HIGHER ORDER LOGIC THEOREM PROVING AND ITS APPLICATIONS

IFIP Transactions A: Computer Science and Technology

International Federation for Information Processing

Technical Committees:

Software: Theory and Practice (TC2)
Education (TC3)
System Modelling and Optimization (TC7)
Information Systems (TC8)
Relationship between Computers and Society (TC9)
Computer Systems Technology (TC10)
Security and Protection in Information Processing Systems (TC11)
Artificial Intelligence (TC12)
Human-Computer Interaction (TC13)
Foundations of Computer Science (SG14)

IFIP Transactions Editorial Policy Board

The IFIP Transactions Editorial Policy Board is responsible for the overall scientific quality of the IFIP Transactions through a stringent review and selection process.

IFIP Transactions Abstracted/Indexed in:
INSPEC Information Services

A-20

HIGHER ORDER LOGIC THEOREM PROVING AND ITS APPLICATIONS

Proceedings of the IFIP TC10/WG10.2 International Workshop on
Higher Order Logic Theorem Proving and its Applications - HOL '92
organized by CHEOPS ESPRIT BRA 3215
sponsored by IMEC and the Commission of the European Communities
Leuven, Belgium, 21-24 September 1992

Edited by

LUC J.M. CLAESEN
Interuniversity Micro-Electronics Center and
Katholieke Universiteit Leuven
Leuven, Belgium

MICHAEL J.C. GORDON
Computer Laboratory
University of Cambridge
Cambridge, England

1993

NORTH-HOLLAND
AMSTERDAM • LONDON • NEW YORK • TOKYO

ELSEVIER SCIENCE PUBLISHERS B.V.
Sara Burgerhartstraat 25
P.O. Box 211, 1000 AE Amsterdam, The Netherlands

Keywords are chosen from the ACM Computing Reviews Classification System, ©1991, with permission.
Details of the full classification system are available from
ACM, 11 West 42nd St., New York, NY 10036, USA.

ISBN: 0 444 89880 8
ISSN: 0926-5473

This book is printed on acid-free paper.

Printed in The Netherlands

Preface

This book contains all the papers submitted to the workshop entitled *Higher Order Logic Theorem Proving and its Applications.* These range from reports on student projects to original research contributions.

The workshop was organized in IMEC, Leuven, Belgium, from 21 to 24 September 1992 by the CHEOPS ESPRIT Basic Research Action in cooperation with IFIP working group 10.2.

The meeting was the fifth in a series of annual workshops dedicated to the topic of higher order logic theorem proving, its usage in the HOL system and its applications. Previous workshops have taken place in Cambridge UK, Aarhus Denmark and Davis California.

The HOL system is a higher order logic theorem proving system implemented at Edinburgh University, Cambridge University and INRIA. It has found many applications, from the verification of hardware designs at all levels to the verification of programs and communication protocols. Contributions and experiences with other systems based on higher order logic, namely Nuprl and LAMBDA, are also presented here.

The papers in this book cover the following topics:

- Mathematical Logic.
- Induction.
- General Modelling and Proofs.
- Formalizing and Modelling of Automata.
- Program Verification.
- Hardware Description Language Semantics.
- Hardware Verification Methodologies.
- Simulation in Higher Order Logic.
- Extended uses of Higher Order Logic.

The workshop was attended by 60 participants (Europe: 45, North America: 15)

Tutorials were presented by Roger B. Jones on *ProofPower* and by Carl Seger and Jeff Joyce on *Hybrid Formal Verification Techniques.*

In addition to regular presentations on the practical application of higher order logic and the HOL system, the workshop included a *HOL users clinic* organized by Tom Melham for in-depth discussion of issues concerning the use of the HOL system.

A special *LAMBDA users meeting* session was organized by Holger Bush.

The next workshop in the series will be organized by Jeff Joyce and Carl Seger at the University of Britisch Columbia in Vancouver Canada.

We are grateful to the Commission of the European Communities for sponsoring this workshop by providing grants for students to participate.

Luc Claesen, Michael Gordon
November 1992

Conference organization.

Workshop Chair:

Luc Claesen
Interuniversity Micro Electronics Center
& Katholieke Universiteit Leuven
Kapeldreef 75,
B-3001 Leuven (Belgium)
e-mail: claesen@imec.be

Workshop Co-Chair:

Michael Gordon
University of Cambridge
Computer Laboratory
Pembroke Street
Cambridge CB2 3QG (U.K.)
email: mjcg@cl.cam.ac.uk

Program Committee

Myla Archer (University of California, Davis, USA)
Graham Birtwistle (University of Calgary, CA)
Holger Bush (Siemens AG, D)
Albert Camilleri (Hewlett-Packard, UK)
Shui-Kai Chin (Syracuse University, USA)
Luc Claesen (IMEC / Kath. Univ. Leuven, B)
Simon Finn (Abstract Hardware Ltd., UK)
Michael Gordon (University of Cambridge, UK)
Elsa L. Gunter (AT&T Bell Labs, USA)
John Herbert (SRI International, UK)
Roger B. Jones (ICL, UK)
Jeff Joyce (University of British Columbia, CA)
Ton Kalker (Philips, NL)
Matt Kaufmann (Computational Logic Inc., USA)
Kurt Keutzer (Synopsys, USA)
Ramayya Kumar (Univ. of Karlsruhe, D)

Miriam Leeser (Cornell University, USA)
Tim Leonard (Digital Equipement Corp., USA)
Paul Loewenstein (Sun Microsystems, USA)
Tom Melham (University of Cambridge, UK)
Carl Seger (University of British Columbia, CA)
David Shepherd (Inmos Ltd., UK)
Gerd Venzl (Siemens AG, D)
Phillip J. Windley (University of Idaho, USA)

Local Organizing Committee.

Catia Angelo Marcondes, Luc Claesen, Geert De Pril, Peter De Vijt, Mark Genoe, Peter Johannes, Wim Ploegaerts, Hans Samsom, Robert Severyns, Annemie Stas, John Tytgat, Jan Vandenbergh, Diederik Verkest, Eric Verlind

Reviewers.

J. Alves-Foss, C. Angelo Marcondes, M. Archer, G. Birtwistle, H. Bush, A. Camilleri, S-K. Chin, L. Claesen, S. Finn, M. Gordon, E.L. Gunter, J. Grundy, J. Herbert, R.B. Jones, J. Joyce, A. Kalker, M. Kaufmann, K. Keutzer, R. Kumar, M. Leeser, T. Leonard, P. Loewenstein, T. Melham, W. Ploegaerts, C. Seger, D. Shepherd, G. Venzl, D. Verkest, Ph.J. Windley

Table of Contents

Chapter 3: General Modelling and Proofs.

Chapter 4: Formalizing and Modelling of Automata.

Chapter 5: Program Verification.

Chapter 6: Hardware Description Language Semantics.

Chapter 7: Hardware Verification Methodologies.

Chapter 8: Simulation in Higher Order Logic.

Chapter 9: Extended uses of Higher Order Logic.

Chapter 1

Mathematical Logic

Higher Order Logic Theorem Proving and its Applications (A-20)
L.J.M. Claesen and M.J.C. Gordon (Editors)
Elsevier Science Publishers B.V. (North-Holland)

The HOL Logic Extended with Quantification over Type Variables

Thomas F. Melham

University of Cambridge Computer Laboratory, New Museums Site, Pembroke Street, Cambridge, CB2 3QG, England.

Abstract

This paper discusses a proposal to extend the primitive basis of the HOL logic with a very simple form of quantification over types. It is shown how certain practical problems with using the definitional mechanisms of HOL would be solved by the additional expressive power gained by making this extension.

Keyword Codes: F.4.1; I.2.3
Keywords: Mathematical Logic; Deduction and Theorem Proving.

1 Introduction

The version of higher order logic mechanized by the HOL system is essentially Church's formulation of simple type theory [2] extended with explicit rules of definition and with object-language polymorphism of the kind developed by Milner for the LCF logic PPλ [4]. In this paper a further extension to the logic is proposed—namely, the addition of a very limited form of object-language quantification over types. The motivation for this extension comes from a particular technical problem that arises when using the definitional mechanisms provided by the HOL logic. This paper explains this problem and shows how the proposed extension solves it.

It is assumed that the reader is familiar with the details of the HOL logic given in the DESCRIPTION volume of the HOL documentation [5].

2 Types and polymorphism in HOL

Type expressions in the HOL logic have the following syntax:

$$\sigma \quad ::= \quad c \quad | \quad \alpha \quad | \quad \sigma_1 \rightarrow \sigma_2 \quad | \quad (\sigma_1, \ldots, \sigma_n)op$$

where σ, σ_1, ..., σ_n range over types, c ranges over type constants, α ranges over type variables, and op ranges over n-ary type operators (for $n \geq 1$). It is the inclusion of type variables in this syntax of types that makes the HOL logic polymorphic. Typing of terms takes place within the context of an assignment of generic types to constants, and a constant is well-typed at any substitution instance of its generic type.

Theorems that contain polymorphic types are also true for any substitution instance of them, so there is a limited form of implicit universal quantification over types in the HOL logic. More precisely, there is an implicit universal quantification over type variables at the level of *sequents*. A sequent $\Gamma \vdash \mathrm{P}$ means that P is provable by natural deduction from the hypotheses Γ. All occurrences of a type variable in such a sequent are identified by the semantics, and there is an implicit universal quantification over the value of this variable whose scope is the entire sequent. This is reflected in the primitive rule of type instantiation shown below.

$$\frac{\Gamma \vdash \mathrm{P}}{\Gamma \vdash \mathrm{P}[\sigma_1,\ldots,\sigma_n \,/\, \alpha_1,\ldots,\alpha_n]} \quad \alpha_1,\ \ldots,\ \alpha_n \text{ not in } \Gamma.$$

where $\mathrm{P}[\sigma_1,\ldots,\sigma_n \,/\, \alpha_1,\ldots,\alpha_n]$ means the result of simultaneously substituting the type σ_i for the type variable α_i for $1 \leq i \leq n$ at every occurrence of α_i in the term P. The side condition on this rule is a consequence of the fact that if a type variable occurs in both the hypotheses and the conclusion of a sequent, then both occurrences are assumed to denote the same set. In fact, from this primitive rule one can derive the more general instantiation rule

$$\frac{\Gamma \vdash \mathrm{P}}{\Gamma[\sigma_1,\ldots,\sigma_n \,/\, \alpha_1,\ldots,\alpha_n] \vdash \mathrm{P}[\sigma_1,\ldots,\sigma_n \,/\, \alpha_1,\ldots,\alpha_n]}$$

which allows one to substitute types for type variables throughout an entire sequent.

There is an additional side condition on both of the rules shown above, namely the condition that no two distinct term variables in P (and, for the second rule, Γ) may become identified as a result of doing the substitution. The HOL system deals with this side condition by renaming variables as required.

3 Motivation

The proposal of this paper is to extend the very simple form of polymorphism described above with limited object-language quantification over types. In particular, we wish to add primitive terms to the HOL logic of the forms $\forall\alpha.\,\mathrm{P}$ and $\exists\alpha.\,\mathrm{P}$, where α is a type variable and P is a boolean term. Informally, the intended interpretation is that $\mathrm{P}[\sigma/\alpha]$ is true for *all* types σ and for *some* type σ, respectively. Note that we are not proposing an extension to the type language of HOL—the quantifications $\forall\alpha.\,\mathrm{P}$ and $\exists\alpha.\,\mathrm{P}$ are new term constructs, and not type constructs of the kind found (for example) in Girard's system F [3]. The extended logic proposed here resembles system Q, a transfinite type theory due to Andrews [1]. It is, however, still much weaker than Andrews' system.

The motivation for this extension originally arose in connection with work on new derived rules of definition for HOL—particularly in work on derived rules for defining abstract data types. The following sections explain this motivational background. Readers primarily interested in the semantics and other details of the proposed extension may wish to skip to section 4.

3.1 Type specifications

In the latest edition of the HOL documentation [5], the following *type specification* rule for the HOL logic is introduced:

$$\frac{\Gamma_1 \vdash \exists x.\, \mathrm{P}\; x, \quad \Gamma_2 \vdash (\alpha \approx \mathrm{P}{:}\sigma \rightarrow bool) \supset \mathrm{Q}}{\Gamma_1 \cup \Gamma_2 \vdash \mathrm{Q}[\nu/\alpha]} \qquad \alpha \text{ not in } \sigma \text{ or P.}$$

where the notation '$\alpha \approx$ P' is an abbreviation for the assertion that there is a bijection between the values of type α and the set of values that satisfy P:

$$\exists f{:}\alpha{\rightarrow}\sigma.\, (\forall a_1\, a_2.\, (f\ a_1 = f\ a_2) \supset (a_1 = a_2)) \wedge (\forall r.\, \mathrm{P}\ r = (\exists a.\, r = f\ a))$$

This (as yet unimplemented) principle of definition is intended to allow new types to be introduced by loose specifications of their properties. The idea is that one finds a non-empty subset P of an existing type σ to represent the values of desired type. One then proves that if the type α is in bijection with this set, then the desired property Q holds of α. If this can be proved, then it is consistent to extend the logic with a new type ν for which $\mathrm{Q}[\nu/\alpha]$ holds.

With the usual HOL rule of type definition, a new type is defined by an axiom that explicitly relates its values to those of a particular subset of an existing type. By contrast, all that is known about a type introduced by the above rule is that it has the properties given by the formula Q. Explicit information about representation is not part of the definition of a type, so types can be only loosely specified by this rule.

A very natural intended application of this type specification rule is in introducing types whose specifications are given by a collection of constants with particular properties—for example, abstract data types with equational specifications. The property by which the new type is specified takes the following general form:

$$\exists v_1 \ldots v_n.\mathrm{Q}[v_1, \ldots, v_n]$$

That is, one shows that there are values v_1, ..., v_n involving the new type (for example, operations over it) for which $\mathrm{Q}[v_1, \ldots, v_n]$ holds. A type having this property is then introduced by a type specification, and the constant specification rule can subsequently be used to introduce constants c_1, ..., c_n that denote these values. The result is a data type characterised by the theorem $\vdash \mathrm{Q}[\mathsf{c}_1, \ldots, \mathsf{c}_n]$ and defined in such a way that consequences of this defining property are all that can be proved about the type.

An example application of this approach is in defining a type *two*, about which all that is known is that it has *at least* two distinct elements. This type can be defined by first proving the theorems

$$\vdash \exists x.\, (\lambda x{:}bool.\mathsf{T})\ x \quad \text{and} \quad \vdash (\alpha \approx (\lambda x{:}bool.\mathsf{T})) \supset \exists x{:}\alpha.\, \exists y{:}\alpha.\neg(x = y)$$

One can then use the type specification rule to introduce the type *two* satisfying

$$\vdash \exists x{:}two.\, \exists y{:}two.\neg(x = y)$$

Finally, two constants **one** and **two** can be introduced for the two values asserted to exist by this theorem, giving the characterising theorem $\vdash \neg(\mathsf{one} = \mathsf{two})$. The result is a type about which we know only that it has at least two distinct values, namely **one** and **two**.

3.2 The problem

The problem that motivates the present proposal to extend the HOL logic is illustrated by the following attempt to use the type definition method described above. Suppose we wish to define in HOL an abstract data type of non-empty bit-vectors, specified algebraically using two unit vectors **t** and **f** and an associative concatenation operation **c**. Suppose further that we wish to use the methodology sketched above—that is, we wish first to prove that the desired characterizing property holds of some appropriate representing set of values and then to introduce the type with this property using a type specification. In this case, we would proceed by first proving

$$\begin{array}{l}\vdash (\alpha \approx (\lambda l{:}(bool)list.\, \neg(l = []))) \supset \\ \quad \exists t{:}\alpha.\, \exists f{:}\alpha.\, \exists c{:}\alpha\to\alpha\to\alpha. \\ \qquad (\forall b_1\, b_2\, b_3.\, c\; b_1\; (c\; b_2\; b_3) = c\; (c\; b_1\; b_2)\; b_3) \wedge \\ \qquad (\forall x{:}\beta.\, \forall y{:}\beta.\, \forall op{:}\beta\to\beta\to\beta. \\ \qquad\quad (\forall x\, y\, z.\, op\; x\; (op\; y\; z) = op\; (op\; x\; y)\; z) \supset \\ \qquad\quad \exists!\, fn{:}\alpha\to\beta.\, (fn\; t = x) \wedge (fn\; f = y) \wedge \\ \qquad\qquad\qquad\qquad (\forall b_1\, b_2.\, fn\; (c\; b_1\; b_2) = op\; (fn\; b_1)\; (fn\; b_2)))\end{array}$$

This theorem says that if α is in bijection with the set of non-empty lists of booleans, then we have an abstract characterization of α as an initial algebra with two elements t and f and an associative concatenation operation c. In particular, we have that there is a unique homomorphism from α to any other type β with two elements x and y and an associative operation op.

Given this theorem, one can use the type specification rule shown in the previous section to introduce a type constant *vect* with the following defining property.

$$\begin{array}{l}\vdash \exists t{:}vect.\, \exists f{:}vect.\, \exists c{:}vect\to vect\to vect. \\ \quad (\forall b_1\, b_2\, b_3.\, c\; b_1\; (c\; b_2\; b_3) = c\; (c\; b_1\; b_2)\; b_3) \wedge \\ \quad (\forall x{:}\beta.\, \forall y{:}\beta.\, \forall op{:}\beta\to\beta\to\beta. \\ \qquad (\forall x\, y\, z.\, op\; x\; (op\; y\; z) = op\; (op\; x\; y)\; z) \supset \\ \qquad \exists!\, fn{:}vect\to\beta.\, (fn\; t = x) \wedge (fn\; f = y) \wedge \\ \qquad\qquad\qquad\qquad (\forall b_1\, b_2.\, fn\; (c\; b_1\; b_2) = op\; (fn\; b_1)\; (fn\; b_2)))\end{array} \tag{1}$$

Having introduced the type *vect* with the above specification, it is natural to wish to use the constant specification rule to give names to the values t, f and c. This would yield the theorem

$$\begin{array}{l}\vdash (\forall b_1\, b_2\, b_3.\, \mathsf{c}\; b_1\; (\mathsf{c}\; b_2\; b_3) = \mathsf{c}\; (\mathsf{c}\; b_1\; b_2)\; b_3) \wedge \\ \quad (\forall x{:}\beta.\, \forall y{:}\beta.\, \forall op{:}\beta\to\beta\to\beta. \\ \qquad (\forall x\, y\, z.\, op\; x\; (op\; y\; z) = op\; (op\; x\; y)\; z) \supset \\ \qquad \exists!\, fn{:}vect\to\beta.\, (fn\; \mathsf{t} = x) \wedge (fn\; \mathsf{f} = y) \wedge \\ \qquad\qquad\qquad\qquad (\forall b_1\, b_2.\, fn\; (\mathsf{c}\; b_1\; b_2) = op\; (fn\; b_1)\; (fn\; b_2)))\end{array} \tag{2}$$

where **t** and **f** are constants of type *vect*, and the concatenation operator **c** is a constant of type *vect*→*vect*→*vect*. This final theorem is the desired algebraic characterization of bit-vectors. The general approach to defining other equationally-specified abstract data types would be similar.

The trouble with the construction sketched above is that the primitive rule of constant specification in HOL does not admit the direct inference of theorem (2) from theorem (1). The rule of constant specification is:

$$\frac{\vdash \exists x_1{:}\sigma_1 \ldots x_n{:}\sigma_n.\, \mathrm{P}}{\vdash \mathrm{P}[\mathsf{c}_1,\ldots,\mathsf{c}_n/x_1,\ldots,x_n]} \quad \mathrm{tyvars}(\sigma_i) = \mathrm{tyvars}(\mathrm{P}) \text{ for } 1 \leq i \leq n.$$

The side condition on this rule requires all the type variables in the defining property P to be included in the types of the constants to be introduced. But in the defining property given by theorem (1) there is a type variable β which does not appear in the type of the existentially quantified variables t, f and c, so the constant specification rule cannot in this case be applied.

The problem illustrated by this example is closely related to the implicit quantification over type variables mentioned in the previous section. Consider a theorem of the form $\vdash \exists x{:}\sigma.\, \mathrm{P}$ where the type variable α appears in P but not in the type σ. Informally, what is meant by such a theorem is:

'for any type τ, we have $\vdash \exists x{:}\sigma.\, \mathrm{P}[\tau/\alpha]$'.

That is, for any type τ, there is a value x such that $\mathrm{P}[\tau/\alpha]$ comes out true. In general, of course, the choice of value x that makes this term true may depend on the type τ. And if the type variable α does not appear in the type of x, then this dependence will be lost if a constant is introduced to denote this value—hence the side-condition on the rule for constant specifications. For example, in the theorem

$$\vdash \exists b.\, b = \forall x{:}\alpha.\, \forall y{:}\alpha.\, x = y$$

the value b must be either true or false, depending on whether the type α denotes a set with exactly one element or not. It is therefore unsound to admit a constant specification

$$\vdash \mathsf{b} = \forall x{:}\alpha.\, \forall y{:}\alpha.\, x = y$$

in which a fixed name b is given to this value, since then by instantiating α in two different ways one can obtain from this theorem both $\vdash \mathsf{b} = \mathsf{T}$ and $\vdash \mathsf{b} = \mathsf{F}$.

Returning to the example, consider again the defining property for the type *vect* given by theorem (1). This contains a type variable β that does not appear in the types of the variables t, f and c, and so the desired constant specification cannot be made. In this particular case, however, the values t, f and c that make the specification true do not, in fact, depend on the set denoted by type variable β; there are three fixed values t, f and c that make this specification hold for *any* type β. This suggests that the defining property for *vect* should be something more like:

$$\begin{array}{l} \vdash \exists t{:}vect.\, \exists f{:}vect.\, \exists c{:}vect\to vect\to vect. \\ \quad (\forall b_1\, b_2\, b_3.\, c\; b_1\; (c\; b_2\; b_3) = c\; (c\; b_1\; b_2)\; b_3) \wedge \\ \quad (\forall \boldsymbol{\beta}.\, \forall x{:}\beta.\, \forall y{:}\beta.\, \forall op{:}\beta\to\beta\to\beta. \\ \quad\quad (\forall x\; y\; z.\, op\; x\; (op\; y\; z) = op\; (op\; x\; y)\; z) \supset \\ \quad\quad \exists!\, fn{:}vect\to\beta.\, (fn\; t = x) \wedge (fn\; f = y) \wedge \\ \quad\quad\quad\quad (\forall b_1\, b_2.\, fn\; (c\; b_1\; b_2) = op\; (fn\; b_1)\; (fn\; b_2))) \end{array} \tag{3}$$

where there is a universal quantification over the type variable β (shown here in boldface) within the scope of the existential quantification over t, f and c.

This new theorem more accurately expresses the initiality property of the type *vect* than theorem (1). Moreover, if the quantification over β is taken to bind the type variable β, and if the function tyvars in the constant specification rule shown above is reinterpreted as returning the *free* type variables in a term, then it would be valid to infer from this theorem the desired constant specification

$$
\begin{aligned}
\vdash\ & (\forall b_1\, b_2\, b_3.\, \mathsf{c}\ b_1\ (\mathsf{c}\ b_2\ b_3) = \mathsf{c}\ (\mathsf{c}\ b_1\ b_2)\ b_3) \wedge \\
& (\forall \beta.\, \forall x{:}\beta.\, \forall y{:}\beta.\, \forall op{:}\beta\to\beta\to\beta. \\
& \quad (\forall x\, y\, z.\, op\ x\ (op\ y\ z) = op\ (op\ x\ y)\ z) \supset \\
& \quad \exists! fn{:}vect\to\beta.\, (fn\ \mathsf{t} = x) \wedge (fn\ \mathsf{f} = y) \wedge \\
& \qquad\qquad\qquad (\forall b_1\, b_2.\, fn\ (\mathsf{c}\ b_1\ b_2) = op\ (fn\ b_1)\ (fn\ b_2)))
\end{aligned}
$$

This is the correct form of algebraic specification for the type *vect*.

4 Universal quantification over types

The proposal of this paper is to extend the HOL logic with universal quantification over types, so that properties like that given by theorem (3) in the previous section can be expressed directly in the logic. With this extension to HOL, the syntax of untyped terms becomes:

$$\mathrm{M} \ ::= \ c \ \mid \ v \ \mid \ \mathrm{M\,N} \ \mid \ \lambda v.\, \mathrm{M} \ \mid \ \forall \alpha.\, \mathrm{M}$$

That is, the term language of HOL (i.e. the simply-typed λ-calculus) is extended with terms of the form $\forall\alpha.\,\mathrm{M}$, where α is a type variable and M is a (boolean) term. The intended meaning of such a term is that the proposition represented by M is true no matter what set is denoted by the type variable α.

More precisely, we take the syntax of types for the extended logic to be

$$\sigma \ ::= \ bool \ \mid \ c \ \mid \ v \ \mid \ \sigma_1 \to \sigma_2 \ \mid \ (\sigma_1, \ldots, \sigma_n)op$$

That is, we assume right from the start that the type expressions include the special type constant *bool*. This is necessary because we wish to form quantifications $\forall\alpha.\,\mathrm{M}$ only for *boolean* terms M. The extended syntax of typed terms is then given by

$$\mathrm{M} \ ::= \ c_\sigma \ \mid \ v_\sigma \ \mid \ (\mathrm{M}_{\sigma_1\to\sigma_2}\ \mathrm{N}_{\sigma_1})_{\sigma_2} \ \mid \ (\lambda v_{\sigma_1}.\, \mathrm{M}_{\sigma_2})_{\sigma_1\to\sigma_2} \ \mid \ (\forall\alpha.\, \mathrm{M}_{bool})_{bool}$$

where (as usual) a constant c_σ is a well-formed term only if σ is a substitution instance of the generic type of c, and where a quantification $\forall\alpha.\,\mathrm{M}$ is well-formed only if for every occurrence of a variable v_σ in M, either the type variable α does not appear in the type σ or the occurrence of v_σ is bound in M by a textually enclosing lambda-abstraction λv_σ. In what follows, it is assumed that all terms are well-formed according to these formation rules.

A quantification $\forall\alpha.\,\mathrm{M}$ binds occurrences of the type variable α in M, so this extended syntax of terms comes with a notion of free and bound type variable, in addition to the usual notion of free and bound (term) variables. Furthermore, the standard notion of a free occurrence of a term in another term must be modified to take into account binding by type quantifiers. Following the definition given by Andrews for his system Q, we say that an occurrence of a term M_σ is *free for* the variable v_σ in a term N if it does not occur after the dot in a subterm of N that has the form $\lambda v_\sigma.\,P$ and if it does not occur after the dot in any subterm of the form $\forall\alpha.\,P$ where α occurs in σ. We define a free occurrence of a variable v_σ in a term N as an occurrence which is free for itself in N. In the presence of the well-formedness condition stated above for type quantifications, this notion of free variable is equivalent to the usual HOL definition: a variable v_σ is free in N if it does not appear after the dot in a subterm of the form $\lambda v_\sigma.\,P$.

In what follows, we assume suitably-defined notions of term and type substitution, with renaming of bound term and type variables as appropriate to avoid variable capture. In particular, we assume for type substitution that if $\sigma s = \sigma_1, \ldots, \sigma_n$ are types and $\alpha s = \alpha_1, \ldots, \alpha_n$ are distinct type variables, then the metalinguistic notation $M[\sigma s/\alpha s]$ stands for the result of simultaneously substituting σ_i for α_i for $1 \leq i \leq n$ at every free occurrence of α_i in the term M, with the condition that no type variable in any σ_i and no free term variable in M becomes bound in the result of the substitution. In practice, an implementation will rename type and term variables in order to satisfy this condition. So, for example,

$$\begin{aligned} (\forall\beta.\,(x{:}\alpha = x{:}\alpha))[\beta/\alpha] &= \forall\beta'.\,(x{:}\beta = x{:}\beta) \\ (\forall x{:}\alpha.\,x{:}bool)[bool/\alpha] &= \forall x'{:}bool.\,x{:}bool \end{aligned}$$

Similar renamings are also carried out in the current HOL implementation; see [5].

4.1 Semantics

Chapters 9 and 10 of the DESCRIPTION volume of the HOL manual [5] contain a set-theoretic semantics of the HOL logic due to Andy Pitts. In this section, we very briefly sketch this semantics and show how it can be extended to cover the quantification over type variables described above.

The semantics of HOL is defined in terms of particular set $\mathcal{U}$ called the *universe*, the elements of which are the sets denoted by the (monomorphic) type expressions. Among other things, the universe is closed under the function space operation and contains the distinguished two-element set $2 = \{1, 0\}$ which is the meaning of the boolean type *bool*.

The semantics of types is given relative to a model M which assigns to each type constant an element of $\mathcal{U}$ and to each n-ary type operator a function $\mathcal{U}^n \to \mathcal{U}$. The notion of a *type-in-context* is also used in defining the semantics of types. A *type context* αs is just a finite list of distinct type variables, and a type-in-context $\alpha s.\sigma$ is a type σ together with a type context αs which contains (at least) all the type variables in σ. The meaning of a type in context $\alpha s.\sigma$, where the context αs is of length n, is then given by a function

$$[\![\alpha s.\sigma]\!]_M : \mathcal{U}^n \to \mathcal{U}$$

which is defined so that for any assignment of sets $Xs = (X_1, \ldots, X_n) \in \mathcal{U}^n$ to the type variables in αs (and hence to the type variables in σ), the element $[\![\alpha s.\sigma]\!]_M(Xs)$ of $\mathcal{U}$ is the corresponding set denoted by σ. The details of the definition of $[\![\_]\!]_M$, which proceeds by induction on the structure of types, can be found in [5].

The notion of a context is also employed in defining the meaning of terms. A *term-in-context* is written 'αs,xs.M' and consists of a term M together with a type context αs and a finite list of variables xs called a *variable context*. The variable context xs of a term-in-context αs,xs.M contains all the variables that occur free in M, and the type context αs contains all the type variables that occur in xs and M. It is evident that this notion of terms in context is easily adapted for the extended term language described above; we simply make the condition that the type context for a term-in-context αs,xs.M contains all the type variables that appear in xs or appear *free* in M. Following [5], we assume that xs contains no variable that appears bound in M—bound term variables may always be renamed to satisfy this condition. Likewise, we assume that the type context αs contains no type variables that are bound in M. This assumption is valid because the well-formedness condition on terms of the form $\forall\alpha.\,$M ensures that M contains no *free* variable v_σ where α occurs in σ. Hence bound type variables can always be renamed so as to be distinct from all type variables in αs.

For the semantics of terms, a model consists of a type model (as described above) together with a function that assigns to each constant c with generic type σ an element of the set of functions

$$\prod_{Xs \in \mathcal{U}^n} [\![\alpha s.\sigma]\!]_M(Xs)$$

where n is the length of the type context αs. For a given model M, the meaning of a term-in-context αs,xs.M, where αs has length n, xs has length m, and M has type τ, is given by a function $[\![\_]\!]_M$ defined by induction on terms such that:

$$[\![\alpha s,xs.\mathrm{M}]\!]_M \in \prod_{Xs \in \mathcal{U}^n} \left(\prod_{j=1}^{m} [\![\alpha s.\sigma_j]\!]_M(Xs)\right) \rightarrow [\![\alpha s.\tau]\!]_M(Xs).$$

where $xs = x_1, \ldots, x_m$ and σ_i is the type of the corresponding variable x_i. The idea is that given an assignment of sets

$$Xs = (X_1, \ldots, X_n) \in \mathcal{U}^n$$

to the type variables in αs (and hence to the free type variables in M) and given an assignment of elements

$$ys = (y_1, \ldots, y_m) \in [\![\alpha s.\sigma_1]\!]_M(Xs) \times \cdots \times [\![\alpha s.\sigma_m]\!]_M(Xs)$$

to the variables in xs (and hence to the variables that occur free in the term M), the result of $[\![\alpha s,xs.\mathrm{M}]\!]_M(Xs)(ys)$ will be an appropriate element of the set $[\![\alpha s.\tau]\!]_M(Xs)$ denoted by the type of M. Again, full details of the definition of $[\![\_]\!]_M$ can be found in [5].

A sequent with hypotheses $\Gamma = \{\mathrm{P}_1, \ldots, \mathrm{P}_p\}$ and conclusion P is *satisfied* by a model M if any assignment of values to free variables that makes all the hypotheses true in M also makes the conclusion true in M. In particular, M satisfies the sequent if for all $Xs \in \mathcal{U}^n$ and all $ys \in [\![\alpha s.\sigma_1]\!]_M(Xs) \times \cdots \times [\![\alpha s.\sigma_m]\!]_M(Xs)$,

$$[\![\alpha s,xs.\mathrm{P}_1]\!]_M(Xs)(ys) = 1, \quad \ldots, \quad [\![\alpha s,xs.\mathrm{P}_p]\!]_M(Xs)(ys) = 1$$

imply that

$$[\![\alpha s,xs.\mathrm{P}]\!]_M(Xs)(ys) = 1,$$

where $\alpha s, xs$ is any valid context for each of P, P_1, ..., P_p with αs of length n, $xs = x_1, \ldots, x_m$, and σ_i the type of the corresponding variable x_i. We write $\Gamma \models_M \mathrm{P}$ to mean that M satisfies the sequent with hypotheses Γ and conclusion P.

It is completely straightforward to extend the definition of $[\![\_]\!]_M$, and hence the notion of satisfaction, to cover type quantifications $\forall\alpha.\,\mathrm{M}$. Suppose $\alpha s,xs.\forall\alpha.\,\mathrm{M}$ is a term-in-context, with αs of length n, and $xs = x_1, \ldots, x_m$ with σ_i the type of x_i. Then, since the term M has type *bool* we have

$$[\![\alpha s,\alpha,xs.\mathrm{M}]\!]_M \in \prod_{Xs \in \mathcal{U}^{n+1}} \left(\prod_{j=1}^{m} [\![\alpha s,\alpha.\sigma_j]\!]_M(Xs) \right) \to 2$$

Note that α is distinct from all the type variables in αs, by the assumption that this type context contains no type variable bound in $\forall\alpha.\,\mathrm{M}$. This ensures that $\alpha s,\alpha,xs.\mathrm{M}$ is in fact a well-formed term-in-context. Now define

$$[\![\alpha s,xs.\forall\alpha.\,\mathrm{M}]\!]_M \in \prod_{Xs \in \mathcal{U}^{n}} \left(\prod_{j=1}^{m} [\![\alpha s.\sigma_j]\!]_M(Xs) \right) \to 2$$

by taking $[\![\alpha s,xs.\forall\alpha.\,\mathrm{M}]\!]_M(Xs)(ys)$ to be 1 if and only if the function in $\mathcal{U} \to 2$ given by $X \mapsto [\![\alpha s,\alpha,xs.\mathrm{M}]\!](Xs, X)(ys)$ yields 1 for all $X \in \mathcal{U}$.

4.2 Primitive inference rules

Having added propositions $\alpha.\,\mathrm{M}$ as primitive terms to HOL, the next step is to extend the logic with new primitive inference rules for reasoning with them. The required rules are the introduction and elimination rules for universal quantification over types shown in figure 1 below. Note that in both rules, it is assumed that $\forall\alpha.\,\mathrm{P}$ is a well formed term—that is, that P contains no free variable v_σ with α occurring in σ.

$$\forall\text{-E:}\ \frac{\Gamma \vdash \forall\alpha.\,\mathrm{P}}{\Gamma \vdash \mathrm{P}[\sigma/\alpha]} \qquad \forall\text{-I:}\ \frac{\Gamma \vdash \mathrm{P}{:}bool}{\Gamma \vdash \forall\alpha.\,\mathrm{P}} \quad \alpha \text{ not free in } \Gamma.$$

Figure 1: Rules for universal quantification over types.

In what follows, we sketch the proof that the elimination rule ∀-E rule for universal type quantification is sound. That is, we show that if a model satisfies the hypothesis of the rule, then it also satisfies the conclusion. The proof depends on the series of lemmas given below.

Lemma 4.1 *Suppose α does not occur in $\alpha s, xs$ and α does not occur in σ, and suppose that $\forall\alpha.\,\mathrm{P}$ is well-formed. Then if $\alpha s, xs$ is a valid context for each of $\mathrm{P}[\sigma/\alpha]$, $\mathrm{P}_1, \ldots, \mathrm{P}_p$, then $\alpha s, xs$ is also a valid context for each of $\forall\alpha.\,\mathrm{P}$, $\mathrm{P}_1, \ldots, \mathrm{P}_p$.*

Proof. Obvious, by the definitions of well-formedness and valid term contexts.[1] □

Lemma 4.2 *If $\alpha s, \alpha.\sigma$ is a type-in-context where α does not occur in σ, then*

$$[\![\alpha s, \alpha.\sigma]\!]_M(Xs, X) = [\![\alpha s.\sigma]\!]_M(Xs)$$

for all $X \in \mathcal{U}$ and $Xs \in \mathcal{U}^n$ where n is the length of αs.

Proof. By structural induction on σ. □

Lemma 4.3 *Suppose $\forall\alpha.\,\mathrm{P}$ is a well-formed term (and hence P contains no free variable v_τ where α occurs in τ). If $\alpha s, xs.\mathrm{P}[\sigma/\alpha]$ is a term-in-context where α does not occur in $\alpha s, xs$ and α does not occur in σ, then for all $Xs \in \mathcal{U}^n$, where n is the length of αs*

$$[\![\alpha s, xs.\mathrm{P}[\sigma/\alpha]]\!]_M(Xs) = [\![\alpha s, \alpha, xs.\mathrm{P}]\!]_M(Xs, [\![\alpha s.\sigma]\!](Xs))$$

Proof. By structural induction on P, using lemma 4.2 to establish that the left- and right-hand sides are functions with the same domain. □

Theorem 4.4 *Rule $\forall$-E is sound. I.e. for any model M and type σ, if $\Gamma \models_M \forall\alpha.\,\mathrm{P}$, then $\Gamma \models_M \mathrm{P}[\sigma/\alpha]$.*

Proof. We may suppose that α appears nowhere in Γ and that σ does not contain α, as the bound type variable α can be renamed if necessary without changing the semantics. Now suppose that $\Gamma \models_M \forall\alpha.\,\mathrm{P}$ with $\Gamma = \{\mathrm{P}_1, \ldots, \mathrm{P}_p\}$. Then for any $\alpha s, xs$ which is a context of each of $\forall\alpha.\,\mathrm{P}$, $\mathrm{P}_1, \ldots, \mathrm{P}_p$ where the length of αs is n, $xs = x_1, \ldots, x_m$, and x_i has type σ_i, we have that for all $Xs \in \mathcal{U}^n$ and $ys \in [\![\alpha s.\sigma_1]\!]_M(Xs) \times \cdots \times [\![\alpha s.\sigma_m]\!]_M(Xs)$, if

$$[\![\alpha s, xs.\mathrm{P}_k]\!]_M(Xs)(ys) = 1 \quad \text{for } 1 \leq k \leq p$$

then $[\![\alpha s, xs.\forall\alpha.\,\mathrm{P}]\!]_M(Xs)(ys) = 1$, which by the semantics of $\forall\alpha.\,\mathrm{P}$ given above means that for all $X \in \mathcal{U}$ we have

$$[\![\alpha s, \alpha, xs.\mathrm{P}]\!]_M(Xs, X)(ys) = 1.$$

Now, suppose that $\alpha s', xs'$ is a context for each of $\mathrm{P}[\sigma/\alpha]$, $\mathrm{P}_1, \ldots, \mathrm{P}_p$. We may assume that $\alpha s', xs'$ does not contain α, since α does not appear in any of $\mathrm{P}[\sigma/\alpha]$, $\mathrm{P}_1, \ldots, \mathrm{P}_p$. Let n' be the length of $\alpha s'$, $xs' = x'_1, \ldots, x'_{m'}$, and σ'_i be the type of the corresponding variable x'_i. Assume that for some $Xs' \in \mathcal{U}^{n'}$ and $ys' \in [\![\alpha s.\sigma'_1]\!]_M(Xs') \times \cdots \times [\![\alpha s'.\sigma'_{m'}]\!]_M(Xs')$ we have for the hypotheses that

$$[\![\alpha s', xs'.\mathrm{P}_k]\!]_M(Xs')(ys') = 1 \quad \text{for } 1 \leq k \leq p.$$

[1] Note that we have assumed that no bound type variables in P have been renamed in $\mathrm{P}[\sigma/\alpha]$. One can always contrive for this to be the case by doing appropriate renamings beforehand.

By lemma 4.1 we have that $\alpha s', xs'$ is a valid context for each of $\forall\alpha.\,\mathrm{P}$, $\mathrm{P}_1, \ldots, \mathrm{P}_p$, so the above assumption implies that for all $X \in \mathcal{U}$

$$[\![\alpha s', \alpha, xs'.\mathrm{P}]\!]_M(Xs', X)(ys') = 1 \tag{4}$$

One then has by lemma 4.3 that

$$[\![\alpha s', xs'.\mathrm{P}[\sigma/\alpha]]\!]_M(Xs')(ys') = [\![\alpha s', \alpha, xs'.\mathrm{P}]\!]_M(Xs', [\![\alpha s'.\sigma]\!](Xs'))(ys')$$

and hence by (4) that $[\![\alpha s', xs'.\mathrm{P}[\sigma/\alpha]]\!]_M(Xs')(ys') = 1$, as required. □

The proof of soundness for the introduction rule ∀-I is left as an exercise for the reader.

4.3 Derived inference rules

Various higher-level inference rules for universal quantification over types are derivable from the primitive introduction and elimination rules given in the previous section. We give the derivations of some of the more important rules below.

The significance of the following congruence rule is that it is needed to extend the implementation of HOL's conversion-based rewriting; see [5] for the details.

Theorem 4.5 *Universal quantification over types satisfies the following congruence rule*

$$\forall\text{-EQ:}\quad \frac{\Gamma \vdash (\mathrm{M}{:}bool) = (\mathrm{N}{:}bool)}{\Gamma \vdash (\forall\alpha.\,\mathrm{M}) = (\forall\alpha.\,\mathrm{N})}\quad \alpha \text{ not free in } \Gamma.$$

Proof. By the following derivation.

1. $\Gamma, \forall\alpha.\,\mathrm{M} \vdash \forall\alpha.\,\mathrm{M}$ [`ASSUME`, `ADD_ASSUM`]
2. $\Gamma, \forall\alpha.\,\mathrm{M} \vdash \mathrm{M}$ [∀-E rule: 1]
3. $\Gamma, \forall\alpha.\,\mathrm{M} \vdash \mathrm{N}$ [`EQ_MP`: hypothesis, 2]
4. $\Gamma, \forall\alpha.\,\mathrm{M} \vdash \forall\alpha.\,\mathrm{N}$ [∀-I rule: 3]
5. $\Gamma \vdash \forall\alpha.\,\mathrm{M} \supset \forall\alpha.\,\mathrm{N}$ [`DISCH`: 4]
6. $\Gamma \vdash \forall\alpha.\,\mathrm{N} \supset \forall\alpha.\,\mathrm{M}$ [similar to steps 1–5]
7. $\Gamma \vdash \forall\alpha.\,\mathrm{M} = \forall\alpha.\,\mathrm{N}$ [`EQ_IMP_RULE`: 5, 6]

Note that the side condition on the ∀-EQ rule is necessary for the steps in this proof that use the introduction rule ∀-I to be valid. □

The following theorem shows that the existing HOL type instantiation rule, described above in section 2, can be derived in the extended logic.

Theorem 4.6 *The following type instantiation rule is derivable.*

$$\forall\text{-INST:}\quad \frac{\Gamma \vdash \mathrm{P}}{\Gamma \vdash \mathrm{P}[\sigma/\alpha]}\quad \alpha \text{ not free in } \Gamma.$$

where there is the additional side condition that no two distinct term variables in P *become identified as a result of doing the substitution (see section 2).*

Proof. Let $x_1, \ldots, x_n$ be all the free variables in P whose types contain α.

1. $\Gamma \vdash \mathrm{P}$ [hypothesis]
2. $\Gamma \vdash \forall x_1. \cdots \forall x_n. \mathrm{P}$ [GENL 1]
3. $\Gamma \vdash \forall \alpha. \forall x_1. \cdots \forall x_n. \mathrm{P}$ [$\forall$-I rule: 2]
4. $\Gamma \vdash (\forall x_1. \cdots \forall x_n. \mathrm{P})[\sigma/\alpha]$ [$\forall$-E rule: 3]
5. $\Gamma \vdash \forall x_1[\sigma/\alpha]. \cdots \forall x_n[\sigma/\alpha]. (\mathrm{P}[\sigma/\alpha])$ [defn. of type substitution: 4]
6. $\Gamma \vdash \mathrm{P}[\sigma/\alpha]$ [SPECL: 5]

Note that step 2 is valid because the side condition that α is not free in Γ ensures that none of $x_1, \ldots, x_n$, each of whose types contains α, is free in Γ. Step 3 is valid because the generalizations in step 2 ensure that '$\forall \alpha. \forall x_1. \cdots \forall x_n. \mathrm{P}$' will be well-formed. Step 5 is valid because the additional side condition mentioned above ensures that no x_i must be renamed to avoid capture by type substitution of other variables free in P. □

It is straightforward use this $\forall$-INST rule to derive the full rule for simultaneous type instantiation shown in section 2. This means that a primitive type instantiation rule, as present in the existing HOL logic, is redundant in the extended system—it can instead be a derived rule. The rule shown in section 2 for instantiation throughout a sequent is also derivable.

Theorem 4.7 *The following alpha-conversion rule holds.*

$$\forall\text{-ACONV: } \quad \frac{}{\vdash (\forall \alpha. \mathrm{M}) = (\forall \beta. \mathrm{M}[\beta/\alpha])} \quad \beta \text{ not free in } \forall \alpha. \mathrm{M}.$$

Proof. By the following derivation.

1. $\forall \alpha. \mathrm{M} \vdash \forall \alpha. \mathrm{M}$ [ASSUME]
2. $\forall \alpha. \mathrm{M} \vdash \mathrm{M}[\beta/\alpha]$ [$\forall$-E rule: 1]
3. $\forall \alpha. \mathrm{M} \vdash \forall \beta. \mathrm{M}[\beta/\alpha]$ [$\forall$-I rule: 2]
4. $\vdash \forall \alpha. \mathrm{M} \supset \forall \beta. \mathrm{M}[\beta/\alpha]$ [DISCH: 3]
5. $\vdash (\forall \beta. \mathrm{M}[\beta/\alpha]) \supset \forall \alpha. \mathrm{M}$ [similar to steps 1–4]
6. $\vdash \forall \alpha. \mathrm{M} = (\forall \beta. \mathrm{M}[\beta/\alpha])$ [EQ_IMP_RULE: 4, 5]

The side condition that β is not free in $\forall \alpha. \mathrm{M}$ is necessary for step 3 to be valid. It also ensures that $\mathrm{M}[\beta/\alpha][\alpha/\beta] = \mathrm{M}$, which is used in step 5. □

5 Existential quantification over types

Adding universal quantification over types to HOL immediately suggests that existential quantification may be possible as well. The simplest approach is to make a term of the form '$\exists \alpha. \mathrm{P}$' an abbreviation for '$\neg \forall \alpha. \neg \mathrm{P}$'. This cannot, of course, be done as an object-language abbreviation using the existing HOL definitional mechanisms. It could, however, be implemented in HOL as a parser-supported metalinguistic abbreviation. Alternatively, the term language could be further extended to

$$\mathrm{M} \ ::= \ c \ | \ v \ | \ \mathrm{M\,N} \ | \ \lambda v. \mathrm{M} \ | \ \forall \alpha. \mathrm{M} \ | \ \exists \alpha. \mathrm{M}$$

with formation and typing rules similar to those given in the previous section for universal

$$\exists\text{-I:}\quad \frac{\Gamma \vdash \mathrm{P}[\sigma/\alpha]{:}\mathit{bool}}{\Gamma \vdash \exists\alpha.\,\mathrm{P}}$$

$$\exists\text{-E:}\quad \frac{\Gamma \vdash \exists\alpha.\,\mathrm{P}, \quad \Gamma, \mathrm{P}[\beta/\alpha] \vdash \mathrm{Q}}{\Gamma \vdash \mathrm{Q}} \quad \beta \text{ not free in } \Gamma \text{ or Q.}$$

Figure 2: Derived rules for existential quantification over types.

quantification. Meaning could then be given to terms of the form '$\exists\alpha.\,\mathrm{M}$' by the addition of the following primitive axiom-scheme (conversion):

$$\exists\text{-CONV:}\quad \frac{}{\vdash (\exists\alpha.\,\mathrm{P}) = \neg\forall\alpha.\neg\mathrm{P}} \quad \beta \text{ not free in } \forall\alpha.\,\mathrm{P}$$

This conversion is the only primitive rule required for existential quantification over types. Appropriate introduction and elimination rules for $\exists\alpha.\,\mathrm{P}$ are easily derived using it.

Theorem 5.1 *The introduction and elimination rules in figure 2 are derivable.*

Proof. For the introduction rule, the derivation is

1. $\Gamma \vdash \mathrm{P}[\sigma/\alpha]$	[hypothesis]
2. $\forall\alpha.\neg\mathrm{P} \vdash \forall\alpha.\neg\mathrm{P}$	[ASSUME]
3. $\forall\alpha.\neg\mathrm{P} \vdash \neg\mathrm{P}[\sigma/\alpha]$	[$\forall$-E rule: 2]
4. $\Gamma, \forall\alpha.\neg\mathrm{P} \vdash \mathsf{F}$	[MP: 3, 1]
5. $\Gamma \vdash \neg\forall\alpha.\neg\mathrm{P}$	[DISCH, NOT_INTRO: 4]
6. $\Gamma \vdash \exists\alpha.\,\mathrm{P}$	[$\exists$-CONV]

For the elimination rule, the derivation is

1. $\Gamma \vdash \neg\forall\alpha.\neg\mathrm{P}$	[$\exists$-CONV: hypothesis]
2. $\Gamma \vdash \mathrm{P}[\beta/\alpha] \supset \mathrm{Q}$	[DISCH: hypothesis]
3. $\Gamma, \neg\mathrm{Q} \vdash \neg\mathrm{P}[\beta/\alpha]$	[CONTRAPOS, UNDISCH: 2]
4. $\Gamma, \neg\mathrm{Q} \vdash \forall\beta.\neg\mathrm{P}[\beta/\alpha]$	[$\forall$-I: 3]
5. $\Gamma, \neg\mathrm{Q} \vdash \forall\alpha.\neg\mathrm{P}$	[$\forall$-E, $\forall$-I: 4]
6. $\Gamma, \neg\mathrm{Q} \vdash \mathsf{F}$	[MP: 1, 5]
7. $\Gamma \vdash \mathrm{Q}$	[DISCH, simplify: 6]

The side condition that β is not free in Γ or Q is needed for the validity of steps 4 and 5 in this second derivation. □

The derivations of congruence and alpha-conversion rules for existential quantification over types are straightforward and will therefore not be given here.

We conclude this section with some preliminary discussion of the idea that extending HOL with existential quantification over types may make it possible to simplify the type specification rule discussed above in section 3.1. In particular, by using existential type quantification the type specification rule may be expressed as

$$\frac{\Gamma \vdash \exists\alpha.\,\mathrm{Q}}{\Gamma \vdash \mathrm{Q}[\nu/\alpha]}$$

That is, if one can prove that there exists a type α with the property given by Q, then it is consistent to extend the logic with a new type expression ν for which $\mathrm{Q}[\nu/\alpha]$ holds. This new formulation of the type specification rule seems somewhat more direct than the one shown in section 3.1, as it does not need the notion of a bijection between sets of values.

With this new formulation, however, the question arises as to whether one can always prove the hypothesis $\Gamma \vdash \exists\alpha.\,\mathrm{Q}$ required for any desired application of the rule. The existential introduction rule $\exists$-I may be used to prove propositions of this form, but this rule requires a witness for the type α asserted to exist. Moreover, this witness must already be expressible in the type language of HOL. It may, therefore, be necessary to add the following additional primitive rule

$$\frac{\Gamma \vdash \exists x:\sigma.\,\mathrm{P}\;x}{\Gamma \vdash \exists\alpha.\,\alpha \approx \mathrm{P}}$$

in order to make the new type specification rule shown above actually useful. This rule states that for any non-empty set of values, there exists a type whose elements are in bijection with this set. There need not, of course, be an actual type expression σ that denotes this set.

6 Implementation

The extensions proposed in the preceding sections have not yet been implemented. Adding terms of the forms $\forall\alpha.\,\mathrm{P}$ and $\exists\alpha.\,\mathrm{P}$ is an extension of the primitive term language of HOL so an implementation based on the existing HOL88 or HOL90 systems would require fairly extensive modifications. The type checker[2] would have to be extended, as would the parser and pretty-printer. The definitions of many of the syntax functions would have to be modified—for example the functions dealing with alpha-conversion, term substitution, type substitution, and free and bound variables. All this should be fairly straightforward, though reimplementing operations like substitution is likely to be error-prone. It is hoped that a prototype system can be be built in the near future, most likely based on HOL90, so that practical experiments can be carried out.

[2]Type checking and type inference for the extended system are still decidable.

Acknowledgements

I am grateful to Andy Pitts for discussions about extending the set-theoretic semantics of HOL, and to the reviewers for helpful comments on the presentation. I have found Andrews' book [1] invaluable in working out many of the technical details of the extension to HOL proposed here. This work was carried out at the University of Cambridge under SERC grant number GR/F 36675.

References

[1] P. B. Andrews, *A Transfinite Type Theory with Type Variables*, Studies in Logic and the Foundations of Mathematics series (North-Holland, 1965).

[2] A. Church, 'A Formulation of the Simple Theory of Types', *The Journal of Symbolic Logic*, Vol. 5 (1940), pp. 56–68.

[3] J.-Y. Girard, 'The System F of Variable Types, Fifteen Years Later', *Theoretical Computer Science*, vol. 45 (1986), pp. 159–192.

[4] M. J. Gordon, A. J. Milner, and C. P. Wadsworth, *Edinburgh LCF: A Mechanised Logic of Computation*, Lecture Notes in Computer Science, vol. 78 (Springer-Verlag, 1979).

[5] University of Cambridge Computer Laboratory, *The HOL System: DESCRIPTION*, revised edition (July 1991).

Higher Order Logic Theorem Proving and its Applications (A-20)
L.J.M. Claesen and M.J.C. Gordon (Editors)
Elsevier Science Publishers B.V. (North-Holland)

A Lazy Approach to Fully-Expansive Theorem Proving

Richard J. Boulton

University of Cambridge Computer Laboratory, New Museums Site, Pembroke Street, Cambridge, CB2 3QG, England.

Abstract

The HOL system is a *fully-expansive* theorem prover: proofs generated in the system are composed of applications of the primitive inference rules of the underlying logic. One can have a high degree of confidence that such systems are sound but they are far slower than theorem provers that exploit meta-theoretic or derived properties.

This paper presents techniques for postponing part of the computation so that the user of a fully-expansive theorem prover can be more productive. The notions of *lazy conversion* and *lazy theorem* are introduced. These not only allow part of the computation to be delayed, but also permit non-local optimisations that are only possible because the primitive inferences are not performed immediately. The approach also opens the way to proof procedures that exploit meta-theoretic properties without sacrificing security; the primitive inferences still have to be performed in order to generate a theorem but during the proof development the user is free of the overheads they entail.

Keyword Codes: F.4.1; I.2.3; D.1.1
Keywords: Mathematical Logic; Deduction and Theorem Proving; Applicative (Functional) Programming

1 Introduction

Proofs generated by a *fully-expansive* theorem prover consist of applications of primitive inference rules to axioms and preproved theorems of the logic. No use is made of meta-theoretic properties which effectively replace portions of the full proof by single steps that are believed to be valid. The HOL system [6] is a fully-expansive theorem prover.

Generating a proof entirely from primitive inferences provides security since, by use of suitable implementation techniques, the critical code of the system can be constrained to the implementation of the primitive inference rules. This also allows the system to be more flexible since users can be given the facilities to write their own derived inference rules without fear of them making the system unsound. The major drawback of the approach is that the system tends to be slow in comparison to theorem provers that exploit meta-theoretic properties.

Tests on the form of terms and on the side-conditions of rules have to be made for each application of a primitive inference rule. A derived rule composed from several primitive inference rules may, in a fully-expansive theorem prover, repeat the same test several times. In a *partially-expansive* theorem prover the derived rule would only make the test once. The potential for inefficiency in fully-expansive theorem provers is obvious. (A certain amount of compromise is possible: for simple derived rules that are used extensively it may be worth adding the derived rule to the critical code. This can improve performance significantly, but with a small price to pay in terms of having more code to verify or trust).

Another possible reason for the superior performance of many theorem provers over a system like HOL is their heavy use of meta-theoretic properties. Many popular decision procedures do not produce (or even follow) a proof in the object logic. One way around this problem is to prove the correctness of the decision procedure and then make it a new primitive inference rule of the system. The verification ensures that the extension is sound. This process is known as *reflection*. The difficulty with it is that in order to verify the implementation of the proof procedure the data-structures and algorithms used in the implementation must be represented and reasoned about in the object logic.

A document was circulated at the Second HOL Users' Meeting which contained an extract from a report by David Musser [8]. In this report Musser proposes the use of a different notion of *tactic* that postpones the justification phase of their operation. This opens up the possibility of proceeding with a proof believing a previous step to be valid when it is not, but does not breach security because the justification has to be performed in order for a theorem to be generated. Musser goes on to mention use of tactics that do not perform the justification step but have been proved to be valid. Howe [7] has done work along these lines in Nuprl. Musser suggests that a mixture of proved and unproved tactics should be allowed with only the unproved tactics having a justification step.

This paper describes a modification to the HOL system that not only allows the justification steps of tactics to be postponed but also permits postponement of the validity checks of inference rules. The technique makes use of *lazy theorems* and *lazy conversions*. Lazy theorems consist of the underlying term structure of a theorem together with a *proof function* capable of generating the theorem for that structure. The security of the system is not compromised. Lazy conversions are a similar idea specialised for equational reasoning.

2 A Simple Form of Lazy Theorem

An ordinary theorem in HOL is an object of ML type `goal` wrapped up as an abstract type `thm`. The type `goal` captures the structure of a theorem and is defined by the following ML type abbreviation:

```
lettype goal = term list × term
```

The list of terms represents the hypotheses of the theorem and the single term is the conclusion.

The simplest representation for a lazy theorem in the HOL system has the following ML type:

```
lettype lazy_thm = goal × (void → thm)
```

Elements of this type are pairs. The first component of such a pair captures the structure of the lazy theorem. The second component is a proof function which takes a dummy argument and returns a theorem. The dummy argument prevents the theorem from being computed until the proof function is applied. In this way the justification steps of a proof can be delayed. The function is essentially the 'lazy theorem', but in order to do anything useful with a lazy theorem, programs need to know its structure without applying the function (computing the theorem).

3 Lazy Conversions

In order to delay the justification steps of a proof for any significant length of time all parts of a proof must be made lazy, including equational reasoning which in HOL is done using *conversions* [9]. The type of conversions is defined by:

```
lettype conv = term → thm
```

If `CONV` is a conversion and t is a term then the application `CONV`(t) returns a theorem of the form:

$\vdash\ t = t'$

The obvious type for lazy conversions is:

```
lettype lazy_conv = term → lazy_thm
```

However, it is known that the theorem will be an equation in which the left-hand side is the original term, so the proof process can be optimised when the conversion does not change the term (i.e., t' is equal to t). This is now explained.

When a conversion is applied to a subterm s within a term t, the theorem:

$\vdash\ t = t[s'/s]$

is constructed, where s' is the result of applying the conversion to s and $t[s'/s]$ denotes t with s replaced by s'. (Strictly, it is only the occurrence of s to which the conversion is applied that is replaced). The construction of the theorem requires the application of significant numbers of primitive inference rules. The deeper the subterm is within the term, the more primitive inferences are required. However, when the conversion has no effect the theorem could be generated by a single primitive inference, namely the reflexivity rule. The rewriting facilities in HOL are an example of where a significant number of conversions are applied with no effect.

3.1 Optimising for Unchanged Subterms

By defining a new ML concrete datatype:

```
type lazy_eq = Lazy_eq_thm of void → thm
             | Lazy_refl
```

and changing the type of lazy conversions to:

```
lettype lazy_conv = term → (goal × lazy_eq)
```

the wasted inferences in building equations for unchanged subterms can be optimised away. When a conversion has no effect the element `Lazy_refl` is returned. The rules for building an equation for the whole term propagate this value. The justification step of the proof normally applies the proof function of a `Lazy_eq_thm` value to generate a theorem for the equation, but if the value is `Lazy_refl` the justification step simply applies the reflexivity rule (stating that a term is equal to itself) to the original term. Hence all the unnecessary inferences in rebuilding unchanged subterms are optimised away.

A similar effect can be achieved for ordinary (non-lazy) conversions by raising an exception when the subterm is not changed. This exception can be captured by the proof functions being applied higher up in the term structure. The depth conversions in HOL88 Version 2.0 use this technique (suggested by Roger Fleming of Hewlett Packard) to significantly improve the performance of rewriting. (Depth conversions are functions which try to apply a conversion to subterms of a term using various term traversal strategies). Use of `Lazy_refl` is not quite as efficient as using exceptions, but applies everywhere conversions are used (provided they are made lazy). `Lazy_refl` is also easier to program with, since programs using exceptions become littered with exception traps.

3.2 Optimising Repeated Dives

A further, though less significant, optimisation can be made in a lazy framework. During equational reasoning the conversion functions may dive down to a subterm several times, possibly diving to another subterm in between. Each dive uses applications of primitive inference rules in order to transform the equation about the subterm into an equation about the whole term. This is because the primitive substitution rule in HOL does not allow substitution inside bindings. The repeated dives to a subterm can often be optimised to a single dive by re-ordering the operations. When theorems are being generated immediately the order of the operations cannot be changed, but in a lazy framework they can be. The ML type can be extended to exploit this:

```
rectype lazy_eq = Lazy_eq_thm of void → thm
                | Lazy_refl
                | Lazy_comb of lazy_eq × lazy_eq
                | Lazy_abs of lazy_eq
```

This type is now recursive. Elements of the type are tree structures similar to terms. There is a branch node for function applications (combinations) and a branch node for λ-abstractions, just as there is in terms. Repeated dives cause parts of the structure (but not theorems) to be rebuilt. Where possible the repeated dives are combined into a single dive. The cost of reconstruction of the `lazy_eq` is small in comparison to the cost of reconstructing a theorem. The reconstruction leaves the old structures as garbage, but the costs of this are compensated for by the reduction in the number of primitive inferences performed afterwards in the justification phase.

The recursive type given above is not sufficiently general to represent all the operations of conversions. It assumes that a subterm into which a dive has taken place retains the same structure throughout the conversion. Consider the following term:

```
(λx. x + 1) (2 + 3)
```

A conversion might dive into the argument of the application and replace the `(2 + 3)` with `5`. This would produce the structure:

```
Lazy_comb (Lazy_refl,Lazy_eq_thm (-))
```

where `(-)` represents a function. The operator, `(λx. x + 1)`, has not been changed. This is represented by the presence of a `Lazy_refl`. The conversion might now perform a β-reduction at the top-level of the term to give:

```
5 + 1
```

How can this be represented? A sequence consisting of the structure above followed by a `Lazy_eq_thm` for the β-reduction is required. The following modification to the type definition captures this:

```
rectype lazy_eq = Lazy_eq_thm of void → thm
                | Lazy_comb of (lazy_eq list) × (lazy_eq list)
                | Lazy_abs of (lazy_eq list)
```

The type of lazy conversions also has to be changed:

```
lettype lazy_conv = term → (goal × (lazy_eq list))
```

Note that `Lazy_refl` values are no longer required. Instead, an empty list represents 'no change'. The example conversion now gives rise to the following structure:

```
[Lazy_comb ([],[Lazy_eq_thm (2+3=5)]); Lazy_eq_thm (β-reduce)]
```

where '*2+3=5*' and '*β-reduce*' are comments indicating what the functions stored at those positions do.

3.3 Transparent Optimisation

So, lazy conversions provide transparent optimisation of equational reasoning. Unnecessary inferences are eliminated in two circumstances:

- When the program builds equations between an unchanged subterm and itself.
- When the program repeatedly dives into a term to perform conversions when one dive would be sufficient.

More subtle programming of the proof procedure could be used to achieve the same results but with lazy conversions the programmer does not have to worry about the inefficiencies. The optimisations are made automatically. The details are hidden from the programmer by re-implementing Paulson's basic conversions and conversionals [9] as lazy conversions. For example, the identity lazy conversion is:

```
let ALL_CONV:lazy_conv = λtm. (([],tm), [])
```

This simply takes a term and returns the same term as the result with an empty hypothesis list and an empty list of changes (`lazy_eq` list). The function for sequencing lazy conversions is `THENC`. It is an infix function:

```
let (lconv1:lazy_conv) THENC lconv2 =
   λtm. let ((hyps1,tm1),leqs1) = lconv1 tm
        in  let ((hyps2,tm2),leqs2) = lconv2 tm1
        in  if (null leqs1) then ((hyps2,tm2), leqs2)
            if (null leqs2) then ((hyps1,tm1), leqs1)
            else ((union hyps1 hyps2,tm2), sequence_lazy_eqs leqs1 leqs2)
```

The lazy conversion generated by sequencing `lconv1` and `lconv2` takes a term and applies `lconv1` to it. `lconv2` is then applied to the resulting term. The function tests to see if either of the `lazy_eq` lists are empty (indicating that the term was not changed). If this is so, the computation can be optimised as indicated above. In particular, if both `lazy_eq` lists are empty then the list returned will be empty. If neither list is empty the set-theoretic union of the hypotheses is formed and the `lazy_eq` lists are combined using the function `sequence_lazy_eqs`. This constructs a new `lazy_eq` list in such a way that any repeated dives into a subterm are avoided. Note that structures are discarded in this process, so the garbage generated due to lazy conversions does not just consist of the structure for the final result but also includes all the intermediate structures.

3.4 Manipulating Subterms from the Root of the Term

The improved performance of rewriting is an obvious benefit of using lazy conversions, but there are other significant examples. A group of examples arise when the issue of integrating other systems with a theorem prover like HOL is considered. Two such external systems are the graphical user interface Centaur [1] and the proof planner CLAM [4]. The interfaces between external systems and a theorem prover are likely to be fairly abstract;

it is undesirable to have an external system concern itself with optimising performance in the theorem prover.

Consider the following term (the step case from the inductive proof of the associativity of addition):

```
(SUC x) + (y + z) = ((SUC x) + y) + z
```

In order to apply the induction hypothesis, the following theorem (part of the definition of addition) must be used three times:

```
⊢ ∀m n. (SUC m) + n = SUC (m + n)
```

If using a graphical user interface to perform this proof, the user might click on the subterm `((SUC x) + y)` and apply the theorem. This subterm can be considered to be at coordinates `[2;1]` of the term (the first argument of the '`+`' that forms the second argument of the '`=`'). Application of the theorem gives rise to the term:

```
(SUC x) + (y + z) = (SUC (x + y)) + z
```

The user might then decide to apply the theorem to the entirety of the left-hand side of the equation (coordinates `[1]`) to produce:

```
SUC (x + (y + z)) = (SUC (x + y)) + z
```

The third application of the theorem is on the entirety of the right-hand side of the equation (coordinates `[2]`). The result is:

```
SUC (x + (y + z)) = SUC ((x + y) + z)
```

There are two things to note here. First, the user interface will tell the theorem prover where to apply the theorem by giving coordinates relative to the root (top) of the term. In the example this is not too significant as regards performance, but when the subterm is deep within the term, the costs of repeatedly building equations for the entire term when multiple manipulations are taking place at the subterm are substantial. The CLAM proof planner also provides instructions to the theorem prover in the form of coordinates relative to the top of the term. The second point is that the user has chosen an order of application of the theorem that is non-optimal because the application on the left-hand side of the equation is made between the two applications on the right-hand side.

Lazy conversions provide optimisation in these situations without the external system or the user having to consider it. There are other ways of optimising the order of equational manipulations, but lazy conversions provide a neat way of doing it within a more generally useful framework. They also allow reasonably efficient equational reasoning procedures to be written in a style in which the programmer can think of subterm positions relative to the top of the term. This should make the code simpler and easier to follow. However, because there is still some overhead in maintaining the `lazy_eq` structures, full optimisation will only be obtained if the procedure is written so that the optimal ordering is used explicitly.

4 Lazy Theorems as an Abstract Type

Lazy conversions not only provide optimisation of equational reasoning, they also allow the justification phase of a proof to be postponed. This is of little use in isolation because most proofs contain blocks of equational reasoning interspersed with more general logical reasoning. To exploit the delay a more general notion, that of a lazy theorem, is required. A simple implementation of lazy theorems is described in Section 2. Unfortunately this implementation has a very serious drawback, as now described.

Suppose the following lazy theorem has been generated:

```
((["x ∧ y"],"y ∧ x"), f)
```

where f is the proof function. The inference rule `CONJ_PAIR` takes a theorem whose conclusion is a conjunction and generates two theorems, one for each conjunct. A lazy version of this rule would when applied to the lazy theorem above produce the following:

```
((["x ∧ y"],"y"), f1)        ((["x ∧ y"],"x"), f2)
```

Both of the functions f_1 and f_2 will contain evaluations of f. So when f_1 is evaluated all the computation involved in evaluating f will be performed. This is also the case when f_2 is evaluated. So, f will be evaluated twice. As other rules are applied to the two new lazy theorems the number of times f can be evaluated may increase further.

One solution to this problem is to make a lazy theorem an updatable (assignable) entity (`ref` types in Standard ML), so that once a lazy theorem has been fully proved by evaluating the proof function, the lazy structure is replaced by a real theorem. This can be achieved by having a concrete representation type for lazy theorems that consists of two possible structures:

```
type lazy_thm_rep = Lazy_thm of goal × (void → thm)
                  | Proved_thm of thm
```

So, the representation is either a lazy theorem as described up to this point, or it is a fully-proved theorem.

ML assignable variables can be used at the top-level of ML to provide the required updatability properties. However in HOL88, they cannot directly be generated within a function and returned from the function. It is normally only the value at return-time that will be returned, not the updatable cell itself. However, the desired effect can be achieved by wrapping the assignable variable in λ-abstractions. The function to generate a new updatable lazy theorem representation is:

```
let lazy_thm_rep_ref (lthrep:lazy_thm_rep) =
   letref x = lthrep
   in ((λlthrep. x := lthrep),(λ():void. x))
```

Within the body of the function an assignable variable is declared and a pair of λ-abstractions is returned. The first component of the pair is a function for storing a new value in the reference cell. The second is a function for obtaining the current value of the cell.

The functions `store` and `fetch` below provide a syntactic sugaring for manipulation of the reference cells:

```
let store cell val = (fst cell) val
and fetch cell = (snd cell) ()
```

These reference cells can be used to implement lazy theorems that will only ever be proved once. In the example with the rule `CONJ_PAIR`, when one of the resulting lazy theorems is proved, the function f will be evaluated and the result used to replace the original lazy theorem. This result will be waiting in the reference cell ready for use when the other lazy theorem is proved.

To ensure that reference cells are always correctly updated, they can be wrapped-up in an abstract datatype:

```
abstype lazy_thm = (lazy_thm_rep → lazy_thm_rep) × (void → lazy_thm_rep)
   with mk_lazy_thm glf = abs_lazy_thm (lazy_thm_rep_ref (Lazy_thm glf))
   and mk_proved_thm th = abs_lazy_thm (lazy_thm_rep_ref (Proved_thm th))
   and goal_of_lazy_thm lth =
      let cell = rep_lazy_thm lth
      in  case (fetch cell)
          of (Lazy_thm (gl,_)) . gl
           | (Proved_thm th) . (dest_thm th)
   and prove_lazy_thm lth =
      let cell = rep_lazy_thm lth
      in  case (fetch cell)
          of (Lazy_thm glf) .
                (let th = apply_proof_fun glf
                 in  store cell (Proved_thm th); th)
           | (Proved_thm th) . th
```

`abs_lazy_thm` and `rep_lazy_thm` are the abstraction and representation-extraction functions for the abstract datatype. `dest_thm` returns the structure of a theorem (a `goal`), and `apply_proof_fun` is described below. The user cannot determine the form of the underlying representation of a lazy theorem. There is no way to know whether it has been proved or not.

The abstract datatype consists of four functions: one to generate a lazy theorem from a goal and a proof function, one to generate a lazy theorem from a proved theorem, one to obtain the structure of a lazy theorem (the goal), and one to obtain a proved theorem from a lazy theorem. It is in a call to the last function that the proof justification is performed using `apply_proof_fun`:

```
let apply_proof_fun (gl,f) =
   let th = f ()
   in  if (dest_thm th = gl)
       then th
       else failwith `apply_proof_fun -- proof does not prove goal`
```

This function takes the proof-function part of the representation of an unproved lazy theorem and applies it to the dummy value '`()`' of type `void`. The result is a theorem, which should have the same structure as the goal part of the lazy theorem representation. However, the user could have made a lazy theorem from one structure together with a function that proves a theorem with a different structure. `apply_proof_fun` therefore checks that this is not the case before it returns the theorem. It fails if the wrong theorem is generated, and of course it is possible for the application of the proof function to fail rather than generate a theorem.

5 Lazy Inference Rules

In the previous section an abstract datatype for lazy theorems was defined. The next step towards exploiting laziness within the HOL system is to define lazy versions of the inference rules.

5.1 Primitive Rules

The primitive inference rules of HOL88 use a function called `mk_thm` to generate a theorem from a goal. When making a lazy version of a primitive inference rule, all the validity checks should be performed within the initial phase of computation, otherwise a lazy theorem may be generated for which the proof function will fail. The only part of the code for the primitive inference rule to be put into the proof function is the call to `mk_thm`. The lazy version of the primitive inference rule `ABS` illustrates this. `ABS` has the following property:

$$\frac{\Gamma \ \vdash \ t_1 = t_2}{\Gamma \ \vdash \ (\lambda x.\ t_1) = (\lambda x.\ t_2)} \ x \text{ is not free in } \Gamma$$

The (not quite correct) lazy implementation is:

```
let ABS x lth =
   let (gamma,eqn) = goal_of_lazy_thm lth
   in  if mem x (freesl gamma)
       then failwith `ABS`
       else let (t1,t2) = dest_eq eqn
            in  let gl = (gamma,mk_eq(mk_abs(x,t1),mk_abs(x,t2)))
            in  mk_lazy_thm (gl,(λ(). mk_thm gl))
```

where `lth` is the lazy theorem on the top of the rule. The function checks that the ML variable `x` (which is bound to the object language variable to be used in the abstraction) does not occur in the list of free variables obtained from the hypotheses (**gamma**) of **lth**. If the check succeeds the underlying structure for the new lazy theorem is built and the lazy theorem is then generated. Note that only the call to **mk_thm** goes into the proof function. The checks are performed during the initial phase of computation. **mk_thm** does actually perform one check itself, namely that the terms used for the hypotheses and conclusion are Boolean valued. In the definition of **ABS** an explicit check is not necessary because the hypotheses are the same as for the original lazy theorem and the conclusion is an equation.

However, there is something seriously wrong with the implementation of the lazy rule above: it does not check that the original lazy theorem actually gives rise to a real theorem. This is a hole in the security of the system because the user could supply a lazy theorem consisting of an invalid statement and a proof function that does not work. (Since the statement it corresponds to is invalid it cannot work). The rule above would then construct a proof function for the new lazy theorem capable of generating a *real* theorem for an invalid statement. The solution to this problem is simple. The original lazy theorem must be proved within the proof function of the new lazy theorem, *before* the **mk_thm** is applied:

```
let ABS x lth =
   let (gamma,eqn) = goal_of_lazy_thm lth
   in  if mem x (freesl gamma)
       then failwith `ABS`
       else let (t1,t2) = dest_eq eqn
            in  let gl = (gamma,mk_eq(mk_abs(x,t1),mk_abs(x,t2)))
            in  mk_lazy_thm (gl,(λ(). prove_lazy_thm lth; mk_thm gl))
```

This is only a problem when **mk_thm** is being used. Derived lazy inference rules will be defined in terms of the primitive rules, so all necessary validity checks will be made, provided the primitive rules make them.

5.2 Derived Rules

Derived rules can be optimised at the risk of generating lazy theorems that will not be provable. If the derived rule is defined in terms of primitive rules then all the checks of the primitive rules will be performed during the initial phase of computation. The inherent repetitive computation will not be delayed. Below is an example of such an implementation for the derived rule that adds a term to the assumptions:

```
let ADD_ASSUM t lth = MP (DISCH t lth) (ASSUME t) ? failwith `ADD_ASSUM`
```

An alternative implementation makes the necessary checks only once during the initial phase of computation:

```
let ADD_ASSUM t lth =
   if (type_of t = bool_ty)
   then let (gamma,conc) = goal_of_lazy_thm lth
        in  let gl = (union [t] gamma,conc)
        in  mk_lazy_thm
               (gl,(λ(). prove_lazy_thm (MP (DISCH t lth) (ASSUME t))))
   else failwith `ADD_ASSUM`
```

Initially this version checks only that the term to be added to the assumptions is Boolean valued. For the purposes of the initial computation this makes the implementation into a primitive rule because a lazy theorem can be generated without all the corresponding primitive inferences being performed. However, the implementation does not use `mk_thm`, so the rule is *derived* with respect to generating real theorems, and so cannot breach the security of the system. For a simple derived rule like `ADD_ASSUM` which only uses three primitive rules, the saving in the time to perform the initial computation may not be significant. However, for derived rules that use tens or hundreds of primitive rules it will be. Note that all the tests necessary to ensure validity are performed in the initial computation, so it is reasonable to expect that a real theorem will be generated when the proof function is applied.

5.3 Laziness Avoids Some Unnecessary Computation

Just as lazy conversions can save inferences without any special coding being required, so can lazy theorems. The function `CONJUNCTS_CONV` takes a pair of Boolean-valued terms and determines whether they are equivalent under associativity, commutativity and idempotency of conjunction. If they are, the function returns a theorem stating the equality. Consider the following pair of terms:

```
("x:bool","x ∧ x")
```

The two terms are equivalent statements. The function proves that `(x = x ∧ x)` by proving `(x ⟹ x ∧ x)` and `(x ∧ x ⟹ x)`. In ordinary HOL, the function performs eight primitive inferences. With lazy rules only seven are performed. The proof of `(x ∧ x ⟹ x)` proceeds by assuming `(x ∧ x)`, i.e., the theorem:

```
x ∧ x ⊢ x ∧ x
```

This is then split into two theorems:

```
x ∧ x ⊢ x     x ∧ x ⊢ x
```

Note that because `x` appears as both arguments of the conjunction the two theorems generated are the same. So, only one of them is required for the proof. The other is never used. In ordinary HOL, this theorem is generated anyway, but with lazy rules only the initial phase of computation is performed. The proof function is discarded and the primitive inference within it is never executed.

6 Three Modes of Operation

It is simple to modify the abstract datatype for lazy theorems so that the theorem prover can operate in different modes. Three modes that have been implemented are *lazy*, *eager* and *draft*. Lazy mode has formed the main topic of this paper so far. Eager mode is the way in which the ordinary HOL system operates; theorems are generated immediately and stored as proved theorems within the representation for lazy theorems. In draft mode the proof functions are not maintained within the lazy theorems. A dummy function is used instead. This allows a proof to be tested without theorem generation taking place, but the lazy theorem produced at the end will not give rise to a real theorem. Lazy mode delays the computation and indeed the computation need never be performed. However, draft mode has the advantage that because it does not retain the proof functions, there is no worry that memory will be exhausted and the garbage collection overheads are reduced.

This system of operating modes allows the lazy version of the theorem prover to be used as an ordinary system (eager mode) with the only change being that the theorems are wrapped-up as lazy theorems. More interestingly, draft mode can be combined with one of the other modes in two passes through a proof. The proof can be developed in draft mode to obtain a lazy theorem. Then if the user is concerned about security the proof can be repeated non-interactively in lazy or eager mode. The proof script developed in draft mode will work without modification in the other modes. The only changes required will be the setting of the mode flag at the beginning of the session, and possibly the generation of a real theorem from the lazy theorem at the end of the session.

7 Results

One of the original motivations for the development of lazy conversions was a proof procedure for a subset of linear arithmetic [2]. This procedure operates in two phases. The first phase normalises the formula using conversions. The second uses more general rules to simplify the normalised term, and hopefully to solve it. The search space for this second phase is quite large. A lazy approach is used to separate the search from the theorem generation. The search can be done very efficiently because after normalisation the formula can be represented in an abstract way using lists of variable names and coefficients. The laziness can only fully be exploited if the normalisation phase is lazy too. Hence the motivation for lazy conversions.

Table 1 compares three implementations of the normalisation phase of the arithmetic procedure. The three implementations use ordinary HOL conversions, conversions optimised using exceptions, and lazy conversions, respectively. Figures are also given for the initial phase of computation in the lazy implementation. This phase obtains a result but does not generate the theorem for it. Table 2 makes similar comparisons for the entire arithmetic procedure including the normalisation.

The tables should be interpreted as follows. 'Run' is the run time in seconds. This does not include garbage collection time. The garbage collection time in seconds is given separately and is labelled 'GC'. 'PInfs' is the number of applications of primitive inference rules used to prove the theorems. The inferences are considered to have taken place only

Table 1: Comparison of Implementations of Normalisation

	Run	GC	PInfs
Ordinary conversions	204.3	78.0	27937
Conversions using exceptions	109.2	42.8	9482
Lazy conversions	106.2	48.8	8433
Lazy conversions (initial computation)	39.2	16.2	0

Table 2: Comparison of Implementations of Arithmetic Proof Procedure

	Run	GC	PInfs
Ordinary conversions	349.4	141.7	49680
Conversions using exceptions	205.1	82.8	21208
Lazy conversions	205.2	95.5	19780
Lazy conversions (initial computation)	43.5	20.5	0

when the real theorems have been generated. The figures given are for a test file containing 54 linear arithmetic formulae and were obtained using HOL88 Version 2.0 built under Franz Lisp and running on a Sun 3/160 with 12 Mbytes of real memory.

The results for the implementation optimised using failure and the lazy implementation are very similar. Slightly fewer theorems are generated in the lazy version thanks to the ability to re-order proof steps. This is offset by the slightly larger garbage collection times due to the use of additional data-structures.

Observe that for the lazy implementation about 50% of the full run time is in the post-normalisation stage, but only about 10% $((43.5 - 39.2)/43.5)$ of the initial computation is involved in this stage. This rapid generation of a result is thanks to the use of lists of variable names and coefficients in place of term structures in finding a solution. In the normalisation stage about 60% $((106.2 - 39.2)/106.2)$ of the computation time is delayed. The overall effect for the proof procedure is for about 80% of the computation to be delayed, i.e., when used lazily the procedure will appear to be 5 times faster than the version optimised using exceptions.

Although the arithmetic procedure (all implementations) makes considerable use of laziness in the post-normalisation stage, lazy theorems as described in Section 4 are not used. The simple notion of lazy theorem described in Section 2 is used instead. This is reasonable because their use is constrained within the proof procedure where the problems can be avoided.

In an attempt to investigate the performance of lazy theorems for a typical proof, the multiplier benchmark for the HOL system has been modified to function in a lazy environment. Part of the proof involves the use of rewriting and some more specialised conversions, but much of it is general tactic-based reasoning. Results for the Franz Lisp version of HOL on a Sun3/160 with 12 Mbytes of real memory are given in Table 3. The results for Allegro Common Lisp on a SPARCstation SLC with 16 Mbytes of real memory can be found in Table 4. 'Run', 'GC' and 'PInfs' are as before. 'Total' is the total execution time (run time plus garbage collection time) in seconds.

Results are given for three systems. The first is the benchmark proved normally using

Table 3: Multiplier Benchmark for Sun3/Franz Lisp

	Run	GC	Total	PInfs
Ordinary theorems in ordinary HOL	344.0	122.5	466.5	17115
Ordinary theorems in lazy HOL	345.8	245.3	591.1	17115
Lazy theorems (total computation)	330.1	457.2	787.3	13560
Lazy theorems (initial computation)	260.5	369.9	630.4	309
Lazy (total) / Ordinary in lazy HOL	95%	186%	133%	
Lazy (initial) / Ordinary in lazy HOL	75%	151%	107%	

Table 4: Multiplier Benchmark for Sun4/Allegro Common Lisp

	Run	GC	Total	PInfs
Ordinary theorems in ordinary HOL	291.3	36.2	327.5	17115
Ordinary theorems in lazy HOL	299.8	41.9	341.7	17115
Lazy theorems (total computation)	326.2	71.7	397.9	13560
Lazy theorems (initial computation)	226.7	51.8	278.5	309
Lazy (total) / Ordinary in lazy HOL	109%	171%	116%	
Lazy (initial) / Ordinary in lazy HOL	76%	124%	82%	

HOL88 Version 2.0. The second is a proof using conventional theorems but performed in a version of HOL extended with support for lazy conversions and lazy theorems. This provides a more realistic comparison with the third system which uses lazy conversions and lazy theorems. This is because garbage collection times increase significantly as the size of the Lisp executable image increases and the addition of the support for laziness causes just such an increase. In a 'production' version, the lazy versions of rules and tactics would replace the ordinary versions, so the increase in image size would be much smaller. The benchmark has had to be modified slightly to perform the tests in the lazy HOL and the same modifications are applied for the tests using ordinary theorems so that a fair comparison can be made. The fourth line of figures in the tables are the initial phase of computation when using lazy theorems, that is the time taken for a lazy theorem to be returned. The 309 primitive inferences performed during the initial computation are due to definitions of constants being made. Since real theorems are stored in the theory files during this process it cannot be made lazy.

The lazy system made use of special primitives, written in Lisp, for updatable cells. The primitives are used in the abstract datatype for lazy theorems in place of the pair of functions (see Section 4). This was necessary because the proof functions within a lazy theorem were not being garbage collected when they had been applied and replaced by the resulting theorem. The reason for this is not certain but is probably a consequence of the way in which the ML compiler deals with λ-abstractions.

The results show that laziness does not dramatically affect the run time. In Franz Lisp the lazy version is a bit faster, but in Allegro it is somewhat slower than using ordinary theorems. However, the garbage collection times are considerably greater for the lazy version. This is more of a problem in Franz Lisp because the resident size of the code plus data has a fixed limit. The Lisp could be rebuilt with a larger value for this limit.

The results show that provided garbage collection doesn't become too dominant (it doesn't in the Allegro run of the benchmark) there is a small advantage to be had from using lazy theorems. Overall computation time is not increased too significantly and lazy theorems can be obtained in about 80% of the time it would take to obtain a real theorem in HOL. Laziness is not being exploited to any great extent in the benchmark. In Section 9 a number of applications in which laziness could be of considerable benefit are discussed, and in Section 10 the issue of very large proofs is considered.

8 A Complete Lazy System

Lazy theorems would not be used everywhere in the theorem prover. In HOL, for example, only theorems can be stored in theory files. This would continue to be the case in a lazy system. Storing lazy theorems would not be allowed. The security of the system would be retained provided users realise that a real theorem, not a lazy theorem, must be generated before they can be sure of the formula's validity.

This raises some issues concerning the interaction between real and lazy theorems. Most of the functions in HOL would be modified to work with lazy theorems. However, those that access the theory files would continue to deal with theorems, as would the functions for defining new logical types and constants because these access the theory as a side-effect. Functions that duplicate the theory file operations but also convert between lazy and real theorems might also be provided. Of course any such function that converts from a lazy to a real theorem might take a long time to execute because it would be performing the justification stage of the proof.

Another significant issue is the interaction between lazy conversions and lazy rules. In ordinary HOL, conversions have ML type:

```
term → thm
```

so they are a form of rule. However, lazy conversions have ML type:

```
term → (goal × (lazy_eq list))
```

rather than:

```
term → lazy_thm
```

so some conversion between these two types would be required wherever conversions are used as rules. This is not difficult but might require significant changes to users' code.

9 Applications

Laziness in a fully-expansive theorem prover will only be worthwhile if it is exploited. In general, the proportion of the computation that is delayed will increase as the code performing the initial computation becomes less like the code for the justification phase. The further the initial computation is from performing primitive inferences, the greater the saving can be. Some possible applications of laziness are discussed below.

9.1 Unsafe Derived Rules

Currently in HOL, derived rules can be made more efficient by adding them to the critical code of the system. This is 'unsafe'. However, in a lazy system the rules operate over lazy theorems so the initial phase of their computation (the time taken to generate a lazy theorem) can be made more efficient without the risk of making the system insecure. The initial phase of computation can employ any available short-cuts while the justification phase (the proof function) continues to perform all the primitive inferences properly. The overall execution time is not reduced but the time to obtain a lazy theorem is.

9.2 Decision Procedures

In Section 7, a proof procedure for a subset of linear arithmetic is mentioned together with results for a lazy implementation of it. The results show that as much as 80% of the computation can be postponed until the justification phase of the lazy theorem. It is likely that similar gains can be made for other proof procedures of this kind.

9.3 Evaluating Constant Natural Number Expressions

The main number system in HOL is a unary representation of natural numbers. Evaluating constant natural number expressions in the logic is very time consuming if done properly by primitive inferences. Lazy rules will be able to produce a result using machine arithmetic in a fraction of the time the inferences take. The lazy theorem produced will however perform the proper primitive inferences during its justification stage.

9.4 Symbolic Compilation and Execution

Fully-expansive theorem provers are often used for formal verification and for mechanising programming language semantics. This may involve symbolic compilation or execution of logical terms representing programs [5]. The numbers of primitive inferences involved in these processes can be huge. Laziness offers the ability to perform the compilation and execution rapidly using an ML or Lisp program while retaining security by checking the result by primitive inference later. This is particularly valuable because a program may be compiled and executed many times before it is perfected, and a theorem is only required for the final run.

9.5 External Theorem Provers and Proof Planners

Various proof tasks (subgoals) could, given a suitable interface, be shipped off to an external theorem prover (such as the Boyer-Moore Theorem Prover [3]) or a proof planner (such as CLAM [4]). For this to ultimately be secure the external system must return not only a result but also a proof of the result that can be checked by the fully-expansive theorem prover. In a lazy system the result could initially be trusted to give the structural part of a lazy theorem while the proof function of the lazy theorem is set up to execute the proof returned by the external system.

Schneider, Kumar and Kropf [10] have combined a first-order theorem prover written in ML with the HOL system so that certain goals in HOL can be solved by the first-order prover. They offer two modes of operation, one safe in which the primitive inference rules are performed, and the other unsafe in which the result of the first-order prover is trusted. Lazy theorems would allow the performance gains of this approach without bypassing the security of the HOL system.

10 Conclusions and Future Work

The lazy approach described in this paper has three advantages:

- Non-local optimisations can be made.
- The justification phase of theorem generation is avoided for theorems that are never used.
- The justification phase is delayed, giving users greater control over their time.

Individual optimisations could be achieved by similar techniques within the existing HOL framework, but a real theorem would have to be generated between each of the optimisations. This would cause many of the benefits to be lost. The optimisations would not build up when proof procedures are combined, and it is only when small delays are chained together into large delays that users gain the freedom to choose when to perform the justification phase of a proof.

The results given in Section 7 suggest that there is little to be gained from a lazy approach for the kinds of proof typically done in HOL. However, some things are not done in HOL because they take too long. Many forms of automatic proof fall into this area, and could be made more practical by the use of lazy theorems. Since it appears that there is little to be lost by taking a lazy approach it is probably worth doing so, so that meta-theoretic properties and short-cuts can be exploited without security being sacrificed. Section 9 listed a number of promising applications of laziness.

It is also worth noting that lazy theorems are not incompatible with reflection mechanisms: those inference rules that can be verified can be used without proof by means of a reflection principle; those rules for which verification is not practical can use unverified code to perform the initial generation of a lazy theorem and primitive inferences in the justification stage.

Laziness does not breach the security of a fully-expansive theorem prover. An invalid formula cannot become a theorem because of the laziness. The worst that can happen is for users to be fooled for a while into thinking they have successfully proved a conjecture when in fact the proof function will not be able to generate a real theorem corresponding to the lazy theorem. Obviously, this would be annoying, but sufficient quality control on the code would make such occurrences rare events. There is no reason why failure of lazy theorems should be any more common than errors in a partially-expansive theorem prover. Furthermore, a failure to generate a theorem would bring to light the error in the code whereas in an insecure system the error might go unnoticed.

The biggest problem with a lazy approach is the extensive use of memory it makes and the extra garbage collection this leads to. It is difficult to do fair tests of the performance until a proper lazy HOL system is built. At the moment all the lazy versions of rules, etc., are additions to the system, extending the code size and reducing the amount of memory available for data. In a proper lazy system most of the functions will be used as replacements for the non-lazy versions. Obviously large memories would make a lazy system more practical, but there will always be proofs so large that garbage collection times become a dominant factor as the memory is filled with the proof functions. However, in these situations a combination of the draft and eager modes of operation described in Section 6 could be used.

One technical difficulty with a lazy approach is the interaction with ML exceptions (failures). In order for the correct behaviour to be retained, lazy conversions and lazy theorems must always be set up so that any exceptions that can be raised will be raised during the initial phase of computation. This is possible but it has to be done explicitly by the programmer and can reduce the potential for optimisation since certain data may have to be computed in the initial phase in order to determine whether or not to raise the exception.

Implementing the HOL system in a lazy functional programming language rather than in ML would not achieve the same effects as described in this paper. This is because as the system is implemented at this time the generation of the structure of a theorem is inextricably linked with performing the primitive inferences that justify it. So, in a lazy language, as soon as one required the structure for a theorem, and so forced the computation of it, the justification would be performed as well. There would of course be the normal savings associated with lazy languages, i.e., any intermediate theorems that were not actually used to obtain the result would not be computed. However, if the structure of a theorem were required in order to determine whether or not to use it, it would still be computed in full, including the justification.

There are many avenues for future work. Extensive testing requires that a proper lazy system be built. At present the lazy versions of functions have different names to the original functions. This means that proof scripts require a lot of tedious editing of function names before they can be used. Replacing the original functions with lazy versions and using the original names for them will relieve this burden substantially.

It would be interesting to see how using general-purpose lazy theorems in the arithmetic procedure would affect the performance (see Section 7). At present a specialised form of lazy theorem is used that is highly optimised for the particular procedure.

Other possible work includes the optimisation of all the derived rules in a lazy system so that they take short-cuts during the initial computation of lazy theorems, and writing a function to sweep HOL's goal stack so that if memory becomes short during an interactive backwards proof, users can instruct the system to convert all the lazy theorems in the goal stack into real theorems while they themselves do some other work. A more radical idea is to provide system support for lazy theorems so that the system could prove them in the background while the user is idle. Finally there are the applications discussed in Section 9 to be investigated.

Acknowledgements

Thanks are due to Mike Gordon for his motivation and guidance, and for delving into Lisp to find and overcome problems with garbage collection. Thanks also to John Van Tassel for help with Lisp and to Jim Grundy, Mike Gordon and the referees for their comments on earlier versions of this paper. This work was supported by the Science and Engineering Research Council of Great Britain.

References

[1] P. Borras, D. Clément, T. Despeyroux, J. Incerpi, G. Kahn, B. Lang, and V. Pascual. CENTAUR: The system. In *SIGSOFT'88, Third Annual Symposium on Software Development Environments*, Boston, November 1988.

[2] R. J. Boulton. On efficiency in theorem provers which fully expand proofs into primitive inferences. Technical Report 248, University of Cambridge Computer Laboratory, February 1992.

[3] R. S. Boyer and J S. Moore. *A Computational Logic Handbook*, volume 23 of *Perspectives in Computing.* Academic Press, 1988. Series edited by W. Rheinboldt and D. Siewiorek.

[4] A. Bundy, F. van Harmelen, J. Hesketh, and A. Smaill. Experiments with proof plans for induction. *Journal of Automated Reasoning*, 7(3):303–324, 1991.

[5] J. Camilleri. Symbolic compilation and execution of programs by proof: A case study in HOL. Technical Report 240, University of Cambridge Computer Laboratory, December 1991.

[6] M. J. C. Gordon. HOL: A proof generating system for higher-order logic. In G. Birtwistle and P. A. Subrahmanyam, editors, *VLSI Specification, Verification and Synthesis*, Kluwer, 1987.

[7] D. J. Howe. Computational metatheory in Nuprl. In *Proceedings of the 9th International Conference on Automated Deduction*, volume 310 of *Lecture Notes in Computer Science*, pages 238–257. Springer-Verlag, 1988.

[8] D. R. Musser. Report on the HOL (Higher Order Logic) proof checker. Computer Science Department, Rensselaer Polytechnic Institute, 1989.

[9] L. Paulson. A higher-order implementation of rewriting. *Science of Computer Programming*, 3:119–149, 1983.

[10] K. Schneider, R. Kumar, and T. Kropf. Automating most parts of hardware proofs in HOL. In K. G. Larsen and A. Skou, editors, *Proceedings of the Third International Workshop on Computer Aided Verification (CAV'91)*, volume 575 of *Lecture Notes in Computer Science*, pages 365–375, Aalborg, Denmark, July 1991.

Higher Order Logic Theorem Proving and its Applications (A-20)
L.J.M. Claesen and M.J.C. Gordon (Editors)
Elsevier Science Publishers B.V. (North-Holland)

Efficient Representation and Computation of Tableaux Proofs

Klaus Schneider, Ramayya Kumar and Thomas Kropf

Institute of Computer Design and Fault Tolerance, University of Karlsruhe
(Prof. Dr. D. Schmid) P.O. Box 6980, W-7500 Karlsruhe, Germany,
e-mail:{schneide,kumar,kropf}@ira.uka.de

Abstract

The current first-order automatic prover FAUST, embedded in HOL, is based on a sequent calculus which is quite slow and memory intensive. In this paper, an improved version of FAUST using a modified form of tableau calculus called *Tableau Graph Calculus* is presented which overcomes the well-known inefficiencies of the traditional tableau calculus to a large extent. This calculus works on a compact representation of analytic tableaux called *tableau graphs* which are obtained by a preprocessing step which covers most of the rule applications of usual tableau calculus. This representation retains the clarity of the input formula and furthermore, its size is linear with respect to the length of the input formula. As a result of this preprocessing, our calculus has only one single rule which is repeatedly applied to obtain a proof. Many optimizations for the rule applications to effectively prune the search space are presented as well and are currently being implemented in a new version of FAUST.

Keyword Codes: I.2.3; F.4.1
Keywords: Deduction and Theorem Proving; Mathematical Logic

1 Introduction

The use of the theorem proving assistant HOL can be greatly enhanced by embedding an automatic first-order prover within it. Especially in the domain of Hardware Verification, the goals to be proven — when appropriate transformations have been undertaken — are mostly solvable using first-order techniques [ScKK91a], [KuKS91a]. To this effect, a prover called FAUST based on a modified form of sequent calculus has been integrated within HOL [ScKK91b], [KuKS91b]. This prover was quite effective in automating the process of Hardware Verification to a large extent. However, when the subgoals which result from the Hardware Verification Environment called MEPHISTO were very large, the proofs were slow and highly memory intensive. This motivated us to improve the efficiency of FAUST. This has been achieved by using a calculus based on a modified tableau calculus, as reported in this paper.

The first-order prover to be embedded should be based on a calculus which lends itself to an easy integration in HOL. Sequent and tableau calculi allow easy translations of obtained proofs to natural deduction on which HOL is based, since their rules reflect the natural semantics of the connectives of the logic. Having experimented with a sequent calculus implementation [ScKK92], we decided to use an alternative approach based on tableau calculus.

Tableau calculi construct a tree labeled with signed formulae of an initial formula set whose inconsistency has to be shown. The rules of tableau calculi which are used to extend paths, can be classified into four types — α, β, γ and δ-rules. α- and β-rules are used to eliminate propositional connectives of the logic. β-rules branch a path but α-rules do not. δ-rules eliminate quantifiers by instantiating an *arbitrary new* variable. Thus α, β and δ-rules can be applied only finitely often.

Since the tableaux as given in [Beth55],[Smul68],[Fitt90] correspond to trees, the number of nodes can grow exponential in terms of the length of the input formula. This effect takes place whenever several β-rules can be applied at the same stage, as shown in figure 1 (left side). After the application of one of those β-rules, e.g. the first one, the other β-rules have to be applied at each open path[1].

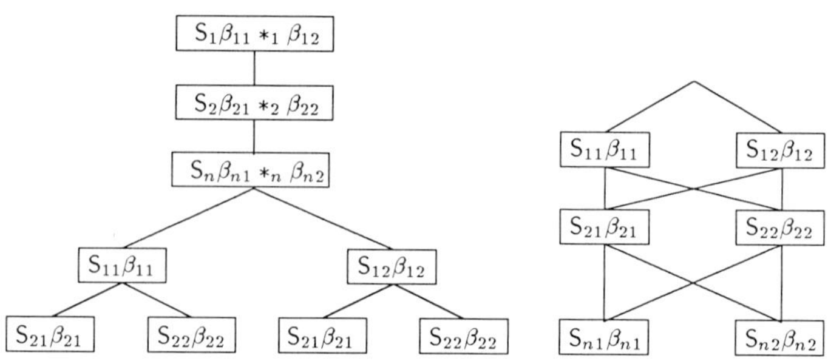

Figure 1: Avoiding redundancy due to β-explosions

However, the application of the first β-rule might have doubled the number of open paths[2] thus enforcing the duplication of the remaining subformulae by further β-rule applications. In fact, the number of these copies can become exponentially large in analytic tableaux. It can however be observed that tableaux need not be stored completely, since they are highly redundant. The redundancy arising from the existence of simultaneously

[1]There are two possibilities how rules can be applied: The first retains the formulae on which the rule is applied in the tableau and generates new subformulae, while the second one removes it from the tableaux (except for γ-rules). Using the second form of rule application, it is necessary to extend *each* open path by copies of subformulae resulting from other rule applications.

[2]The number of open paths is not always doubled, since one of the paths could have been closed after the application of a β-rule.

applicable β-rules can be avoided by referring to the original subformulae instead of using copies as illustrated in figure 1 (right side).

In this paper, we define a calculus which is similar to tableau calculus, but works on such efficient representations of formulae and proofs. As this representation is even more flexible than usual tableau trees, further enhancements which speed up the proof process are possible. Some of these enhancements stem from connection calculi ([Andr81],[Bibe87]) to which the presented calculus is also related to. Thus, our calculus combines advantages of tableau calculus — namely the clarity and naturalness — and connection calculus — namely its efficient implementations. As the various features presented here are mainly independent from each other, one can choose an appropriate subset to implement a suitably efficient prover.

The paper is organized as follows: First, we describe an improvement of the representation of analytic tableaux developed by [Beth55],[Smul68] on which our tableau graph calculus is based. The definitions and the related soundness and completeness proofs are also given. The next section describes various algorithms for computing proofs within this calculus. Implementational details, mainly concerning extension-rules are briefly sketched in section 4. A short section concerning the implementation of our calculus in HOL90 with experimental results then follows. Finally, the paper is concluded with a summary.

2 The Tableau Graph Calculus $\mathcal{C}_{TG}$

In this section, we define the tableau graph calculus $\mathcal{C}_{TG}$, which uses graphs as shown in figure 1 as the underlying representation of the input formula. Before we formalize these ideas, we list the notation which is used (for detailed definitions see any logic textbook, for example [Fitt90]):

V	set of object variables
T_Σ	set of first order terms over a signature $\Sigma = (P_\Sigma, F_\Sigma, \alpha_\Sigma)$, where P_Σ, F_Σ denote the set of predicate symbols and functions symbols, respectively and $\alpha_\Sigma : P_\Sigma \cup F_\Sigma \to \mathbb{N}$ is the arity function
For_Σ	set of first order formulae over a signature $\Sigma = (P_\Sigma, F_\Sigma, \alpha_\Sigma)$
T,F	signs of tableau nodes
true,false	truth values
$SFor_\Sigma$	set of signed formulae over a signature $\Sigma = (P_\Sigma, F_\Sigma, \alpha_\Sigma)$
$[\varphi]_x^\tau$	substitution of the variable x in the term or formula φ by τ
ξ_x^d	modified assignment: $\xi_x^d(x) := d$ and $\xi_x^d(y) := \xi(y)$ if $x \neq y$
$\omega^{D,I,\xi}$	evaluation function of an interpretation (D, I) and an assignment ξ
$\wp(M)$	set of all subsets of a set M
mgu	abbreviation for most general unifier
id	the identity substitution
$\circ$	composition of functions, that is $(f \circ g)(x) := f(g(x))$

The tableau graph calculus $\mathcal{C}_{TG} = (\mathcal{SS}_\Sigma, \mathcal{CSS}_\Sigma, \{\vdash^\mathcal{E}\})$ consists of a formal language $\mathcal{SS}_\Sigma$, a subset $\mathcal{CSS}_\Sigma \subset \mathcal{SS}_\Sigma$ of it called the axioms and a rule $\vdash^\mathcal{E}$ which maps elements of the language $\mathcal{SS}_\Sigma$ to other elements of it. In general, a proof in $\mathcal{C}_{TG}$ will proceed as follows:

First, for the given formula the corresponding element of $\mathcal{SS}_\Sigma$ has to be computed, then the rule $\vdash^{\mathcal{E}}$ has to be applied until an element of $\mathcal{CSS}_\Sigma$ is obtained. Implementations of $\mathcal{C}_{TG}$ however work more sophisticatedly than described above.

First, we define the language $\mathcal{SS}_\Sigma$ of our calculus:

Definition 2.1 (Structure Sets $\mathcal{SS}_\Sigma$) *The following rules define the set of structure sets $\mathcal{SS}_\Sigma$ over a given signature Σ:*

1. $\{\} \in \mathcal{SS}_\Sigma$
2. $\{\varphi\} \cup S \in \mathcal{SS}_\Sigma$, *where* $\varphi \in SFor_\Sigma$ *and* $S \in \mathcal{SS}_\Sigma$
3. $\{(S_1, S_2)_\beta\} \cup S \in \mathcal{SS}_\Sigma$, *where* $S_1, S_2, S \in \mathcal{SS}_\Sigma$
4. $\{(S_1)^\mu_\gamma\} \cup S \in \mathcal{SS}_\Sigma$, *where* $S_1, S \in \mathcal{SS}_\Sigma$

$(S_1, S_2)_\beta$ is called a β-structure and $(S)^\mu_\gamma$ is called a γ-structure, where μ is an arbitrary metavariable[3]*. Elements of a structure set are called rows.*

The next definition shows how structure sets can be computed for signed formulae :

Definition 2.2 (Structure Set of a signed formula) *$\Im : SFor_\Sigma \to \mathcal{SS}_{\Sigma_{sk}}$ assigns to each signed formula $\Phi \in SFor_\Sigma$, a structure set $\Im(\Phi)$ according to the following rules*[4] *(Σ_{sk} is obtained by the introduction of new skolem function symbols in Σ):*

$$\begin{array}{ll}
\Im(\mathsf{T}P(\tau_1,\ldots,\tau_n)) := \{\mathsf{T}P(\tau_1,\ldots,\tau_n)\} & \Im(\mathsf{F}P(\tau_1,\ldots,\tau_n)) := \{\mathsf{F}P(\tau_1,\ldots,\tau_n)\} \\
\Im(\mathsf{T}\neg\varphi) := \Im(\mathsf{F}\varphi) & \Im(\mathsf{F}\neg\varphi) := \Im(\mathsf{T}\varphi) \\
\Im(\mathsf{T}\varphi \wedge \psi) := \Im(\mathsf{T}\varphi) \cup \Im(\mathsf{T}\psi) & \Im(\mathsf{F}\varphi \wedge \psi) := \{(\Im(\mathsf{F}\varphi), \Im(\mathsf{F}\psi))_\beta\} \\
\Im(\mathsf{T}\varphi \vee \psi) := \{(\Im(\mathsf{T}\varphi), \Im(\mathsf{T}\psi))_\beta\} & \Im(\mathsf{F}\varphi \vee \psi) := \Im(\mathsf{F}\varphi) \cup \Im(\mathsf{F}\psi) \\
\Im(\mathsf{T}\varphi \to \psi) := \{(\Im(\mathsf{F}\varphi), \Im(\mathsf{T}\psi))_\beta\} & \Im(\mathsf{F}\varphi \to \psi) := \Im(\mathsf{T}\varphi) \cup \Im(\mathsf{F}\psi) \\
\Im(\mathsf{T}\forall x.\varphi) := \{(\Im(\mathsf{T}[\varphi]^\mu_x))^\mu_\gamma\} & \Im(\mathsf{F}\forall x.\varphi) := \Im(\mathsf{F}[\varphi]^{f(\mu_1,\ldots,\mu_n)}_x) \\
\Im(\mathsf{T}\exists x.\varphi) := \Im(\mathsf{T}[\varphi]^{f(\mu_1,\ldots,\mu_n)}_x) & \Im(\mathsf{F}\exists x.\varphi) := \{(\Im(\mathsf{F}[\varphi]^\mu_x))^\mu_\gamma\}
\end{array}$$

In the rules above, $\{\mu_1, \ldots, \mu_n\}$ is the set of metavariables of φ, f is a new skolem-function symbol and μ is a new metavariable.

Structure sets are a set-representation of the skolemized negation normal form of the given formula. Thus, provided that no equivalence or antivalence operator occurs, the following lemma holds:

Lemma 2.1 (Size of Structure Sets) *The number of atomic formulae of the structure set $\Im(\varphi)$ of a each formula $\varphi \in SFor_\Sigma$ is less or equal to the number of atomic formulae in φ.*

---

[3]Unification can be introduced in tableau calculus by using *Metavariables* (see [Fitt90], [Reev87], [ScKK91b], [ScKK92]). They can be seen as placeholders which are introduced by a γ-rule and are instantiated with suitable terms found with unification later on. Substitutions which assign terms to metavariables are called *metasubstitutions*.

[4]Here we use the liberalized δ-rule presented in [HaSc92] which takes only the free variables of the formula as arguments for the skolem function, instead of the free variables of a considered path.

Proof:
Easy by induction with Definition 2.2. As $\Im(\mathsf{T}\psi \wedge \psi) = \Im(\mathsf{T}\psi)$, the number ot atomic formulae of the structure set $\Im(\varphi)$ can also be less than the number of atomic formulae of the input formula.

■

Example 2.1 (Structure Sets) *1. Let Φ_1 be the formula:*

$$\neg(\neg h \wedge (\neg[g \wedge m] \rightarrow \neg[h \vee g]) \wedge ([\{g \rightarrow m\} \wedge \{\neg z \rightarrow h\}] \vee z) \\ \wedge(z \rightarrow [\neg g \wedge m]) \wedge (\neg h \rightarrow g))$$

$\mathsf{F}\Phi_1$ has the following structure set $\Im(\mathsf{F}\Phi_1)$:

$$\{\mathsf{F}h, (\{\mathsf{T}g, \mathsf{T}m\}, \{\mathsf{F}h, \mathsf{F}g\})_\beta, (\{(\{\mathsf{F}g\}, \{\mathsf{T}m\})_\beta, (\{\mathsf{T}z\}, \{\mathsf{T}h\})_\beta\}, \{\mathsf{T}z\})_\beta, \\ (\{\mathsf{F}z\}, \{\mathsf{F}g, \mathsf{T}m\})_\beta, (\{\mathsf{T}h\}, \{\mathsf{T}g\})_\beta, \}$$

2. Let Φ_2 be the formula:

$$\neg([\forall x.(S(x,a) \vee R(b)) \wedge (Q(c,a) \vee U(x))] \\ \wedge[(R(b) \vee [S(a,a) \wedge S(b,a)]) \rightarrow (\neg Q(c,a) \wedge \neg[U(a) \wedge U(b)])])$$

The structure set $\Im(\mathsf{F}\Phi_2)$ contains a γ-structure and a β-structure at the topmost level:

$$\{(\{(\{\mathsf{T}S(\mu,a)\}, \{\mathsf{T}R(b)\})_\beta, (\{\mathsf{T}Q(c,a)\}, \{\mathsf{T}U(y)\})_\beta\})^\mu_\gamma, \\ (\{\mathsf{F}R(b), (\{\mathsf{F}S(a,a)\}, \{\mathsf{F}S(b,a)\})_\beta\}, \quad \{\mathsf{F}Q(c,a), (\{\mathsf{F}U(a)\}, \{\mathsf{F}U(b)\})_\beta\})_\beta\}$$

To define the set of axioms $\mathcal{CSS}_\Sigma$, we need the notion of the *path set* of a given structure set. To define it, the following operator is introduced: $\{p_1, \ldots, p_n\} \uplus p := \{p_1 \cup p, \ldots, p_n \cup p\}$. $\uplus$ extends all paths which are the elements of its left argument with the path which is its right argument.

Definition 2.3 (Path Set of a structure set) *$\mathcal{PS} : \mathcal{SS}_\Sigma \rightarrow \wp(\wp(SFor_\Sigma))$ assigns to each structure set $S \in \mathcal{SS}_\Sigma$ a set of paths $\mathcal{PS}(S)$. $\mathcal{PS}$ is defined as follows:*

$$\mathcal{PS}(\{\}) := \{\{\}\}$$
$$\mathcal{PS}(\{\varphi\} \cup S) := \{\{\varphi\} \cup p \mid p \in \mathcal{PS}(S)\}$$
$$\mathcal{PS}(\{(S_1, S_2)_\beta\} \cup S) := \{\mathcal{PS}(S_1) \uplus p \mid p \in \mathcal{PS}(S)\} \cup \{\mathcal{PS}(S_2) \uplus p \mid p \in \mathcal{PS}(S)\}$$
$$\mathcal{PS}(\{(S_1)^\mu_\gamma\} \cup S) := \{\mathcal{PS}(S_1) \uplus p \mid p \in \mathcal{PS}(S)\}$$

Each element of $\mathcal{PS}(S)$ is called a path of the structure set. Given that $\Phi \in SFor_\Sigma$, then each element of $\mathcal{PS}(\Im(\Phi))$ is called a path of the formula Φ.

Note that the path set of a structure set is a set of structure sets. The number of paths $\Pi(S)$ of a structure set S can be computed as follows:

$$\Pi(\{\}) := 1$$
$$\Pi(\{\varphi\} \cup S) := \Pi(S)$$
$$\Pi(\{(S_1, S_2)_\beta\} \cup S) := (\Pi(S_1) + \Pi(S_2)) \cdot \Pi(S)$$
$$\Pi(\{(S_1)^\mu_\gamma\} \cup S) := \Pi(S_1) \cdot \Pi(S)$$

Example 2.2 (Path Sets) *Some paths of $\Im(\Phi_1)$ (see Example 2.1) are:*

$\{\mathsf{F}h, \mathsf{T}g, \mathsf{T}m, \mathsf{F}g, \mathsf{T}z, \mathsf{F}z, \mathsf{T}h\}$
$\{\mathsf{F}h, \mathsf{T}g, \mathsf{T}m, \mathsf{F}g, \mathsf{T}z, \mathsf{F}z, \mathsf{T}g\}$
$\{\mathsf{F}h, \mathsf{T}g, \mathsf{T}m, \mathsf{F}g, \mathsf{T}h, \mathsf{F}g, \mathsf{T}m, \mathsf{T}h\}$

The set of axioms $\mathcal{CSS}_\Sigma$ of our calculus is characterized by the following definition:

Definition 2.4 (Closed Structure Set $\mathcal{CSS}_\Sigma$) *A structure set S is closed under the metasubstitution σ if and only if for each path $p \in \mathcal{PS}(S)$ there exist signed formulae $\mathsf{T}\varphi_p, \mathsf{F}\psi_p \in p$ such that $\sigma(\varphi_p) = \sigma(\psi_p)$. $(\mathsf{T}\varphi_p, \mathsf{F}\psi_p)$ is then called a complementary pair under σ. The set of all closed structure sets of $\mathcal{SS}_\Sigma$ is denoted by $\mathcal{CSS}_\Sigma$.*

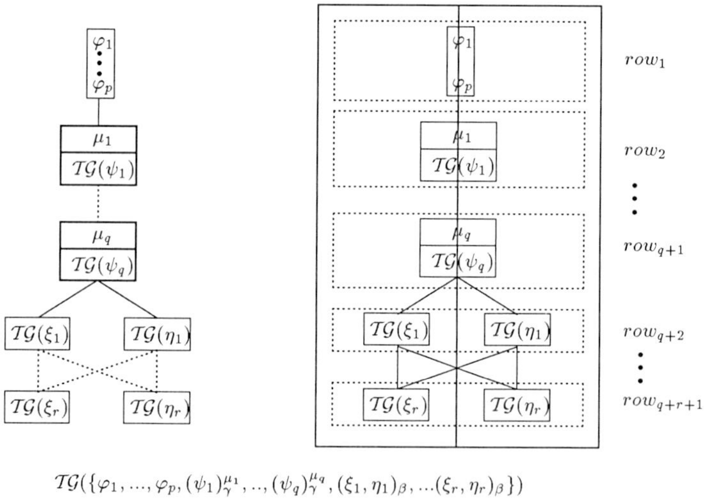

Figure 2: Mapping of structure sets into graphs and the resulting rows and columns

Structure sets can be pictorially represented by graphs called **tableau graphs**. Therefore a function $\mathcal{TG}$ has to be defined which assigns a graph to a given structure set. To define $\mathcal{TG}$, a given structure set $S \neq \{\}$ is split in the subsets $\{\varphi_1, \ldots, \varphi_p\}$ of signed formulae, $\{(\psi_1)^{\mu_1}_\gamma, \ldots, (\psi_q)^{\mu_q}_\gamma\}$ of γ-structures and $\{(\xi_1, \eta_1)_\beta, \ldots, (\xi_r, \eta_r)_\beta\}$ of β-structures (which is always possible due to definition 2.1). Then figure 2 (left side) shows such a pictorial representation of a tableau graph $\mathcal{TG}(S)$ of S. If S is empty, then $\mathcal{TG}$ assigns the empty graph to it. It should be noted that the nodes corresponding to γ- and β-structures in the tableau graph themselves contain tableau graphs.

The mapping given in figure 2 (left side) is not well-defined, since structure sets provide no fixed order on their elements. By grouping all tableau graphs which can be built from a structure set S with a specific ordering, one can define an equivalence relation such that all graphs in that group form an equivalence class. All graphs of an equivalence class differ only in the order of their literals, their γ-structures and their β-structures. An arbitrary element of such an equivalence class corresponding to a structure set of a formula Φ is called a tableau graph for Φ.

In a tableau graph representation, there are at most two columns resulting from the β-structures and arbitrarily many rows corresponding to the rows of the underlying structure set as shown in figure 2 (right side). Rows have to be interpreted as conjunctively connected and columns represent disjunctive connections. The truth signs T and F represent positive and negative literals, respectively.

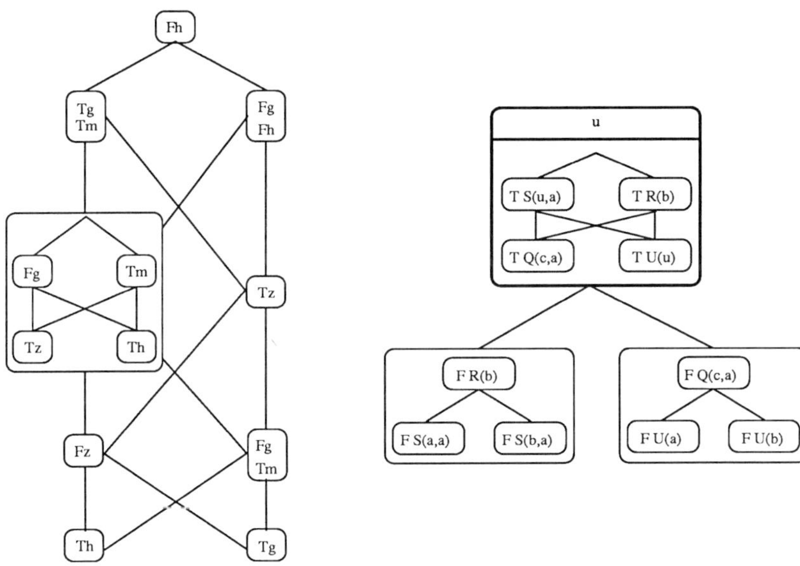

Figure 3: Examples of tableau graphs

Example 2.3 (Examples of tableau graphs) *Figure 3 shows possible tableau graphs for* $\mathsf{F}\Phi_1$ *and* $\mathsf{F}\Phi_2$*. The graph for* Φ_1 *has 40 paths and the graph for* Φ_2 *has 16 paths. The left graph is closed under id, thus it is an axiom and the formula* Φ_1 *is proven. The graph* *for* $\mathsf{F}\Phi_2$ *cannot be closed though* Φ_2 *is valid, too. Here some rule applications of the only rule* $\vdash^{\mathcal{E}}$ *defined below are necessary.*

Definition 2.5 (Extension Rule) *The extension relation* $\vdash^{\mathcal{E}}$ *on the set of all structure sets* $\mathcal{SS}_\Sigma$ *is defined by:*

1. $\{(S_1)^{\mu}_{\gamma}\} \cup S \vdash^{\mathcal{E}} \{(S_1)^{\mu}_{\gamma}, ([S_1]^{\overline{\mu}}_{\mu})^{\overline{\mu}}_{\gamma}\} \cup S$, *where* $\overline{\mu}$ *is a new metavariable*
2. $\{(S_1, S_2)_\beta\} \cup S \vdash^{\mathcal{E}} \{(\overline{S_1}, S_2)_\beta\} \cup S$, *if* $S_1 \vdash^{\mathcal{E}} \overline{S_1}$
3. $\{(S_1, S_2)_\beta\} \cup S \vdash^{\mathcal{E}} \{(S_1, \overline{S_2})_\beta\} \cup S$, *if* $S_2 \vdash^{\mathcal{E}} \overline{S_2}$

In other words, an extension of a structure set includes a variant of an arbitrary γ-structure of it, obtained by replacing the bound metavariable μ by $\overline{\mu}$. This γ-structure variant can be generated from a γ-structure which is not necessarily an element of the structure set, but can also be generated from a γ-structure within nested structure sets due to item 2. and 3. of the definition. One extension step in $\mathcal{C}_{TG}$ corresponds to the application of a γ-rule and all the subsequent $\alpha-, \beta-$ and δ-rules applicable to the obtained formula in tableau calculus.

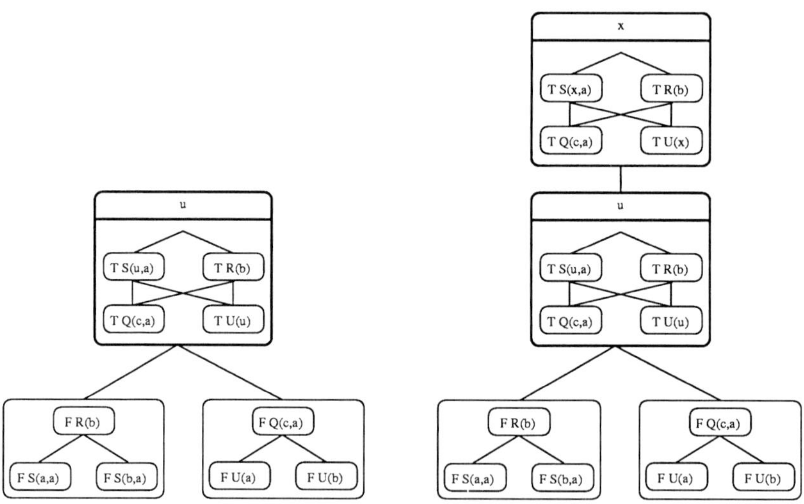

Figure 4: Extension of a tableau graph

Definition 2.6 (Derivation in $\mathcal{C}_{TG}$) *A derivation of* S_1 *in* $\mathcal{C}_{TG}$ *is a finite list* $S_1, \ldots, S_n$ *of structure sets, such that the following conditions hold:*

- *There is a metasubstitution* σ *such that* S_n *is closed under* σ

- $S_i \vdash^{\mathcal{E}} S_{i+1}$, *for all* $i \in \{1, \ldots, n-1\}$

n is called the length of the derivation.

It should be noted that for each formula of propositional logic, the structure set does not contain a γ-structure, thus no extension rule can be applied. Hence, proofs for propositional logic have just to check if the original path set itself is closed under *id*.

In order to prove soundness and completeness results, we have to define the semantics of structure sets. Lemma 2.2 lists the essential properties of structure sets, which are needed to show the correctness.

Definition 2.7 (Semantics of Structure Sets) *Given an interpretation (D, I) and an assignment $\xi : V \to D$, $\Omega^{D,I,\xi} : \mathcal{SS}_\Sigma \to \{\mathsf{true}, \mathsf{false}\}$ assigns each structure set a truth value according to the following rules:*

1. $\Omega^{D,I,\xi}(\{\}) := \mathsf{true}$

2. $\Omega^{D,I,\xi}(\{\mathsf{T}\,\varphi\} \cup S) := \begin{cases} \mathsf{true}: & \omega^{D,I,\xi}(\varphi) = \mathsf{true} \textit{ and } \Omega^{D,I,\xi}(S) = \mathsf{true} \\ \mathsf{false}: & \textit{otherwise} \end{cases}$

3. $\Omega^{D,I,\xi}(\{\mathsf{F}\,\varphi\} \cup S) := \begin{cases} \mathsf{true}: & \omega^{D,I,\xi}(\varphi) = \mathsf{false} \textit{ and } \Omega^{D,I,\xi}(S) = \mathsf{true} \\ \mathsf{false}: & \textit{otherwise} \end{cases}$

4. $\Omega^{D,I,\xi}(\{(S_1, S_2)_\beta\} \cup S) := \begin{cases} \mathsf{true}: & (\Omega^{D,I,\xi}(S_1) = \mathsf{true} \textit{ and } \Omega^{D,I,\xi}(S) = \mathsf{true}) \\ & \textit{or } (\Omega^{D,I,\xi}(S_2) = \mathsf{true} \textit{ and } \Omega^{D,I,\xi}(S) = \mathsf{true}) \\ \mathsf{false}: & \textit{otherwise} \end{cases}$

5. $\Omega^{D,I,\xi}(\{(S_1)^\mu_\gamma\} \cup S) := \begin{cases} \mathsf{true}: & \Omega^{D,I,\xi^d_\mu}(S_1) = \mathsf{true} \textit{ for each } d \in D \\ & \textit{and } \Omega^{D,I,\xi}(S) = \mathsf{true} \\ \mathsf{false}: & \textit{otherwise} \end{cases}$

Lemma 2.2 (Properties of Structure Sets) *For each $\varphi \in For_\Sigma, S \in \mathcal{SS}_\Sigma$ and each Interpretation D, I with assignment ξ the following holds:*

1. *$\Im(\mathsf{T}\varphi)$ is satisfiable iff φ is satisfiable.*

2. *$\Im(\mathsf{F}\varphi)$ is unsatisfiable iff φ is valid.*

3. $\Omega^{D,I,\xi}(S) = \mathsf{true}$ *implies that for each* σ $\omega^{D,I,\xi}(\bigvee_{i=1}^{n} \bigwedge_{j=1}^{l_i} \sigma(sgn(\varphi_{i,j}))) = \mathsf{true}$ *where*

   $\mathcal{PS}(S) = \{\{\varphi_{i,1}, \ldots, \varphi_{i,l_i}\} \mid i \in \{1, \ldots, n\}\}$ *and* $sgn(\varphi) := \begin{cases} \psi & : \textit{if } \varphi = \mathsf{T}\psi \\ \neg\psi & : \textit{if } \varphi = \mathsf{F}\psi \end{cases}$

4. $\Omega^{D,I,\xi}(S) = \mathsf{false}$ *holds if there is a substitution σ such that for each path $p \in \mathcal{PS}(S)$* $\Omega^{D,I,\xi}(\sigma(p)) = \mathsf{false}$ *holds.*

5. *S is unsatisfiable if there is a substitution σ such that S is closed under σ*

6. $S_1 \vdash^{\mathcal{E}} S_2$ *implies* $\Omega^{D,I,\xi}(S_1) = \Omega^{D,I,\xi}(S_2)$.

Proof sketch:
1 is obtained by the usual skolemization theorem and structural induction. 2 is equivalent to 1. 3 can be easily proven by structural induction with definition 2.7 and a substitution lemma. 4 follows from 3.

If a structure set S is closed under a substitution σ then each path p of it is unsatisfiable[5]. Hence 5 follows from 4. The proof of 6 is based on bounded renaming and idempotency of conjunction.

■

Theorem 2.1 (Soundness and Completeness) *φ is valid iff there is a derivation for $\Im(\mathsf{F}\varphi)$ in $\mathcal{C}_{TG}$.*

Proof sketch:(The numbers used below refer to lemma 2.2)
Correctness: Assume there is a derivation $\Im(\mathsf{F}\varphi) \vdash^{\mathcal{E}} S_1 \vdash^{\mathcal{E}} \ldots \vdash^{\mathcal{E}} S_n$, then S_n is closed under a substitution σ and because of 5 it is unsatisfiable. By repeated use of 6 one can conclude, that $\Im(\mathsf{F}\varphi)$ is also unsatisfiable, and by 2, that φ has to be valid.
Completeness: The completeness follows from the theorem of Gödel-Herbrand-Skolem and the compactness theorem of first-order logic, but can also obtained from the completeness of tableau calculus in the following way: Without loosing completeness, we remove the signed formula on which a rule is applied from the tableau unless it is a γ-formula. Moreover, allowing the tableau only to be atomically closed also retains the completeness in tableau calculus. Now we define a homomorphism Θ which assigns each tableau a tableau graph by the following rules: The tableau graph $\Theta(T_1)$ for the initial tableau T_1 is an arbitrary tableau graph built from the structure set of the input formula. Applications of α-,β-, δ- and the first applications of γ-rules do not change the assigned tableau graph. Further applications of γ-rules in tableau calculus correspond to extension steps in $\mathcal{C}_{TG}$. Then we can show, that if the tableau is atomically closed under a substitution σ, the corresponding tableau graph is also closed under σ. As for each valid formula φ a closed tableau described as above for $\mathsf{F}\varphi$ exists (since tableaux calculus is complete), the completeness is proven.

■

3 Algorithmic aspects

In order to prove the validity of a given formula $\varphi \in For_\Sigma$ in $\mathcal{C}_{TG}$, one has to compute the structure set $\Im(\mathsf{F}\varphi)$ and to check if it can be closed under a certain substitution σ. If it can be closed, the proof is finished, if not, an extension rule has to be applied and again the closure has to be checked and so on. Theorem 2.1 guarantees, that if the extension

[5]Assuming the satisfiability of p means that there must be D, I such that $\Omega^{D,I,\xi}(\{\varphi\}) = \mathsf{true}$ for each $\varphi \in p$ under each assignment ξ. Using an arbitrary assignment ξ with the property $\xi(\mu) := \xi_I(\sigma(\mu))$ leads then to a contradiction.

rule is applied in a fair manner, the process terminates yielding in a finite derivation of the valid formula.

While the computation of the structure set is deterministic and can be done in a time linear to the length of the input formula (if no equivalence or antivalence connective occurs) (cf. lemma 2.1), checking the closure of the structure set is much more complex (because of the $\mathcal{NP}$-completeness of the whole problem). Moreover, the extension rule introduces an indeterminism, since the γ-structure from which a variant should be included is not specified. In this section we describe, how checking the closure can be enhanced by precomputation of complementary pairs and reordering of rows in the tableau graph[6]. In the next section we give some improvements in order to remove some of the indeterminism and the complexity of extension steps.

Several proof procedures can be obtained by combining some of the following items which are explained later:

- depth-first or breadth-first path-traversal
- backtracking or collecting all possible unifiers in a set $\mathcal{U}$
- reordering of rows or using a fixed order

Let us first describe a non-backtracking depth-first-traversal without reordering of rows: The procedure starts with a tableau graph using an arbitrary, but fixed order and the unifier set $\mathcal{U}_0 := \{id\}$. The paths are traversed in a depth-first manner from left to right until the considered path p, instantiated with a unifier $\sigma \in \mathcal{U}_n$, contains a complementary pair under ϱ. Then the path is left and the unifier set is updated by composing the unifiers found so far with the unifiers which cause the path to become complementary as follows:

$$\mathcal{U}_{n+1} := \{\varrho \circ \sigma \mid \sigma \in \mathcal{U}_n \wedge \exists \mathsf{F}\varphi, \mathsf{T}\psi \in p. \varrho = mgu\{\sigma(\varphi), \sigma(\psi)\}\}.$$

In other words, all possibilities of closing the path p proceeding from the unifier set $\mathcal{U}_n$ are saved in $\mathcal{U}_{n+1}$. If there is a unifier $\sigma' \in \mathcal{U}_n$ which has no refinement ϱ' but other unifiers of $\mathcal{U}_n$ can be refined, then σ' is omitted in $\mathcal{U}_{n+1}$.

Instead of collecting the most general unifiers which close the path set traversed so far, one can also choose a unifier among them. If this unifier can be refined to a substitution under which the whole path set is closed, a proof is obtained, otherwise one has to backtrack after a given number of extensions and choose another unifier (backtracking depth-first-traversal without reordering of rows).

Instead of a depth first traversal of the paths, one can also use a breadth-first traversal: Beginning with the root node of the tableau graph all paths are traversed simultaneously until all paths can be closed under a substitution σ. Unlike the depth-first approaches presented above, this procedure is complete, but it turns out to be too slow for practical use since *all* paths are traversed further even if some of them could already be closed (breadth-first traversal without reordering of rows).

[6]Precomputation of complementary pairs and reordering are concepts used in *connection calculi* [Bibe87], but can also be found in tableau calculi [Wrig84].

While the procedures described so far use a given ordering of the tableau graph, one can *reorder the tableau graph* during the traversal in order to close paths as soon as possible. Since a complementary pair indicates that each path containing it is complementary, a single complementary pair can close many paths. In general, the closer the complementary pair lies to the root, the larger is the number of paths it closes, thus the rows should be reordered[7]. As the complementary pairs now control the path traversal, reordering requires that all complementary pairs must be known before the paths are traversed. This precomputation of complementary pairs can be done in polynomial time[8]. In this step, the tableau graph is extended by links labeled with a most general unifier which makes the formulae connected by that link complementary (this is usually done in connection calculi, too).

Having computed all complementary pairs and the unifiers which make them complementary, we reorder the rows of the tableau graph in such a manner that complementary nodes clash together as soon as possible.

For example consider the depth-first traversal without backtracking using reordering: Again we start with the unifier set $\mathcal{U}_0 := \{id\}$ and a leftmost path p in an arbitrary row R_1 of the tableau graph which contains at least one signed formula φ which is part of a complementary pair (φ, ψ). If $\psi \in p$, too, then the path can be closed, otherwise there is another row R_2 in the graph which contains ψ. Thus, we exchange rows such that R_2 immediately follows R_1 in order to bring the complementary pair φ, ψ as near as possible to each other. As R_2 contains again a structure set which has rows itself, the reordering has to proceed recursively in R_2 such that the row which contains ψ will be the first one in R_2, and so on.

A possible proof of the formula $\mathsf{F}\Phi_1$ given in section 2 is shown in figure 5. First, the root node is taken into account. Two links are found, one connecting the root node with the node labeled with $\mathsf{T}h$ in the third row and another connecting the root node with the node labeled with $\mathsf{T}h$ in the fifth row. We decided to use the fifth row and therefore we have to exchange the second and the fifth row of the graph. This situation is given in the leftmost graph in figure 5. As one of the two paths leaving from the root is immediately closed, the number of paths to consider is halved after this step. Taking the remaining path $\{\mathsf{F}h, \mathsf{T}g\}$ into account, we find a link $(\mathsf{T}g, \mathsf{F}g)$ between the second and the fourth row in the leftmost graph in figure 5, thus we exchange the third and the fourth row to obtain the graph as shown in the middle of figure 5. The contradiction on the closed path $\{\mathsf{F}h, \mathsf{T}g, \mathsf{F}g, \mathsf{T}m\}$ halves again the remaining path set, thus the number of paths is reduced to a fourth by these two steps. In the graph in the middle of figure 5, we choose

---

[7]The reordering of rows is correct since rows are conjunctively connected and the conjunction is an associative and commutative operation.

[8]Assume the structure set of a formula of length n contains p atomic formulae (nested arbitrarily deep) and $l_1, \ldots, l_p$ are the lengths of these atomic formulae, thus $\sum_{i=1}^{p} l_i \leq n$ and $p \leq n$. If all literals are unified with each other, and the unification of l_i and l_k requires a time bounded by $c \cdot (l_i + l_k)$, the time required for all unifications is

$$\sum_{i=1}^{p}\sum_{k=1}^{p} c \cdot (l_i + l_k) = \sum_{i=1}^{p} c \cdot p \cdot l_i \cdot \left(\sum_{k=1}^{p} l_k\right) = c \cdot p \cdot \left(\sum_{i=1}^{p} l_i\right) \cdot \left(\sum_{k=1}^{p} l_k\right) \leq c \cdot p \cdot n \cdot n \leq c \cdot n^3$$

the fourth row as the next row to be visited, thus no exchanges have to be performed. The rightmost graph in figure 5 shows which paths are finally visited. It has to be noted that just 6 of the 40 paths of the structure set of the formula have been visited !

Moreover, it has to be pointed out, that the last row which stems from the formula $\{g \wedge m\} \vee \{\neg h \wedge \neg g\}$ is not needed for the proof. The tableau graph calculus makes use of this fact since the unnecessary branches are not considered.

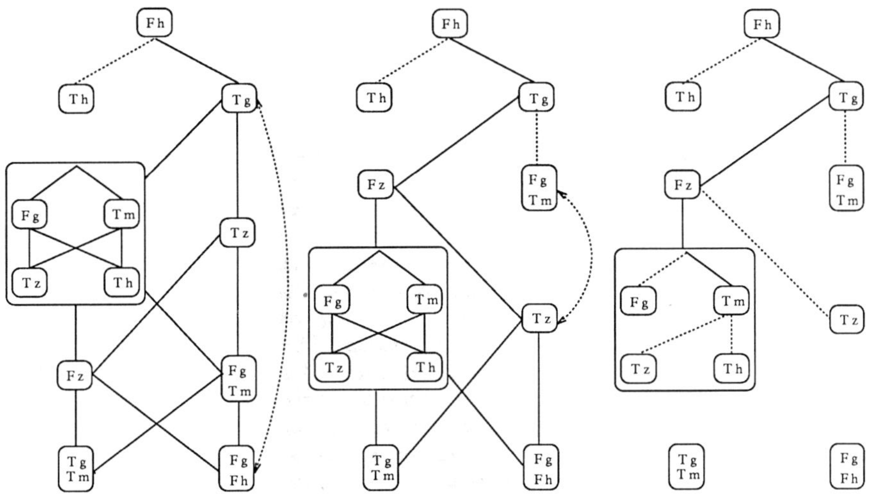

Figure 5: A proof in tableau graph calculus using reordering

To summarize, the proof process is split into three processes:

1. Construction of the tableau graph, i.e. the structure set
2. Computation of all complementary pairs and the corresponding unifiers
3. Checking the closure using the reordering technique, eventually apply the extension rule and check again.

As rule application in tableau calculus corresponds with path traversal in tableau graph calculus, this *reordering of rows* can be interpreted as a *postponed rule application* of those β-rules in tableau calculus which could have been applied at the same situation, in order to optimize the tableau length. Of course, this strategy will not always find a shortest tableau, but the proof length is reduced drastically.

Paths of analytic tableau can be closed during their construction, so the complementary pair need not necessarily be atomic. Therefore, some big formulae of simple structure (for example $\Theta \vee \neg\Theta$) can be proved in constant time regardless of the length of the formula. Since non-atomic closure is not available in tableau graph calculus, the tableau graph calculus can not p-simulate ([Eder92]) the tableau calculus. However, definitional forms ([Eder92]) can simulate the non-atomic closure in $\mathcal{C}_{TG}$, as each subformula is replaced by a new atomic formula which can then become part of a complementary pair. On the other hand, it can be observed, that the path set of the definitional form can be exponentially larger than the path set of the original formula.

4 Implementational details concerning extensions

In tableau calculus γ-rules are treated in a special manner. Unlike the other rules, γ-rules can be applied arbitrarily often. Therefore, some special treatment of these rules is necessary in order to compute short proofs. As the γ-rule of tableau calculus corresponds with the extension rule of $\mathcal{C}_{TG}$, the same holds for the extension rule. We have implemented a prover based on tableau graph calculus, using various improvements mainly concerning the extension-rule, which are given in this section.

Fairness. After a finite number of extensions of a γ-structure, other γ-structures have to be extended. In other words, no γ-structure can be preferred infinitely often.

Unification. The γ-rule in tableau calculus allows the instantiation of an arbitrary term, but usually only specific terms allow the closure of the tableau path. Therefore tableau calculi with unification have been defined (see [Fitt90, Reev87, ScKK92]). Mainly there are two approaches: in Fitting's approach, the δ-quantifiers are eliminated by the introduction of skolem-functions, while in the other two approaches, the δ-rules introduce new eigenvariables[9]. Since the δ-rule of tableau calculus requires that the introduced eigenvariable has to be *new*, each metavariable m has to be attributed with a restriction list fs_m for calculating the terms which are allowed for substitution. In other words, a term τ can be instantiated for a metavariable m if and only if no variable of fs_m occurs in τ. The substitution is then called *allowed*.

The two approaches are dual in sense, but the second one can be implemented more efficiently. If we do not use liberalized δ-rules, then it can be seen, that each restriction list fs_m contains all eigenvariables which are introduced after the metavariable m. If the eigenvariables are encountered in the order of their introduction, then each restriction list has the form $\{y_{r_m}, y_{r_m+1}, \ldots\}$, where r_m is the index of the eigenvariable which are introduced immediately after m. Thus the only information which is relevant for restrictions is the *restriction index* r_m which can be seen as an attribute of m. Using restriction indices, a term is called allowed iff the maximum index of all eigenvariables occuring in it is less than the *restriction index* of the metavariable.

In Fitting's approach the dependency of the metavariables and eigenvariables is expressed by using a skolem function. The allowedness of a term is detected by the occurcheck of the used unification algorithm. Since the terms instantiated by δ-rules can become arbitrary large and the occur check is known to be a frequently used operation, we suppose that the approach using restrictions is more efficient.

[9]The variables introduced by δ-rules are called eigenvariables.

Both approaches carry over to tableau graph calculus. In section 2 we gave the definitions of the calculus with Fitting's approach, the definition of the calculus with the other approach is easily obtained from [Reev87], [ScKK92].

γ-templates. In general several extensions of a γ-structure are necessary to close the structure set.

If all γ-structures have been extended once, but the tableau graph is not closed under a substitution, another extension of a γ-structure is required. Since the subgraph of this extension has exactly the same structure as the original one, only a *template* is stored for all extensions which is parametrized with the metavariable to be introduced (cf. figure 4). For this reason, a new sort of node is introduced, namely the γ-templates, which represent all subgraphs which stem from extensions of the same γ-structure. These nodes are parametrized with some offset informations needed for making extensions. Thus, applying further extensions this way reduces storage requirements, but of course, γ-templates do not decrease the number of paths.

Multiple γ-templates. If two or more γ-quantifiers immediately follow each other (for example $\mathsf{T}\forall x.\forall y.\varphi(x,y)$), then only one γ-template will be created for them. Of course, this does not affect the completeness of our calculus, since applying an extension rule on the outermost γ-structure implies applying an extension on the inner γ-structure as well. Hence this is in principle not one extension-rule, but a *multiple* one. The advantage is, that the number of applicable extension rules decreases and this way the search space is pruned. Since applying extensions on outermost γ-structures does not require more effort than applying them on inner ones as their subgraph is exactly the same, neither the number of nodes nor the number of paths is increased by using these *multiple γ-templates*.

Antiprenexing. Each time an extension is made, the number of paths to close is decreased. Thus, the scopes of all quantifiers should be as small as possible in order to obtain γ-templates with a minimal number of paths. For this reason, we use in our prover a technique called *antiprenexing* which is used in other provers as well.

Management of extension-rules in a stack. Since the repeated use of an extension-rule on a γ-structure is simulated by using offsets of the corresponding template graph, all extensions of a γ-structure are , at a first glance, forced to immediately follow each other. It is however possible that the extension is required only for the closure of one path, but now the extension is 'visible' for each path running through the γ-template. Therefore the number of paths running through the γ-template is multiplied by the number of paths of the subgraph inside the γ-template, each time an extension is made.

To get rid of this unsatisfying situation, the following strategy is used: After a path p which requires a further extension is detected (this can only be the case at the end of p), the extension-rule application is simulated by adding an appropriate offset in a γ-template on p. The traverse control now dives into the γ-template as if this template would be at the end of p. Assume that the considered paths (the γ-template can branch p into more paths which have to be closed now) can now be closed, the extension will be removed by simply deleting the offset information in the template. This way the extension-rule application is actually done at the end of p although the template is somewhere else. As several extensions can be required, and the γ-templates added to p are managed in a FIFO-manner, it turns out, that the γ-template traversals are similar to subprogram calls and can be handled similarly by a stack.

Local metavariables and their scopes. As a metavariable occurs only in the γ-template which introduced it, the γ-template is the scope of the metavariable. After leaving an instance of a γ-template, the instances of the metavariables introduced there can be removed of each unifier of the unifier set $\mathcal{U}_n$ found so far, since they are not relevant for the other paths to be searched through. This has the effect, that different unifiers in the set $\mathcal{U}_n$ may become equal and therefore the number of unifiers shrinks down. Thus, treating metavariables locally within their scopes has drastic effects on the time needed for the proof process.

Choice of next extension rule. If a path cannot be closed, the question arises, which one of the γ-templates on the path might be a good candidate for a further instantiation. Fairness has to be guaranteed, but it is inefficient to use all γ-templates in a queue, since the number of extensions of each γ-template is then exactly the same whether an extension of it is useful or not. However, this is not required by fairness.

If a path cannot be closed, then the information found so far may give some hints which can be used to choose the next γ-template for an extension. For example, if a unifier was deleted due to restrictions or an occur check, the γ-template which introduced the metavariable whose restriction was violated is a good candidate for a further instantiation, because an extension introduces a new metavariable $\overline{\mu}$ which has no restriction, thus the deleted unifier may become allowed with $\overline{\mu}$ instead of μ. Other good candidates are γ-templates, for which different unifiers suggest different instantiations of the introduced metavariable, though one might fail with this suggestion also, e.g. in $(\exists x.Px)\wedge(\exists y.Qy) \rightarrow (\exists z.Pz \wedge Qz)$.

Some of the information needed for the decisions described above can be precomputed in the second step: For example, unifiers which are not allowed are not thrown away, but stored with a flag, so that they are available for extensions in which they may become allowed.

Stopping criteria. Predicate logic is only semidecidable, but special formula fragments are nevertheless decidable. In general, some specific stopping criteria can be formulated. For example, if no unifier has been deleted since it was not allowed, it will be of no use to work further at all. Another stopping criteria is, that unifiers suggest only terms which are already instantiated in the considered γ-rule.

Merging/Lemma generation. The enhancements usually called *merging* and *lemma generation* [Brod91] will also be integrated in our calculus. Lemma generation is based on the asymmetric β-rules given below:

$$\frac{\beta}{\beta_1 \mid \begin{array}{c}\beta_2\\ \neg\beta_1\end{array}} \qquad\qquad \frac{\beta}{\begin{array}{c}\beta_1\\ \neg\beta_2\end{array} \mid \beta_2}$$

Both rules are correct due to the absorption theorems $(\varphi \lor \psi) \leftrightarrow (\varphi \lor (\neg\varphi \land \psi))$ and $(\varphi \lor \psi) \leftrightarrow ((\varphi \land \neg\psi) \lor \psi)$.

Merging is done be removing each path p_2 which contains another path p_1. This is correct since closing p_1 implies closing of the larger path p_2. Both techniques can influence the length of shortest proofs drastically: There are examples which can be proven in a time of $O(2^k)$ without merging or lemma generation, while proofs of length $O(k)$ exist when merging or lemma generation is used.

5 Embedding C_{TG} in HOL

The procedures corresponding to the various steps in the proof process (cf. section 3) are being implemented in SML within HOL90. Similar to the previous implementation of FAUST, there is a fast modus in which the theorems are smuggled into HOL using 'mk_thm' and a slower modus in which the proofs are automatically translated into a HOL-tactic or a HOL-rule. The tactic generation procedure was earlier implemented using a backward proof technique and a new implementation present in the public-domain version of FAUST uses a forward proof technique, which is much faster than the former. Additionally, we are currently implementing another version which converts the original higher-order goal into a propositional goal and then solves it using a propositional proof tactic. The propositional goal can be obtained by substituting the appropriate instantiations (this can be more than one for a quantifier) for the quantified variables, thereby removing the quantifications.

	$\mathcal{F}_{new}$	$\mathcal{F}_{old}$		$\mathcal{F}_{new}$	$\mathcal{F}_{old}$		$\mathcal{F}_{new}$	$\mathcal{F}_{old}$		$\mathcal{F}_{new}$	$\mathcal{F}_{old}$
P_1	3	20	P_{13}	6	40	P_{25}	22	120	P_{36}	21	800
P_2	2	< 10	P_{14}	8	50	P_{26}	29	39	P_{37}	43	720
P_3	1	< 10	P_{15}	4	10	P_{27}	15	220	P_{38}	387	8850
P_4	3	10	P_{16}	1	< 10	P_{28}	18	210	P_{39}	5	20
P_5	2	20	P_{17}	10	10	P_{29}	27	370	P_{40}	17	6410
P_6	1	< 10	P_{18}	4	40	P_{30}	10	90	P_{41}	22	1670
P_7	1	< 10	P_{19}	10	140	P_{31}	9	100	P_{42}	18	200
P_8	1	10	P_{20}	12	110	P_{32}	18	160	P_{43}	* * *	* * *
P_9	5	60	P_{21}	11	160	P_{33}	21	220	P_{44}	16	450
P_{10}	7	80	P_{22}	8	70	P_{34}	305	11710	P_{45}	153	3590
P_{11}	1	< 10	P_{23}	7	50	P_{35}	3	70	P_{46}	113	106850
P_{12}	19	150	P_{24}	27	70						

Table 1: Runtimes of the Pelletier benchmarks

Table 1 compares the proof times of the theorems of [Pell86] required by the old FAUST ($\mathcal{F}_{old}$) and the new version ($\mathcal{F}_{new}$). All times are given in milliseconds and are an average value of ten proofs.

6 Summary

A variant of the tableau calculus has been presented in this paper, which is based on an efficient representation obtained by a preprocessing step. Since most of the rules of traditional calculus are implicitly applied during this step, only a single rule is necessary for obtaining proofs. Efficient strategies for applying this rule have also been described.

The underlying concepts retain the advantages of tableau calculus – retention of the input formula structure and the clarity of proofs – and additionally yield more efficient implementations. Since the size of the representation used for proofs is linear in terms of the length of the formula, more complex problems can be handled efficiently.

The implementational aspects of the calculus, especially the handling of extension-rules have been elaborated. New improvements, made possible due to the underlying graph representation, have also been embedded in our calculus.

References

[Andr81] Andrews P.B., *Theorem Proving via General Matings*, Journal of the ACM, vol. 28, no. 2, pp 193-214, April 1981

[Beth55] Beth E.W., *Semantic entailment and formal derivability*, Mededlingen der Koninklijke Nederlandse Akademie van Wetenschappen, 18(13):309-342, 1955, reprinted in *Logik-Texte. Kommentierte Auswahl zur Geschichte der modernen Logik*, Berka & Kreiser, eds., pp 262-266, Akademie-Verlag, Berlin 1986

[Bibe87] Bibel W., *Automated Theorem Proving*, Artificial Intelligence, Vieweg, Braunschweig/Wiesbaden, second edition, 1987

[Brod91] Broda K., *The Application of Semantic Tableaux with Unification to Automated Deduction*, PhD thesis, Department of Computing, Imperial College, 1991

[Eder92] Eder E.,*Relative Complexities of First Order Calculi*, Vieweg Verlag, 1992

[Fitt90] Fitting M.C., *First-Order Logic and Automated Theorem Proving*, Springer Verlag, New York, 1990

[HaSc92] Haehnle R., Schmitt P.H., *The Liberalized δ-rule in Free Variable Semantic Tableaux*, to appear in Journal of the automated reasoning, 1991

[KuKS91a] Kumar R., Kropf T., Schneider K., *First Steps towards Automating Hardware Proofs in HOL*, in Proc. HOL'91, Davis, IEEE Press

[KuKS91b] Kumar R., Kropf T., Schneider K., *Integrating a First-Order Automatic Prover in the HOL Environment*, in Proc. HOL'91, Davis, IEEE Press

[MuRo87] Murray N.V., Rosenthal E., *Path dissolution: a strongly rule of inference*, in Proc. of the 6th National Conference on Artificial Intelligence, Seattle, pp 161-166, July 1987

[Pell86] F.J. Pelletier. *Seventy-Five Problems for Testing Automatic Theorem Provers*, Journal of Automated Reasoning, Vol. 2, pp 191-216,1986

[ScKK91a] Schneider K., Kumar R., Kropf T., *Structuring Hardware Proofs: First steps towards Automation in a Higher-Order Environment*, in Proc. VLSI '91, Edinburgh, North-Holland

[ScKK91b] Schneider K., Kumar R., Kropf T., *Automating most parts of Hardware Proofs in HOL* , in Proc. of Conf. on Computer Aided Verification, CAV-91, Aalborg, Denmark, 1991, LNCS-575, Springer Verlag, Berlin 1992

[ScKK92] Schneider K., Kumar R., Kropf T., *The FAUST-prover*, in Proc. of Conf. on Automated Deduction, CADE-11, Albany, New York, 1992

[Reev87] Reeves S.V., *Semantic tableau as a framework for automated theorem proving.*, Department of Computer Science and Statistics, Queen Mary College, Univ. of London, 1987

[Smul68] Smullyan R.M.,*First Order Logic*, Springer Verlag Berlin, 1968

[Wrig84] Wrightson G., *Semantic tableaux, unification and links*, technical report, Department of Computer Science, University of Wellington, New Zealand, 1984

Higher Order Logic Theorem Proving and its Applications (A-20)
L.J.M. Claesen and M.J.C. Gordon (Editors)
Elsevier Science Publishers B.V. (North-Holland)

A Note on Interactive Theorem Proving with Theorem Continuation Functions

Ching-Tsun Chou[1] (chou@cs.ucla.edu)

Computer Science Department, University of California at Los Angeles,
Los Angeles, CA 90024, U.S.A.

Abstract

A simple technique for using theorem continuation functions interactively with HOL's subgoal package is presented. An interesting aspect of the technique is that it hinges on the existence of assignable variables in ML.

Keyword Codes: F.4.1; I.2.3
Keywords: Mathematical Logic; Deduction and Theorem Proving.

1 An Example

Suppose we wish to prove `a /\ b ==> b /\ a` using HOL's subgoal package [1]:

```
#g "a /\ b ==> b /\ a" ;;                                        1
"a /\ b ==> b /\ a"
```

The natural proof is to assume the antecedent `a /\ b` to be true and then deduce the succedent `b /\ a` from the assumed antecedent, which is called an *assumption*. In HOL there are, broadly speaking, *two* methods by which assumptions are manipulated. The first method is to add the assumption to the *assumption list* of the goal:

```
#expand( DISCH_TAC );;                                           2
OK..
"b /\ a"
    [ "a /\ b" ]
```

and then prove the the *conclusion* of the goal using the assumption list:

```
#expand( ASM_REWRITE_TAC [ ] );;                                 3
OK..
goal proved
. |- b /\ a
|- a /\ b ==> b /\ a
```

[1]Supported by IBM Graduate Fellowship.

where ASM_REWRITE_TAC [ ] rewrites the conclusion with the assumptions as rewrite rules.

The second method is used when one does not want to add an assumption to the assumption list. One such situation is when the new assumption can be used in a way not involving existing assumptions. Then it may be preferable to use the new assumption directly without first adding it to the assumption list, which may already contain many assumptions. Another such situation is when, if added to the assumption list, it will become difficult to get hold of and manipulate the new assumption in intricate ways. The main reason for this is that the assumption list of a goal is better viewed as an unordered set than as an ordered list, since the order in which assumptions appear on the assumption list often depends on implementation details which are best left unspecified. Hence it may be desirable to 'intercept' an assumption before it is added to the assumption list and not have to later worry about how to get hold of it in an unordered set. Whatever the reason, in order to avoid adding an assumption to the assumption list, one has to understand of the notion of *theorem continuation functions*, which is explained below.

A *theorem continuation*[2] is an ML function of type thm -> tactic (abbreviated as thm_tactic), which converts a theorem into a tactic for continuing the (goal-oriented backward) proof. A simple example is:

```
#let ttac : thm_tactic =                                          1
#   \ t . let (l,r) = CONJ_PAIR t in
#         CONJ_TAC THENL [ACCEPT_TAC r; ACCEPT_TAC l] ;;
ttac = - : thm_tactic
```

Theorem continuation ttac splits the input theorem t (which must be a conjunction) into its left conjunct l and right conjunct r using CONJ_PAIR, then reduces the goal (which must also be a conjunction) into two subgoals corresponding to its two conjuncts using CONJ_TAC, and finally solves the left (right) subgoal with theorem r (l) using ACCEPT_TAC. *Theorem continuation functions* are higher-order ML functions that take theorem continuations as arguments, an example of which is:

```
#DISCH_THEN ;;                                                    2
- : (thm_tactic -> tactic)
```

When applied to an implicative goal, DISCH_THEN removes the antecedent from the goal, creates a theorem by assuming the antecedent, produces a tactic by applying its first argument (which is a theorem continuation) to that theorem, and reduces the succedent of the original goal using the resulting tactic. Schematically, if

```
  ?- u
========   ttac (t |- t)
  ?- v
```

(read: under the assumption t, to prove u, it suffices to prove v), then

```
  ?- t ==> u
==============   DISCH_THEN ttac
      ?- v
```

[2]There is a resemblance between theorem continuations in HOL and continuations in Denotational Semantics, but the reader need not know the latter in order to understand the former or the technique presented in this note.

(read: to prove `t ==> u`, it suffices to prove `v`). The point here is that the assumption `t` is not added to the assumption list, but rather it is used directly by the theorem continuation `ttac`. This is in contrast to how `DISCH_TAC` works:

```
   ?- t ==> u
==============   DISCH_TAC
     t ?- u
```

Plugging `ttac` into `DISCH_THEN` produces a tactic that can solve `a /\ b ==> b /\ a` in one step:

```
#g "a /\ b ==> b /\ a" ;;                                             3
"a /\ b ==> b /\ a"

#expand( DISCH_THEN ttac );;
OK..
goal proved
|- a /\ b ==> b /\ a
```

The pattern of inference effected by `(DISCH_THEN ttac)` corresponds closely to the intuitive argument one uses to prove `a /\ b ==> b /\ a`. Expanding with the definition of `ttac`, the tactic `(DISCH_THEN ttac)` becomes:

```
DISCH_THEN \ t .
  let (l,r) = CONJ_PAIR t in
  CONJ_TAC THENL
  [ ACCEPT_TAC r ;
    ACCEPT_TAC l ]
```

which can be read line-for-line as expressing the following informal proof:

Assume the antecedent `t = a /\ b` is true.
1. Hence both `l = a` and `r = b` are true.
2. To prove `b /\ a`, it suffices to prove both `b` and `a`.
   2.1. Assumption `r` proves `b`.
   2.2. Assumption `l` proves `a`.

All built-in theorem continuation functions in HOL88 (*viz.*, those ML functions with names ending with '`_THEN`', '`_THENL`' or '`_THEN2`') afford equally intuitive interpretations.

2 The Technique

In the above example the goal `a /\ b ==> b /\ a` is so simple that it is trivial to figure out what the theorem continuation argument `ttac` of `DISCH_THEN` should be. Indeed, there is no need to use `DISCH_THEN` at all; the more conventional solution with `DISCH_TAC` is actually simpler. But what if our proof goal is more complicated and, for one reason or another, we have decided to use a theorem continuation function like `DISCH_THEN` and hence must somehow construct a complex theorem continuation argument?

In HOL a complex tactic is constructed by *interactively* building, traversing and sometimes backtracking over a proof tree using the *subgoal package* [1]. It is crucial to be able to perform the proof search interactively, for theorem proving is computationally too hard to be fully automated, and to have a tool like the subgoal package to do all the bookkeeping, for it is too tedious and error-prone for a human to keep track of all the details. But can we construct complex theorem continuations, not just tactics, interactively using the subgoal package?

A theorem continuation `ttac` has type `thm -> tactic`. Obviously, without knowing the value of its theorem argument (call it `t`), `ttac` (more precisely, the part of `ttac` that has been constructed) cannot be interactively tested using the subgoal package. The problem is: How does the user generate the correct value of `t` during an interactive session?

When a theorem continuation `ttac` is used as an argument of a theorem continuation function `tcl`, `ttac`'s input theorem `t` is produced by `tcl` either from the goal to be proved via the inverse of an introduction rule (*e.g.*, when `tcl` is `DISCH_THEN`), or from an already generated theorem via an elimination rule (*e.g.*, when `tcl` is `CHOOSE_THEN`; see the example in Section 3), or from a combination of both. It is possible, in principle at least, to generate the correct value of `t` by mimicking `tcl` manually. But this approach is as tedious and error-prone as doing interactive proofs manually without the help of the subgoal package. The contribution of this note is a technique for overcoming this difficulty.

The basic idea behind the technique is very simple: Let `ttac` assign the value of its theorem argument to an *assignable variable* which is global and hence can be accessed from outside `ttac`.

```
#letref t = ARB_THM ;;     % Initialize t to some arbitrary theorem %      1
t = |- $= = $=

#let ttac : thm_tactic = ( \t' . t := t' ; ALL_TAC ) ;;
ttac = - : thm_tactic
```

It is essential that `t` be an assignable variable, since non-assignable variables (*i.e.*, those variables declared with `let` instead of `letref`) cannot be re-assigned a new value. Let us examine the behavior of `ttac` by re-doing the previous example:

```
#g "a /\ b ==> b /\ a" ;;                                                   2
"a /\ b ==> b /\ a"

#expandf( DISCH_THEN ttac );;
OK..
"b /\ a"

#t ;;
a /\ b |- a /\ b
```

Thus the theorem that `DISCH_THEN` feeds `ttac` with has been 'captured' and stored in `t`, which can now be accessed from anywhere. Continuing with the proof:

```
#let (l,r) = CONJ_PAIR t ;;                                                 3
l = a /\ b |- a
r = a /\ b |- b
```

4

```
#expandf( CONJ_TAC THENL [ACCEPT_TAC r; ACCEPT_TAC l] );;
OK..
goal proved
. |- b /\ a
|- a /\ b ==> b /\ a
```

Notice that we must use `expandf` instead of `expand` in the last step:

5

```
#backup () ;;
"b /\ a"

#expand( CONJ_TAC THENL [ACCEPT_TAC r; ACCEPT_TAC l] );;
OK..
evaluation failed     Invalid tactic
```

The reason why `expand` fails is that since we have used `DISCH_THEN`, the last goal has no assumption at all. But theorems `l` and `r` both have the assumption `a /\ b`, which causes the *validity check* of `expand` to fail. Since our technique results in a proof style which is often incompatible with the default validity check of `expand`, we will in the sequel use `expandf` exclusively during interactive construction of theorem continuations and adopt the following abbreviation:

1

```
#let f = expandf ;;
f = - : (tactic -> void)
```

But we will continue to use `expand` to test the final tactics for solving top-level goals.

The basic technique can be refined. The assignable variable to which `ttac` assigns the value of its theorem argument does *not* really have to be global. It is sufficient to have a local and anonymous assignable variable to hold the 'captured' theorem, which is then returned as a function value. Furthermore, instead of writing a special piece of code for (the initial skeleton of) each theorem continuation that we might want to plug into `DISCH_THEN`, we can define a *uniform* transformation for all theorem continuation functions of type `thm_tactic -> tactic`. Below the prefix '`f_`' arises (obviously) from our abbreviating `expandf` as `f`.

2

```
#let f_ttac_tac (ttac_tac : thm_tactic -> tactic) : void -> thm =
#  letref th = ARB_THM in
#  let ttac : thm_tactic = ( \ th' . th := th' ; ALL_TAC ) in
#  ( \ () . f (ttac_tac ttac) ; th )
#;;
f_ttac_tac = - : ((thm_tactic -> tactic) -> void -> thm)

#let f_DISCH_THEN = f_ttac_tac DISCH_THEN ;;
f_DISCH_THEN = - : (void -> thm)
```

Now we can have an interactive proof which is almost identical to the previous one:

```
#g "a /\ b ==> b /\ a" ;;                                              3
"a /\ b ==> b /\ a"

#let t = f_DISCH_THEN () ;;
OK..
"b /\ a"

t = a /\ b |- a /\ b

#let (l,r) = CONJ_PAIR t ;;
l = a /\ b |- a
r = a /\ b |- b

#f( CONJ_TAC THENL [ACCEPT_TAC r; ACCEPT_TAC l] );;
OK..
goal proved
. |- b /\ a
|- a /\ b ==> b /\ a
```

3 Another Example

The theorem continuation function

```
CHOOSE_THEN : thm_tactic -> thm -> tactic
```

can be described schematically as follows. If

```
   ?- u
========   ttac (t[x'/x] |- t[x'/x])
   ?- v
```

then

```
   ?- u
========   CHOOSE_THEN ttac (|- ?x.t)
   ?- v
```

where `x'` is a variant of `x` chosen not to be free in the assumption list of the goal. In other words, `CHOOSE_THEN` uses an existentially quantified theorem by instantiating it to a particular but arbitrary witness. Analogous to `f_ttac_tac` and `f_DISCH_THEN`, we can define:

```
#let f_ttac_ttac (ttac_ttac : thm_tactic -> thm -> tactic)             4
#     : void -> thm -> thm =
#   letref th = ARB_THM in
#   let ttac : thm_tactic = ( \ th' . th := th' ; ALL_TAC ) in
#   ( \ () t . f (ttac_ttac ttac t) ; th )
#;;
f_ttac_ttac = - : (thm_tactical -> void -> thm -> thm)

#let f_CHOOSE_THEN = f_ttac_ttac CHOOSE_THEN ;;
f_CHOOSE_THEN = - : (void -> thm -> thm)
```

Furthermore we shall suppress the printing of the assumption lists of theorems, since they can be very long:

5

```
#top_print print_thm ;;
- : (thm -> void)
```

Now consider the goal:

6

```
#g "(?n1. !n. n >= n1 ==> P1 n) /\ (?n2. !n. n >= n2 ==> P2 n) ==>
#    (?n3. !n. n >= n3 ==> P1 n /\ P2 n)" ;;
"(?n1. !n. n >= n1 ==> P1 n) /\ (?n2. !n. n >= n2 ==> P2 n) ==>
 (?n3. !n. n >= n3 ==> P1 n /\ P2 n)"
```

Since the goal is implicative, we can strip and assume the antecedent:

7

```
#let p = f_DISCH_THEN () ;;
OK..
"?n3. !n. n >= n3 ==> P1 n /\ P2 n"

p = . |- (?n1. !n. n >= n1 ==> P1 n) /\ (?n2. !n. n >= n2 ==> P2 n)

#let (p1,p2) = CONJ_PAIR p ;;
p1 = . |- ?n1. !n. n >= n1 ==> P1 n
p2 = . |- ?n2. !n. n >= n2 ==> P2 n
```

Now we have two existentially quantified theorems `p1` and `p2` which we can use by means of `CHOOSE_THEN`:

8

```
#let p1' = f_CHOOSE_THEN () p1 ;;
OK..
"?n3. !n. n >= n3 ==> P1 n /\ P2 n"

p1' = . |- !n. n >= n1 ==> P1 n

#let p2' = f_CHOOSE_THEN () p2 ;;
OK..
"?n3. !n. n >= n3 ==> P1 n /\ P2 n"

p2' = . |- !n. n >= n2 ==> P2 n
```

Now we are ready to solve the existential goal. A suitable witness for `n3` is `n1 + n2`:

9

```
#f( EXISTS_TAC "n1 + n2" );;
OK..
"!n. n >= (n1 + n2) ==> P1 n /\ P2 n"

#f( GEN_TAC );;
OK..
"n >= (n1 + n2) ==> P1 n /\ P2 n"

#let q = f_DISCH_THEN () ;;
OK..
"P1 n /\ P2 n"

q = . |- n >= (n1 + n2)
```

Suppose the following theorems have already been proved:

```
#(th1,th2) ;;                                                    10
(|- !n1 n2 n. n >= (n1 + n2) ==> n >= n1,
 |- !n1 n2 n. n >= (n1 + n2) ==> n >= n2)
: (thm # thm)
```

Then some forward reasoning will generate suitable theorems to finish the proof:

```
#let q1 = itlist MATCH_MP [p1'; th1] q                           11
#and q2 = itlist MATCH_MP [p2'; th2] q ;;
q1 = .. |- P1 n
q2 = .. |- P2 n

#f( ACCEPT_TAC (CONJ q1 q2) );;
OK..
goal proved
... |- P1 n /\ P2 n
.. |- n >= (n1 + n2) ==> P1 n /\ P2 n
.. |- !n. n >= (n1 + n2) ==> P1 n /\ P2 n
.. |- ?n3. !n. n >= n3 ==> P1 n /\ P2 n
.. |- ?n3. !n. n >= n3 ==> P1 n /\ P2 n
. |- ?n3. !n. n >= n3 ==> P1 n /\ P2 n
|- (?n1. !n. n >= n1 ==> P1 n) /\ (?n2. !n. n >= n2 ==> P2 n) ==>
   (?n3. !n. n >= n3 ==> P1 n /\ P2 n)
```

Finally, the whole proof session can be condensed into a single tactic, which we use `expand` to test:

```
#g "(?n1. !n. n >= n1 ==> P1 n) /\ (?n2. !n. n >= n2 ==> P2 n) ==>   12
#   (?n3. !n. n >= n3 ==> P1 n /\ P2 n)" ;;
"(?n1. !n. n >= n1 ==> P1 n) /\ (?n2. !n. n >= n2 ==> P2 n) ==>
 (?n3. !n. n >= n3 ==> P1 n /\ P2 n)"

#expand(
#  DISCH_THEN \ p .
#    let (p1,p2) = CONJ_PAIR p in
#    CHOOSE_THEN ( \ p1' .
#    CHOOSE_THEN ( \ p2' .
#      EXISTS_TAC "n1 + n2" THEN
#      GEN_TAC THEN
#      DISCH_THEN \ q .
#        let q1 = itlist MATCH_MP [p1'; th1] q
#        and q2 = itlist MATCH_MP [p2'; th2] q in
#        ACCEPT_TAC (CONJ q1 q2)
#    ) p2
#    ) p1
#);;
OK..
goal proved
|- (?n1. !n. n >= n1 ==> P1 n) /\ (?n2. !n. n >= n2 ==> P2 n) ==>
   (?n3. !n. n >= n3 ==> P1 n /\ P2 n)
```

Appendix: The Code

Analogous to `f_ttac_tac` and `f_ttac_ttac`, we can define a uniform transformation for each type of built-in theorem continuation functions in HOL88. Notice that these definitions are needed only during interactive construction of theorem continuation arguments of theorem continuation functions. Once a proof is completed, the record of interaction can be condensed into a single tactic containing no '`f_...`' functions, as demonstrated in the last example. Also notice that our technique applies, *mutatis mutandis*, to other LCF-style systems, such as Cambridge LCF [2], as well.

```
let f = expandf ;;


let f_ttac_tac (ttac_tac : thm_tactic -> tactic)
    : void -> thm =
  letref th = ARB_THM
  in
  let ttac : thm_tactic = ( \ th' . th := th' ; ALL_TAC )
  in
  ( \ () . f (ttac_tac ttac) ; th )
;;

let f_DISCH_THEN                = f_ttac_tac DISCH_THEN
and f_INDUCT_THEN (th : thm)    = f_ttac_tac (INDUCT_THEN th)
and f_RES_THEN                  = f_ttac_tac RES_THEN
and f_STRIP_GOAL_THEN           = f_ttac_tac STRIP_GOAL_THEN
and f_SUBGOAL_THEN (t : term)   = f_ttac_tac (SUBGOAL_THEN t)
;;


let f_ttac_ttac (ttac_ttac : thm_tactic -> thm -> tactic)
    : void -> thm -> thm =
  letref th = ARB_THM
  in
  let ttac : thm_tactic = ( \ th' . th := th' ; ALL_TAC )
  in
  ( \ () t . f (ttac_ttac ttac t) ; th )
;;

let f_ALL_THEN                            = f_ttac_ttac ALL_THEN
and f_ANTE_RES_THEN                       = f_ttac_ttac ANTE_RES_THEN
and f_CHOOSE_THEN                         = f_ttac_ttac CHOOSE_THEN
and f_CONJUNCTS_THEN                      = f_ttac_ttac CONJUNCTS_THEN
and f_DISJ_CASES_THEN                     = f_ttac_ttac DISJ_CASES_THEN
and f_FREEZE_THEN                         = f_ttac_ttac FREEZE_THEN
and f_IMP_RES_THEN                        = f_ttac_ttac IMP_RES_THEN
and f_NO_THEN                             = f_ttac_ttac NO_THEN
and f_STRIP_THM_THEN                      = f_ttac_ttac STRIP_THM_THEN
and f_X_CASES_THEN (xll: term list list) = f_ttac_ttac (X_CASES_THEN xll)
and f_X_CHOOSE_THEN (x : term)            = f_ttac_ttac (X_CHOOSE_THEN x)
;;
```

```
let f_ttac_ftac (ttac_ftac : thm_tactic -> term -> tactic)
    : void -> term -> thm =
  letref th = ARB_THM
  in
  let ttac : thm_tactic = ( \ th' . th := th' ; ALL_TAC )
  in
  ( \ () x . f (ttac_ftac ttac x) ; th )
;;

let f_FILTER_DISCH_THEN = f_ttac_ftac FILTER_DISCH_THEN
and f_FILTER_STRIP_THEN = f_ttac_ftac FILTER_STRIP_THEN
;;


let f_ttac_ttac_ttac (ttac_ttac_ttac : thm_tactic -> thm_tactic ->
                                       thm -> tactic)
    : void -> void -> thm -> (thm # thm) =
  letref th1 = ARB_THM and th2 = ARB_THM
  in
  let ttac1 : thm_tactic = ( \ th1' . th1 := th1' ; ALL_TAC )
  and ttac2 : thm_tactic = ( \ th2' . th2 := th2' ; ALL_TAC )
  in
  ( \ () () t . f (ttac_ttac_ttac ttac1 ttac2 t) ; (th1,th2) )
;;

let f_CONJUNCTS_THEN2  = f_ttac_ttac_ttac CONJUNCTS_THEN2
and f_DISJ_CASES_THEN2 = f_ttac_ttac_ttac DISJ_CASES_THEN2
;;


let f_ttacl_ttac (ttacl_ttac : thm_tactic list -> thm -> tactic)
    : void list -> thm -> thm list =
  letref thl = [ ] : thm list
  in
  let ttacl : int -> thm_tactic list =
      letrec ttacl' (m : int) =
        if (m = 0) then [ ]
        else ( \ th' . thl := thl @ [th'] ; ALL_TAC ).(ttacl' (m - 1))
      in
      ( \ n . thl := [ ] ; ttacl' n)
  in
  ( \ vl t . f (ttacl_ttac (ttacl (length vl)) t) ; thl )
;;

let f_CASES_THENL                             = f_ttacl_ttac CASES_THENL
and f_DISJ_CASES_THENL                        = f_ttacl_ttac DISJ_CASES_THENL
and f_X_CASES_THENL (xll: term list list) = f_ttacl_ttac (X_CASES_THENL xll)
;;
```

Acknowledgements

Tom Melham reminded me of the existence of assignable variables in ML. Ray Toal prompted me to write this note and also read an early draft. Peter Homeier, Sara Kalvala, Phil Windley and the anonymous referee made suggestions which greatly improved the presentation of this note. I am grateful to all of them.

References

[1] DSTO and SRI International, *The HOL System: DESCRIPTION*, (1991).

[2] L. C. Paulson, *Logic and Computation: Interactive Proof with Cambridge LCF*, Cambridge Tracts in Theoretical Computer Science 2 (Cambridge University Press, 1987).

A Sequent Formulation of a Logic of Predicates in HOL

Ching-Tsun Chou[1] (chou@cs.ucla.edu)

Computer Science Department, University of California at Los Angeles,
Los Angeles, CA 90024, U.S.A.

Abstract

By a *predicate* we mean a term in the HOL logic of type `* -> bool`, where `*` can be any type. Boolean connectives, quantifiers and sequents in the HOL logic can all be *lifted* to operate on predicates. The lifted logical operators and sequents form a *Logic of Predicates* (LP) whose behavior resembles closely that of the unlifted HOL logic. Of the applications of LP we describe two in some detail: (1) a semantic embedding of Lamport's *Temporal Logic of Actions*, and (2) an alternative formulation of set theory. The main contribution of this paper is a systematic approach for *lifting* tactics that works in the unlifted HOL logic to ones that works in LP, so that one can enjoy the rich proof infrastructure of HOL when reasoning in LP. The power of this approach is illustrated by examples from modal and temporal logics. The implementation technique is briefly described.

Keyword Codes: F.4.1; I.2.3
Keywords: Mathematical Logic; Deduction and Theorem Proving.

1 A Logic of Predicates

By a *predicate* we mean a term in the HOL [3] logic of type `* -> bool`, where `*` is called the *domain* of the predicate and can be any type. Boolean connectives and quantifiers in the HOL logic can all be *lifted* to operate on predicates with the following definitions:

```
         (TT)(x)  =  T
         (FF)(x)  =  F
       (~~ P)(x)  =  ~ P(x)
  (P //\\ Q)(x)  =  P(x) /\ Q(x)
  (P \\// Q)(x)  =  P(x) \/ Q(x)
  (P ==>> Q)(x)  =  P(x) ==> Q(x)
  (P  ==  Q)(x)  =  (P(x) = Q(x))
       (!! R)(x)  =  ! i . (R i)(x)
       (?? R)(x)  =  ? i . (R i)(x)
```

[1]Supported by IBM Graduate Fellowship.

where `P` and `Q` are predicates of type `* -> bool` and `R : ** -> (* -> bool)` is an indexed family of predicates. Notice our notational convention of 'doubling' the symbols of the original operators to form those of the lifted operators.

The last two definitions above need a little more explanation. The lifted quantifiers `!!` and `??` are actually implemented as *binders* so that for any predicate `R[i] : * -> bool` which may contain `i : **` among its free variables, the following equations hold:

```
(!! i . R[i])(x)  =  ! i . R[i](x)
(?? i . R[i])(x)  =  ? i . R[i](x)
```

These equations have to be derived from the definitions of `!!` and `??` using β-conversion for each `R[i]`.

Just as logical operators can be lifted, so can sequents. The *lifted sequent* with *assumptions* `P1`, ..., `Pn` and *conclusion* `Q` is defined as:

```
[ P1 ; ... ; Pn ] |= Q  =  ! x . P1(x) /\ ... /\ Pn(x) ==> Q(x)
```

Notice that, while it does not make sense to say whether a predicate is true or false, a lifted sequent *is* either true or false.

The lifted logical operators and sequents form a *Logic of Predicates* (LP) whose behavior resembles closely that of the unlifted HOL logic, in the sense that theorems, inference rules and tactics in the unlifted HOL logic all have lifted counterparts in LP. For example, *Modus Ponens* can be lifted:

```
   [ ] |= (P ==>> Q)     [ ] |= P
---------------------------------
            [ ] |= Q
```

so can `DISCH_TAC`:

```
   [ ] ?= (P ==>> Q)
=====================
      [ P ] ?= Q
```

where `?=` indicates that the lifted sequents containing it have not yet been proved. Indeed, the main contribution of this paper is a systematic approach for *lifting* tactics that works in the unlifted HOL logic to ones that works in LP, so that one can enjoy the rich proof infrastructure of HOL when reasoning in LP.

It should be noted that the success of our approach depends crucially upon our adopting *sequents* instead of *formulas* as the basis of our logic of predicates. Had we chosen to reason about the validities of individual predicates, instead of sequents of predicates, by defining:

```
|= Q  =  !x. Q(x)
```

we would have obtained a lifted logic obeying a Hilbert style calculus of formulas (*i.e.*, predicates), which is incompatible with the natural deduction style calculus of sequents upon which the unlifted HOL logic is based. Such a mismatch would render impossible the relatively straightforward lifting of HOL tactics described in this paper, thus incurring a great deal of unnecessary work in the process of mechanizing the lifted logic.

2 Applications of LP

Two applications of LP are discussed in this section: §2.1 contains a typical example of the semantic embeddings of other logics in HOL, and §2.2 briefly describes the 'predicates as sets' formulation of set theory using LP.

2.1 A Temporal Logic of Actions in HOL

A common method of semantically embedding various logics, such as programming logics [4] and modal logics [2], in HOL is to use *predicates* in the HOL logic to represent *propositions* in the embedded logic. The logical operators of the embedded logic are naturally represented by the lifted logical operators of LP. The non-logical operators, such as modal operators, are embedded by referring to the semantics of the domains of predicates. These ideas are nicely illustrated by the semantic embedding of Lamport's *Temporal Logic of Actions* (TLA) [5] described below, which is an on-going project of the author's.

In TLA there are three domains on which predicates are needed:

states	:	`*state`
transitions	:	`*state # *state`
behaviors	:	`num -> *state`

That is, a state can be anything (usually a tuple of values of program variables), a transition is a pair of states (representing a step of program execution), and a behavior is an infinite sequence of states (representing an infinite program execution). Predicates on states are called *state predicates* or simply *predicates*, predicates on transitions *actions*, and predicates on behaviors *temporal properties*.

In addition to the lifted logical operators, there are two kinds of non-logical operators in TLA: (*type*) *coercion operators* and *temporal operators*. Coercion operators are all defined by specializing the *inverse image operator*. Let `f : *1 -> *2` be any mapping. For any predicate `P : *2 -> bool`, the inverse image of `P` under `f`, `(inv f P) : *1 -> bool`, is defined by:

```
(inv f P)(x)   =   P(f x)
```

(The more familiar notation for `(inv f P)` is $f^{-1}(P)$.) For instance, let `map_b_s` be the projection that maps each behavior to the first state in that behavior:

```
map_b_s (b : num -> *state)   =   b(0) : *state
```

Then, by specializing the `f` in `(inv f)` to `map_b_s`:

```
b_s  =  inv map_b_s : (*state -> bool) -> ((num -> *state) -> bool)
```

we obtain a coercion operator that 'coerces' a state predicate into a temporal property by evaluating the predicate at the first state of a behavior:

```
(b_s P)(b)   =   P(b(0))
```

There are several other coercion operators in TLA that allow one to view a predicate as an action, an action as a temporal property, and so on.

The advantage of defining all coercion operators by specializing `(inv f)` is that one can prove properties of coercion operators simply by specializing properties of `inv`, which have to be proved only once. For instance, one can prove that `(inv f)` distributes over the lifted implication[2]:

```
! f . ! P Q . (inv f)(P ==>> Q)  =  (inv f)(P) ==>> (inv f)(Q)
```

As a special case, `b_s` distributes over the lifted implication:

```
! P Q . b_s(P ==>> Q)  =  b_s(P) ==>> b_s(Q)
```

and so do all other coercion operators. All these facts can be proved by specializing the above distributivity theorem for `(inv f)`.

In TLA there are two basic modal operators `[]` (read: *box*) and `<>` (read: *diamond*) on temporal properties which formalize the notions of, respectively, '*always*' and '*eventually*' (where 'eventually' includes 'now'):

```
([] G)(b : num -> *state)  =  ! n . G(suffix n b)
(<> G)(b : num -> *state)  =  ? n . G(suffix n b)
```

where

```
suffix n b  =  \ m . b(m + n)
```

denotes the `n`-th suffix of behavior `b`. In other words, a temporal property `G` is always (eventually) true of a behavior `b` if and only if it is true of the `n`-th suffix of `b` for all (some) `n`. Other temporal operators in TLA are defined in terms of `[]` and `<>`. For example,

```
(G ~~> H)  =  [](G ==>> <> H)
```

(read: `G` *leads to* `H`) expresses the notion that whenever `G` is true, `H` will eventually be true.

In TLA, not only the temporal properties of programs, but programs themselves are also expressed as predicates on behaviors. A program `Prog` which starts in a state satisfying the initial condition `Init`, henceforth takes only steps allowed by action `Next`, and meets the fairness condition `Fair`, is formalized as:[3]

```
Prog  =  b_s(Init) //\\ [](b_t(Next)) //\\ Fair
```

The statement that program `Prog` satisfies temporal property `Spec` is expressed by:

```
[ ] |= Prog ==>> Spec
```

(Do not confuse `[]`, the modal operator, with `[ ]`, the empty list!) Such a statement is to be proved by a mixture of HOL and TLA reasoning. On the one hand, reasoning about individual predicates and actions, such as proving a particular action preserves a particular invariant, is application-specific and handled directly in HOL. On the other hand, temporal reasoning follows common patterns and is handled by TLA inference rules, which are actually proved as HOL theorems. For instance, the transitivity of `~~>`:

---

[2]As a matter of fact, `(inv f)` distributes over *all* lifted logical operators.

[3]This is an oversimplification, since TLA formulas takes the so-called *stuttering* into account; see [5].

```
! P Q R : (num -> *state) -> bool .
  [ ] |= ((P ~~> Q) //\\ (Q ~~> R)) ==>> (P ~~> R)
```

can be proved as an HOL theorem and instantiated with particular temporal properties `P`, `Q` and `R` when needed.

Comparison with Another Embedding of TLA

In [7] and [8], von Wright and Långbacka describe another semantic embedding of TLA in HOL which is very similar to ours. As far as the main thrust of this paper is concerned, there are two major differences between their embedding and ours. Firstly, von Wright and Långbacka use curried functions to represent actions. For example, assuming the state consists only of two numeric variables `x` and `y`, the action `Inc_x` that increments `x` by 1 but keeps `y` unchanged would be formalized in their system as:

```
Inc_x (x,y) (x',y')  =  (x' = x + 1) /\ (y' = y)
```

while in our system it would be formalized as:

```
Inc_x ((x,y), (x',y'))  =  (x' = x + 1) /\ (y' = y)
```

As a consequence, they do not have a uniform treatment of logical operators at predicate, action and temporal levels, nor can they define the coercion operators uniformly by specializing the inverse image operator. Thus their work suffers from an unnecessary proliferation of similar but slightly different definitions.

Secondly, von Wright and Långbacka's embedding is based on a Hilbert style calculus of the validities of single predicates, so their system does not enjoy the easy lifting of HOL reasoning to TLA reasoning which our system does (see the last paragraph of Section 1).

2.2 Predicates as Sets

By identifying sets with their characteristic functions, predicates can be viewed as sets. Thus viewed, a predicate `P : * -> bool` *is* the set of elements of type `*` which satisfies property `P`:

```
P  =  { x : * | P(x) }
```

An extensive HOL library for predicates as sets, called the `pred_sets` library, has been written by Melham [6] (based on earlier work of Kalker). Many operations and relations on sets have logical interpretations if sets are identified with predicates. Using the notation of `pred_sets`, all of the following are theorems:

```
        UNIV  =  TT
       EMPTY  =  FF
   P INTER Q  =  P //\\ Q
   P UNION Q  =  P \\// Q
   P DIFF  Q  =  P //\\ ~~Q
  P SUBSET Q  =  [ ] |= P ==>> Q
DISJOINT P Q  =  [ ] |= (P //\\ Q) == FF
```

Furthermore, the lifted quantifiers `!!` and `??` provide, respectively, union and intersection over an indexed family of sets, which are not available in `pred_sets`. The upshot is that the proof technique described below can be used to reason about sets as well.

3 Lifting Tactics and Theorem Tactics

In principle one can always prove statements in an embedded logic, such as LP or the LP-based embedding of TLA, by expanding the embedded operators with their definitions and reasoning directly in HOL. But doing so defeats the very purpose of embedding: if all the reasoning is to be done directly in HOL, then why bother with the embedding in the first place? An embedded logic provides its user with not only more concise and elegant notations, but also (conceptually) larger inference steps, than available in plain HOL. Hence it seems reasonable to accept as a general principle that the user of an embedded logic should perform as much reasoning as possible in the embedded logic. This is not to say that the actual inference steps executed by the HOL system should contain few expansions of embedded operators. To the contrary, the technique described below involves a lot of translating back and forth between LP and plain HOL. The point is that the user should be shielded from the implementation details and be able to imagine that she or he is doing proofs directly in the embedded logic.

In the rest of this section we shall describe, by means of a series of examples drawn from propositional, modal and temporal logics, several *tactic transformers* which can lift tactics in plain HOL to tactics in LP, thus supporting the illusion of 'doing proofs directly in LP'. Due to limited space, the implementation technique can only be outlined.

In the HOL sessions shown below, in order not to have to supply explicit type information too often, we use anti-quotations with the following ML binding:

```
#let VALID = " ($|= [ ]) : (* -> bool) -> bool " ;;                    1
VALID = "$|= [ ]" : term
```

Also, output from HOL is edited for ease of reading.

3.1 Tautology Checking

The tactic `pred_TAUT_TAC` is the lifted tautology checking tactic for LP, which is based on Boulton's tautology checker [1] for HOL.

```
#g" ^VALID( ((X ==>> Y) ==>> X) ==>> X ) ";;                            2
"[ ] |= ( ((X ==>> Y) ==>> X) ==>> X )"

#e( pred_TAUT_TAC );;
OK..
goal proved
|- [ ] |= ( ((X ==>> Y) ==>> X) ==>> X )

#g" ^VALID( !! k : ** . (Z k) \\// ~~(Z k) ) ";;
"[ ] |= ( !! k. (Z k) \\// ~~(Z k) )"

#e( pred_TAUT_TAC );;
OK..
goal proved
|- [ ] |= ( !! k. (Z k) \\// ~~(Z k) )
```

Incidentally, the examples above are known as the *Peirce's Law* and the (indexed) *Law of Excluded Middle*, both of which are intuitionistically *in*valid.

3.2 Lifting Tactics

Consider the following proof in the unlifted HOL logic:

3

```
#g" p ==> q ==> (p /\ q) ";;
"p ==> q ==> p /\ q"

#e( REPEAT DISCH_TAC );;
OK..
"p /\ q"
    [ "p" ]
    [ "q" ]

#e( ASM_REWRITE_TAC [ ] );;
OK..
goal proved
.. |- p /\ q
|- p ==> q ==> p /\ q
```

Now consider the same proof *lifted*:

4

```
#g" ^VALID( P ==>> Q ==>> (P //\\ Q) ) ";;
"[ ] |= ( P ==>> Q ==>> (P //\\ Q) )"

#e( pred_TCL (REPEAT DISCH_TAC) );;
OK..
"[ Q ; P ] |= ( P //\\ Q )"

#e( pred_TCL (ASM_REWRITE_TAC [ ]) );;
OK..
goal proved
|- [ Q ; P ] |= ( P //\\ Q )
|- [ ] |= ( P ==>> Q ==>> (P //\\ Q) )
```

The tactical `pred_TCL : tactic -> tactic` converts a tactic that works on unlifted sequents into one that works on lifted sequents in exactly the same way.

Admittedly the above proof is unnecessarily arduous: calling the tautology checker is much easier. But it is done to demonstrate, in as simple a setting as possible, how general tactics are lifted. The reason why the lifting of general tactics is desirable is that when LP is used to embed other logics, it may be necessary to mix logical reasoning with special-purpose, non-logical reasoning, such as modal or temporal reasoning. Also, one needs to reason about lifted quantifiers as well, which in general cannot be handled by tautology checking. Examples for all these appear in the next subsection, which also introduces the lifting of theorem tactics.

The last result is needed later, so we bind it with an ML identifier:

5

```
#let Lemma_0 = top_thm () ;;
Lemma_0 = |- [ ] |= ( P ==>> Q ==>> (P //\\ Q) )
```

3.3 Lifting Theorem Tactics

A broad class of modal logics are the so-called *normal logics* [2], which are characterized by *Schema K* and the *Rule of Necessitation*:

```
#K_Schema ;;                                                    6
|- !P Q. [ ] |= ( [](P ==>> Q) ==>> ([] P ==>> [] Q) )

#Necessitation_Rule ;;
|- !P. [ ] |= ( P ) ==> [ ] |= ( [] P )
```

For example, TLA is a normal logic and both `K_Schema` and `Necessitation_Rule` can be proved as theorems of our embedding. But for our present purpose, it suffices to regard them as axioms for the uninterpreted modal operator `[] : (* -> bool) -> (* -> bool)`.

Now consider the following goal:

```
#g" ^VALID( ([] P //\\ [] Q) ==>> [](P //\\ Q) ) ";;            7
"[ ] |= ( ([] P //\\ [] Q) ==>> [](P //\\ Q) )"
```

The first step is to strip the antecedent:

```
#e( pred_TCL STRIP_TAC );;                                      8
OK..
"[ [] Q ; [] P ] |= ( [](P //\\ Q) )"
```

The following key lemma is derived from the result of the previous example using the Rule of Necessitation:

```
#let Lemma_1 = MATCH_MP Necessitation_Rule Lemma_0 ;;           9
Lemma_1 = |- [ ] |= ( [](P ==>> Q ==>> (P //\\ Q)) )
```

To add `Lemma_1` to the assumption list of the current goal, we need to lift `ASSUME_TAC`. But `ASSUME_TAC` is a theorem tactic, which is lifted by `pred_TTCL` rather than `pred_TCL`:

```
#e( pred_TTCL ASSUME_TAC Lemma_1 );;                            10
OK..
"[ [](P ==>> Q ==>> (P //\\ Q)) ; [] Q ; [] P ] |= ( [](P //\\ Q) )"
```

Finally, applying the lifted `IMP_RES_TAC` twice with `K_Schema` as the implicative theorem finishes the proof:

```
#e( pred_TTCL IMP_RES_TAC K_Schema );;                          11
OK..
"[ [](Q ==>> (P //\\ Q)) ; [](P ==>> Q ==>> (P //\\ Q)) ; [] Q ; [] P ] |=
  ( [](P //\\ Q) )"

#e( pred_TTCL IMP_RES_TAC K_Schema );;
OK..
goal proved
|- [ [](Q ==>> (P //\\ Q)) ; [](P ==>> Q ==>> (P //\\ Q)) ; [] Q ; [] P ] |=
   ( [](P //\\ Q) )
|- [ [](P ==>> Q ==>> (P //\\ Q)) ; [] Q ; [] P ] |= ( [](P //\\ Q) )
|- [ [] Q ; [] P ] |= ( [](P //\\ Q) )
|- [ ] |= ( ([] P //\\ [] Q) ==>> [](P //\\ Q) )
```

The next example shows that lifted quantifiers can also be handled:

```
#Reflexivity ;;                                                          12
|- !P. [ ] |= ( P ~~> P )

#Transitivity ;;
|- !P Q R. [ ] |= ( ((P ~~> Q) //\\ (Q ~~> R)) ==>> (P ~~> R) )

#g" ^VALID( (!! n . R(SUC n) ~~> R(n)) ==>> (!! n . R(n) ~~> R(0)) ) ";;
"[ ] |= ( (!! n . R(SUC n) ~~> R(n)) ==>> (!! n . R(n) ~~> R(0)) )"

#e( pred_TCL (DISCH_TAC THEN INDUCT_TAC) );;
OK..
2 subgoals
"[ R(n) ~~> R(0) ; !! n . R(SUC n) ~~> R(n) ] |= ( R(SUC n) ~~> R(0) )"

"[ !! n . R(SUC n) ~~> R(n) ] |= ( R(0) ~~> R(0) )"

#e( pred_TTCL MATCH_ACCEPT_TAC Reflexivity );;
OK..
goal proved
|- [ !! n . R(SUC n) ~~> R(n) ] |= ( R(0) ~~> R(0) )

Previous subproof:
"[ R(n) ~~> R(0) ; !! n . R(SUC n) ~~> R(n) ] |= ( R(SUC n) ~~> R(0) )"

#e( pred_TCL (FIRST_ASSUM (ASSUME_TAC o SPEC "n:num")) );;
OK..
"[ R(SUC n) ~~> R(n) ; R(n) ~~> R(0) ; !! n . R(SUC n) ~~> R(n) ] |=
 ( R(SUC n) ~~> R(0) )"

#e( pred_TTCL IMP_RES_TAC Transitivity );;
OK..
goal proved
|- [ R(SUC n) ~~> R(n) ; R(n) ~~> R(0) ; !! n . R(SUC n) ~~> R(n) ] |=
   ( R(SUC n) ~~> R(0) )
|- [ R(n) ~~> R(0) ; !! n . R(SUC n) ~~> R(n) ] |= ( R(SUC n) ~~> R(0) )
|- [ ] |= ( (!! n . R(SUC n) ~~> R(n)) ==>> (!! n . R(n) ~~> R(0)) )
```

4 Outline of Implementation

Roughly speaking, the effect of applying the tactic (pred_TCL tac) to a goal is achieved in three steps:

1. *Unfold* to eliminate lifted operators in the goal;
2. Apply tac;
3. *Fold* to re-introduce lifted operators.

For instance, this is what happens when (pred_TCL STRIP_TAC) is applied to a goal:

```
#g" ^VALID( (P //\\ Q) ==>> (P \\// Q) ) ";;                    13
"[ ] |= ( (P //\\ Q) ==>> (P \\// Q) )"

#e( pred_SEQ_UNFOLD_TAC "x:*" );;
OK..
"P x /\ Q x ==> P x \/ Q x"

#e( STRIP_TAC );;
OK..
"P x \/ Q x"
    [ "P x" ]
    [ "Q x" ]

#e( pred_SEQ_FOLD_TAC "x:*" );;
OK..
"[ Q ; P ] |= ( P \\// Q )"
```

The tactic (pred_TTCL ttac th) follows a similar pattern except that the theorem th must also be unfolded, while pred_TAUT_TAC does not need the folding phase at all.

Acknowledgements

The author is grateful to the members of HOL Seminar at UCLA for interesting discussions and to Dr. von Wright for sending him [7] and [8].

References

[1] R. J. Boulton, *The HOL* taut *Library*, Univ. of Cambridge Computer Lab., (1991).

[2] R. Goldblatt, "Logics of Time and Computation", CSLI Lecture Notes 7, (1987).

[3] M. J. C. Gordon, "HOL: A Proof Generating System for Higher-Order Logic", in *VLSI Specification, Verification and Synthesis*, edited by G. Birtwistle and P. A. Subrahmanyam, (Kluwer Academic Publishers, 1988), pp. 73–128.

[4] M. J. C. Gordon, "Mechanizing Programming Logics in Higher-Order Logic", in *Current Trends in Hardware Verification and Automated Theorem Proving*, edited by G. Birtwistle and P. A. Subrahmanyam, (Springer-Verlag, 1989), pp. 387–439.

[5] L. Lamport, "The Temporal Logic of Actions", DEC SRC technical report 79, (1991).

[6] T. F. Melham, *The HOL* pred_sets *Library*, Univ. of Cambridge Computer Lab., (1992).

[7] J. von Wright, "Mechanising the Temporal Logic of Actions in HOL", in *Proc. of the HOL Tutorial and Workshop*, (1991).

[8] J. von Wright and T. Långbacka, "Using a Theorem Prover for Reasoning about Concurrent Algorithms", in *Proc. of Workshop on Computer-Aided Verification*, (1992).

Higher Order Logic Theorem Proving and its Applications (A-20)
L.J.M. Claesen and M.J.C. Gordon (Editors)
Elsevier Science Publishers B.V. (North-Holland)

A Classical Type Theory with Transfinite Types

Garrel Pottinger[1]

Mathematical Sciences Institute, 409 College Avenue, Cornell University, Ithaca, New York 14850, U.S.A.

Abstract
This paper defines a system of classical logic that includes transfinite types formed by means of the general Cartesian product operation. The consistency of the system is established by means of a semantical argument.

Keyword Codes: F.4.1; I.2.3
Keywords: Mathematical Logic; Deduction and Theorem Proving.

1 Introduction

This paper defines CT^3 (Classical Transfinite Type Theory), a classical logic with complex type symbols of the form $\prod x : X.Y$. Such type symbols denote the general Cartesian product determined by taking the type denoted by X as an index set and associating with each index the type denoted by Y when that index is assigned as the value of the variable x.

Models of CT^3 include a type of types denoted by U that has the type of truth values as a member and is closed under function space formation. The type denoted by U is a member of the type of types denoted by O, which is also closed under function space formation and is a member of the type of types denoted by T_1. Both the type denoted by U and the type denoted by O are subsets of the type denoted by T_1. The type denoted by T_1 is closed under product formation and is both a member and a subset of the type of types denoted by T_2, which, in turn, is closed under product formation. Every type mentioned so far is both a member and a subset of the universal type denoted by Ω. Empty types are permitted, and the system includes quantification and definite descriptions for each of the types that is a member of the type denoted by T_1.

The semantics for CT^3 defined below establishes the consistency of the system.[2] Although the models introduced in section 5 are adequate for proving that CT^3 is consistent, it will

[1]Partly supported by the U.S. Army Research Office through the Mathematical Sciences Institute of Cornell University.

I am grateful to ORA Corporation and the Cornell University Department of Computer Science for allowing me to use their computer systems in preparing this document. Paul Howard proved a theorem that is essential for understanding what is required in order to prove the consistency of the system described in this paper, and earlier drafts were improved due to comments from Peter Andrews, Thierry Coquand, Jim Hook, and Tom Melham.

I, of course, am responsible for any remaining errors and infelicities.

[2]See section 14.

be necessary to pass, in the manner of [Hen50], to a larger class of models in order to prove completeness. This is so because the Peano axioms can be expressed in CT^3, and they can be formulated so as to be categorical in the class of models considered below.

CT^3 is descended from the HOL logic [Gor85, HOL], which, in turn, is descended from Church's type theory [Chu40, Hen50, And86]. But CT^3 differs from both the HOL logic and Church's type theory in very basic ways — the λ-operator has three operands, rather than two; in general, terms, as such, do not have type symbols; and some type symbols may denote the empty set.

In the HOL logic and Church's type theory, each variable x_σ has a fixed type symbol, σ, and abstraction terms have the form $\lambda\, x_\sigma.Y$. In CT^3, on the other hand, variables do not carry type symbols with them, and abstraction terms have the form $\lambda\, x : X.Y$.[3] Since each constant of the HOL logic is assigned a fixed type scheme, each constant of Church's type theory is assigned a fixed type symbol, and type symbols cannot be affected by λ-conversion in these systems, terms of the HOL logic, including terms that contain type variables, and terms of Church's type theory have fixed type symbols. In contrast with this, terms of CT^3, in general, have type symbols only relative to environments that assign type symbols to their free variables, and λ-conversion can affect type symbols.

The HOL logic extends Church's type theory by adding type variables, but no way of binding them is included. Consequently, the HOL logic includes a primitive rule allowing substitution of type symbols for type variables.

The work that led to formulating CT^3 developed out of proving a Henkin-style completeness theorem for the HOL logic [Pot92]. In the course of working on this completeness result, two problems that stem from the way type variables are handled in the HOL logic came to light.

Problem 1: The combination of fixed type symbol assignment and the type symbol substitution rule makes the proof theory of the HOL logic quite unpleasant, for the following reasons. A primitive substitution rule generally involves restrictions on the variables to which it can be applied, and dealing with these restrictions usually involves proving that all applications of such a rule can be pushed up to the top of derivations. This, in fact, had to be done in the work reported in [Pot92]. Furthermore, in order achieve the correspondence between substitutions and assignments of values to variables involved in Henkin's completeness argument, it was necessary to go beyond a substitution operation that replaces type variables with type symbols and deal with a substitution operation that replaces x_σ with terms that depend on both x_σ and the type symbol σ' obtained by applying the substitution to σ.

Problem 2: Because of my experience with various systems of type symbol assignment for constructive type theory and the λ-calculus,[4] saying, as one does in the HOL logic, that the type scheme of the equality constant has the form $\alpha \to (\alpha \to o)$, where α is a type variable and o denotes the type of truth values, struck me as an inadequate way of expressing what is really going on — clearly, it seemed, the equality constant should be assigned an appropriate type *symbol* formed by means of $\prod$.

The basic idea that led, in the end, to defining CT^3 was to overcome both of these problems by modifying the HOL logic to include type symbols of the form $\prod x : X.Y$,[5] abstraction terms

[3] [Bru68] is the earilest publication using this form of λ-abstraction that has come to my attention. I am indebted to Jonathan Seldin for this reference.

[4] For a survey of systems of the latter kind, see [Sel87]. The system of [Pot79] applies type symbol assignment techniques to relevance logic. My experience with Coquand's theory of constructions [Pot87, Pot88a, Pot88b] was especially important in the work that led to the definition of CT^3. Accounts of the theory of constructions and references to the relevant literature are given in [Coq90], [Sel87], and [Pot87, Pot88a, Pot88b].

[5] Bourbaki may have introduced this notation. I owe this suggestion to Per Martin-Löf and to Bengt Nordström,

of the form $\lambda x : X.Y$, and application of terms to type symbols, thereby both enhancing the expressivness of the language and, *via* λ-conversion, avoiding a primitive substitution rule for type symbols. Once this was done, it turned out to be easy to avoid assuming that all types are non-empty.

Andrews' system $\mathcal{Q}$ of transfinite theory [And65] is also a descendant of Church's type theory. $\mathcal{Q}$ includes quantification over a set of finite types similar to the CT^3 type denoted by U, but this set is not counted as a type and λ-abstraction with respect to type variables is not permitted. In fact, the type structures in which $\mathcal{Q}$ is interpreted [And65, pp. 8–9] include neither functions taking types as arguments nor functions returning types a values. Since the type structures used to interpret CT^3 do include such functions and abstraction terms denoting them are included in the system, CT^3 seems, *prima facie*, to be quite different from $\mathcal{Q}$.

On the other hand, the type structures used to interpret $\mathcal{Q}$ include a transfinite type that is the union of the finite types, and these structures are closed under function space formation. Therefore, finite types can be represented as functions from the transfinite type that is the union of the finite types to the type of truth values, functions mapping finite types to finite types can be represented as relations between functions of the kind just described, and so on. Consequently, although the availability of λ-abstraction with respect to type variables in CT^3 is a practical advantage, from an abstract point of view, the differences between CT^3 and $\mathcal{Q}$ may not be as radical as they seem. A subsequent detailed comparison of CT^3 and $\mathcal{Q}$ will be necessary in order to sort this out.

2 The Metalanguage

The metalanguage is Morse-Kelley set theory with atoms,[6] the *axiom of universal choice*, which asserts that there is a bijection from the class of sets and atoms to the ordinals, and the *axiom of inaccessible cardinals*, which asserts that, for every ordinal α, there is an inaccessible cardinal κ such that $\alpha < \kappa$. The objects countenanced by the metalanguage are *classes*, *sets*, which are those classes that are members of classes, and *atoms*, which are not classes. It is assumed that every condition on classes and atoms determines the class of all sets and atoms satisfying the condition in question. Classes that are not sets are *proper* classes.

Where u and v are sets or atoms, the ordered pair (u, v) is $\{\{u\}, \{u, v\}\}$. A *relation* is a class of ordered pairs of sets and atoms, and u is a *function* if, and only if, u is a relation and, for all v, w_1, w_2, if $(v, w_1), (v, w_2) \in u$, then $w_1 = w_2$. From now on, $f, g, h, f_1, \dots$ are functions.

Where u is a set and v is a class, define: $u \xrightarrow{\epsilon} v = \{f | \mathit{Dom}\, f = u \text{ and } \mathit{Rng}\, f \subseteq v\}$. "$\xrightarrow{\epsilon}$" associates to the right, when parentheses are omitted.

Define: (1) $u \downarrow v$ if, and only if, u is a function and $v \in \mathit{Dom}\, u$, and (2) for $n \geq 1$, $u \downarrow v_1, \dots, v_n$ if, and only if, $u \downarrow v_1$ and $\cdots$ and $u \downarrow v_n$.

Where u is a class and $u \subseteq \mathit{Dom}\, f$, define: (1) $\lambda\, u\, f = \{(v, fv) | v \in u\}$, and (2) $\lambda\, v \in u.fv = \lambda\, u\, f$.

Where u is a set and f is a set-valued function such that $u \subseteq \mathit{Dom}\, f$, define: (1) $\prod u\, f = \{g | \mathit{Dom}\, g = u \text{ and, for all } v \in u, gv \in fv\}$, and (2) $\prod v \in u.fv = \prod u\, f$.

In what follows, the second form of the metalanguistic λ and $\prod$ notations is used without bothering to transform the part following the dot into a function application.

who passed it along from Martin-Löf to me.

[6] Morse-Kelley with atoms and choice is the system obtained by omitting the restrictions of [Rub67, p. 31, notes 7 and 8]. For an account of Morse-Kelley without atoms, see [Kel55, pp. 251–281].

The axiom of choice is required in order to show that $\prod u\, f$ is non-empty, if, for every $v \in u$, $f\,v \neq \emptyset$. Since this theorem is needed in order to show that CT^3 type structures behave as expected, the axiom of choice is indispensable for the semantical metatheory of CT^3.

The truth values $\boldsymbol{t}$ and $\boldsymbol{f}$ are atoms, and so are the terms, assertions, and sequents considered in the sequel.

Where u and v are classes, $u \times v = \{(u_1, v_1) | u_1 \in u \text{ and } v_1 \in v\}$, and, where u is a class and $n \geq 1$, $u^1 = u$ and $u^{n+1} = u^n \times u$. Also, for all $u_1, \ldots u_n, u_{n+1}$ ($n \geq 1$), $(u_1) = u_1$ and $(u_1, \ldots, u_n, u_{n+1}) = ((u_1, \ldots, u_n), u_{n+1})$.

Let *Atm* be the class of atoms, and let $Tupl_\omega$ be the result of closing *Atm* under pairing. The function *Flat*, which maps $Tupl_\omega$ to $Atm^\omega = \bigcup_{n \geq 1} Atm^n$ is defined by: (1) If $v \in Atm$, then $Flat\, v = v$ and $Flat\,(u, v) = (Flat\, u, v)$, and (2) $Flat\,(u, (v, w)) = Flat\,((u, v), w)$.

Where u is a class, define u^*, the class of finite sequences of members of u, by: $u^* = \bigcup_{n \in \omega} n \xrightarrow{\epsilon} u$. Define *Items* on tuples and finite sequences by letting $Items\,(v_1, \ldots, v_n)$ ($n \geq 1$) be $\{v_1, \ldots, v_n\}$ and, for a finite sequence σ, letting $Items\, \sigma$ be $Rng\, \sigma$.

Let $[u \mapsto v]f$ be $\lambda\, w \in (Dom\, f) \cup \{u\}.\text{if } w = u \text{ then } v, \text{ else } f w$.

3 Terms, Assertions, Environments, and Sequents

The set *Var* of *variables* is denumerable, and *Con*, the class of *constants*, has $Req = \{\Omega, \mathrm{T}_2, \mathrm{T}_1, O, U, o, \simeq, \iota\}$ as a subset and is proper. The members of *Req* are the *required constants.*

From now on, $x, y, z, x_1, \ldots$ are members of *Var*, and $a, b, c, a_1, \ldots$ are members of *Con*.

The class *Term* of *terms* is disjoint from $\{\boldsymbol{t}, \boldsymbol{f}\}$, and its members are formed from variables and constants by means of the *term forming operations* $(\_2\_3), (\lambda\_1 : \_2.\_3)$, and $(\prod\_1 : \_2.\_3)$ of *application, abstraction*, and *production.* Argument 1 of the term forming operations is a variable, and arguments 2 and 3 are terms.

From now on, $X, Y, Z, X_1, \ldots$ are terms, $(Op\_1 : \_2.\_3), (Op_1\_1 : \_2.\_3), \ldots$ are abstraction or production, and parentheses are dropped with the understanding that application associates left and the other term forming operations get the largest scope possible. The convention for dropping parentheses also applies to metalinguistic application and metalinguistic variable binders involving dots.

The scope of abstraction and production is discontinuous — in a term of the form $Op\, x : X.Y$, the outermost term forming operation binds x in Y, but it binds nothing in X. Also, it is assumed that the term forming operations behave so that terms that would ordinarily be counted as being the same up to alphabetic change of bound variables are, in fact the same term.

The *free variable operation* FV maps X to the set of variables free in X, and $Sub[Y/x/X]$ is the result of substituting Y for x in X. It is assumed that *Sub* avoids variable capture.

The term X is an *immediate subterm of* the term Y if, and only if, Y has one of the forms $Y_1\, Y_2$ or $Op\, x : Y_1.Y_2$ and X is Y_1 or Y_2. Let $\prec$ be the transitive closure of the relation immediate subterm. Given a formal version of the intuitive account of terms contained in the preceding four paragraphs, it can be shown that $\prec$ is well-founded and $\{Y | Y \prec X\}$ is a finite or denumerable set.

A *lexicon* is a subset of *Con* that has the required constants as members. From now on, $L, L_1, \ldots$ are lexicons, and $Term_L$ is the smallest set of terms that contains L and *Var* and is closed under the term forming operations. The *required constants of* L are the required constants, and the *optional constants of* L are those members of L that are not required constants.

In what follows, it is assumed that distinct notations for variables denote distinct variables.

The class *Assert* of *assertions* is disjoint from $\{\boldsymbol{t}, \boldsymbol{f}\}$ and *Term*, and its members are formed from terms by means of the *assertion forming operations* $\_1 : \_2$ and $\_1 \cong \_2$. Assertions of the form $X : Y$ are *typing assertions*, assertions of the form $X \cong Y$ are *equality assertions*, and the set $Assert_L$ is $\{X : Y | X, Y \in Term_L\} \cup \{X \cong Y | X, Y \in Term_L\}$.

From now on, $\_1\ Op\ \_2$, $\_1\ Op_1\ \_2$, ... are assertion forming opperations. Also, ψ, ψ_1, ... are terms or assertions, and, where $\psi = X\ Op\, Y$, $FV\ \psi = (FV\, X) \cup (FV\, Y)$.

A finite sequence of typing assertions $x_0 : X_0, \ldots, x_{n-1} : X_{n-1}$ $(n \geq 0)$ has the *matrioshka property*[7] if, and only if, for all i $(0 \leq i \leq n-1)$: (1) $FV\, X_i \subseteq \{x_0, \ldots, x_{i-1}\}$ and (2) $x_i \notin \{x_0, \ldots, x_{i-1}\}$. An *environment* is a finite sequence of typing assertions that has the matrioshka property. From now on, $\gamma, \delta, \theta, \xi, \gamma_1, \ldots$ are environments. Also, *Env* is the class of environments, $Env_L = \{\gamma | Items\, \gamma \subseteq Assert_L\}$, and, where $\gamma = x_0 : X_0, \ldots, x_{n-1} : X_{n-1}$, $FV\, \gamma = FV\, x_0 : X_0 \cup \cdots \cup FV\, x_{n-1} : X_{n-1}$.

The pair (γ, ψ) has the *matrioshka property* if, and only if, $FV\, \psi \subseteq FV\, \gamma$. The class *Seq* of *sequents* is disjoint from $\{\boldsymbol{t}, \boldsymbol{f}\}$, *Term*, and *Assert*, and its members are formed from pairs (γ, ψ) that have the matrioshka property by means of the *sequent forming operation* $\_1 \gg \_2$. In what follows, $s, s_1, \ldots$ are sequents, $\zeta, \zeta_1, \ldots$ are sets of sequents, $Seq_L = \{\gamma \gg \psi | \gamma \in Evn_L, \psi \in Assert_L \cup Term_L$, and (γ, ψ) has the matrioshka property$\}$, and $FV\, \gamma \gg \psi = (FV\, \gamma) \cup (FV\, \psi)$.

Where $\gamma = x_0 : X_0, \ldots, x_{n-1} : X_{n-1}$, let $LV\, \gamma = \{x_0, \ldots, x_{n-1}\}$. Note that, due to the matrioshka property, $FV\, \gamma = LV\, \gamma$ and $FV\, \gamma \gg \psi = LV\, \gamma$.

4 Set Theoretic Closure Conditions

Define: (1) u is *closed under* $\overset{\epsilon}{\rightarrow}$ if, and only if, u is a class and, for all $v, v_1 \in u$, if v and v_1 are classes, then $v \overset{\epsilon}{\rightarrow} v_1 \in u$, and (2) u is *closed under* $\prod$ if, and only if, u is a class and, for all v and f, if $v \in u$, v is a class, $f \in v \overset{\epsilon}{\rightarrow} u$, and every member of $Rng\, f$ is a class, then $\prod v\, f \in u$.

Consider the system of set theory obtained from the system described in section 2 by omitting the axiom of inaccessible cardinals. In the diminished system, it can be shown that for every set w, there is a set u such that $w \subseteq u$ and u is closed under $\prod$ if, and only if, for every ordinal α, there is an inaccessible cardinal κ such that $\alpha < \kappa$.[8] Since closure under $\prod$ is an essential component of the semantical apparatus defined in section 5, it should be clear that the axiom of inaccessible cardinals is needed in order to deal with the semantics of CT^3. This remark is refined in sections 5 and 14.

5 Structures, Models, Denotation, Satisfaction, Validity, and Consequence

An L *structure* is a function $\mathcal{S}$ such that $Dom\, \mathcal{S} = L$ and the following conditions hold: (1) Each of $\mathcal{S}\, U$, $\mathcal{S}\, O$, $\mathcal{S}\, T_1$, and $\mathcal{S}\, T_2$ is a set of sets. (2) $\mathcal{S}\, o$ is a non-empty set, and $\mathcal{S}\, o \in \mathcal{S}\, U$. (3) $\mathcal{S}\, U \in \mathcal{S}\, O$, $\mathcal{S}\, U \subseteq \mathcal{S}\, T_1$, $\mathcal{S}\, O \in \mathcal{S}\, T_1$ and $\mathcal{S}\, O \subseteq \mathcal{S}\, T_1$, and $\mathcal{S}\, T_1 \in \mathcal{S}\, T_2$ and $\mathcal{S}\, T_1 \subseteq \mathcal{S}\, T_2$. (4) $\mathcal{S}\, U$ and $\mathcal{S}\, O$ are closed under $\overset{\epsilon}{\rightarrow}$, and $\mathcal{S}\, T_1$ and $\mathcal{S}\, T_2$ are closed under $\prod$. (5) $\mathcal{S} \simeq\ \in \prod u \in \mathcal{S}\, T_1 . u \overset{\epsilon}{\rightarrow} u \overset{\epsilon}{\rightarrow} \mathcal{S}\, o$. (6) $\mathcal{S}\, \iota \in \prod u \in \mathcal{S}\, T_1 . u \overset{\epsilon}{\rightarrow} (u \overset{\epsilon}{\rightarrow} \mathcal{S}\, o) \overset{\epsilon}{\rightarrow} u$. (7) $\mathcal{S}\, \Omega = (\bigcup(\mathcal{S}\, T_2)) \cup (\mathcal{S}\, T_2) \cup \{\mathcal{S}\, T_2\}$. (8) If c is an optional constant of L, then there is a u such that either $u \in \mathcal{S}\, T_2$ and $\mathcal{S}\, c \in u$ or $u \in (\mathcal{S}\, \Omega) \cup \{\mathcal{S}\, \Omega\}$ and $\mathcal{S}\, c = u$.

An individual or class u is a *structure* if, and only if, for some L, u is an L structure. From now on, $\mathcal{S}, \mathcal{S}_1, \ldots$ are structures, and $L_{\mathcal{S}} = Dom\, \mathcal{S}$. The *types of* $\mathcal{S}$ are the members of

---

[7] "*matrioshka*" is the Russian word for sets of dolls that nest inside one another.

[8] The proof of the right to left half is straightforward. The left to right half was proved by Paul Howard.

$(\mathcal{S}\,\mathrm{T}_2) \cup \{\mathcal{S}\,\mathrm{T}_2\} \cup \{\mathcal{S}\,\Omega_2\}$. The *entities of* $\mathcal{S}$ are the members of $(\mathcal{S}\,\Omega) \cup \{\mathcal{S}\,\Omega\}$, and the class of structures is *Str*.

It follows from the theorem stated in section 4 that the full force of the axiom of inaccessible cardinals is needed to insure that, for every pair of sets u_U and v_o satisfying the conditions (1) and (2), above, there is a structure $\mathcal{S}$ such that $\mathcal{S}\,U = u_U$ and $\mathcal{S}\;o = v_o$.

The set $Val_{\mathcal{S}}$ of $\mathcal{S}$ *valuations* is $\mathrm{Var} \xrightarrow{c} \bigcup(\mathcal{S}\,\mathrm{T}_2)$. From now on, $V, V_1, \ldots$ are $\mathcal{S}$ valuations.

A *model* is a structure such that: (1) $\mathcal{S}\,o = \{\boldsymbol{t}, \boldsymbol{f}\}$, (2) $\mathcal{S} \simeq = \lambda\, u \in \mathcal{S}\,\mathrm{T}_1.\lambda\, v \in u.\lambda\, w \in u.\mathrm{if}\, v = w$ then $\boldsymbol{t}$, else $\boldsymbol{f}$, and (3) for $u \in \mathcal{S}\,\mathrm{T}_1$, $v \in u$, and $f \in u \xrightarrow{c} \{\boldsymbol{t}, \boldsymbol{f}\}$, if there is a $v_1 \in u$ such that $\{w \in u | f w = \boldsymbol{t}\} = \{v_1\}$, then $\mathcal{S}\,\iota\, u\, v\, f$ is the $v_1 \in u$ such that $f v_1 = \boldsymbol{t}$.

In what follows, $\mathcal{M}, \mathcal{M}_1, \ldots$ are to be models, and *Mod* is the class of models.

For members of $Term_{L_{\mathcal{S}}}$, define $D_{\mathcal{S}}$, the *conditional denotation function over* $\mathcal{S}$, by the following recursion on $\prec$:

(1) $D_{\mathcal{S}}\,x = \lambda\, V \in Val_{\mathcal{S}}.V\,x$

(2) $D_{\mathcal{S}}\,c = \lambda\, V \in Val_{\mathcal{S}}.\mathcal{S}\,c$

(3) Where $X = Y\,Z$, $u = (Dom\,(D_{\mathcal{S}}\,Y)) \cap (Dom\,(D_{\mathcal{S}}\,Z))$, $v = \{V \in u | D_{\mathcal{S}}\,Y\,V$ is a function$\}$, and $w = \{V \in v | D_{\mathcal{S}}\,Y\,V \downarrow D_{\mathcal{S}}\,Z\,V\}$: $D_{\mathcal{S}}\,X = \lambda V \in w.(D_{\mathcal{S}}\,Y\,V)(D_{\mathcal{S}}\,Z\,V)$.

(4) Where $X = \lambda\, x : Y.Z$, $u = Dom\,(D_{\mathcal{S}} Y)$, $v = \{V \in u | D_{\mathcal{S}}\,Y\,V$ is a set$\}$, and $w = \{V \in v |$for all $v_1 \in D_{\mathcal{S}}\,Y\,V, D_{\mathcal{S}}\,Z \downarrow [x \mapsto v_1]V\}$: $D_{\mathcal{S}}\,X = \lambda\, V \in w.\lambda v_1 \in D_{\mathcal{S}}\,YV.D_{\mathcal{S}}\,Z\,[x \mapsto v_1]V$.

(5) Where $X = \prod x : Y.Z$, $u = Dom\,(D_{\mathcal{S}} Y)$, $v = \{V \in u | D_{\mathcal{S}}\,Y\,V$ is a set$\}$, $w_1 = \{V \in v |$for all $v_1 \in D_{\mathcal{S}}\,Y\,V, D_{\mathcal{S}}\,Z \downarrow [x \mapsto v_1]V\}$, and $w_2 = \{V \in w_1 |$for all $v_1 \in D_{\mathcal{S}}\,Y\,V, D_{\mathcal{S}}\,Z\,[x \mapsto v_1]V$ is a set$\}$: $D_{\mathcal{S}}\,X = \lambda\, V \in w_2.\prod v_1 \in D_{\mathcal{S}}\,Y\,V.D_{\mathcal{S}}\,Z[x \mapsto v_1]V$.

It can be shown by induction on $\prec$ that:

(1) If V_1 and V_2 agree on $FV\,X$, then $D_{\mathcal{S}}\,X \downarrow V_1$ if, and only if, $D_{\mathcal{S}}\,X \downarrow V_2$, and if $D_{\mathcal{S}}\,X \downarrow V_1, V_2$, then $D_{\mathcal{S}}\,X\,V_1 = D_{\mathcal{S}}\,X\,V_2$.

(2) If $D_{\mathcal{S}}\,X \downarrow V$, $D_{\mathcal{S}}\,Y \downarrow V$, $V' = [x \mapsto D_{\mathcal{S}}\,Y\,V]V$ and $X' = Sub[Y/x/X]$, then $D_{\mathcal{S}}\,X' \downarrow V$, $D_{\mathcal{S}}\,X \downarrow V'$, and $D_{\mathcal{S}}\,X'\,V = D_{\mathcal{S}}\,XV'$.

Result (1) is the *agreement principle*, and result (2) is the *substitution principle*.

The term X is *closed* if, and only if, $FV\,X = \emptyset$. Otherwise, X is *open*. It follows from the agreement principle that if X is closed and $Dom\,(D_{\mathcal{S}}\,X) \neq \emptyset$, then $Dom\,(D_{\mathcal{S}}\,X) = Val_{\mathcal{S}}$ and, for all $V_1, V_2 \in Val_{\mathcal{S}}$, $D_{\mathcal{S}}\,X\,V_1 = D_{\mathcal{S}}\,X\,V_2$. Where $u_{\mathcal{S}}$ is the intersection of the class of closed terms with $\{X \in Term_{L_{\mathcal{S}}} | Dom\,(D_{\mathcal{S}}\,X) \neq \emptyset\}$, define $[\![\_]\!]_{\mathcal{S}}$, the *unconditional denotation function over* $\mathcal{S}$, by: $[\![\_]\!]_{\mathcal{S}} = \lambda\, v \in u_{\mathcal{S}}.D_{\mathcal{S}}\,v\,\lambda\, w \in Var.\boldsymbol{t}$. In contexts where $\mathcal{S}$ is fixed or one structure serves as well as another, the subscript is omitted from the notation just defined.

By the substitution principle, if $D_{\mathcal{S}}\,(\lambda\, x : X.Y)\,Z \downarrow V$ then $D_{\mathcal{S}}\,Sub[Z/x/Y] \downarrow V$ and $D_{\mathcal{S}}\,(\lambda\, x : X.Y)\,Z\,V = D_{\mathcal{S}}\,Sub[Z/x/Y]\,V$. This is the semantical version of β-conversion.

Besides making it possible to define $[\![\_]\!]_{\mathcal{S}}$, the agreement principle also implies that if $D_{\mathcal{S}}\,\lambda\, x : X.Y\,x \downarrow V$, $x \notin FV\,Y$, and $Dom\,(D_{\mathcal{S}}\,Y\,V) = D_{\mathcal{S}}\,X\,V$, then $D_{\mathcal{S}}\,Y \downarrow V$ and $D_{\mathcal{S}}\,(\lambda\, x : X.Y\,x)\,V = D_{\mathcal{S}}\,Y\,V$. This is the semantical version of η-conversion.

Define *satisfaction* for terms and assertions and *validity in a structure* for sequents and sets of sequents as follows:

(1) $\models^V_{\mathcal{S}} X$ if, and only if, $D_{\mathcal{S}} \downarrow X$, $D_{\mathcal{S}}\,X \downarrow V$, and $D_{\mathcal{S}}\,X\,V = \boldsymbol{t}$.

(2) $\models^V_{\mathcal{S}} X : Y$ if, and only if, $D_{\mathcal{S}} \downarrow X, Y$, $D_{\mathcal{S}}\,X \downarrow V$, $D_{\mathcal{S}}\,Y \downarrow V$, $D_{\mathcal{S}}\,Y\,V$ is a type of $\mathcal{S}$, and $D_{\mathcal{S}}\,X\,V \in D_{\mathcal{S}}\,Y\,V$.

(3) $\models^V_{\mathcal{S}} X \cong Y$ if, and only if, $D_{\mathcal{S}} \downarrow X, Y$, $D_{\mathcal{S}}\,X \downarrow V$, $D_{\mathcal{S}}\,Y \downarrow V$, $D_{\mathcal{S}}\,X\,V$ and $D_{\mathcal{S}}\,Y\,V$ are entities of $\mathcal{S}$, and $D_{\mathcal{S}}\,X\,V = D_{\mathcal{S}}\,Y\,V$.

(4) $\models_{\mathcal{S}} \gamma \gg \psi$ if, and only if, $\gamma \gg \psi \in Seq_{L_{\mathcal{S}}}$ and, for every $V \in Val_{\mathcal{S}}$, either there is an item ψ' of γ such that $\not\models^{V}_{\mathcal{S}} \psi'$ or $\models^{V}_{\mathcal{S}} \psi$.

(5) $\models_{\mathcal{S}} \zeta$ if, and only if, for every $s \in \zeta$, $\models_{\mathcal{S}} s$.

Define *structural consequence* for sequents and sets of sequents, *structural validity* for sequents, and *structural validity* for terms and assertions as follows:

(6) $\zeta \models^{\clubsuit} s$ if, and only if, for every $\mathcal{S} \in Str$ such that $\zeta \cup \{s\} \subseteq Seq_{L_{\mathcal{S}}}$, if $\models_{\mathcal{S}} \zeta$, then $\models_{\mathcal{S}} s$.

(7) $\models^{\clubsuit} s$ if, and only if, $\emptyset \models^{\clubsuit} s$.

(8) $\models^{\clubsuit} \psi$ if, and only if, $\models^{\clubsuit} \gg \psi$.

And define *consequence* for sequents and sets of sequents, *validity* for sequents, and *validity* for terms and assertions as follows:

(9) $\zeta \models s$ if, and only if, for every $\mathcal{S} \in Mod$ such that $\zeta \cup \{s\} \subseteq Seq_{L_{\mathcal{S}}}$, if $\models_{\mathcal{S}} \zeta$, then $\models_{\mathcal{S}} s$.

(10) $\models s$ if, and only if, $\emptyset \models s$.

(11) $\models \psi$ if, and only if, $\models \gg \psi$.

Structural consequence and validity are appropriate for dealing with typing rules and typing axioms, and consequence and validity are appropriate for dealing with boolean rules and boolean axioms.

6 Abbreviations

Abbreviations are enhancements of the metalanguage — they provide new ways of talking about terms, but they do *not* introduce new terms.

Abbreviate:

Abbr $\stackrel{\wedge}{\to}$: $X \stackrel{\wedge}{\to} Y = \prod z : X.Y$, where $z \notin FV\,Y$.

"$\stackrel{\wedge}{\to}$" associates to the right.

Abbr $\simeq$: $X \simeq_Z Y = \ \simeq Z\,X\,Y$

Abbr $\underline{\Leftrightarrow}$1: $\underline{\Leftrightarrow} = \ \simeq o$

Abbr $\underline{\Leftrightarrow}$2: $X \underline{\Leftrightarrow} Y = \underline{\Leftrightarrow} X\,Y$

Abbr $\underline{\Leftrightarrow}$2 makes the metalanguage ambiguous — at this point, it is not clear whether "$X \underline{\Leftrightarrow} Y$" denotes the result of applying X to $\underline{\Leftrightarrow}$ and then applying the term so formed to Y or, instead, the result of applying $\underline{\Leftrightarrow}$ to X and then applying the term so formed to Y. This ambiguity is resolved by stipulating that, in parsing metalinguistic expressions, abbreviations take precedence over primitive notations. Thus, "$X \underline{\Leftrightarrow} Y$" denotes the result of applying $\underline{\Leftrightarrow}$ to X and then applying the term so formed to Y.

Abbr $\underline{T}$: $\underline{T} = \ \Leftrightarrow \simeq_{o \stackrel{\wedge}{\to} o \stackrel{\wedge}{\to} o} \Leftrightarrow$

Abbr $\underline{\forall}$1: $\underline{\forall} = \lambda\, z : \mathrm{T}_1.\lambda x : z \stackrel{\wedge}{\to} o.(\lambda\, y : z.\underline{T}) \simeq_{z \stackrel{\wedge}{\to} o} x$

Abbr $\underline{\forall}$2: $\underline{\forall} x : Z.X = \underline{\forall}\, Z\, \lambda\, x : Z.X$

Abbr ι: $\iota\, x : Z\,[Y].X = \iota\, Z\, Y\, \lambda\, x : Z.X$

Abbr $\underline{F}$: $\underline{F} = \underline{\forall} x : o.x$

Abbr $\underline{\neg}$: $\underline{\neg} = \ \simeq o\, \underline{F}$

Abbr $\underline{d}$: $\underline{d} = \lambda\, x : o.\lambda\, y : o.\lambda\, z : o \stackrel{\wedge}{\to} o \stackrel{\wedge}{\to} o.z\, x\, y$

Abbr $\underline{\Rightarrow}$1: $\underline{\Rightarrow} = \lambda\, x : o.\lambda\, y : o.\underline{\neg}(\underline{d}\, x\, y \simeq_{o \stackrel{\wedge}{\to} o \stackrel{\wedge}{\to} o} \underline{d}\, \underline{T}\, \underline{F})$

Abbr $\underline{\Rightarrow}$2: $X \underline{\Rightarrow} Y = \underline{\Rightarrow} X\, Y$

Abbr $\underline{d}$ and, consequently, **Abbr** $\underline{\Rightarrow}$**1** may seem mysterious. It can be shown that, for all $u_1, u_2, v_1, v_2 \in [\![o]\!]$, if $[\![\underline{d}]\!]\, u_1\, u_2 = [\![\underline{d}]\!]\, v_1\, v_2$, then, $u_1 = v_1$ and $u_2 = v_2$. Therefore, $[\![\underline{d}]\!]$ is a pairing operation on $[\![o]\!]$, and in a model, for all $u, v \in [\![o]\!]$, $[\![\underline{\Rightarrow}]\!]\, u\, v$ is if $u = \boldsymbol{t}$ and $v = \boldsymbol{f}$ then $\boldsymbol{f}$, else $\boldsymbol{t}$. It follows that **Abbr** $\underline{\Rightarrow}$**1** is a correct way of introducing the truth functional conditional.

Abbr $\underline{Void}$: $\underline{Void} = \lambda\, x : \mathrm{T}_1 . \underline{\forall}\, y : x . \underline{\neg}\, y \simeq_x y$

Rules for restoring omitted parentheses: (1) "$\underline{\forall}$" and "ι" written as a variable binders get the largest scope possible. (2) Infixed "$\simeq$" gets a smaller scope than infixed "$\underline{\Leftrightarrow}$" and "$\underline{\Rightarrow}$" and a smaller scope than "$\underline{\neg}$". (3) Infixed "$\underline{\Leftrightarrow}$" and "$\underline{\Rightarrow}$" and prefixed "$\underline{\neg}$" get decreasing scopes in the order in which they have just been listed. (4) Infixed "$\underline{\Leftrightarrow}$" and "$\underline{\Rightarrow}$" associate to the right. (5) With the exception of notations for applications headed by "$\underline{\neg}$", notations for applications have a smaller scope than infixes.

7 Inferences, Rules of Inference, Derivations, and Derivability

The class *Inf* of *inferences* is $\bigcup_{n \geq 1} Seq^{n+1}$, and, in what follows, I, I_1, ... are inferences.

A *rule of inference* is a class of inferences. From now on, R, R_1, ... are rules of inference.

Let $\gamma \gg \psi_1; \ldots; \psi_n$ $(n \geq 1)$ be $(\gamma \gg \psi_1, \ldots, \gamma \gg \psi_n)$. Where $m, n_1, \ldots, n_m \geq 1$, the scheme:

$$\frac{\gamma^1 \gg \psi^1_1; \ldots; \psi^1_{n_1} \qquad \ldots \qquad \gamma^m \gg \psi^m_1; \ldots; \psi^m_{n_m}}{\gamma \gg \psi}$$

depicts the class of inferences of the form:

$$(Flat\,(\gamma^1 \gg \psi^1_1; \ldots; \psi^1_{n_1}, \ldots, \gamma^m \gg \psi^m_1; \ldots; \psi^m_{n_m}), \gamma \gg \psi)$$

Substitution instances of schemata of this form, possibly with provisos, are used to define rules of inference.

Define derivations reccursively by: (1) s is a *derivation.* (2) If $\boldsymbol{D}_1, \ldots, \boldsymbol{D}_n$ $(n \geq 1)$ are derivations, then $(\boldsymbol{D}_1, \ldots, \boldsymbol{D}_n, s)$ is a *derivation.*

From now on, $\boldsymbol{D}$, $\boldsymbol{E}$, $\boldsymbol{F}$, $\boldsymbol{D}_1$, ... are derivations.

The *endsequent of* $\boldsymbol{D}$ is $\boldsymbol{D}$, if $\boldsymbol{D}$ is a sequent, and, otherwise, it is the second component of the ordered pair $\boldsymbol{D}$. In what follows, $s_{\boldsymbol{D}}$ is the endsequent of $\boldsymbol{D}$.

If $\boldsymbol{D}$ is $(\boldsymbol{D}_1, \ldots, \boldsymbol{D}_n, s)$ $(n \geq 1)$, then $I_{\boldsymbol{D}}$ is $(s_{\boldsymbol{D}_1}, \ldots, s_{\boldsymbol{D}_n}, s)$, the *last inference of* $\boldsymbol{D}$ is $I_{\boldsymbol{D}}$, and the *immediate subderivations of* $\boldsymbol{D}$ are $\boldsymbol{D}_1, \ldots, \boldsymbol{D}_n$. If $\boldsymbol{D}$ is a sequent, then no derivation is an *immediate subderivation of* $\boldsymbol{D}$.

The relation *proper subderivation* is the transitive closure of the relation immediate subderivation, and the relation *subderivation* is the reflexive closure of the relation proper subderivation.

If $\boldsymbol{D}$ is a sequent, then $Top\,\boldsymbol{D} = \{\boldsymbol{D}\}$, and if $\boldsymbol{D}$ is not a sequent and the set of immediate subderivations of $\boldsymbol{D}$ is $\{\boldsymbol{D}_1, \ldots, \boldsymbol{D}_n\}$, then $Top\,\boldsymbol{D} = (Top\,\boldsymbol{D}_1) \cup \cdots \cup (Top\,\boldsymbol{D}_n)$. The members of $Top\,\boldsymbol{D}$ are the *top sequents of* $\boldsymbol{D}$.

The derivation $\boldsymbol{D}$ is an *R derivation* if, and only if, for every subderivation $\boldsymbol{E}$ of $\boldsymbol{D}$, if $\boldsymbol{E}$ is not a sequent, then $I_{\boldsymbol{E}}$ is a member of R. The sequent s is *R derivable from* ζ if, and only if, for some R derivation $\boldsymbol{D}$, $Top\,\boldsymbol{D} \subseteq \zeta$ and $s_{\boldsymbol{D}} = s$. Also, where $n \geq 1$, $(\zeta_1, \ldots, \zeta_n) \sqsupset_R s$ holds if, and only if s is R derivable from $\zeta_1 \cup \cdots \cup \zeta_n$.

The inference $I = (s_1, \ldots, s_n, s)$ is *structurally sound* if, and only if, $\{s_1, \ldots, s_n\} \models^{\clubsuit} s$, and I is *sound* if, and only if, $\{s_1, \ldots, s_n\} \models s$. The rule of inference R is *structurally sound* if, and only if, every member of R is a structurally sound inference, and R is *sound* if, and only if, every member of R is a sound inference. It should be clear by inspection that the rules of sections 8–11 are structurally sound, the axioms of section 12 are structurally valid, and the rules of section 13 are sound.

8 The Monotonicity Rule

The rule [Reit] allows assertions and terms derived in a given environment to be used in extensions of that environment.[9]

$$\frac{\gamma \gg \psi \qquad \gamma, \psi_1 \gg \psi_2}{\gamma, \psi_1 \gg \psi} \quad \text{[Reit]}$$

9 Closure Rules

The rules [U Cl] and [O Cl] correspond to the semantical conditions (i) $[\![U]\!]$ is closed under $\stackrel{\epsilon}{\to}$ and (ii) $[\![O]\!]$ is closed under $\stackrel{\epsilon}{\to}$.

$$\frac{\gamma \gg X : U\,;\, Y : U}{\gamma \gg X \stackrel{\wedge}{\to} Y : U} \quad [U\ \text{Cl}] \qquad \frac{\gamma \gg X : O\,;\, Y : O}{\gamma \gg X \stackrel{\wedge}{\to} Y : O} \quad [O\ \text{Cl}]$$

The rule [T_1 Cl] corresponds to the semantical conditions (i) $[\![U]\!] \subseteq [\![\mathrm{T}_1]\!]$, (ii) $[\![O]\!] \subseteq [\![\mathrm{T}_1]\!]$, and (iii) $[\![\mathrm{T}_1]\!]$ is closed under $\prod$.

$$\frac{\gamma \gg X : U}{\gamma \gg X : \mathrm{T}_1} \quad [\mathrm{T}_1\ \text{Cl}\ U] \qquad \frac{\gamma \gg X : O}{\gamma \gg X : \mathrm{T}_1} \quad [\mathrm{T}_1\ \text{Cl}\ O]$$

$$\frac{\gamma \gg Z_1 : \mathrm{T}_1 \qquad \gamma, x : Z_1 \gg Z_2 : \mathrm{T}_1}{\gamma \gg \prod x : Z_1.Z_2 : \mathrm{T}_1} \quad [\mathrm{T}_1\ \text{Cl}\ \textstyle\prod]$$

Define: $[\mathrm{T}_1\ \text{Cl}] = [\mathrm{T}_1\ \text{Cl}\ U] \cup [\mathrm{T}_1\ \text{Cl}\ O] \cup [\mathrm{T}_1\ \text{Cl}\ \prod]$

The rule [T_2 Cl] corresponds to the semantical conditions (i) $[\![\mathrm{T}_1]\!] \subseteq [\![\mathrm{T}_2]\!]$ and (ii) $[\![\mathrm{T}_2]\!]$ is closed under $\prod$.

$$\frac{\gamma \gg X : \mathrm{T}_1}{\gamma \gg X : \mathrm{T}_2} \quad [\mathrm{T}_2\ \text{Cl}\ \mathrm{T}_1] \qquad \frac{\gamma \gg Z_1 : \mathrm{T}_2 \qquad \gamma, x : Z_1 \gg Z_2 : \mathrm{T}_2}{\gamma \gg \prod x : Z_1.Z_2 : \mathrm{T}_2} \quad [\mathrm{T}_2\ \text{Cl}\ \textstyle\prod]$$

Define: $[\mathrm{T}_2\ \text{Cl}] = [\mathrm{T}_2\ \text{Cl}\ \mathrm{T}_1] \cup [\mathrm{T}_2\ \text{Cl}\ \prod]$

In the context of the other rules and the axioms of section 12, the rule [Ω Cl] expresses the semantical condition $(\bigcup [\![\mathrm{T}_2]\!]) \cup [\![\mathrm{T}_2]\!] \subseteq [\![\Omega]\!]$.

$$\frac{\gamma \gg X : Z}{\gamma \gg X : \Omega} \quad [\Omega\ \text{Cl}]$$

10 Conversion Rules

Taken together with the axiom for Ω stated in section 12,[10] the rules [ρ] through [η] insure that $\cong$ has the properties of $\beta\eta$-conversion. The names of these rules are variations on the standard names of the rules used to analyze conversion in the type-free λ-calculus [CF58].

$$\frac{\gamma \gg X : \Omega}{\gamma \gg X \cong X} \quad [\rho] \qquad \frac{\gamma \gg X \cong Y}{\gamma \gg Y \cong X} \quad [\sigma] \qquad \frac{\gamma \gg X \cong Y\,;\, Y \cong Z}{\gamma \gg X \cong Z} \quad [\tau]$$

---

[9] In the name of the rule, "Reit" is short for "Reiteration". The name is taken from that of a rule used in the system of [Fit52].

[10] See the definition of [Ω] at the beginning of that section.

$$\frac{\gamma \gg Z\,X : \Omega\,;\, X \cong Y}{\gamma \gg Z\,X \cong Z\,Y}\ [\mu] \qquad \frac{\gamma \gg X\,Z : \Omega\,;\, X \cong Y}{\gamma \gg X\,Z \cong Y\,Z}\ [\nu]$$

$$\frac{\gamma \gg Op\,x : Z.X : \Omega \qquad \gamma, x : Z \gg X \cong Y}{\gamma \gg Op\,x : Z.X \cong Op\,x : Z.Y}\ [\xi] \qquad \frac{\gamma \gg Op\,x : X.Z : \Omega\,;\, X \cong Y}{\gamma \gg Op\,x : X.Z \cong Op\,x : Y.Z}\ [\xi.]$$

$$\frac{\gamma \gg (\lambda\,x : Z_1.X)Y : \Omega}{\gamma \gg (\lambda\,x : Z_1.X)Y \cong Sub[Y/x/X]}\ [\beta] \qquad \frac{\gamma \gg \lambda\,x : Z_1.X\,x : \Omega\,;\, X : \prod x : Z_1.Z_2}{\gamma \gg \lambda\,x : Z_1.X\,x \cong X}\ [\eta]$$

Provided $x \notin LV\,\gamma$.

The rules [:≅] and [≅:] extend the replacement properties of ≅ guaranteed by the rules [μ] through [ξ.] to include replacement in the context $\_\_1 : \_\_2$.

$$\frac{\gamma \gg Z : X\,;\, X \cong Y}{\gamma \gg Z : Y}\ [:\cong] \qquad \frac{\gamma \gg X : Z\,;\, X \cong Y}{\gamma \gg Y : Z}\ [\cong:]$$

11 Assignment Rules

The rule [Hyp] assigns type symbols to variables,[11] and the rules [∏ E] and [∏ I] assign type symbols to terms formed by means of application and abstraction.

$$\frac{\gamma \gg Z : \mathrm{T}_2}{\gamma, x : Z \gg x : Z}\ [\mathrm{Hyp}] \qquad \frac{\gamma \gg X : \prod x : Z_1.Z_2\,;\, Y : Z_1}{\gamma \gg X\,Y : Sub[Y/x/Z_2]}\ [\textstyle\prod \mathrm{E}]$$

Provided $x \notin LV\gamma$.

$$\frac{\gamma \gg Z_1 : \mathrm{T}_2 \qquad \gamma, x : Z_1 \gg X : Z_2\,;\, Z_2 : \mathrm{T}_2}{\gamma \gg \lambda\,x : Z_1.X : \prod x : Z_1.Z_2}\ [\textstyle\prod \mathrm{I}]$$

Let $R_{\Vdash\clubsuit}$ be the union of the structural rule, closure rules, and assignment rules.

12 Morphologies

Define: $[\Omega] = \{\gg \Omega \cong \Omega\}$, $[\mathrm{T}_2] = \{\gg \mathrm{T}_2 : \Omega\}$, $[\mathrm{T}_1] = \{\gg \mathrm{T}_1 : \mathrm{T}_2\}$, $[O] = \{\gg O : \mathrm{T}_1\}$, $[U] = \{\gg U : O\}$, $[o] = \{\gg o : U\}$, $[\simeq] = \{\gg \simeq : \prod x : \mathrm{T}_1.x \stackrel{\wedge}{\to} x \stackrel{\wedge}{\to} o\}$, and $[\iota] = \{\gg \iota : \prod x : \mathrm{T}_1.x \stackrel{\wedge}{\to} (x \stackrel{\wedge}{\to} o) \stackrel{\wedge}{\to} x\}$. Let $[Req]$ be the union of these sets. The members of $[Req]$ correspond to the semantical conditions on structures that are not covered by the rules of sections 9 and 11.

A *morphology over* L is a set of sequents M satisfying the following conditions:

(1) Every member of M has the form $\gg c\,Op\,X$, where c is an optional constant of L.

(2) If $\gg c\,Op_1\,X_1 \in M$ and $\gg c\,Op_2\,X_2 \in M$, then $\_\_1Op_1\_\_2 = \_\_1Op_2\_\_2$ and $X_1 = X_2$.

(3) If c is an optional constant of L, then some member of M has the form $\gg c\,Op\,X$.

(4) There is a well-ordering w of the optional constants of L such that, where c is an optional constant of L and M_c is the set of members $\gg c_1\,Op_1\,X_1$ of M such that $(c_1, c) \in w$, the following conditions hold:

[11] In the name of the rule, "Hyp" is short for "Hypothesis". As in the case of [Reit], the name is taken from that of a rule used in the system of [Fit52].

(a) If $\gg c : X \in M$ then $M_c \cup [Req] \sqsupset_{R_{\Vdash^{\clubsuit}}} \gg X : \mathrm{T}_2$.

(b) If $\gg c \cong X \in M$ then $M_c \cup [Req] \sqsupset_{R_{\Vdash^{\clubsuit}}} \gg X : \Omega$ or $M_c \cup [Req] \sqsupset_{R_{\Vdash^{\clubsuit}}} \gg X \cong \Omega$.

Let $M^{\clubsuit} = M \cup [Req]$. The sequent s is a *morphological theorem over* u if, and only if, for some L and M, M is a morphology over L, $u = M$, and $M^{\clubsuit} \sqsupset_{R_{\Vdash^{\clubsuit}}} s$. Also, $u \Vdash^{\clubsuit} s$ holds if, and only if, s is a morphological theorem over u.

The constant ι is used to represent definite descriptions [Rus56]. In view of [And86, pp. 165 and 182], the expected the type symbol of ι would be $\prod x : \mathrm{T}_1.(x \stackrel{\wedge}{\rightarrow} o) \stackrel{\wedge}{\rightarrow} x$, but giving ι this type symbol rules out empty types. Giving ι the more complicated type symbol $\prod x : \mathrm{T}_1.x \stackrel{\wedge}{\rightarrow} (x \stackrel{\wedge}{\rightarrow} o) \stackrel{\wedge}{\rightarrow} x$ allows empty types, at the expense of complicating the terms used to represent descriptions. Inuitively, giving ι the type symbol $\prod x : \mathrm{T}_1.x \stackrel{\wedge}{\rightarrow} (x \stackrel{\wedge}{\rightarrow} o) \stackrel{\wedge}{\rightarrow} x$ amounts to making it impossible to say "The thing of type monarch that is presently king of France is bald", but allowing us to say, instead, "The thing of type monarch, containing Elizabeth II, that is presently king of France is bald".

Note that, although type symbols may denote empty types, every member of $Term_L$ that gets a type symbol in the empty environment *via* $\Vdash^{\clubsuit}$ and a morphology M over L will have a denotation in every $\mathcal{S}$ such that $\models^{\clubsuit}_{\mathcal{S}} M^{\clubsuit}$. So CT^3 is a free logic [Fra66, LF72], because it does not require all domains of quantification to be non-empty. But, unlike many systems of free logic, CT^3 does not permit well-typed closed terms that lack denotations.

The constants permitted by condition (4)(a) allow for addition of new primitives needed in applications of CT^3, and the constants permitted by condition (4)(b) provide for equational definitions. Due to the presence of the λ operator and the strong closure conditions on types built into the system, equational definitions are very powerful in CT^3.

For example, it is possible to define a constant corresponding to each of the underlined symbols introduced by the abbreviations of section 6, and, in applications, it is best to do this. For the purposes of this paper, however, it is better to proceed *via* abbreviations, even in cases where a corresponding definition can be given — this is the only way to make it clear what the primitive basis of the system is.

13 Boolean Rules

The rules [$\underline{\Rightarrow}$ E] through [$\underline{\neg}$ E] are adaptations of a familiar set of postulates for classical propositional logic, and the rules [$\cong$ E] through [DT ξ'] insure that an appropriate version of $\beta\eta$-conversion applies to $\simeq$ and, also, that $\simeq$ has the replacement properties it should. Rules the names of which contain "DT" play a special role in proving the deduction theorem for CT^3.

$$\frac{\gamma \gg X \underline{\Rightarrow} Y \,;\, X}{\gamma \gg Y} \; [\underline{\Rightarrow}\,\mathrm{E}] \qquad \frac{\gamma \gg X \underline{\Rightarrow} Y \underline{\Rightarrow} X : o}{\gamma \gg X \underline{\Rightarrow} Y \underline{\Rightarrow} X} \; [\mathrm{DT\,Null}]$$

$$\frac{\gamma \gg (X \underline{\Rightarrow} Y \underline{\Rightarrow} Z) \underline{\Rightarrow} (X \underline{\Rightarrow} Y) \underline{\Rightarrow} X \underline{\Rightarrow} Z : o}{\gamma \gg (X \underline{\Rightarrow} Y \underline{\Rightarrow} Z) \underline{\Rightarrow} (X \underline{\Rightarrow} Y) \underline{\Rightarrow} X \underline{\Rightarrow} Z} \; [\mathrm{DT} \underline{\Rightarrow} \mathrm{E}]$$

$$\frac{\gamma \gg (X \underline{\Leftrightarrow} Y) \underline{\Rightarrow} X \underline{\Rightarrow} Y : o}{\gamma \gg (X \underline{\Leftrightarrow} Y) \underline{\Rightarrow} X \underline{\Rightarrow} Y}\ [\underline{\Leftrightarrow}\,\mathrm{E1}] \qquad \frac{\gamma \gg (X \underline{\Leftrightarrow} Y) \underline{\Rightarrow} Y \underline{\Rightarrow} X : o}{\gamma \gg (X \underline{\Leftrightarrow} Y) \underline{\Rightarrow} Y \underline{\Rightarrow} X}\ [\underline{\Leftrightarrow}\,\mathrm{E2}]$$

Define: $[\underline{\Leftrightarrow}\,\mathrm{E}] = [\underline{\Leftrightarrow}\,\mathrm{E1}] \cup [\underline{\Leftrightarrow}\,\mathrm{E2}]$

$$\frac{\gamma \gg (X \underline{\Rightarrow} Y) \underline{\Rightarrow} (Y \underline{\Rightarrow} X) \underline{\Rightarrow} (X \underline{\Leftrightarrow} Y) : o}{\gamma \gg (X \underline{\Rightarrow} Y) \underline{\Rightarrow} (Y \underline{\Rightarrow} X) \underline{\Rightarrow} (X \underline{\Leftrightarrow} Y)}\ [\underline{\Leftrightarrow}\,\mathrm{I}]$$

$$\frac{\gamma \gg (\underline{\neg} X \underline{\Rightarrow} \underline{F}) \underline{\Rightarrow} X : o}{\gamma \gg (\underline{\neg} X \underline{\Rightarrow} \underline{F}) \underline{\Rightarrow} X}\ [\underline{\neg}\,\mathrm{E}] \qquad \frac{\gamma \gg X \simeq_Z Y : o;\ X \cong Y}{\gamma \gg X \simeq_Z Y}\ [\cong\ \mathrm{E}]$$

$$\frac{\gamma \gg X \simeq_{Z_1} Y \underline{\Rightarrow} Z\,X \simeq_{Z_2} Z\,Y : o}{\gamma \gg X \simeq_{Z_1} Y \underline{\Rightarrow} Z\,X \simeq_{Z_2} Z\,Y}\ [\mu'] \qquad \frac{\gamma \gg X \simeq_{Z_1} Y \underline{\Rightarrow} X\,Z \simeq_{Z_2} Y\,Z : o}{\gamma \gg X \simeq_{Z_1} Y \underline{\Rightarrow} X\,Z \simeq_{Z_2} Y\,Z}\ [\nu']$$

The rule $[\xi']$ is the version of $[\xi]$ appropriate to $\simeq$. But there is no version of $[\xi.]$ for $\simeq$ — $\simeq$ does *not* support replacement to the right of the colon.

$$\frac{\gamma \gg (\lambda\, x : Z.X) \simeq_{Z_2} \lambda\, x : Z.Y : o \qquad \gamma, x : Z \gg X \simeq_{Z_1} Y}{\gamma \gg (\lambda\, x : Z.X) \simeq_{Z_2} \lambda\, x : Z.Y}\ [\xi']$$

$$\frac{\gamma \gg (\underline{\forall}\, x : Z.X_1 \underline{\Rightarrow} X \simeq_{Z_1} Y) \underline{\Rightarrow} X_1 \underline{\Rightarrow} (\lambda\, x : Z.X) \simeq_{Z_2} \lambda\, x : Z.Y : o}{\gamma \gg (\underline{\forall}\, x : Z.X_1 \underline{\Rightarrow} X \simeq_{Z_1} Y) \underline{\Rightarrow} X_1 \underline{\Rightarrow} (\lambda\, x : Z.X) \simeq_{Z_2} \lambda\, x : Z.Y}\ [\mathrm{DT}\ \xi']$$

Provided $x \notin LV\,\gamma$.

The rule $[\iota]$ is the only postulate needed for the description operator. The rule $[\,\underline{\mathit{Void}}\,]$ asserts that there is at most one empty type, which is true because CT^3 types are sets.

$$\frac{\gamma \gg \iota\, Z\, X\, (\simeq Z\, Y) \simeq_Z Y : o}{\gamma \gg \iota\, Z\, X\, (\simeq Z\, Y) \simeq_Z Y}\ [\iota] \qquad \frac{\gamma \gg \underline{\mathit{Void}}\, X \underline{\Rightarrow} \underline{\mathit{Void}}\, Y \underline{\Rightarrow} X \simeq_Z Y : o}{\gamma \gg \underline{\mathit{Void}}\, X \underline{\Rightarrow} \underline{\mathit{Void}}\, Y \underline{\Rightarrow} X \simeq_Z Y}\ [\,\underline{\mathit{Void}}\,]$$

Let $R_{\Vdash}$ be the union of $R_{\Vdash^\clubsuit}$ and the boolean rules.

14 Bases, Theoremhood, and Consistency

A *basis over* L is a pair of sets $B = (M, A)$ such that M is a morphology over L and every member of A has the form $\gamma \gg X$, where $M \Vdash^\clubsuit \gamma \gg X : o$.

Let $B^+ = (M^\clubsuit, A)$. The sequent s is a *theorem over* u if, and only if, for some L and some basis B over L, $u = B$ and $B^+ \sqsupset_{R_{\Vdash}} s$. Also, $u \Vdash s$ holds if, and only if, s is a theorem over u.

Consider a basis B of the form $(M, \emptyset)$. It can be shown that, for some $\mathcal{M}$, $\models_{\mathcal{M}} M^\clubsuit$. Since the typing rules of CT^3 are structurally sound, the boolean rules of CT^3 are sound, and for all $\mathcal{M}$, $\not\models_{\mathcal{M}} \gg \underline{F}$, it follows that $\gg \underline{F}$ is not a theorem over $(M, \emptyset)$. In other words, CT^3 is consistent.

Analysis of the proof of the theorem stated in section 4 shows that, even in the case of the minimal basis $(\emptyset, \emptyset)$, it must be assumed that there are at least two inaccessible cardinals in order carry out the argument relied on in the preceding paragraph.

15 Concluding Remarks

CT^3 is *very* powerful, and the system seems nice to work with, even by hand, at least for someone accustomed to typed λ-calculi. Furthermore, I think the metatheory of CT^3 will go smoothly. It should be possible to prove a completeness theorem for CT^3 by adapting the techniques Henkin used to prove the completeness of Church's type theory. Also, it should also be possible to prove normalization and Church-Rosser theorems for the reduction relation associated with $\cong$, thereby providing a basis for defining type checking algorithms for CT^3.

If these conjectures are correct, then CT^3 is an excellent candidate for tactic-style mechanization [GMW79, C+86, HOL].

References

[And65] Peter B. Andrews. *A Transfinite Type Theory with Type Variables.* North-Holland, 1965.

[And86] Peter B. Andrews. *An Introduction to Mathematical Logic and Type Theory: To Truth through Proof.* Academic Press, New York, 1986.

[Bru68] N. G. de Bruijn. Automath, a language for mathematics. Technical Report T.H.-Report 68-WSK-05, Department of Mathematics, Technical University of Einghoven, November 1968.

[C+86] Robert L. Constable et al. *Implementing Mathematics with the Nuprl Proof Development System.* Prentice-Hall, Inc., Englewood Cliffs, New Jersey, 1986.

[CF58] Haskell Brooks Curry and Robert Feys. *Combinatory Logic*, volume 1. North-Holland Publishing Company, Amsterdam, 1958. Reprinted 1968 and 1974.

[Chu40] Alonzo Church. A formulation of the simple theory of types. *Journal of Symbolic Logic*, 5:56–68, 1940.

[Coq90] Thierry Coquand. Metamathematical investigations of a calculus of constructions. In P. Odifreddi, editor, *Logic and Computer Science.* Academic Press, 1990. Apic Series 31.

[Fit52] Fredric Brenton Fitch. *Symbolic Logic.* The Ronald Press Company, New York, 1952.

[Fra66] Bas C. van Fraassen. The completeness of free logic. *Zeitschrift für Mathematische Logik und Grundlagen der Mathematik*, 12:219–234, 1966.

[GMW79] M. J. Gordon, J. Milner, and C. P. Wadsworth. *Edinburgh LCF: A Mechanized Logic of Computation.* Springer Verlag, 1979. Lecture Notes in Computer Science 78.

[Gor85] M. Gordon. HOL: A machine oriented formulation of higher-order logic. Technical Report 68, University of Cambridge Computer Laboratory, July 1985. Revised version.

[Hen50] Leon Henkin. Completeness in the theory of types. *Journal of Symbolic Logic*, 15:81–91, 1950.

[HOL] Cambridge Research Center of SRI International. *The HOL System Description.* Distributed with version 1.12 of the HOL system.

[Kel55] J. L. Kelley. *General Topology.* Van Nostrand, 1955.

[LF72] Karel Lambert and Bas C. van Fraassen. *Derivation and Counterexample: An Introduction to Philosophical Logic.* Dickenson Publishing Company, Encino, California, and Belmont, California, 1972.

[Pot79] Garrel Pottinger. On analysing relevance constructively. *Studia Logica*, 38:171–185, 1979.

[Pot87] Garrel Pottinger. Strong normalization for terms of the theory of constructions. Technical Report TR 11-7, ORA Corporation, February 1987.

[Pot88a] Garrel Pottinger. Ulysses: Logical and computational foundations of the primitive inference engine. Technical Report TR 11-8, ORA Corporation, January 1988.

[Pot88b] Garrel Pottinger. Ulysses: Logical foundations of the definition facility. Technical Report TR 11-9, ORA Corporation, January 1988.

[Pot92] Garrel Pottinger. Completeness for the HOL logic, preliminary report. Posted by electronic mail to info-hol, the mailing list for the HOL interest group, January 1992.

[Rub67] Jean E. Rubin. *Set Theory for the Mathematician.* Holden-Day, 1967. Axiom **A6**, pp. 66 and 81, should be replaced by $(\mathrm{Func}(F) \,\&\, \mathrm{S}(u)) \rightarrow \mathrm{S}(F''u)$. This will not affect the set of theorems of the whole system, including axiom **A10**, p. 85, but will make the arguments of pp. 66ff correct. For example, the proof of theorem 3.3.8, p. 66, which is circular in the context of version of **A6** given in the book, is correct for the modified version stated in this note.

[Rus56] Bertrand Russell. On denoting. In Robert C. Marsh, editor, *Logic and Knowledge*, pages 41–56. George Allen & Unwin, London, 1956.

[Sel87] Jonathan P. Seldin. MATHESIS: The mathematical foundation for ULYSSES. Technical Report RADC-TR-87-223, Rome Air Development Center, November 1987.

Chapter 2

Induction

Higher Order Logic Theorem Proving and its Applications (A-20)
L.J.M. Claesen and M.J.C. Gordon (Editors)
Elsevier Science Publishers B.V. (North-Holland)

Unification-Based Induction

Holger Busch

Siemens AG, Corporate Research and Development, Munich

Abstract

The experiences of large case studies show that there is a definite need for far better efficiency of machine-supported proving. The integration of automatic special-purpose systems with interactive theorem provers promises better performance. By means of interactive proof systems problems at higher abstraction levels can be transformed into manageable - in terms of complexity and degree of abstraction - subtasks for fully-automatic tools.

Interactive theorem provers themselves have the potential for more automation, with the advantage to attain areas beyond the domains of fully automatic tools. The *Boyer-Moore* system [3] impressively demonstrates that for a quantifier-free first-order logic mechanized induction and other heuristics significantly reduce the amount of user interaction. While this system uses routines fully coded in *LISP*, systems such as *LAMBDA* [17] allow induction schemes incarnated in rules. The expressiveness of higher-order logic entails additional difficulties, but, in conjuction with higher-order resolution and other *LAMBDA* features, it also opens up more flexibility.

This paper is a presentation of ideas for an induction tool in *LAMBDA*.

Keyword Codes: F.4.1; B.7.1; I.2.3
Keywords: Mathematical Logic; Integrated Circuits, Types and Design Styles; Deduction and Theorem Proving.

1 Introduction

Despite promising research results and applications it is still difficult to transfer formal proof methods into industrial design departments. The educational barrier is not the only reason. Even skilled proof experts are scarcely able to achieve results in time for being useful in other than safety- and security-critical projects. Since these safety-critical projects are less in number, theorem proving techniques must still improve to become more relevant in industry.

Automatic verification tools have reached a state-of-the-art where they are applied to some important verification tasks in real design, e.g., the comparison of finite-state machines [9]. These tools lack the versatility of interactive systems. Interactive theorem provers could be used to formalize most if not all design aspects. Generally, however, they are better suited for reasoning about functional behaviour at higher levels of abstraction than for treating with reasonable effort other issues which designers consider to be as important as logic functionality, e.g., timing.

The integration of efficient automatic tools with interactive theorem provers [18, 20] seems to be a viable solution. This way, more automation can be gained from both parts, i.e., extending automatic tools and enhancing interactive proof systems with more automation.

The expressiveness of a formalism influences the degree of automation. The more expressive the language the wider is the range of problems to be dealt with, and the more difficult it becomes to develop algorithms for automatic tools. Nevertheless, there are good reasons for higher-order logic[12]. In particular, higher-order logic enables more powerful reasoning procedures than less expressive formalisms. Reuse of proofs by means of genericity [16, 27] and transformational reasoning [8] are examples.

The proof system *LAMBDA* [17] comprises an induction tool for structural induction, by means of user-supplied tactics. In practice, it turns out that often more sophisticated strategies are required to determine appropriate induction approaches, for instance, substitutions for induction variables. This paper presents ideas to furnish *LAMBDA* with a more flexible induction tool, exploiting higher-order unification and other features of *LAMBDA*. The graphical frontend *DIALOG* to *LAMBDA* [10] demonstrates that a combination of design rules and programmed ML-tactics to carry out scheduling and allocation is an excellent way to mechanize proof and design heuristics. *LAMBDA* with its existing and extendible functionality therefore is a good basis for developing an induction tool comparable to that of other systems [3, 15].

In practice, especially in hardware applications, structural induction is often sufficient. The induction approach presented in this paper is based on well-founded induction, which enables generic induction schemes for better adaptation to individual proof goals.

At definition time of functions, recursion analysis is performed to generate tailored induction rules which match their recursion patterns. For conjectures with recursive terms, a generic induction rule with meta-variables is generated and applied to the goal. The meta-variables are gradually instantiated on the basis of the induction schemes of recursive subterms and other available information, until, possibly after some manipulations, the induction assertions can be *fertilized* with the induction hypotheses. The resulting induction premises are further manipulated by means of induction lemmas that are either available or can be constructed from remaining proof obligations. New induction proofs have to be initiated for induction lemmas, but usually these secondary goals are less complex than the primary goal.

The motivation for this work is the experience that most of the harder proofs include induction. This is not only true for conjectures about parameterized hardware [7]. Even when parameters are fixed, word-lengths of 32 or more exclude unfolding expressions in order to apply basic proof procedures. Specifications originally written in VHDL [24] lead to induction, because this language provides iterations and recursive definitions, apart from the need to formalize simulation cycles [23]. A large amount of proof goals can hopefully be delegated to fully-automatic tools [18, 20], but many goals still have to be proved by induction in more powerful systems.

The outline of this paper is as follows. Section 2 is a summary of the most important *LAMBDA* features in the context of this paper. Section 3 treats the formalization of well-founded induction in *LAMBDA*. The determination of an appropriate induction approach is the subject of Section 4. The manipulation of inductive step premises is discussed in Section 5. Section 6 lists further problems.

2 The Proof System *LAMBDA*

LAMBDA consists of a prover kernel with rewriting, unification, subgoaling, and many other functions available in state-of-the-art systems, structured libraries of rules and common logic definitions, and a user-interface which includes a browser and the graphics tool *DIALOG* [10, 11]. The system is implemented and runs in Poly/ML. The *LAMBDA* logic is axiomatized as a classical higher-order predicate logic with polymorphic types [17]. The syntax of *LAMBDA* terms is kept close to ML, enhanced with the common operators of higher-order logic.

Notation In this paper, examples are written in the ASCII syntax of *LAMBDA*. The representation of rules is modified. Commas are used as separators of hypotheses. The bottom symbols `G,H` for the construction of hypotheses lists [17] are omitted.

Operators and quantifiers of higher-order predicate logic are written as `NOT`, `/\`, `\/`, `->>`, `==`, `forall`, `exists`, `fn x =>`, `any` for negation, conjunction, disjunction, implication, equivalence, universal and existential quantification, lambda abstractions, and Hilbert-operator, respectively.

Syntactic functions provide a meta-level for logical expressions to specify arbitrary contexts with occurrences of subterms that appear as syntactic function arguments. They are explained by an example. The expression `x == y + z` can be described by means of a *context* expression `P#(x,y + z)`, where the symbol `P` is the syntactic function `lam a,b. a == b`. Syntactic functions are always curried. The symbol `lam` is analogous to the λ-operator for object-level functions. In fact, the expression above could equivalently be represented by `p x (y + z)` with `p := fn a => fn b => a == b`. Users may define fixed syntactic functions as macros, called *abbreviations*. The concept of syntactic functions is essential for the abstraction of rules from irrelevant details. The conjectures to be proved are typically unified with generic context expressions of induction rules.

Rules consist of conclusion, and 0 to n premises, both in sequent form, and separated by a horizontal line. A rule without premises is called *theorem*.[1] Each sequent consists of a list of hypotheses, and, separated by the turnstile, an assertion.

Higher-order unification and resolution are available in *LAMBDA* similarly to the system *ISABELLE* [19]. Resolution denotes unification of two sequents of different rules. Pattern matching is not sufficient, because especially in synthesis, but also in the

[1] For theorems, the horizontal line is sometimes omitted in his paper.

induction approach described in the following sections, variables in both sequents have to be instantiated.

Meta-Variables [19], also called *flexible variables* in *LAMBDA*, can be specialized in the course of a proof. Free (unrestricted) variables in an applied rule are flexible, while the free variables in the goal are usually rigid. *LAMBDA* has a facility to control the flexibility of free variables. Flexible variables are also introduced by inference rules with free and unrestricted variables in premises that do not occur in the conclusion. *LAMBDA* has a textual instantiation facility for meta-variables.

Restricted variables are introduced to specify new objects that do not occur as subterms in the sequent at the time of the introduction. Consequently, flexible variables that existed before the introduction of the restricted variable must not be instantiated later to terms containing the restricted variable. Restricted variables therefore carry information about restrictions concerning instantiation of flexible variables. This way, flawed proofs are avoided, such as for the assertion `|- exists x. forall y. x == y`. Restricted variables cannot be made flexible. They are generated by eliminating universal quantification in assertions and existential quantification in hypotheses.

Recursive Definitions Users can indirectly extend the axiomatization of the logic without compromising the consistency by including axioms that are automatically generated by *LAMBDA* from user definitions.

Recursive datatypes can be defined in *LAMBDA* along ML conventions, with few restrictions to avoid paradox definitions. For each datatype definition, a couple of axioms are generated, such as uniqueness and induction rules. For the definition of the natural numbers,

```
datatype natural = 0 | S of natural;
```

the following axioms are generated.

```
---------------------      ----------------------------
|- (S x = 0) == false      |- (S x == S y) == (x == y)

2: P#(r') |- P#(S r')
1: |- P#(0)
   ------------------
   |- forall x. P#(x)
```

Recursive functions are defined as in ML[21], i.e., all patterns of recursion including mutually recursive functions are allowed. The only difference is the exclusion of non-overlapping case patterns. As the logic is classical, functions must terminate. The solution is to restrict the evaluation of expressions to unfold equations for recursive clauses that have been proved to terminate. For this purpose, *LAMBDA* generates unfold rules with additional termination hypotheses, which in fact have to be discharged before actually using the unfold rule. For primitively recursive definitions, which are recognized statically, the termination property is not generated. For all other definitions, termination needs to be proved in order to obtain usable unfold equations. The definition

```
fun gcd 0 y = y | gcd (S x) 0 = S x
|   gcd (S x) (S y) = if y<=x then gcd (x-y)(S y)
                                        else gcd (S x) (y-x);
```

causes the generation of the following rules.

```
----------------------        ---------------------
|- gcd 0 y == y               |- gcd (S x) 0 == S x

------------------------      -------------------------
|- DOM'gcd#(0,y) == TRUE      |- DOM'gcd#(1'x,0) == TRUE

----------------------------------------------------------
DOM'gcd#(S x,S y)
|- gcd (S x) (S y) ==
   if y <= x then gcd (x - y) (S y) else gcd (S x) (y - x)

-------------------------------------------------
|- DOM'gcd#(S x,S y) ==
   ((y <= x == true ->> DOM'gcd#(x - y,S y))
    /\ (y <= x == false ->> DOM'gcd#(S x,y - x)))
```

The termination of such definitions can only be shown by means of induction proofs.

This short introduction to *LAMBDA* has necessarily been restricted to few details relevant in this paper. More information is given in *LAMBDA* users manuals [17].

3 Well-Founded Induction in *LAMBDA*

Structural induction can be reduced to natural induction over the depth of the elements of a construction sequence for a tuple of recursively constructed objects. There are examples, however, where it is difficult or impossible to specify such a depth.

The recursive arguments of the Ackermann function are lexicographically descending, but cannot be assigned a depth to allow for structural induction.

```
fun ack 0      n      = n + 1
|   ack (S m) 0       = ack m 1
|   ack (S m) (S n) = ack m (ack (S m) n);
```

Well-founded or Noetherian induction is able to cope with such functions, and what is important for the approach taken in this paper, provides the freedom to determine the actual induction approach gradually as more and more information is available. In addition, generic induction rules that incorporate the essentials of the Noetherian induction approach allow for uniform procedures to perform the distinct phases of induction proofs.

Well-founded induction has been successfully mechanized in the *NQTHM* or *Boyer-Moore* system [4], *Oyster-CLAM* [5], *INKA* [2], and *Nuprl* [15].

3.1 Well-Founded Sets

Noetherian induction is based on the notion of a *well-founded set*. A well-founded set has at least one minimal element in each non-empty subset with respect to a partial order, which is then called a *well-founded relation*. An equivalent property of a well-founded set is that there is no infinite descending sequence of elements with respect to the partial order. In order to prove an assertion on a well-founded set by Noetherian induction, the assertion has to be proved in the base case for all minimal elements, and it has to be shown that the assertion holds for all non-minimal elements, if the assertion holds for all smaller elements.

In *LAMBDA* such a well-founded relation is defined in terms of the following function which establishes a lexicographic order between tuples of natural numbers, i.e., here lists are used in order to make the function independent of the number of measured arguments.

```
fun lex [] [] = FALSE | lex (h1::t1) [] = FALSE | lex [] (h2::t2) = FALSE
|   lex (h1::t1) (h2::t2) = ((h1 < h2) == true /\ (length t1 == length t2)) \/
                            ((h1 == h2) /\ lex t1 t2);
```

Recursive arguments are then mapped onto tuples of natural numbers by means of measure functions.

In *LAMBDA* it has been proved that the function `lex` along with an appropriate measure function `me` defines a well-founded relation

```
-------------------------------------------
W#(x) /\ length (me x) == S n
|- exists z. W#(z) /\ length (me z) == S n /\
     forall y. lex (me y) (me z) ->> NOT W#(y)
```

For instance, if the well-founded set is given as pair of lists, a possible instantiation of `me` is the measure function

```
fun m2Lists (x,y) = [length y,length x];
```

3.2 A Generic Induction Rule

Predicates to distinguish the recursive cases according to the definition clauses and destructor functions to reduce recursive arguments are the essential constituents of induction rules. The destructor functions define the substitution of induction variables in induction hypotheses. The most basic induction rule for well-founded induction is given as the following.

```
4: |- 0 < length (me x) == true
3: |- forall x. Q#(x) ->> lex (me (d x)) (me x))
2: |- forall x. NOT (Q#(x)) ->> P#(x)
1: |- forall x. Q#(x) /\ P#(d x) ->> P#(x)
--------------------------------------------
   |- P#(x)
```

Premise 4 excludes degenerate cases. The measure function me is variable, i.e., implicitly all-quantified. The same is true for the destructor function d, the context variable P, [2] and the predicate Q, which tests whether recursion occurs. The base case of the induction corresponds to premise 2, and the step case to premise 1.

3.3 Adapted Induction Rules

The generic induction rule given above is restricted to one destructor function and one case. Most times, however, there will be several cases and different destructor functions. Hence it is necessary to adapt the induction rule to the expressions to be dealt with.

An extended induction rule matches the number of cases in the Ackermann function.

```
7: |- 0 < length (me (m,n)) == true
6: |- (forall x,y. Q2#(x,y) ->> lex (me d22#(x,y)) (me (x,y)))
5: |- (forall x,y. Q2#(x,y) ->> lex (me d21#(x,y)) (me (x,y))e)
4: |- (forall x,y. Q1#(x,y) ->> lex (me d1#(x,y)) (me (x,y)))
3: |- (forall x,y. NOT (Q1#(x,y) \/ Q2#(x,y)) ->> P#(x,y))
2: |- (forall x,y. Q2#(x,y) /\ P#(d21#(x,y)) /\ P#(d22#(x,y)) ->> P#(x,y))
1: |- (forall x,y. Q1#(x,y) /\ P#(d1#(x,y)) ->> P#(x,y))
   ----------------------------------------------------------------------
   |- P#(m,n)
```

This induction rule is polymorphic: it can be instantiated to range over arbitrarily typed tuples, which depend on the number of variables to be included in the induction approach.

With the instantiations

```
Q1  := lam x,y. 0 < x == true /\ y == 0           d1  := lam x,y. (x - 1,1)
Q2  := lam x,y. 0 < x == true /\ 0 < y == true    d21 := lam x,y. (x - 1,ack x (y - 1))
                                                  d22 := lam x,y. (x,y - 1)
me  := fn (x,y) => [x,y]
```

a tailored induction rule for terms with occurrences of the Ackermann function is obtained.

```
3: |- (forall n.   P#(0,n))
2: |- (forall m,n. P#(m,ack (S m) n) /\ P#(S m,n) ->> P#(S m,S n))
1: |- (forall m.   P#(m,1) ->> P#(S m,0))
   ---------------------------------------------------------------
   |- P#(m,n)
```

It is now straightforward to prove automatically

```
|- 0 < ack m n == true
```

and other goals by applying the induction rule and rewriting the resulting premises.

The question how to determine appropriate instantiations is subject of the next section.

[2]Note that a syntactic function conveniently represents the induction skeleton introduced by Hutter [14].

4 Finding an Induction Approach

The generic induction rules above do not uniquely specify the cases and corresponding substitutions of induction variables in inductive hypotheses. The effort and the success of an induction approach critically depend on a good instantiation of these parameters, which allow *fertilization* of the induction conclusion with the induction hypothesis [4].

The definitions of the functions give hints how to derive promising induction schemes. Meta-variables in *LAMBDA* additionally allow one to specify an induction approach gradually, i.e., not all parameters need to be instantiated at the same time. A term may immediately suggest particular instantiations for some induction variables, while not referring to other ones. In *LAMBDA* it is possible to start proofs for the individual induction premises in any order and perform instantiations only as far as needed to prove the individual premise. Partially instantiated terms are manipulated until the resulting terms suggest further instantiations. For such manipulations induction lemmas are used as discussed in Section 5; thus induction lemmas are another important source for finding suitable instantiations.

4.1 Recursion Analysis

In order to obtain appropriate induction schemes, recursion analysis is performed whenever a function is defined. Recursion analysis [4] essentially consists of extracting the cases and corresponding destructors from function definitions. For this purpose, the generated unfold rules for each function are decomposed by means of rule-destruction functions available in *LAMBDA*.

The unfold rules for each function are analysed with respect to the number of recursive calls which determines the number of destructor functions for each case. With this information, a generic induction rule as shown in the previous section can already be generated automatically. Then the cases and destructor functions are identified and extracted. They are used to instantiate generic induction rules to ones matching the recursive reductions.

Measure functions are determined on the basis of the datatype definitions for the types of the recursive arguments. An appropriate order of the arguments ensures well-foundedness of the lexicographic relation. Often the measure function simply takes the argument values in the same order as they occur. As the depth of recursively constructed datatypes can be measured, it is possible to construct measure functions mechanically for use in the premise that asserts well-founded relations with the lexicographic order function `lex`.

Conditional expressions Within function bodies, `case`- or `if`-expressions may occur. It is favourable to assign different cases `Q` at this level. For the function `gcd` of Section 2 there are two cases, with corresponding destructors.

```
Q1#(m,n) = 0 < m /\ 0 < n /\ n <= m
    d1 = fn (m,n) => (m - n, n + 1)

Q2#(m,n) = 0 < m /\ 0 < n /\ m < n
```

```
d2 = fn (m,n) => (m + 1, n - m)
```

If a branching condition itself contains a recursive call, however, it cannot be included in the set of cases `Qi`.

4.2 Non-Exhaustive Patterns

In the constructive logic of the previous *LAMBDA* versions it was possible to define partial functions. Although functions are total in the classical logic of the renewed *LAMBDA* system, function definitions need not be exhaustive. Function applications with arguments that do not match a pattern of the function definition are treated as irreducible expressions. Irreducible terms must be avoided, if a goal cannot be proved after abstraction of the irreducible terms into free variables. Such cases must be excluded by supplying additional hypotheses.[3]

If for partially defined functions the recursive case is still distinguished by predicates `Q`, the negation of their disjunction does no longer define the base case. In particular, it is not possible to discharge the non-recursive induction premise

```
|- forall x. NOT (Q#(x)) ->> P#(x)
```

Hence, the induction term must be extended by a constraint `R#(x) ->> P#(x)`. One could naïvely assume that the constraint might simply be the negation of the irreducible case, say a constraint `R1#(x)`. Unfortunately, this constraint is not generally sufficient to prove the inductive step.

```
|- forall x. Q#(x) /\ (R1#(d x) ->> P#(d x)) ->> (R1#(x) ->> P#(x))
```

In fact, as the constraint `Q#(x)` by definition does not include the undefined case, it simply implies the constraint `R1#(x)`, and the inductive step degenerates to

```
|- forall x. Q#(x) /\ (R1#(d x) ->> P#(d x)) ->> P#(x))
```

without a chance to satisfy `R1#(d x)`. A global constraint must be added that excludes the undefined case through all recursions.

```
|- forall x. Q#(x) /\ R1#(d x) /\ R#(d x) /\ P#(d x) ->>
             (R1#(x) /\ R#(x) ->> P#(x))
```

The global constraint must have the properties

```
|- forall x. R#(x) ->> R1#(x)
```

```
|- forall x. Q#(x) /\ R#(x) ->> R#(d x)
```

The base case then reads

[3]The retiming approach of [7] has to be modified, since now `|- delay (fn (t:time) => f (x t)) 0 == f (delay x 0)` is no longer provable, but relates two irreducible expressions instead of `UNDEFINED === UNDEFINED` for strict functions `f`.

```
|- forall x. NOT (Q#(x)) ->> (R1#(x) ->> P#(x))
```

The goal of recursion analysis is the generation of an induction scheme that is suitable for a context of the analysed function without other assumptions about this context. Therefore, the reducibility constraint is better made an explicit component of the induction rule. In addition, the local reducibility constraints `R1` can be determined automatically, using similar procedures as the ML compiler, which issues warnings for non-exhaustive matches.

There are good chances to construct the global constraints `R` automatically as well, using the defining properties above and information about the relevant datatypes.

Alternatively to an explicit construction, the premise relating initial and the unknown global condition can simply be made part of the induction rule. The global condition can be instantiated at any time during the proof of the induction premises.[4]

```
7: |- R#(x)
6: |- 0 < length (me x) == true
5: |- forall x. R#(x) ->> R1#(x)
4: |- forall x. Q#(x) /\ R#(x) ->> R#(d x)
3: |- forall x. Q#(x) ->> lex (me (d x)) (me x))
2: |- forall x. NOT Q#(x) /\ R1#(x) ->> P#(x)
1: |- forall x. Q#(x) /\ R#(x) /\ P#(d x) ->> P#(x)
----------------------------------------------------
   |- P#(x)
```

For illustration, the following recursive list function and its local reducibility constraint are considered.

```
fun before 0 l = []
|   before (S n) (h::t) = h :: before n t;

R1#(n,l) := 0 < n == true ->> l = [] == false
```

The global constraint is characterized by the properties

```
R#(S n,h::l) |-  R#(n,l)

R#(n,l) /\ 0 < n == true |-  l = [] == false
```

The second property can be replaced by

```
|-  R#(S n,[]) ==  FALSE
```

It is straightforward to choose the appropriate condition

```
R#(n,l) := n <= length l == true
```

The following induction rule is obtained.

```
3: |- n <= length l == true
2: |- P#(0,[])
1: P#(n',t') /\ n' <= length t' == true |- P#(S n',h'::t')
-----------------------------------------------------------
   |- P#(n,l)
```

[4]The condition `R` in Premise `1` is kept, for it could be needed for further manipulations.

4.3 Merging Induction Schemes

A proof goal may contain several occurrences of recursive subterms. Individual induction schemes may be available for the recursive subterms. Based on the available information the cases and destructors pertaining to individual functions are merged [4, 22]. In *LAMBDA* these schemes can be used to instantiate the generic induction approach. As *LAMBDA* provides higher-order resolution, the instantiation is performed by applying auxiliary rules that are derived from the individual induction schemes. The auxiliary rules may embed repeated substitutions.

Incompatibility of induction schemes is detected by mismatches between terms already instantiated by some induction schemes and the substitutions required by other ones.

The procedure is illustrated with the example of the **ge**-function.

```
fun ge 0     (S y) = false
|   ge x     0     = true
|   ge (S x) (S y) = ge x y;
```

The corresponding ge-induction rule is

```
2: |- (forall m,n. m == 0 \/ n == 0 ->> P#(m,n))
1: |- (forall m,n. 0 < m == true /\ 0 < n == true /\
                          P#(m - 1,n - 1) ->> P#(m,n))
   -------------------------------------------------------
   |- P#(m,n)
```

For being able to prove the transitivity of the **ge**-function, merging should yield an induction rule with the step premise

```
|- (forall m,n,l. 0 < m == true /\ 0 < n == true /\ 0 < l == true /\
                  P#(m - 1,n - 1,l - 1) ->> P#(m,n,l))
```

Starting from a generic induction rule, the initial premise

```
|- forall m,n,l. Q1#(m,n,l) /\ TRUE /\
                 P#(d1#(m,n,l),d2#(m,n,l),d3#(m,n,l)) ->> P#(m,n,l)
```

is instantiated in the first step to

```
|- forall m,n,l. Q2#(m,n,l) /\ (0 < m == true /\ 0 < n == true /\ TRUE) /\
                 P#(m - 1,n - 1,d3#(m,n,l)) ->> P#(m,n,l)
```

This first instantiation is obtained by resolution with the rule

```
|- forall m,n,l. Q2#(m,n,l) /\ (0 < m == true /\ 0 < n == true /\ Q#(m,n,l)) /\
                 P#(m - 1,n - 1,d3#(m,n,l)) ->> P#(m,n,l)
-------------------------------------------------------------------------
|- forall m,n,l. (Q2#(m,n,l) /\ 0 < m == true /\ 0 < n == true) /\ Q#(m,n,l) /\
                 P#(m - 1,n - 1,d3#(m,n,l)) ->> P#(m,n,l)
```

which is generated from the ge-induction rule. The second instantiation is obtained analogously.

For repeated destruction, a subsumed scheme[22] is generated by repeatedly applying the induction rule.

```
|- forall m,n,l. (Q1#(m,n,l) /\(1 < m == true /\\ 1 < n == true)) /\ Q#(m,n,l) /\
                 P#(m - 2,n - 2,d3#(m,n,l)) ->> P#(m,n,l)
```

In general such repeated schemes can be generated by tentatively applying individual induction schemes and evaluating the resulting instantiations. Incompatible induction schemes are detected by failure of the application of the generated auxiliary rule.

This example demonstrates how in the process of merging induction schemes meta-functions are gradually instantiated until the induction approach is completely determined.

Generalization of Constants and Terms In some cases it is necessary to generalize constants or expressions in order to be able to apply induction schemes. An example is given in Section 5.2.

Implicitly recursive functions are not recursive by definition but contain in their bodies applications of genuinely recursive functions. The recursion analysis as discussed does not always lead to useful induction rules. For proving conjectures about terms that contain implicitly recursive subterms it is more efficient to avoid excessive unfolding by supplying induction rules for implicitly recursive induction rules as well. They are derived similarly to the treatment of terms.

4.4 Destructor vs. Constructor Induction

The destructor form of step premises is often less appropriate for manipulation by induction lemmas, because function definition clauses and derived induction lemmas are more handy in constructor formulations. The destructor formulation is more general, though, because not all function definitions can be captured by constructor formulations. The definition of the `gcd` -function is such an example.

The approach taken here is to start from destructor formulations and transform those into constructor versions. Once an induction approach is determined, it is often possible to transform terms of induction step premises from destructor into constructor form. For this purpose, rules such as

```
                                           |- P#(S n)
---------------------------------          ----------------------
|- NOT (m == 0) == (0 < m == true)         0 < m == true |- P#(m)


------------------------
|- (S m - 1) == m
```

are used.

4.5 Case and Destructor Splitting

If an initial induction approach turns out to be too coarse in a proof process, it is necessary to distinguish more cases or to provide more destruction functions. Inductive hypotheses, for instance, may be used for several fertilization steps with different instantiations, which corresponds to providing different destructor functions.

The first example rule is used to split a case `Q1` with one destructor function, while the second one introduces another destructor without splitting the case.

```
------------------------------------------------------------
forall x. Q11#(x) /\ P#(Q11 x) ->> P#(x),
forall x. Q12#(x) /\ P#(Q12 x) ->> P#(x),
forall x. Q11#(x) ->> lex (m (Q11 x)) (m x),
forall x. NOT (Q11#(x) \/ Q12#(x) \/ Q2#(x)) ->> P#(x),
forall x. Q12#(x) ->> lex (m (Q12 x)) (m x)
|- forall x. NOT (Q11#(x) \/ Q12#(x) \/ Q2#(x)) ->> P#(x)

--------------------------------------------------------------------------------
forall x. Q1#(x) ->> lex (m (d11 x)) (m x),
forall x. Q1#(x) ->> lex (m (d12 x)) (m x),
forall x. Q1#(x) /\ P#(d11 x) /\ P#(d12 x) ->> P#(x)
|- forall x1. Q1#(x1) /\ P#((fn x =>
     if (P#(d11 x) ->> P#(d12 x)) = TRUE then d11 x else d12 x) x1) ->> P#(x1)
```

Destructor splitting is closely related to multiple fertilization [6], i.e., the multiple use of the inductive hypothesis with distinct substitutions for manipulating the inductive assertion.

5 The Completion of Induction Proofs

After determination of a suitable induction scheme, the induction proof is to be accomplished by proving the generated induction premises.

This section is a discussion of ways to automate parts of these proofs.

5.1 Manipulating Induction Premises

Even if the set of available lemmas is complete it is not always straightforward to prove induction premises by rewriting. Advanced rewriting techniques that include completion procedures should be investigated for this purpose [1]. Many induction lemmas are unfortunately conditional or implicative.

As induction step premises match particular patterns, it is possible and recommended to exploit this fact to reduce the search space [6, 14]. The notion of *wavefronts* and manipulations called *rippling* are successfully used to guide and mechanize induction proofs[6], in particular to prepare the fertilization of induction assertions with hypotheses.

In *LAMBDA*, wavefronts could be treated at the level of expressions, i.e., wavefronts could be explicit parts of terms moved by rule applications. This would mean that induction lemmas would have to contain explicit wavefront expressions as well.

Alternatively, wavefronts could be handled implicitly by supplying conditional tactics which analyse and record the evolving syntactical structure of expressions. It should be investigated to what extent such analyses would affect the efficiency and how to manage the implementation.

5.2 Construction of Induction Lemmas

In this section it is shown that induction lemmas can be constructed by adding flexible hypotheses over induction variables. These hypotheses are instantiated to remaining proof goals that cannot be discharged completely by means of available lemmas.

The procedure is illustrated with the list reversion function

```
fun rev [] a = a | rev (h::t) a = rev t (h::a);
```

and the goal

```
|- rev (a@b) []== rev b [] @ rev a []
```

Generalization is carried out and an obligation is introduced.

```
Q#(a,b,x,y,z)|- rev (a@b) x == rev b y @ rev a z
```

Based on recursion analysis the induction approach is chosen as follows. (The flexible variables `b,x,y,z` correspond to the destructor expressions `di#(a',b',x',y',z')`.)

```
Q#(a',b,x,y,z) ->> rev (a'@b) x == rev b y @ rev a' z
|- Q#(h'::a',b',x',y',z') ->> rev ((h'::a')@b') x' == rev b' y' @ rev (h'::a') z'
```

For discharging the base case,

```
|- forall b,x,y,z. Q#([],b,x,y,z) ->> rev b x == rev b y @ rev [] z
```

the meta-variable is instantiated accordingly and a second meta-variable is introduced that represents other still unknown obligations.

```
Q := lam a,b,x,y,z. Q1#(b,x,y,z) /\ rev b x == rev b y @ rev [] z
```

Manipulation of the inductive step suggests instantiations like

```
Q1 := lam b,x,y,z. Q2#(b,y,z) /\ forall h. rev (h :: b) y @ z == rev b y @ (h :: z)
```

The arity of the newly generated syntactic functions is reduced, for they only need to refer to the variables in the remaining expressions.

After the proof of the extended induction rule has been accomplished, the rule is applied to the original proof goal, where now the missing induction lemmas appear as obligations. The remaining goals

```
|- rev b [h] == rev b [] @ [h]
|- rev b [h] @ z == rev b [] @ (h::z)
```

have to be proved, the first essentially being a special case of the second one. Although another induction proof has to be performed, the complexity of the induction compared to the original goal is reduced.

6 Additional Problems

The logic of *LAMBDA* is more expressive than the *Boyer-Moore* logic. For this and other reasons additional questions arise. In Section 4.2 the problem of incomplete function definitions has been discussed. In addition, existential quantification calls for different instantiation procedures. Not all induction approaches are found by recursion analysis (6.2). (This problem is not specifically related to higher-order logic.) *LAMBDA* allows terms with partial function applications. Their treatment in the context of induction is discussed in Section 6.3. We have not yet investigated function datatypes, but at first sight they do not appear to pose further problems (Section 6.4).

6.1 Existential Quantification

The approach discussed so far only refers to free variables. Functions in *LAMBDA* may defined as logical formulas that include existential quantification. *Skolemization* is a common method to deal with existential quantification in induction proofs [2]. Explicitly constructing a unique object to demonstrate existence is not always necessary and could be awkward. In higher-order logic, the Hilbert-operator (`any`) can be used, which is axiomatized in *LAMBDA* in such a way that it satisfies

```
|- (exists x. P#(x)) == P#(any x. P#(x))
```

This operator is used to instantiate **skolem** meta-functions and later infer the desired properties. The following definition is considered as an example.

```
fun cfoldL [] (cin,[]) y = y == cin
|   cfoldL (C::Cs) (cin,h::t) y = exists z. C (cin,h) z /\
                                            cfoldL Cs (z,t) y;
```

It has been used for specifying accumulative cascades of hardware components [7]. Neglecting length conditions, a goal is to be proved by induction. It states that one initialized cascade may replace another one.

```
exists u. A u /\ cfoldL Bs (u,x) y |- exists u. A u /\ cfoldL Cs (u,x) y
```

Writing destructor expressions as meta-variables without primes, the induction step reads

```
exists u. (A u /\\ cfoldL Bs' (u,x') y')  /\
exists u. (A u /\ cfoldL Bs (u,x) y) ->> exists u. (A u /\\ cfoldL Cs (u,x) y)
|- exists u. (A u /\\ cfoldL Cs' (u,x') y')
```

Recall that meta-variables must not be instantiated to terms with occurrences of restricted variables that are introduced after the creation of the meta-variables. For this reason, we cannot introduce existentially quantified hypotheses, for which simplification, i.e., removal of existential quantification, would generate terms with such restricted variables not to be used for instantiation.

In order to allow fertilization, the first predicate has to be split in such a way that a part of it matches the flexible fertilization precondition. (Of course, a direct unification would not yield a well-founded destruction.) The splitting need not be specified immediately. By introducing additional premises which are to be discharged later

```
|- U#(Bs',Cs',x',y')

U#(Bs',Cs',x',y') /\ exists u. (A u /\\ cfoldL Bs' (u,x') y')
|- exists u. (A u /\ cfoldL (dB#(Bs',Cs',x',y')) (u,dx#(...)) (dy#(...)) /\
             rB#(...) (dy#(...),rx#(...)) y')
```

the **modus ponens** rule can be applied. Similarly the cascade `cfoldL Cs' (u,x') y')` is split.

```
exists u. (A u /\ cfoldL (dB#(...)) (u,dx#(...)) (dy#(...)) /\
          rB#(...) (dy#(...),rx#(...)) y')  /\
exists u. (A u /\\ cfoldL Cs (u,dx#(...)) (dy#(...)))
|- exists u. (A u /\\ cfoldL (dC#(...))(u,dx#(...)) (dy#(...))) /\
          rC#(...) (dy#(...),rx#(...)) y')  /\
```

Now fertilization can take place leaving the goal

```
exists u. (A u /\ cfoldL (dB#(...)) (u,dx#(...)) (dy#(...)) /\
          rB#(...) (dy#(...),rx#(...)) y')  /\
exists u. (A u /\\ cfoldL Cs (u,dx#(...)) (dy#(...)))
|- rC#(...) (dy#(...),rx#(...)) y')
```

As the cascade components can define relations with degrees of freedom [7], it is not adequate in general to specify `dy#(...)` as a unique function. For this reason the **Hilbert**-operator `any` is used to instantiate

```
dy = lam Bs,Cs,x,y. any z.
                      exists u. (A u /\ cfoldL (dB#(...)) (u,dx#(...)) z /\
                                rB#(...) (z,rx#(...)) y')
```

The `any`-expressions are eliminated with the rule

```
P#(x') |- Q#(x')
-------------------------------------
P#(any x. P#(x)) |- Q#(any x. P#(x))
```

The remaining proof goals are discharged by instantiating the meta-variables and properties of the components. The selection of appropriate destructors is discussed in Section 6.2.

6.2 Special Destructors

The destructors available from recursive definitions are not always the most favourable ones. In [7], a couple of inductions are given, where the destructor has to reduce arguments in a different way than they are reduced in the recursive function definition.

The induction proof started in Section 6.1 is based on destruction at the tail of the list arguments, corresponding to the unfold rule

```
|- cfoldL (Cs@[C]) (cin,x@[l]) y ==
   exists z. cfoldL Cs (cin,x]) z /\ C (z,l) y
```

The permutation function

```
fun permute x il = map (fn i => el i x) il;
```

requires the induction approach

```
2: |- forall a1,a2,n. eqm (iter 0 n) (a1 @ a2) /\ P#(a1 @ a2,n) ->>
                        P#(a1 @ (n :: a2),S n)
1: |- P#([],0)
   ---------------------------------------------------------------
   eqm a (iter 0 n) |- P#(a,n)
```

where the function `el` returns list elements, `iter` generates a list of natural numbers in the given range, and `eqm` tests whether two lists have the same members.

In these and other examples it can hardly be imagined to derive such induction schemes automatically from recursion analysis. In the second case, there are even auxiliary functions `iter` and `eqm` to be supplied to construct a useful induction scheme. There is hope, however, that once such schemes and unfold rules have been derived, they can be selected and applied mechanically.

6.3 Partial Function Application

The *LAMBDA* logic allows higher-order functions. The best alternative to induction proofs is to reason about them at function level by means of reduced equations [7] if those are available. The usual procedure to derive reduced equations with partial applications of recursive functions is applying function extension, eta-expansion, followed by induction.

In order to determine induction schemes, functions with partial recursive applications are just considered as far as the depth of application.

```
fun mult 0 f n = f n
|   mult (S m) f n = (S m) * mult m f n;

fun multfac x 0 n = n | multfac x (S m) n = mult x (multfac (S x) m) (S n);
```

The recursion scheme for the function `multfac` is

```
Q#(x,m,n) /\ P#(S x,m - 1,d#(x,m,n)) ->> P#(x,m,n)
```

The third argument is irrelevant for the well-foundedness. If the well-foundedness depends on the argument not applied explicitly, it is more difficult to conceive a sound induction approach. Further investigations will have to be carried out.

In the context of permutation proofs [7], the following goal was encountered, which has been proved by applying the permutation induction scheme. Again, this induction approach could have hardly been found automatically.

```
eqm a b /\ b == iter 0 (length a)
|- map (fn i => P#(el (el i a))) b == map (fn i => P#(el i)) a
```

6.4 Datatypes

Functional datatypes are legal in *LAMBDA* according to ML conventions, with few additional restrictions to avoid flawed definitions [17].

```
datatype 'a fundt = empty | cons of 'a * 'a fundt | fcons of unit -> 'a fundt;
```

A function to determine the construction depth of elements of this datatype can be produced in the same way as for first-order datatypes.

```
fun depth_fundt empty = 0
|   depth_fundt (cons (x,f) = 1 + depth_fundt (x,f)
|   depth_fundt (fcons (f())) = 1 + depth_fundt f;
```

Particular problems have not been encountered yet.

LAMBDA supports subtypes. It has not yet been investigated how induction approaches have to be extended to deal with subtypes, but the predicate defining the subtype, and the abstraction function to obtain the parent type will have to be taken into account.

7 Conclusions

Future success of formal methods will depend on the ability to provide more automation. Theorem provers will continue to play an important role on top of automatic verification tools. The proofs remaining in the domain of interactive theorem provers will exploit the expressiveness of predicate logic. Such proofs frequently include induction. Reports about applications of the *Boyer-Moore* prover agree that mechanized induction is an essential support for interactively guided proofs [13, 25]. The study discussed in this paper is a step in the direction to supply *LAMBDA* with a tool for increasing the degree of automation for induction proofs in *LAMBDA*.

In further case studies heuristics for efficiently determining instantiations of destructor functions and for discharging induction premises have to be investigated. Many examples have led to the perception that the decisive step in an induction proof consists in identifying an appropriate induction lemma that has to be selected from an existing database or has to be proved by induction. Such an induction proof is often an order of magnitude less in complexity than the original conjecture. Measures for the complexity are the number of variables or the number of recursive terms in the conjecture.

Most of the investigations and proofs for the examples of this paper have been carried out interactively in LAMBDA, but many ideas for automatic ML-functions have been obtained already. Abstraction and generalization as means to obtain successful induction approaches [4], but also for coping with the complexity of expressions [8] in the generated induction premises, is an important topic to be tackled in this context. The systematic application [6, 14] and construction of induction lemmas [26] are essential for future efforts to automate inductions.

Novices in logic, in particular hardware designers, typically have considerable initial difficulties to devise appropriate induction strategies. From this point of view, mechanical support for induction proofs is particularly important, besides the general gain in efficiency of interactive theorem proving. If the construction and proof of induction goals can be mechanized in most relevant cases, the practical benefits for large industrial projects will be significant.

References

[1] Leo Bachmair, 'Canonical Equational Proofs', Birkhäuser, Boston, 1991.

[2] S. Biundo, 'Automated Synthesis of Recursive Algorithms as a Theorem Proving Tool', ECAI, 1988.

[3] R.S. Boyer and J.S. Moore, 'The User's Manual for A Computational Logic', Computational Logic, Inc., Austin, Texas, February 1988.

[4] R.S. Boyer and J.S. Moore, 'A Computational Logic', Academic Press, ACM Monograph Series, 1979.

[5] A. Bundy, F. van Harmeln, C. Horn, and A. Smaill, 'The Oyster-CLAM System', 10th International Conference on Automated Deduction, 1990.

[6] A. Bundy, A. Stevens, F. van Harmelen, A. Irelan, and A. Smaill, 'Rippling: A Heuristic for Guiding Inductive Proofs', University of Edinburgh, DAI Research Paper No. 567, December 1991.

[7] H. Busch, 'Hardware Design by Proven Transformations', PhD thesis, Brunel University of West London, 1991.

[8] H. Busch, 'Transformational Design in a Theorem Prover', TPCD Workshop, June 22-24, 1992, Nijmegen.

[9] T. Filkorn, 'A Method for Symbolic Verification of Synchronous Circuits', in *Proceedings of IFIP WG 10.2 10th International Symposium on CHDLs, Marseille*, edited by D. Borrione and R. Waxman, 1991.

[10] S. Finn, M. Fourman, M. Francis, and R.Harris, 'Formally Based System Design - Interactive Synthesis Based on Computer-Assisted Formal Reasoning', in *Formal VLSI Specification and Synthesis - VLSI Design Methods-I*, edited by L. Claesen, (North-Holland, 1990), pp. 139–152.

[11] M.P. Fourman and E. Mayger, 'Formally Based System Design - Interactive Hardware Scheduling', in *Proceedings of the International Conference on VLSI, 1989*, edited by G. Musgrave and U. Lauther, pp. 101–112.

[12] M.J.C. Gordon, 'Why higher-order logic is a good formalism for specifying and verifying hardware', in *Formal Aspects of VLSI Design*, edited by G. J. Milne and P. A. Subrahmanyam, North-Holland, 1986.

[13] W.A. Hunt Jr., 'FM8501: A Verified Microprocessor', PhD thesis, The University of Texas at Austin, 1985.

[14] D. Hutter, 'Guiding Induction Proofs', 10th International Conference on Automated Deduction, 1990.

[15] P. Jackson, '*Nuprl* and its Use in Circuit Design', TPCD Workshop, June 22-24, 1992, Nijmegen.

[16] J.J. Joyce, 'Generic Specification of Digital Hardware', in *: Proceedings of the Workshop on Designing Correct Circuits, Oxford, September 1990*, edited by G. Jones and M. Sheeran (Springer).

[17] E. Mayger, B. Harris, S. Finn, M. Fourman, and M. Francis, Abstract Hardware Limited Version 4.2 *LAMBDA* Documentation, 1992.

[18] R. Kumar, T. Kropf, and K. Schneider, 'Integrating a First-Order Automatic Prover in the HOL Environment', in *International Tutorial and Workshop on the HOL Theorem Proving System and its Applications*, Davis, August 27-30, 1991.

[19] L.C. Paulson, 'Introduction to Isabelle', Computer Laboratory, University of Cambridge, 1992.

[20] C.-J. Seger and J.J. Joyce, 'A Two-Level Formal Verification Methodology using HOL and COSMOS', in *:Computer Aided Verification '91, Aalborg, Denmark*, July 1-4, 1991.

[21] R. Harper, R. Milner, and M. Tofte, 'The Definition of Standard ML, Version 3', University of Edinburgh,LFCS Report Series, ECS-LFCS-89-81, May 1989, May.

[22] A. Stevens 'A Rational Reconstruction of Boyer and Moore's Technique for Constructing Induction Formulas', in *Proceedings the 8th ECAI*, 1988.

[23] G. Umbreit 'Providing a VHDL-Interface for Proof Systems', EURO-VHDL '92, September 7 - 10, 1992, Hamburg.

[24] IEEE Standard VHDL Language Reference Manual, 1988.

[25] D. Verkest, L. Claesen, and H. De Man, 'On the use of the *Boyer-Moore* theorem prover for correctness proofs of parameterized hardware modules', in: *Formal VLSI Specification and Synthesis - VLSI Design Methods-I*, edited by L. Claesen, (North-Holland, 1990), pp. 99–116.

[26] C. Walther 'Computing Induction Axioms', Proceedings LPAR-92, St. Petersburg, 1992.

[27] P.J. Windley, 'The Formal Verification of Generic Interpreters', Research Report CSE-90-22, July 1990.

Higher Order Logic Theorem Proving and its Applications (A-20)
L.J.M. Claesen and M.J.C. Gordon (Editors)
Elsevier Science Publishers B.V. (North-Holland)

Introducing well-founded function definitions in HOL

Mark van der Voort

University of Twente, Department of Computer Science, P.O. Box 217, 7500 AE Enschede, The Netherlands

Abstract

Reasoning about functional programs is rapidly becoming a hot topic in the field of automated theorem-proving. This paper contributes to the field by introducing a set of theorem proving tools to facilitate reasoning about functional programs using the HOL system. It extends the existing introduction schemes so as to allow a wider class of definitions to be accepted automatically. Furthermore it provides definition specific induction which allows theorem proving to take place on the level of abstraction the user had in mind when writing down the definitions in the functional language.

Keyword Codes: D.2.4; F.3.1; I.2.3
Keywords: Program Verification; Specifying and Verifying and Reasoning about Programs; Deduction and Theorem Proving

1 Introduction

One of the main strong points of functional programming is the ease with which reasoning about programs or expressions can be performed. In this field much effort has been directed towards the development of a calculus for functional expressions [4],[5]. This paper more or less complements this approach by providing a tool to facilitate direct inductive reasoning about functional expressions

The HOL system as it is shows to be limited for this activity. Lots of functions whose existence is trivially clear cannot be introduced directly. This is because a function-definition effectively specifies a constant of a certain (functional) type, for which certain laws should hold. In order to maintain consistency HOL requires an existence theorem. In this way every constant symbol is certified to denote an object in the theory being constructed.

For some classes of constant definitions the existence theorem can be derived automatically. Nonrecursive definitions which rely solely on predefined objects form one class. Primitive recursive functions, be it on numbers, lists or arbitrary types formed with the type-definition package form another such class.

Theoretically every provably total recursive function is primitive in a higher order system. This knowledge however is of limited practical use for HOL users. Using HOL one would like an easy to use introduction mechanism without having to bother about inventing an equivalent definition which *is* acceptable. The mechanisms existing to date greatly restrict the class of allowable definitions, by imposing a strict adherence to a certain pattern of definition.

This paper presents a means to enlarge the class of allowable function definitions. A mechanism is presented which allows the user to define recursive functions on finite arguments without imposing the restriction to the schema of primitive recursion. The mechanism makes an ordinary recursive definition in a Miranda-like language acceptable.

The idea as such is not novel. It is mentioned in [3], and an implementation of it exists in the Boyer-Moore theorem-prover [2]. In contrast with the Boyer-Moore implementation the HOL version introduces definitions by formal proof.

Interesting developments from within the HOLcommunity are those described by Ploegaerts *et al.* [6] and Melham [7]. Our approach differs from that of Ploegaerts in that definitions do not rely on the existence of constructors. The advantage of this is that functions with multiple recursive calls such as quicksort can be introduced.

In this respect we do not go beyond the capabilities of the system of Melham's. Melham describes a way in which inductively defined relations can be introduced. Our system is geared towards the definitions of functions. Using Melham's system to do this requires a uniqueness-proof for the value returned plus an existence proof showing on what values the function is defined.

2 Concepts and definitions

Strictly spoken there is a clear distinction between a function and its definition. A function is a mathematical object which can be modelled in several ways.

One of the ways of modelling a function is by means of a (recursive) definition. Such a definition states equalities between applications of the function for different argument combinations.

Since we are working in a symbolic environment with a strong syntactic bias, we will take the liberty to talk about a function when we formally mean a function definition. So we might say that " a function f can be constructed " where we mean that one can give a construction of a definition of the function f. The intended meaning of the word function will be clear from the context in which it appears.

In this paper the phrase *provably well-founded function* denotes any total first order function, which can be defined in such a way that a provably strictly decreasing measure on the arguments can be given. A *measure* on arguments is a function from the arguments to the natural numbers. A measure m is said to be *strictly decreasing* for a definition of a function f if and only if for all (combinations of) arguments the measure on the arguments in each recursive call in the definition of f is smaller than the measure on the original arguments.

A measure m is said to be *provably strictly decreasing* for f if and only if a proof has been established stating the fact that m is strictly decreasing for f.

A function *definition* consists of a number of clauses, each guarded by a condition. The function name itself does not occur in these conditions[1]. A clause which does not mention the function name is called non-recursive, otherwise a clause is recursive. The function *calls* itself in a recursive clause, this is a recursive *call*. The conditions which guard the non-recursive clauses specify the base case of the function.

Notice that standard definitional mechanisms of functional languages provide two ways of guarding a clause. The first is by pattern matching, the second by explicitly stating a condition. Both impose a condition to be fulfilled, and both will be called guards.

A clause is only guarded by its guard. It will however only be taken when all the guards of previous clauses are false, while its own guard should be true. So we will introduce the *case* for a clause to be the conjunction of the actual guard and the negations of all previous guards.

3 The introduction scheme

In order to preserve consistency of any HOL theory the introduction of a constant should always be accompanied by a proof of existence. For a limited class of definitions, this existence proof can be given automatically. In this article we propose an introduction mechanism for recursive functions. It operates by creating another primitive recursive function definition which can be accepted. The actual definition of the constant will be nonrecursively in terms of this already accepted function, which means that it can consistently added.

The intended definition is seen as a number of clauses together uniquely specifying the function. It must be shown that the constant as defined obeys the intended properties. This equivalence proof can also be performed mechanically.

Central to the idea of creating an equivalent definition lies the observation that any function definition can be extended with an extra parameter which stands for the number of steps the computation may take. This yields a total function, which can be expressed by a primitive recursive definition.

A computation step will be taken to mean *one* unfolding of the function at hand. As this number of computation steps actually depends on the arguments we will denote it by $\mathcal{C}_f\, a_1 \ldots a_n$

We can now set about to define a function f' which satisfies the lemma:

$$m\, a_1 \ldots a_n \geq (\mathcal{C}_f\, a_1 \ldots a_n) \Rightarrow f'\,(m\, a_1 \ldots a_n)\, a_1 \ldots a_n = f\, a_1 \ldots a_n \tag{1}$$

A function f' can be constructed as follows:

> When called with a zero first argument this function f' behaves as f whenever a base case for f has been reached. If the latter is not the case an arbitrary value is yielded.
>
> When called with a non-zero first argument, say $k+1$, f' mimics the clauses of f with the one difference that each occurrence of f is replaced by $f'\,k$.

[1]This can be checked directly as mutually recursive functions are not allowed

This f' is obviously primitive recursive in its first argument, and its existence can be proven by existing introduction mechanisms in HOL.

The construction given satisfies the stated requirements. The first argument is the number of steps which can be taken during the computation of the function. When f' is called with the exact number of steps needed to reach a result for f, f' will reach the same result, as every step coincides with exactly one decrement of the number of remaining steps.

When called with a surplus number of steps the base case situation for f will be reached with a certain number k of remaining steps. Since $f'\ k$ exactly mimics f in this case, the value returned by $f'\ k$ will be the value returned by f.

This proof sketch clarifies an alternative definition f' can be proven equivalent. To aid the formal proof somewhat more detail is needed. The function f' enjoys some nice properties which make it suitable for automated introduction. For the base cases of the function f, $\mathcal{C}_f\, a_1 \ldots a_n = 0$ which allows us to specialize lemma 1 into

$$\forall k.\ \mathit{base\_case} \Rightarrow f'\ k\ a_1 \ldots a_n = f'\ 0\ a_1 \ldots a_n \tag{2}$$

furthermore, since it is generally hard to give an expression for $\mathcal{C}_f\, a_1 \ldots a_n$ we can deduce

$$\forall kl.\ k \geq l \geq (\mathcal{C}_f\, a_1 \ldots a_n) \Rightarrow f'\ k\, a_1 \ldots a_n = f'\ l\, a_1 \ldots a_n \tag{3}$$

The definition f' mentions an extra clause in which an arbitrary value is yielded. This can only happen when the arguments do not specify a base case and the added first parameter to f' is zero. To rule this out one needs the following lemma:

$$\neg \mathit{base\_case} \Rightarrow 0 < \mathcal{C}_f\ a_1 \ldots a_n \tag{4}$$

Shown is how an equivalent definition f' can be constructed when defined in terms of f. The actual construction will be the other way round. Given is a definition of f, which cannot be accepted directly and a strictly decreasing measure for f.

It now is necessary to deduce (slight variants of) the lemmas 2,3 and 4 as these suffice to prove the equivalence.

For these proofs induction is needed which is why a strictly decreasing measure is needed instead of an arbitrary upperbound.

Any strictly decreasing measure m on the arguments $a_1 \ldots a_n$ is indeed an upperbound on the number of calculation steps: $m\, a_1 \ldots a_n \geq (\mathcal{C}_f\, a_1 \ldots a_n)$ as is shown by the following argument.

The only requirement for a base case of f is that this measure is greater than or equal to zero. This holds trivially since the value of an application of the measure yields a natural number. Now suppose that the measure is not suitable. Then there exist arguments for which $m\ a_1 \ldots a_n < \mathcal{C}_f\ a_1 \ldots a_n$. But then since the measure is strictly decreasing, it decreases at least as fast as the number of computation steps needed to reach the result. So if the measure is smaller for the particular arguments mentioned above, it will also be smaller for all intermediate arguments during the computation, and in particular it will be smaller than the number of computation steps to calculate the base case, which is zero. This is a contradiction so every strictly decreasing measure is suitable.

It is the responsibility of the user to come up with a suitable measure, and proof for its monotonicity.

4 Examples

Following are two examples which form a recurring theme in this article.

4.1 Exponential

The exponential function is defined by the following primitive recursive definition:

```
m EXP       0 = 1
m EXP (SUC n) = m * (m EXP n)
```

Another implementation, which is computationally less expensive, is

```
let fastexp_DEF_term =                                                    1
   "!m n. FASTEXP m n = ( (n=0)        => 1
                        | (EVEN_NUM n) => let a = (FASTEXP m (n DIV 2))
                                          in a * a
                        |                 (m * (FASTEXP m (n-1))))";;
```

This definition doesn't follow the pattern for primitive recursive definitions and hence is not acceptable. Using the mechanism described above an equivalent definition can be constructed provided that some measure can be found.

As a measure function we will take `Mfastexp m n = n`. To aid the introduction the proof obligations can be calculated:

```
#let Mfastexp = "Mfastexp n:num = n";;                                    2
Mfastexp = "Mfastexp n = n" : term

#monotonicity_proof_obligations(fastexp_DEF_term,Mfastexp);;
Introducing definition of Mfastexp.
"!n.
  (~(n = 0) ==> EVEN_NUM n ==> (Mfastexp(n DIV 2)) < (Mfastexp n)) /\
  (~(n = 0) ==> ~EVEN_NUM n ==> (Mfastexp(n - 1)) < (Mfastexp n))"
: term
```

As a side effect this function introduces the measure function as a constant in the current theory. To know exactly what variables are to be measured it is important to use the same names for the variables as in the term defining the function.

4.2 Quicksort

A quicksort routine on lists of numbers can also be defined. Interesting about this definition is that the one recursive clause contains two different recursive calls.

```
let filter_DEF =                                                          3
    new_list_rec_definition(`filter_DEF`,
        "(filter P [] = []) /\
         (filter P (CONS x xs) = (P (x:*)) => (CONS x (filter P xs))
                                         |    filter P xs)");;
let qs_DEF_term =
   "!xs. QS xs = ( (xs=[]) => []
                          |let h = HD xs
                           and t = TL xs
                           in   (APPEND (APPEND (QS (filter (\y. y < h) t))
                                                [h])
                                        (QS (filter (\y. ~(y < h)) t))))";;
```

The length of the argument list is a perfect measure.

```
   let Mqs = "Mqs (xs:(num)list) = LENGTH xs";;                           4

   monotonicity_proof_obligations qs_DEF_term Mqs;;
   Introducing definition of Mqs.
   "!xs.
     (~(xs = []) ==> (Mqs(filter(\y. y < (HD xs))(TL xs))) < (Mqs xs)) /\
     (~(xs = []) ==> (Mqs(filter(\y. ~y < (HD xs))(TL xs))) < (Mqs xs))"
   : term
```

These obligations are easy to prove knowing that every filtered list is at most as long as the original list.

5 Formalisation

It is important for the function to be total. This requirement can be checked syntactically since at present the system only accepts definitions which are written as possibly nested conditionals. So for all combinations of arguments an appropriate unfolding exists. Termination is guaranteed by the proof of the monotonicity of the measure.

Generality comes at a price. There is no single closed pattern a function should comply with. Consequently, there is no single theorem expressing existence for all provably well-founded function definitions. The existence of each function has to be proven separately.

With the aid of some extra notation a general scheme can be formulated. This notation will also be of use in the display of proofs

5.1 Notation used

Throughout the rest of this document the following abbreviations will be used

- *args* denotes a list of variables. *rec_args* denotes a list of terms.
- *selected* denotes a function which when given a list of terms selects those terms on the positions of the measured variables.

- *measure* is the measure function
- an application of *selected* will be abbreviated by $\_\epsilon$
- an application of *measure* · *selected* will be abbreviated by $\|\_\epsilon\|$
- the name of the function will consistently be F, whereas *prepreF* is the name of the function that is introduced by the mechanism.

Furthermore some conventions from functional programming are adopted. In the presentation of lemma's heavy use is made of λ-abstractions, applications and the higher-order function `map`, denoted as $\_^* \_$. Each such expression is an abbreviation for the value it represents. At the language level these terms are not abbreviations for their values, although they can be proven to be equal to their values.

5.2 An intermediate data-structure

The aim of this paper is to demonstrate a rather general principle which holds for a class of function definitions. However not every part of the definition is equally important for this introduction scheme. As the main interest is in the structure of the recursion we try to capture this structure in an abstract fashion. Consider for instance the following definition of the fibonacci function:

```
fib n = (  (n=0) ==> 1
         | (n=1) ==> 1
         |           (fib (n-1)) + (fib (n-2)))
```

This type of presentation can be abstracted from, reducing further proof to simple formula manipulation.

A function definition as above effectively specifies a set of conditional rewrite rules together uniquely determining the same function as the definition. Another abstraction on this view can be performed yielding yet another presentation of the fibonacci function.

```
fib n = [<n=0, (λ dummy. 1) []>
        ;<[¬(n=0);n=1], (λ dummy. 1) []>
        ;<[¬(n=0);¬(n=1)], (λ [rc1;rc2]. rc1 + rc2) (fib*  [[n-1];[n-2]])>
        ]
```

The definition has now been reduced to a list of clauses, each specifying a conditional rewrite rule as a pair, whose first element specifies a set of conditions and whose second element specifies the right hand side of a rewrite rule written as an application. Each element consists of a series of conditions consisting of the guard of a clause together with the negations of the guards of the preceding clauses. Following Boyer and Moore this series will be called the *governing terms* of a clause. A further abstraction can be applied. Let $\mathcal{C}$ be a datastructure which acts as a list of clauses, where each clause *cl* permits the following operations

- $\mathcal{G}_{cl}$ denotes the governing terms of the clause. In those case where an explicit argument is given the subscript will be dropped.
- $\mathcal{T}_{cl}$ denotes the term structure of the clause, the λ-abstraction in the above example. It takes a list of recursive calls as parameter. In its definition it may mention any previously introduced constants and all of the arguments to the function. This structure is of no importance for the introduction mechanism.
- $\mathcal{R}_{cl}$ denotes the list of lists of arguments to the various recursive calls. In the example if *cl* is the last clause $\mathcal{R}_{cl}$ = `[[n-1];[n-2]]`. The example function only takes one argument but one should allow for a function to take an arbitrary number of parameters. $\mathcal{R}_{cl}$ depends on *args* in that each of the variables mentioned in *args* may be used to form the terms in $\mathcal{R}_{cl}$.

Furthermore the structure $\mathcal{C}$ as a whole permits the following two operations

- *base_cases* $\mathcal{C}$ denotes the list of clauses which do not contain a recursive call.
- *non_base_cases* $\mathcal{C}$ denotes the list of clauses which do contain at least one recursive call.

A restriction on the form of acceptable definitions can now be stated concisely

- $(\mathit{base\_cases}\,\mathcal{C}) +\!\!+ (\mathit{non\_base\_cases}\,\mathcal{C}) = \mathcal{C}$, expressing the assumption that all base cases must precede the non base cases.

One last abbreviation is used

- The condition *base_case* states the cases in which a base case can be taken. It is formally expressed as $\bigvee \mathcal{G}^* \,(\mathit{base\_cases}\,\mathcal{C})$

Its negation, none of the base cases, is by totality of the definition equivalent to any of the non base cases. Or formally: $\neg \mathit{base\_case} = \bigvee \mathcal{G}^* \,(\mathit{non\_base\_cases}\,\mathcal{C})$

5.3 An equivalent prim-rec definition

The whole mechanism of introducing a provably well-founded definition relies on the fact that an equivalent primitive-recursive definition can be automatically constructed and proven equivalent.

Using the recently introduced notation any acceptable function F can be expressed as

$$(\lambda cl.\ \langle \mathcal{G}_{cl}, \mathcal{T}_{cl}\,(F^*\,\mathcal{R}_{cl})\rangle)^*\,\mathcal{C}$$

The equivalent definition will be constructed as follows:

$$
\begin{array}{l}
\forall args. \\
\qquad \mathit{prepreF}\ \mathit{0}\ \mathit{args} \\
\qquad\quad = (\lambda cl.\ \langle \mathcal{G}_{cl}, \mathcal{T}_{cl}\,[\,]\rangle)^*\,(\mathit{base\_cases}\,\mathcal{C}) \\
\qquad\qquad +\!\!+ [\langle \text{"}\mathit{True}\text{"}, \mathit{any\_value}\rangle] \\
\wedge \\
\qquad \mathit{prepreF}\ (\mathrm{SUC}\ n)\ \mathit{args} \\
\qquad\quad = (\lambda cl.\ \langle \mathcal{G}_{cl}, \mathcal{T}_{cl}\,[\,]\rangle)^*\,(\mathit{base\_cases}\,\mathcal{C}) \\
\qquad\qquad +\!\!+ (\lambda cl.\ \langle \mathcal{G}_{cl}, \mathcal{T}_{cl}\,((\mathit{prepreF}\ n)^*\,\mathcal{R}_{cl})\rangle)^*\,(\mathit{non\_base\_cases}\,\mathcal{C})
\end{array}
$$

For the `fib` example reversing the abstraction yields the somewhat more usual definition:

$$
\begin{array}{llll}
\multicolumn{4}{l}{\texttt{preprefib}\ 0\ a} \\
= (& (a=0) & \rightarrow & 1 \\
\mid & (a=1) & \rightarrow & 1 \\
\mid & & & \text{any value}) \\
\multicolumn{4}{l}{\texttt{preprefib}\ (\mathrm{SUC}\ n)\ a} \\
= (& (a=0) & \rightarrow & 1 \\
\mid & (a=1) & \rightarrow & 1 \\
\mid & & & (\texttt{preprefib}\ n\ (a-1)) + (\texttt{preprefib}\ n\ (a-2)))
\end{array}
$$

5.4 Proving the applicability of the measure

It is the obligation of the user to provide a theorem stating that the measure introduced is in fact a correct measure for the definition at hand. Examples of this proof obligation are given in section 4. For the coming proofs a formal description is needed. The obligation is a conjunction of lemmas. For each non base case clause *cl* and then again for each of the recursive calls $rec\_args \in \mathcal{R}_{cl}$ there is a lemma:

$$| - \forall args.\ \mathcal{G}_{cl} \Rightarrow\ \|rec\_args_{\epsilon}\|\ <\ \|args_{\epsilon}\| \tag{5}$$

In the above lemma *args* is the list of quantified arguments of the original definition. The list *rec_args* is a list of terms built up only from known constants and variables out of *args*. It also has the same length as *args*.

This is the monotonicity proof obligation from which the lemmas 2, 3 and 4 can be deduced.

5.4.1 Measure is greater than zero for non base cases

This theorem is needed during further proof to ensure that the added case, which yields any value, will never be taken. It is called `measure_zero_imp_base_case`. It equals lemma 4 except that it is proven for the measure instead of the actual number of computation steps needed.

$$\forall args.\ \neg base\_case \Rightarrow \mathit{0} <\ \|args_{\epsilon}\|$$

which equals by expansion of $\neg base\_case$

$$\forall args.\ \bigwedge(\mathcal{G}^{*}\ (non\_base\_cases\ \mathcal{C}) \Rightarrow 0 <\ \|args_{\epsilon}\|\ )$$

which is trivially solved as lemma 5 shows $\|rec\_args_{\epsilon}\|\ <\ \|args_{\epsilon}\|$

5.4.2 measure is suitable for base cases

The lemma is `base_case_measure_is_enough` is equal to lemma 2. It states that for base cases the number of allowed computation steps simply doesn't matter:

$$| - \forall args.\ base\_case \Rightarrow \forall n.\ (prepreF\ n\ args) = (prepreF\ \mathit{0}\ args)$$

Simple unfolding of both definitions, followed by a case analysis does the job.

5.4.3 measure is suitable in all cases

In order to prove that the measure is applicable for the whole of the definition we need to prove the lemma `measure_is_enough`: This is the lemma corresponding to lemma 3.

$$|- \forall n.\ \forall args.\ \|args_\epsilon\| \leq n \Rightarrow (prepreF\ n\ args) = (prepreF\ \|args_\epsilon\|\ args)$$

This proof will be by general induction on n. In the step case a the function definition dictates a case split. If the governing terms dictate a base case rewriting with the lemma `base_case_measure_is_enough` yields an equality. Remains only the following proof obligation:

$$\textbf{Case:} \left[\begin{array}{l} \text{induction step case} \\ \text{induction hypothesis:} \\ \forall m\ args.\ \|args_\epsilon\| \leq m < n \Rightarrow (prepreF\ m\ args) = (prepreF\ \|args_\epsilon\|\ args) \\ \neg base\_case \end{array}\right]$$

$$\begin{array}{ll} & \forall args.\ \|args_\epsilon\| \leq n \Rightarrow (prepreF\ n\ args) = (prepreF\ \|args_\epsilon\|\ args) \\ = & \left\{\begin{array}{l} \text{using unfolding } prepreF \\ \text{for each } cl \in non\_base\_cases\ \mathcal{C} \end{array}\right. \\ & \|args_\epsilon\| \leq n \Rightarrow \\ & \quad (\mathcal{T}_{cl}\ ((\lambda\ as \cdot prepreF\ (\text{PRE}\ n)\ as)^*\ \mathcal{R}_{cl})) \\ & \quad = \\ & \quad (\mathcal{T}_{cl}\ ((\lambda\ as \cdot prepreF\ (\text{PRE}\ \|args_\epsilon\|\ )\ as)^*\ \mathcal{R}_{cl})) \\ = & \left\{\begin{array}{l} \text{by monotonicity of } \|\_\epsilon\|\text{, for each } rec\_args \in \mathcal{R}_{cl} \text{ both} \\ \|rec\_args_\epsilon\| \leq (\text{PRE}\ n) < n \text{ and } \|rec\_args_\epsilon\| \leq (\text{PRE}\ \|args_\epsilon\|\ ) < n. \\ \text{Hence the induction hypothesis can be applied} \end{array}\right. \\ & \|args_\epsilon\| \leq n \Rightarrow \\ & \quad (\mathcal{T}_{cl}\ ((\lambda\ as \cdot prepreF\ \|as_\epsilon\|\ as)^*\ \mathcal{R}_{cl})) \\ & \quad = \\ = & \quad (\mathcal{T}_{cl}\ ((\lambda\ as \cdot prepreF\ \|as_\epsilon\|\ as)^*\ \mathcal{R}_{cl})) \\ & true \end{array}$$

5.5 Proving the equivalence

The actual introduction of the constant F is

$$|- \forall args.\ F\ args = prepreF\ \|args_\epsilon\|\ args$$

The last proof obligation now is to show that this constant is effectively the constant which has been specified in the first place. That is, does the following hold

$$\forall args.\ F\ args = (\lambda cl.\ \langle \mathcal{G}_{cl}, \mathcal{T}_{cl}\ (F^*\ \mathcal{R}_{cl})\rangle)^*\ \mathcal{C}$$

where $\mathcal{C}$ is the abstract set of clauses as derived from the intended definition.

This proof proceeds by case analysis where the cases are suggested by the definition of *prepreF*. Whenever *base_case* holds lemma `base_case_measure_is_enough` can be used followed by an unfolding of the occurrence of *prepreF*. If *base_case* doesn't hold there are two cases to consider. Use `measure_zero_implies_base_case` to rule out the case that the measure is zero Now only one goal is left, which can be proven using monotonicity and the lemma `measure_is_enough`

Case: $[\neg base\_case \wedge \neg \|args_\epsilon\| = 0]$

$$
\begin{array}{ll}
 & F\ args \\
= & \\
 & prepreF\ \|args_\epsilon\|\ args \\
= & \quad \text{Since } \neg \|args_\epsilon\| = 0 \\
 & prepreF\,(\text{SUC}(\text{PRE}\,\|args_\epsilon\|\,))\ args \\
= & \\
 & (\lambda cl.\ \langle \mathcal{G}_{cl}, \mathcal{T}_{cl}\,((\lambda as.\ prepreF\,(\text{PRE}\,\|args_\epsilon\|\,)\ as)^*\ \mathcal{R}_{cl})\rangle)^*\ (non\_base\_cases\ \mathcal{C}) \\
= & \begin{cases} \text{by } (\lambda rec\_args.\ (\text{PRE}\,\|args_\epsilon\|\,) \geq \|rec\_args_\epsilon\|\,)^*\ \mathcal{R}_{cl} \\ \text{and } \texttt{measure\_is\_enough} \end{cases} \\
 & (\lambda cl.\ \langle \mathcal{G}_{cl}, \mathcal{T}_{cl}\,((\lambda as.\ prepreF\ \|as_\epsilon\|\ as)^*\ \mathcal{R}_{cl})\rangle)^*\ (non\_base\_cases\ \mathcal{C}) \\
= & \quad \text{definition } \forall as.\ prepreF\ \|as_\epsilon\|\ as = F\ as \\
 & (\lambda cl.\ \langle \mathcal{G}_{cl}, \mathcal{T}_{cl}\,((\lambda as.\ F\ as)^*\ \mathcal{R}_{cl})\rangle)^*\ (non\_base\_cases\ \mathcal{C}) \\
= & \quad \text{Using assumption } \neg base\_case \text{ and polishing up} \\
 & (\lambda cl.\ \langle \mathcal{G}_{cl}, \mathcal{T}_{cl}\,(F^*\ \mathcal{R}_{cl})\rangle)^*\ \mathcal{C}
\end{array}
$$

5.6 Derivation of induction

Merely having an introduction mechanism as explained above won't be very useful. The acceptance of a function expresses that the recursion employed uses an inductive definition of the domain it operates on. The same view on the domain can be used to construct an induction mechanism. Introduction of a function yields both a definition of this function and a theorem stating the induction rule which corresponds to the recursion scheme of the function.

6 Examples continued

This section continues the examples started in section 4. There the monotonicity proof obligations were calculated. Assume that these have been proven.

```
Mfastexp_thm =                                                          5
|-!n.
   (~(n = 0) ==> EVEN_NUM n ==> (Mfastexp(n DIV 2)) < (Mfastexp n)) /\
   (~(n = 0) ==> ~EVEN_NUM n ==> (Mfastexp(n - 1)) < (Mfastexp n))
Mqs_thm =
|- !xs.
    (~(xs = []) ==> (Mqs(filter(\y. y < (HD xs))(TL xs))) < (Mqs xs)) /\
    (~(xs = []) ==> (Mqs(filter(\y. ~y < (HD xs))(TL xs))) < (Mqs xs))
```

The actual definition will be performed now. Also shown is the time taken to deduce the existence. Time is measured on a SUN SparcStation 1, which is being shared among several people.

```
let (fastexp_DEF,fastexp_ind_thm) =                                    6
    new_pwf_recursive_definition(fastexp_DEF_term,Mfastexp,Mfastexp_thm);;
fastexp_DEF =
|- !m n.
    FASTEXP m n =
    ( (n = 0) =>
      1 |
      (EVEN_NUM n =>
      let a = FASTEXP m(n DIV 2) in a * a |
      m * (FASTEXP m(n - 1))))
fastexp_ind_thm =
|- !Q.
    (!m n.
      ((n = 0) ==> Q m n) /\
      (~(n = 0) ==> EVEN_NUM n ==> Q m(n DIV 2) ==> Q m n) /\
      (~(n = 0) ==> ~EVEN_NUM n ==> Q m(n - 1) ==> Q m n)) ==>
    (!m n. Q m n)
: (thm # thm)
Run time: 29.9s
Intermediate theorems generated: 1940
```

The aim is to show that the implementation FASTEXP is equivalent to the exponential function which is already defined in HOL. The goal can be matched perfectly against the conclusion of the induction theorem.

```
# g !a b . FASTEXP a b = EXP a b";;                                    7
"!a b. FASTEXP a b = a EXP b"
```

For all induction theorems deduced an appropriate induction tactic can be generated by the function mk_pwf_induct_tac. This tactic only splits up the goal in subgoals according to antecedent of the induction theorem. Induction hypotheses are not moved into the assumptions, hence the unfold_TAC. Besides discharging of the assumptions this tactic unfolds any applications of the function, followed by a crude case analysis.

```
# let unfold_TAC def extras = REPEAT STRIP_TAC                         8
#                              THEN ONCE_REWRITE_TAC [def]
#                              THEN ASM_REWRITE_TAC extras;;
unfold_TAC = - : (thm -> thm list -> tactic)
```

Now everything has been set up for the application of the induction tactic

```
#e(mk_pwf_induct_tac fastexp_ind                                    9
#  THEN unfold_TAC fastexp_DEF []);;
OK..
3 subgoals
"x * (x EXP (n - 1)) = x EXP n"
    [ "~(n = 0)" ]
    [ "~EVEN_NUM n" ]
    [ "FASTEXP x(n - 1) = x EXP (n - 1)" ]

"(let a = x EXP (n DIV 2) in a * a) = x EXP n"
    [ "~(n = 0)" ]
    [ "EVEN_NUM n" ]
    [ "FASTEXP x(n DIV 2) = x EXP (n DIV 2)" ]

"1 = x EXP 0"
    [ "n = 0" ]

() : void
```

The resultant subgoals do no longer make mention of the term `FASTEXP`. For their proofs lemmas about `EXP` and `DIV` will be sufficient.

The quicksort definition is equally acceptable:

```
#let (qs_DEF,qs_ind_thm) =                                          10
#    new_pwf_recursive_definition(qs_DEF_term,Mqs,Mqsthm);;
qs_DEF =
|- !xs.
    QS xs =
    ((xs = []) =>
     [] |
     APPEND
     (APPEND(QS(filter(\y. y < (HD xs))(TL xs)))[HD xs])
     (QS(filter(\y. y >= (HD xs))(TL xs))))
qs_ind_thm =
|- !Q.
    (!xs.
      ((xs = []) ==> Q xs) /\
      (~(xs = []) ==>
       Q(filter(\y. y < (HD xs))(TL xs)) ==>
       Q(filter(\y. y >= (HD xs))(TL xs)) ==>
       Q xs)) ==>
    (!xs. Q xs)
: (thm # thm)
Run time: 29.2s
Intermediate theorems generated: 1300
```

An example goal is the fact that the length of a quicksorted list equals the length of the original list

```
#g "!xs. LENGTH (QS xs) = LENGTH xs";;                                11
"!xs. LENGTH(QS xs) = LENGTH xs"

#e(mk_induct_tac qs_ind_thm);;
OK..
2 subgoals
"~(xs = []) ==>
 (LENGTH(QS(filter(\y. y < (HD xs))(TL xs))) =
  LENGTH(filter(\y. y < (HD xs))(TL xs))) ==>
 (LENGTH(QS(filter(\y. y >= (HD xs))(TL xs))) =
  LENGTH(filter(\y. y >= (HD xs))(TL xs))) ==>
 (LENGTH(QS xs) = LENGTH xs)"

"(xs = []) ==> (LENGTH(QS xs) = LENGTH xs)"
```

The base case is solved by rewriting with `xs = []`. The step case contains a nasty antecedent. After proving that it is equal to the fact there exist `h` and `t` such that `xs = (CONS h t)` and further simplification the following subgoal is returned:

```
OK..                                                                  12
"(LENGTH(QS(filter(\y. y < h)t)) = LENGTH(filter(\y. y < h)t)) ==>
 (LENGTH(QS(filter(\y. y >= h)t)) = LENGTH(filter(\y. y >= h)t)) ==>
 (LENGTH(QS(CONS h t)) = LENGTH(CONS h t))"

#e(unfold_TAC qs_DEF [NOT_CONS_NIL; HD; TL; LENGTH_CLAUSES]);;
OK..
"SUC((LENGTH(filter(\y. y < h)t)) + (LENGTH(filter(\y. y >= h)t))) =
 SUC(LENGTH t)"
    [ "LENGTH(QS(filter(\y. y < h)t)) = LENGTH(filter(\y. y < h)t)" ]
    [ "LENGTH(QS(filter(\y. y >= h)t)) = LENGTH(filter(\y. y >= h)t)" ]
```

Again the remaining subgoal doesn't rely on any properties of quicksort. Obvious from this latter example is the fact the induction hypothesis really is very well suited for theorems about the introduced function. The example also demonstrates the need for automated support to massage the goal into a form in which the induction hypotheses can be applied.

7 Conclusions

It has been shown that a large class of functions can be introduced in a purely definitional fashion. Furthermore do these function definitions suggest a new induction principle for the domains they act upon. After some initial proof effort by the user the introduction and the derivation of powerful tools can be performed mechanically.

Although the mechanism as such are powerful there is still room for extensive research since the tools are not always directly applicable.

References

[1] DSTO, The University of Cambridge, and SRI International, *The HOL System: DESCRIPTION* (1991).

[2] R. S. Boyer and J S. Moore, 'A computational Logic' (Academic Press, 1979).

[3] Stephen Cole Kleene, Mathematical logic (Wiley, 1967).

[4] L. Meertens, 'Constructing a calculus of programs', in J.L.A. van de Snepscheut, editor, *Mathematics of Program Construction*, Lect. Notes in Comp. Sc., pages 66–90. (Springer Verlag, 1989)

[5] M. M. Fokkinga, 'Law and Order in Algorithmics', PhD thesis, (University of Twente, 1992)

[6] W. Ploegaerts, L. Claesen and H. De Man 'Defining Recursive Function in HOL', in M. Archer, J. J. Joyce, K. N. Levitt and P. J. Windley, editors, *Proceeding of the 1991 International Workshop on the HOL Theorem Proving System and its Applications, August 1991*, (IEEE Computer Society Press, 1992)

[7] T. F. Melham 'A Package for Inductive Relation Definitions in HOL', in M. Archer, J. J. Joyce, K. N. Levitt and P. J. Windley, editors, *Proceeding of the 1991 International Workshop on the HOL Theorem Proving System and its Applications, August 1991*, (IEEE Computer Society Press, 1992)

Higher Order Logic Theorem Proving and its Applications (A-20)
L.J.M. Claesen and M.J.C. Gordon (Editors)
Elsevier Science Publishers B.V. (North-Holland)

Boyer-Moore Automation for the HOL System

Richard J. Boulton

University of Cambridge Computer Laboratory, New Museums Site, Pembroke Street, Cambridge, CB2 3QG, England.

Abstract

Boyer and Moore's heuristics for their first-order logic without quantifiers have been re-implemented in the HOL system. They have been adapted to behave as an automatic prover for the subset of higher-order logic that roughly corresponds to the Boyer-Moore logic. This paper describes the modifications required and presents some initial results and conclusions from the exercise.

Keyword Codes: F.4.1; I.2.3
Keywords: Mathematical Logic; Deduction and Theorem Proving

1 Introduction

During the late 1970's two quite different theorem proving systems emerged, both of which have been popular within the verification community in their original or some derived form. The Boyer-Moore theorem prover [1] is a largely automatic system originally for a quantifier-free first-order logic. LCF is a proof assistant developed over several years at various locations. There are currently a number of systems derived from LCF or based on its methodology. This includes the HOL system [3], a derivative of LCF for classical higher-order logic.

The Boyer-Moore theorem prover provides extensive automation, but its logic is not as expressive as that of the HOL system. The HOL system is also more flexible than the Boyer-Moore theorem prover in that users can write their own proof procedures in the meta-language of the system, ML (a strongly-typed functional programming language).

However, the lack of automation in the HOL system is a serious drawback. The effort required to prove certain trivial theorems is so great that many people find the system too frustrating to use. The work described here is an investigation into the feasibility of providing Boyer-Moore automation in the HOL system for a first-order subset of its higher-order logic. The aim has been to implement in the HOL system the proof procedures described in *A Computational Logic* [1], adapting them appropriately. The extensions to the Boyer-Moore prover during the 1980's, in particular the incorporation of a decision procedure for arithmetic, have not been considered.

The description below assumes that the reader is reasonably familiar with the workings of the Boyer-Moore theorem prover as described in *A Computational Logic.* Boyer and Moore's theorem prover is referred to as 'the Boyer-Moore prover' and the HOL implementation of their heuristics is called either 'the automatic prover' or 'the HOL implementation'.

2 Comparison between the Boyer-Moore Prover and the HOL Implementation

2.1 The Logic

There is a subset of the logic of the HOL system that corresponds closely to the logic described by Boyer and Moore in *A Computational Logic.* The Boyer-Moore logic features the usual propositional connectives, conditionals (if-then-else), an equality relation, and has (first-order) function application. HOL has corresponding notions. Since the Boyer-Moore logic is first-order, there are no λ-abstractions, no function-valued variables, and functions cannot be curried. HOL functions that are curried must therefore be fully applied in any formula on which the automatic prover is to be used.

One of the main features of the Boyer-Moore logic is the shell principle. This allows what are essentially recursive datatypes to be introduced together with constructor, destructor (accessor) and discriminator (recogniser) functions. The recursive types package [4] in HOL serves a similar function, though destructor and discriminator functions are defined separately in HOL. The Boyer-Moore logic also has a definitional principle which allows new functions to be introduced. These can be recursive provided they satisfy some suitable notion of well-foundedness for the 'shells' over which they are defined.

2.2 The Form of Definitions

In the Boyer-Moore system definitions are equations in which the left-hand side is an application of the function being defined and the arguments are variables. On the right-hand side of the equation the variables may appear with destructor functions applied to them. In HOL, the usual form of definitions is for constructors to appear on the left-hand side in one of the argument positions and for the variables used on the left to appear without destructors on the right. For example, the function `LENGTH` to compute the length of a list is defined in the Boyer-Moore system by:

```
(LENGTH X) =
(IF (LISTP X)
    (ADD1 (LENGTH (CDR X)))
    0)
```

and in HOL by:

```
(LENGTH [] = 0) /\
(LENGTH (CONS h t) = SUC (LENGTH t))
```

Definitions of the Boyer-Moore form can be repeatedly applied as a rewrite rule to a term. This rewriting can continue to build up more and more applications of the destructor in the term being rewritten. This is undesirable and the Boyer-Moore theorem prover has heuristics to control how much the definition is used. In HOL the definitions can only be applied if the constructor is present in the term being rewritten, so the number of rewrites is limited by the number of constructors present in the term. The controlling heuristics used by the Boyer-Moore prover are therefore not required in the HOL system.

2.3 Generating a HOL Theorem

The HOL system requires a proof to be given in terms of primitive inferences of the logic in order for a theorem to be generated. Derived rules can be written in ML that automate much of this process. However, the derived rules must ultimately execute functions which implement the primitive inferences. In particular, meta-theoretic properties cannot be used. It was not clear at the outset of the investigation as to whether the Boyer-Moore prover used meta-theoretic properties, and therefore whether the algorithms given in *A Computational Logic* could be used in HOL. It turns out that the logical properties used by the Boyer-Moore prover, and required for the HOL implementation, have corresponding object-level proofs in HOL, and most of these proofs are straightforward. Despite this, there is extra work to do in the HOL system. The Boyer-Moore prover processes and breaks up a conjecture until it is satisfied of the validity. In HOL, it is also necessary to apply the inference rules associated with the processing, so that a theorem is generated.

3 System-Wide Features

3.1 The ML Datatype of Heuristics

Each *heuristic* in the HOL implementation is meant to take a term (in clausal form) as the argument and return a list of new clauses to be proved together with a proof of the original clause from these new clauses. (In this respect, heuristics resemble the *tactics* of the HOL system). A heuristic fails if it can do nothing with its input. The ML type used for heuristics is:

```
(term # bool) -> ((term # bool) list # (thm list -> thm))
```

The Boolean value paired with each term is a flag used to indicate whether the term is a step case of an induction.

3.2 Clausal Form

The prover takes the formula to be proved and converts it to clausal form. In the HOL implementation the clausal form normaliser is included as a 'heuristic' so that the other heuristics do not have to concern themselves with returning terms in clausal form.

3.3 The Waterfall and Pool

Boyer and Moore use a waterfall metaphor to describe their prover. The formula to be proved is converted to clausal form and the clauses produced are poured over the top of a 'waterfall' of heuristics. Each clause proceeds down until one of the heuristics can do something with it. The clause then gives rise to new clauses which are themselves poured over the waterfall. The heuristics are ordered so that the ones tried first are those that are least likely to generate an invalid subclause from a valid clause.

The implementation of the waterfall uses a new concrete recursive datatype for holding partly processed clauses. A clause is either still to be proved, has been proved, or can be proved once subclauses have been. The datatype is defined in ML by:

```
rectype clause_tree = Clause of term # bool
                    | Clause_proved of thm
                    | Clause_split of clause_tree list # (thm list -> thm);;
```

The clauses which cannot be processed by any of the heuristics in the waterfall collect in a 'pool' at the bottom. This pool is processed to remove clauses that are substitution instances of others. The most general clauses are then passed to the induction heuristic which is not included in the waterfall. The induction heuristic will generate new clauses for the base cases and step cases of the induction. These are poured over the waterfall and the whole process begins again.

3.4 Shells

The shell principle of the Boyer-Moore theorem prover is implemented in HOL using the recursive types package [4]. However, by default, the recursive types package only generates and stores a theorem characterising the new type. The sort of properties required for automatic proof have to be requested explicitly and the system does not control their naming or storage. Passing the theorems capturing such properties as arguments to the Boyer-Moore procedure would be tedious and error prone. Hence, the automatic prover incorporates an environment containing all the necessary details about the types to be used. The environment is implemented using an ML assignable variable. This can be updated by the user when new types are created, and the automatic proof procedures can access the information they require. The basic system of the automatic prover includes information for natural numbers and lists.

3.5 Definitions and Rewrite Rules

The automatic proof procedures require access to the definitions of functions appearing in the conjecture. This could be achieved by searching the theory hierarchy, but it is more efficient to restrict attention to definitions currently in use. Furthermore, definitions as stored by HOL require some processing before they can be used by the automatic proof procedures. For these reasons, the user is required to add to an environment the definitions to be used.

As for definitions, the user can make rewrite rules available to the automatic prover. In the Boyer-Moore theorem prover, the user specifies when a lemma is proved whether it is to be used as a rewrite rule. In HOL, lemmas (theorems) can be proved interactively so the selection of rewrite rules is controlled separately from their proof.

4 The Heuristics

This section notes the major differences between the implementations of the individual heuristics in HOL and their equivalents as described in *A Computational Logic*. The subsections correspond to chapters in *A Computational Logic*.

4.1 Using Type Information

The Boyer-Moore prover has heuristics for performing type reasoning. This includes reasoning about true and false. In HOL, only the reasoning about true and false is required; the other reasoning is unnecessary because it is implicit in the HOL system's typed logic.

4.2 Using Rewrite Rules

An important consideration when rewriting automatically is avoiding infinite looping. Infinite looping can arise because of a permutative rewrite rule or because of conditional rewriting. An equation is *permutative* if the left-hand and right-hand sides are instances of one another. In *conditional rewriting*, the rewrite rule is an implication, the consequent of which is an equation. The left-hand side of the equation is used as the pattern. The rule is applicable if the antecedant of the implication can be satisfied after it has been instantiated through the pattern matching. If conditional rewriting is used in the attempt to satisfy the antecedant, the process can loop due to a rule matching part or all of its own or some other rule's condition. The following rule has this property:

$$m < (\texttt{PRE}\ n) \implies m < n = \texttt{T}$$

The automatic prover in HOL uses a straightforward implementation of some of the algorithms described in *A Computational Logic*, but only detects infinitely repeated applications of conditional rewrite rules in very simple cases.

4.3 Using Definitions

Since only constructor-style definitions are currently admitted in the HOL implementation, the process of opening up definitions, even recursive ones, cannot loop. So, for simplicity, the HOL implementation opens up definitions as far as possible. The Boyer-Moore prover has to be more delicate because of the different style of definitions.

4.4 Rewriting Terms and Simplifying Clauses

The HOL implementation of the simplification heuristic calls subfunctions based on the form of the term to be rewritten. Variables are left unchanged by the rewriting functions, as are constant-valued expressions (explicit values) except that numeric constants are expanded into applications of the successor function to zero. For a conditional statement $(b \Rightarrow x \mid y)$, the x and y are rewritten under the assumption that b is true or false respectively. Function applications are dealt with as described in *A Computational Logic.*

The rewriting of equalities is of most interest. In addition to their other techniques for rewriting equalities, Boyer and Moore give a number of situations in which it can be shown that an equality is false. If two different expressions of a recursive or structured type contain no variables, they can be shown to be distinct by using the injectivity and distinctness theorems for the constructors of the type. It may also be possible to show that two structures are unequal if they contain the same variable. Consider the equation:

$$\texttt{CONS}\ x\ (\texttt{CONS}\ y\ l)\ =\ l$$

Since `CONS` is a constructor function the left-hand side is obviously 'bigger' than the right, so they cannot be equal. Boyer and Moore appear to take this meta-theoretic 'size' approach, but in HOL it is necessary to *prove* that the equation is false using the injectivity of `CONS` and induction.

Clauses are simplified by rewriting each literal in turn under a list of assumptions consisting of the negations of the other literals. If a simplified literal contains conditionals they are eliminated using the theorem:

$$\vdash (b \Rightarrow x \mid y) = (\neg b \lor x) \land (b \lor y)$$

Boyer and Moore describe three discrepancies between this process and their actual implementation. The latter two are to do with subsumption of clauses and these are not considered by the HOL implementation. The first, however, is implemented. Before beginning to simplify the literals of a clause the prover looks for a literal of the form $\neg(x = t)$ where x is a variable and t is a term not containing x. If there is such a literal, all occurrences of x in the clause are replaced by t and the literal is deleted.

4.5 Eliminating Destructors

Since the HOL implementation does not currently admit definitions in a destructor style, the heuristics for eliminating destructors have not yet been implemented.

4.6 Using Equalities

Following an induction it is often necessary to use and eliminate the induction hypothesis. This increases the chance of a further induction proving the term. The HOL implementation of this heuristic closely follows the description in *A Computational Logic* with only minor modifications.

4.7 Generalisation

The generalisation heuristic is as described in *A Computational Logic*. The only significant difference is that the Boyer-Moore prover adds type restrictions when generalising if it can. In HOL these restrictions are inherent in the logic. Generalisation lemmas are kept in an environment much as rewrite rules are.

4.8 Eliminating Irrelevance

This heuristic is another form of generalisation. It is used to obtain a stronger conjecture prior to induction. This is useful because the induction hypothesis will be stronger and so the induction is more likely to succeed. However, the heuristic is limited in its ability to detect irrelevance and it is not clear that the techniques described by Boyer and Moore are particularly useful in the HOL context, though they have been implemented.

4.9 Induction

Much of Boyer and Moore's work is concerned with induction. For HOL, the only induction schemes considered are the standard induction theorems for each recursive type. More sophisticated inductions are possible, but at this time the form of automatic recursive function definitions in HOL is quite restricted, and the standard induction scheme often suffices. For this reason, most of the heuristics for induction described in *A Computational Logic* have not been implemented in HOL. If a more flexible function definition system such as the one proposed by Ploegaerts *et al.* [5] becomes available, it will be worth implementing the induction heuristics in full. One difficulty with this is that in order to use an induction scheme in HOL a theorem must be proved for it. The existing heuristic does make some choices, specifically concerning which variable to perform induction on.

5 The Automatic Prover in Use

This section gives an example to illustrate the use of the automatic prover in HOL. The proof begins by starting a new theory and defining a function, `REVERSE`, over lists:

```
#new_theory `reverse`;;
() : void

#let REVERSE =
# new_list_rec_definition
#  (`REVERSE`,
#   "(REVERSE ([]:(*)list) = []) /\
#    (REVERSE (CONS (x:*) l) = APPEND (REVERSE l) [x])");;
REVERSE =
|- (REVERSE[] = []) /\ (!x l. REVERSE(CONS x l) = APPEND(REVERSE l)[x])
```

For the conjecture `"REVERSE (REVERSE l) = (l:(*)list)"` the automatic proof procedure will fail at this point because it does not have the definitions of `REVERSE` and `APPEND`. Before rectifying this, proof printing is activated so that the proof can be seen in detail:

```
#proof_printer true;;
false : bool
```

This printing facility is intended for debugging. It has nothing like the sophistication of the Boyer-Moore prover's English commentary. However, the displayed terms are informative and compact. When a clause is split into several new clauses by the procedure the indentation is increased by one space. Blank lines are inserted between the proofs of each new clause. The proof can be seen below:

```
#new_def APPEND;;
() : void

#new_def REVERSE;;
() : void

#BOYER_MOORE "REVERSE (REVERSE l) = (l:(*)list)";;

"REVERSE(REVERSE l) = l"

 "REVERSE(REVERSE[]) = []"

 "(REVERSE(REVERSE t) = t) ==> (REVERSE(REVERSE(CONS h t)) = CONS h t)"
 "~(REVERSE(REVERSE t) = t) \/ (REVERSE(REVERSE(CONS h t)) = CONS h t)"
 "~(REVERSE(REVERSE t) = t) \/
  (REVERSE(APPEND(REVERSE t)[h]) = CONS h t)"
 "F \/ (REVERSE(APPEND(REVERSE t)[h]) = CONS h(REVERSE(REVERSE t)))"
 "REVERSE(APPEND(REVERSE t)[h]) = CONS h(REVERSE(REVERSE t))"
 "REVERSE(APPEND l[h]) = CONS h(REVERSE l)"

  "REVERSE(APPEND[][h]) = CONS h(REVERSE[])"
  "APPEND(REVERSE[])[h] = [h]"

  "(REVERSE(APPEND t[h]) = CONS h(REVERSE t)) ==>
   (REVERSE(APPEND(CONS h' t)[h]) = CONS h(REVERSE(CONS h' t)))"
  "~(REVERSE(APPEND t[h]) = CONS h(REVERSE t)) \/
   (REVERSE(APPEND(CONS h' t)[h]) = CONS h(REVERSE(CONS h' t)))"
  "~(REVERSE(APPEND t[h]) = CONS h(REVERSE t)) \/
   (APPEND(REVERSE(APPEND t[h]))[h'] = CONS h(APPEND(REVERSE t)[h']))"
  "F \/ (APPEND(CONS h(REVERSE t))[h'] = CONS h(APPEND(REVERSE t)[h']))"
  "APPEND(CONS h(REVERSE t))[h'] = CONS h(APPEND(REVERSE t)[h'])"

|- REVERSE(REVERSE l) = l
```

The prover first performs an induction. The base case can be proved trivially using the definition of `REVERSE`. The step case is converted to clausal form, some definitions are opened up, and the induction hypothesis is used by substitution. The prover then generalises `(REVERSE t)` to `l`, and performs a second induction to prove the generalised clause. Both cases are successfully proved by opening up definitions and, for the step case, by use of the new induction hypothesis.

The automatic prover is also available as a tactic.

6 Results

The tables below list theorems proved by the HOL implementation of the Boyer-Moore heuristics and the theorems used as definitions. For each theorem there are two sets of results. The 'Cumulative' figures give run times in seconds (labelled 'Run') and number of intermediate theorems generated (labelled 'Thms') for the proof of the theorems with all the preceding theorems in the table as rewrite rules. The 'Separate' figures are for the same theorems proved in an empty rewrite-rule environment. The proofs were performed in HOL88 Version 2.0 built under Franz Lisp and running on a Sun 3/160 with 12 Mbytes of real memory.

Theorem	Cumulative		Separate	
	Run	Thms	Run	Thms
$\vdash$ (x $\Rightarrow$ T $\mid$ (T $\land$ x)) $\lor$ $\neg$((x $\land$ y) $\land$ (z $\lor$ F))	2.5	272	2.5	272
$\vdash$ n + 0 = n	4.0	437	4.0	437
$\vdash$ m + (SUC n) = SUC(m + n)	5.0	561	5.0	561
$\vdash$ m + (n + p) = n + (m + p)	5.8	644	10.6	1208
$\vdash$ m + n = n + m	4.4	477	12.8	1436
$\vdash$ (m + n) + p = m + (n + p)	2.3	250	5.7	668
$\vdash$ n * 0 = 0	3.8	415	4.2	481
$\vdash$ m * (n + p) = (m * n) + (m * p)	11.1	1203	24.9	2883
$\vdash$ m * (SUC n) = m + (m * n)	8.9	952	17.6	2013
$\vdash$ m * n = n * m	5.0	520	35.0	3991
$\vdash$ m * (n * p) = n * (m * p)	12.6	1318	35.3	4073
$\vdash$ (m * n) * p = m * (n * p)	2.4	250	31.6	3616
$\vdash$ (m * n = 0) = (m = 0) $\lor$ (n = 0)	34.1	3474	26.8	2881
$\vdash$ $\neg$m < p $\land$ n < p $\Longrightarrow$ $\neg$m < n	15.6	1547	15.0	1547
$\vdash$ $\neg$m < p $\land$ m < n $\Longrightarrow$ $\neg$n < p	1.5	167	15.3	1569
$\vdash$ n < p $\land$ m < n $\Longrightarrow$ m < p	1.2	128	14.0	1424
$\vdash$ APPEND(APPEND l1 l2)l3 = APPEND l1(APPEND l2 l3)	7.4	745	6.5	745
$\vdash$ APPEND l[] = l	4.9	502	4.5	502
$\vdash$ REVERSE(APPEND l1 l2) = APPEND(REVERSE l2)(REVERSE l1)	8.3	768	19.7	2225
$\vdash$ REVERSE(REVERSE l) = l	9.7	889	13.0	1480
$\vdash$ REVERSE(APPEND l[x]) = CONS x(REVERSE l)	3.5	318	8.4	958
$\vdash$ LENGTH(APPEND l1 l2) = (LENGTH l1) + (LENGTH l2)	10.1	878	6.3	701
$\vdash$ LENGTH(REVERSE l) = LENGTH l	11.1	961	12.2	1344
$\vdash$ LENGTH(APPEND l1 l2) = LENGTH(APPEND l2 l1)	1.5	121	17.6	1967

Definitions
$\vdash$ ($\forall$n. 0 + n = n) $\land$ ($\forall$m n. (SUC m) + n = SUC (m + n))
$\vdash$ ($\forall$n. 0 * n = 0) $\land$ ($\forall$m n. (SUC m) * n = (m * n) + n)
$\vdash$ ($\forall$n. $\neg$(n < 0)) $\land$ ($\forall$m n. m < (SUC n) = ((m = n) $\lor$ m < n))
$\vdash$ ($\forall$l. APPEND [] l = l) $\land$ ($\forall$l1 l2 h. APPEND (CONS h l1) l2 = CONS h (APPEND l1 l2))
$\vdash$ (REVERSE [] = []) $\land$ ($\forall$x l. REVERSE (CONS x l) = APPEND (REVERSE l) [x])
$\vdash$ (LENGTH [] = 0) $\land$ ($\forall$h t. LENGTH (CONS h t) = SUC (LENGTH t))

The cumulative times tend to be smaller because the system can make use of rewrite rules to short-cut the proofs. However, searching the list of rewrites takes longer as more lemmas are added to it. The performance could be improved in this respect by using a discrimination net (as used by the HOL system's rewriting facilities) instead of a list. Little attempt has been made to minimise the number of intermediate theorems generated.

7 Remarks

The investigation described in this paper has shown that automation in the style of Boyer and Moore can be achieved within the HOL system. However, the question remains as to whether this is a good approach. An alternative is to interface HOL to another system such as the Boyer-Moore theorem prover or the proof planner CLAM [2].

Interfacing to other systems has the advantage that advances in those systems can more easily be exploited in HOL than if they have to be re-implemented. In many cases the advances could become available in HOL with no extra work. A possible disadvantage with interfacing is that definitions and formulae have to be passed between the systems. The work described in this paper suggests that there is a logically sound correspondence between the Boyer-Moore logic and a subset of the HOL logic, but nothing has been done to confirm this formally.

Possible avenues for future work include extending the variety of induction schemes that can be used, incorporating decision procedures into the prover, improving efficiency, and extending the prover to deal with higher-order terms. Any extension is likely to be difficult as the prover is very sensitive to change. A small change in one heuristic can affect the operation of another; one may modify the prover to work for a new example at the cost of it ceasing to prove an earlier example.

References

[1] R. S. Boyer and J S. Moore. *A Computational Logic.* ACM Monograph Series. Academic Press, New York, 1979.

[2] A. Bundy, F. van Harmelen, J. Hesketh, and A. Smaill. Experiments with proof plans for induction. *Journal of Automated Reasoning*, 7(3):303–324, 1991.

[3] M. J. C. Gordon. HOL: A proof generating system for higher-order logic. In G. Birtwistle and P. A. Subrahmanyam, editors, *VLSI Specification, Verification and Synthesis*, Kluwer, 1987.

[4] T. F. Melham. Automating recursive type definitions in higher order logic. In G. Birtwistle and P. A. Subrahmanyam, editors, *Current Trends in Hardware Verification and Automated Theorem Proving.* Springer-Verlag, 1989.

[5] W. Ploegaerts, L. Claesen, and H. De Man. Defining recursive functions in HOL. In *Proceedings of the 1991 International Workshop on the HOL Theorem Proving System and its Applications.* IEEE Computer Society Press, 1992.

Chapter 3

General Modelling and Proofs

Higher Order Logic Theorem Proving and its Applications (A-20)
L.J.M. Claesen and M.J.C. Gordon (Editors)
Elsevier Science Publishers B.V. (North-Holland)

Constructing the real numbers in HOL

John Harrison

University of Cambridge Computer Laboratory, New Museums Site, Pembroke Street, Cambridge, CB2 3QG, England.

Abstract

We describe a construction of the real numbers in the HOL theorem-prover by strictly definitional means using a version of Dedekind's method. We also outline the theory of mathematical analysis that has been built on top of it, and discuss current and potential applications in verification and computer algebra.

Keyword Codes: F.4.1; I.2.3
Keywords: Mathematical Logic; Deduction and Theorem Proving

1 Constructing the real numbers

We do not address any philosophical issues, but merely take a formal view that the real numbers are an algebraic structure obeying the following axioms:

1. $1 \neq 0$
2. $\forall x\ y.\ x + y = y + x$
3. $\forall x\ y\ z.\ x + (y + z) = (x + y) + z$
4. $\forall x.\ 0 + x = x$
5. $\forall x.\ (-x) + x = 0$
6. $\forall x\ y.\ xy = yx$
7. $\forall x\ y\ z.\ x(yz) = (xy)z$
8. $\forall x.\ 1x = x$
9. $\forall x.\ (x \neq 0) \Longrightarrow (x^{-1}x = 1)$
10. $\forall x\ y\ z.\ x(y + z) = xy + xz$

11. $\forall x\ y.\ (x = y) \vee x < y \vee y < x$

12. $\forall x\ y\ z.\ x < y \wedge y < z \Longrightarrow x < z$

13. $\forall x.\ x \not< x$

14. $\forall y\ z.\ y < z \Longrightarrow \forall x.\ x + y < x + z$

15. $\forall x\ y.\ 0 < x \wedge 0 < y \Longrightarrow 0 < xy$

16. $\forall S.\ (\exists x.\ x \in S)\ \wedge (\exists M.\ \forall x \in S.\ x \leq M)$
    $\Longrightarrow \exists m.\ (\forall x \in S.\ x \leq m)\ \wedge (\forall m' < m.\ \exists x \in S.\ x > m'))$

In the above, we use the normal notation for ordered fields without prejudice as to what 0, say, actually denotes. Note that juxtaposition represents 'multiplication' here; context and spacing should disambiguate it from function application later on.

The axiom that sets the reals apart from other ordered fields is the last of the above, the so-called *supremum property* which states that every nonempty set which has an upper bound has a *least upper bound* (LUB) or *supremum*. For example, the two sets $\{x \mid x^2 < 2\}$ and $\{x \mid x^2 \leq 2\}$ *both* have a supremum of $\sqrt{2}$, although one of the sets contains $\sqrt{2}$ and the other does not.

1.1 *The* real numbers?

An analagous axiomatization to the above in first-order logic cannot incorporate the necessary quantification (over sets or properties of real numbers) to express the last axiom properly; this can be got around by using a schema of axioms, but this still admits of nonstandard models.

However in higher-order logic, the above axioms do determine the reals uniquely up to isomorphism, *provided* the inverse is defined only on the subset of nonzero reals. (For a proof, see [4], [6] or [16].) Since HOL functions are total, this is not convenient in our formalization. We therefore have an extra 'degree of freedom' that does not exist traditionally: we can define 0^{-1} to be whatever we like. To avoid surprises we keep it 'undefined' (effectively $\varepsilon x.\ F$), but it is important to realize that this is not the same as true undefinedness. For example we can prove $\forall x.\ 0x = 0$ and so in particular, assuming we define $x/y = xy^{-1}$ (as we do),

$$0/0 = 0$$

Conversely, a theorem true classically which is *not* true in our framework is:

$$\forall x \in \mathbb{R}.\ (tan(x) = 0) \Longrightarrow \exists n \in \mathbb{Z}.\ x = n\pi$$

because we cannot exclude the possibility that $cos(x)^{-1}$ is zero at odd multiples of $\pi/2$, in other words that $0^{-1} = 0$. There is even something to be said for *defining* $0^{-1} = 0$ since this makes the inverse into a bijection, so things like the following are true universally:

$$\forall x.\ (x^{-1})^{-1} = x$$

$$\forall x.\ 0 < x \equiv 0 < x^{-1}$$

We feel these issues are unlikely to present problems in practice, because division by zero is normally treated as a special case anyway, but one should be aware of them.

1.2 The classical construction

There are a variety of different ways of constructing the real numbers from a simple basis. One interesting recent method is the use of Conway Games explained in [7]. Others favour approaches using positional representations — see for example [1], [4] or [8].

However the two classical methods both start from the rational numbers. In their turn, these are normally constructed from the integers, which are themselves constructed from the naturals. We look at each of these steps in more detail.

1.2.1 From naturals ($\mathbb{N}$) to integers ($\mathbb{Z}$)

There are various possible representations of the integers in terms of the natural numbers, such as:

- A pair consisting of a boolean 'sign bit' and a natural number. For example $(true, 1)$ might represent $+1$ and $(false, 2)$ represent -2.
- A pair of natural numbers, where one imagines (m, n) standing for $m - n$ in the integers. Thus $(1, 0)$ represents $+1$ and $(5, 7)$ represents -2.

The main problem with both the above is non-uniqueness. Manifestly, every number has an infinity of representatives in the second case; $+1$ could equally well be represented by $(2, 1)$, $(3, 2)$, $(4, 3)$ and so on. Less egregious is the first case, but there are still two representations of zero, $(false, 0)$ and $(true, 0)$. There are two natural ways round this problem:

- Consider only a minimal set of representatives, which are in some sense canonical. For example one might in the first case exclude $(false, 0)$, and in the second insist that one or both numbers of the pair be zero.
- Define an equivalence relation expressing the effective identity of sets of terms, and use the equivalence classes under this relation, rather than the representatives themselves, to construct the new type.

1.2.2 From integers ($\mathbb{Z}$) to rationals ($\mathbb{Q}$)

This stage is a particular case of a well-known construction in abstract algebra, constructing the *field of fractions of an integral domain*, an integral domain being a nontrivial commutative ring with the property that

$$\forall x\ y.\ (xy = 0) \Longrightarrow (x = 0) \ \lor \ (y = 0)$$

The procedure consists of considering pairs of integers, which one thinks of as the numerator and denominator of a fraction; it is necessary to exclude 0 from the possible denominators. Then one uses equivalence classes of this subset of pairs under the obvious 'cross multiplication' equivalence relation

$$(x, y) \sim (x', y') \equiv (xy' = x'y)$$

As with the path from $\mathbb{N}$ to $\mathbb{Z}$, we have the option of eschewing equivalence classes in favour of choosing canonical elements. The natural choice of canonical form would be to insist that the pair of elements be coprime, i.e. represent a cancelled fraction (though this is not available in a general integral domain).

1.2.3 From rationals ($\mathbb{Q}$) to reals ($\mathbb{R}$)

This is the most difficult stage. There are two well-established classical methods, which were published independently by Cantor and Dedekind, both in 1872. (Cantor's method was largely anticipated by Meray in 1869, and one can find precursors of Dedekind's method as far back as Eudoxus with his theory of proportion.)

Cantor's Method This method identifies a real number with the set of all rational sequences which converge to it. To say that a sequence (s_n) converges to s, written $s_n \longrightarrow s$ means:

$$\forall \epsilon > 0.\ \exists N.\ \forall n \geq N.\ |s_n - s| < \epsilon$$

This is no good as a definition, because it contains the limit itself, which may not be rational. However the following variant is equivalent (we prove this later after constructing our real numbers):

$$\forall \epsilon > 0.\ \exists N.\ \forall m \geq N, n \geq N.\ |s_m - s_n| < \epsilon$$

(It does not matter that we will restrict ϵ to rational values, since $\mathbb{Q}$ is dense in $\mathbb{R}$, i.e. between any two distinct reals there is a rational.) A sequence which satisfies this property is called a *Cauchy sequence.*

The fact that two series (s_n) and (t_n) converge to the same limit can also be expressed without using the limit itself:

$$\forall \epsilon > 0.\ \exists N.\ \forall n \geq N.\ |s_n - t_n| < \epsilon$$

It is easy to see that the above defines an equivalence relation on Cauchy sequences, and the real numbers can be defined as its equivalence classes. The arithmetic operations can be inherited from those of the rationals in a fairly natural way, although the supremum presents slightly more difficulty. A complete treatment is given by Thurston [17].

Dedekind's Method Dedekind's method identifies a real number with the set of all rational numbers less than it. Once again this is not immediately satisfactory as a definition, but it is possible to give an equivalent definition of such a set without referring to the bounding real number. We shall call such a set a *cut*. The four properties required of a set C for it to be a cut are as follows:

1. $\exists x.\ x \in C$
2. $\exists x.\ x \notin C$
3. $\forall x \in C.\ \forall y < x.\ y \in C$
4. $\forall x \in C.\ \exists y > x.\ y \in C$

These state respectively that a cut is not empty, is not $\mathbb{Q}$ in its entirety, is a 'down set', and has no greatest element. Again the arithmetic operations can be inherited from $\mathbb{Q}$ in a natural way, and the supremum of a set of cuts is simply the union of that set.

1.3 The lattice of number systems

It is harder than it might appear at first sight to make the above rigorous. Defining the integers as sign/magnitude pairs excluding $(false, 0)$ means that even proving the associative law of addition for integers represents a considerable amount of work, because there are eight different cases according to the sign of each of the three integers. (Ostensibly at any rate; avoiding the case split requires careful exploitation of symmetry and the distributivity of negation over addition.)

Defining the rationals as pairs of integers is even worse. Using cancelled fractions seems a bad idea because the proofs of the elementary theorems require a fair few lemmas about coprimality and divisibility. It seems that the use of equivalence classes is better: essentially we would have to prove similar theorems anyway in the case of cancelled fractions, with the added complexity of canonicalization. However the use of equivalence classes is slightly harder than intuition would suggest, particularly since the equivalence relation is *not* an equivalence relation over the whole type, but only on the subset with nonzero denominators. To see this, observe that

$$(1,2) \sim (0,0) \ \wedge \ (0,0) \sim (1,3)$$

But clearly $(1,2) \not\sim (1,3)$, so transitivity fails globally. Therefore many of the proofs are hedged with conditions about membership of the admissible subset, which renders them more complicated. (An alternative which is slightly cleaner, but still tedious, is to define the 'subtype' of nonzero integers and prove the closure of the arithmetic operators under certain conditions.) Finally, when it comes to constructing the reals from the rationals, neither of the classical methods is very appealing.

- Cauchy's method requires the generation of a significant body of 'analytical' lemmas about the rational numbers. Moreover, the problem of defining equivalence relations over subsets of types would have to be faced, which as noted above is tiresome.

- When proving the axioms for the structure, Dedekind's method requires several case-splits according to the sign of variables. This is mainly because the product of two negative rationals is positive, so the natural definition of multiplication on cuts

  $$XY = \{xy \mid x \in X \ \wedge \ y \in Y\}$$

  does not work. The two cuts X and Y extend to $-\infty$, so there will exist products of these large and negative numbers which are arbitrarily large and positive. Therefore the set on the right is not a cut.

  This difficulty is usually noted in sketch proofs given in books, but to carry through in detail the complicated case splits they gloss over is far from easy.

A rather more sophisticated approach is suggested by Conway. Although the novel method for constructing the reals explained in [7] did not seem to be possible to formalize in HOL, his incidental discussion of the classical methods is perceptive. He considers the following lattice of number systems:

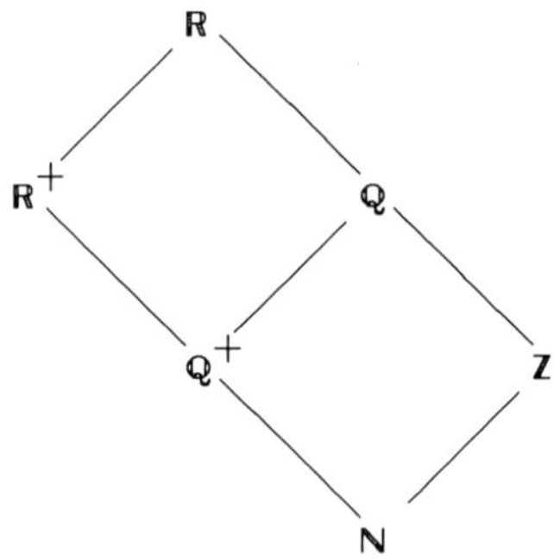

Here $\mathbb{Q}^+$ represents the positive rationals and $\mathbb{R}^+$ the positive reals (we do not worry yet about whether these include 0). Firstly, Conway emphasizes the difficulty of constructing $\mathbb{R}$ from $\mathbb{Q}$ by Dedekind cuts:

> Nobody can seriously pretend that he has ever discussed even eight cases in such a theorem — yet I have seen a presentation in which one theorem actually had 64 cases ... Of course an elegant treatment will manage to discuss several cases at once, but one has to work very hard to find such a treatment.

He advocates instead following the path through $\mathbb{Q}^+$ and $\mathbb{R}^+$, at least if Dedekind's method is to be used. This avoids the case splits (otherwise it is essentially the same as the signed case presented above), and as we shall see, has other advantages as well.

One apparent drawback of using this path is that we lose the potentially useful intermediate types $\mathbb{Z}$ and $\mathbb{Q}$. However this is not really so, because the method used to construct $\mathbb{R}$ from $\mathbb{R}^+$ can be used almost unchanged (and this is where a computer theorem prover scores over a human) to construct $\mathbb{Z}$ and $\mathbb{Q}$ from their positive-only counterparts.

Landau's book [14] is one of the few to present Dedekind's construction in more or less full detail. This also, significantly, uses the path through $\mathbb{Q}^+$ and $\mathbb{R}^+$. Landau makes the step from $\mathbb{R}^+$ to $\mathbb{R}$ by introducing zero and negative numbers, which is similar to the sign/magnitude representation, and with the same problems of case splitting. Conway anticipates this and prefers instead the use of equivalence classes of pairs of numbers. In fact there is then quite a close analogy with the construction of the rationals, with addition taking the role of multiplication.

1.4 The theory of semirings

To journey along the path through $\mathbb{Q}^+$ and $\mathbb{R}^+$, we must know which algebraic laws for these structures we are going to need to get $\mathbb{R}$ out at the other end, and whether we are going to include 0. We will refer to structures like $\mathbb{N}$, $\mathbb{Q}^+$ and $\mathbb{R}^+$ as *semirings*, although to some authors [9] this implies that they contain a zero.

Semirings can be characterized by relatively few axioms. Ordering does not have to be primitive: we can *define* an ordering as follows, whether or not the structure contains a zero:

$$x < y \equiv (x \neq y) \wedge (\exists d.\ y = x + d)$$

Investigation reveals that the following set of axioms is sufficient to allow the derivation of a nontrivial ordered ring by the method of equivalence classes of pairs of numbers under $(x, y) \sim (x', y') \equiv (x + y' = x' + y)$. They are chosen to be true whether or not the semiring contains a zero, which explains why the penultimate one looks a bit peculiar.

1. $\forall x\ y.\ x + y = y + x$
2. $\forall x\ y\ z.\ x + (y + z) = (x + y) + z$
3. $\forall x\ y.\ xy = yx$
4. $\forall x\ y\ z.\ x(yz) = (xy)z$
5. $\forall x\ y\ z.\ x(y + z) = xy + xz$
6. $\forall x.\ 1x = x$
7. $\forall x\ y.\ (x = y)\ \vee (\exists d.\ x = y + d)\ \vee (\exists d.\ y = x + d)$
8. $1 + 1 \neq 1$
9. $\forall x\ y\ z.\ (x + (y + z) = x) \Longrightarrow (x + y = x)$
10. $\forall x\ y\ z.\ (x + y = x + z) \Longrightarrow (y = z)$

It is not hard to see that the last three axioms can all be derived from the single axiom

$$\forall x\ y.\ x + y \neq x$$

This is a strong argument for not including a zero in the structure: we will have to prove fewer axioms as primitive for the semirings we construct. Further, it allows certain theorems such as the field axiom and the Archimedean property (see below) to be written in a simpler form.

There is unfortunately a problem: the standard HOL theory of natural numbers does contain a zero. We could have defined a new type of nonzero natural numbers, but that seemed rather wasteful, so instead we wrote a procedure which works whether or not the semiring contains a zero. The procedure requires the full list of axioms above, although where possible it is easier to prove $\forall x\ y.\ x + y \neq x$ explicitly and derive the others from that.

There are a few extra axioms we need for particular semirings. To get from $\mathbf{R}^+$ to $\mathbf{R}$, we require a form of the supremum property for $\mathbf{R}^+$. And to get from $\mathbf{Q}^+$ to $\mathbf{R}^+$ we need to prove for $\mathbf{Q}^+$ both the field axiom

$$\forall x.\ x^{-1}x = 1$$

and also a form of the *Archimedean property.* This states that if we define a function

$$addn\ n\ x = x + \ldots + x$$

where there are n terms in the sum (of course if we are sloppy with types, $addn\ n\ x$ is just nx), then the following is true

$$\forall x\ y.\ \exists n.\ addn\ n\ x > y$$

Note that neither of the above would be true if $\mathbf{Q}^+$ contained a zero; they would both be consequent on $x \neq 0$.

1.5 Equivalence relations

Every step we are to take in the lattice, with the sole exception of the line from $\mathbf{Q}^+$ to $\mathbf{R}^+$, involves constructing a set of equivalence classes. To make this easier, we wrote a procedure to automate it, given:

- A name for the new type
- A theorem asserting that a (2-place curried) relation (say R) is an equivalence relation, in the following simple form:

  $$\forall x\ y.\ xRy \equiv (Rx = Ry)$$

  Supposing we are constructing the integers from pairs of natural numbers as explained above, the relation would be defined by

  $$(x_1, y_1)R(x_2, y_2) = (x_1 + y_2 = x_2 + y_1)$$

- A list of operations on the representatives together with the desired name of the corresponding operators over the equivalence classes. For example, we might give it an addition operation defined on pairs of numbers as follows:

$$(x_1, y_1) + (x_2, y_2) = (x_1 + x_2, y_1 + y_2)$$

- A list of theorems asserting that the operations on representatives are all well-defined, in the sense that, taking addition as our example again:

$$(x \; R \; x') \; \wedge \; (y \; R \; y') \Longrightarrow (x + y) \; R \; (x' + y')$$

  When the relevant argument or result is not of the representing type, equality takes the place of R.

- A list of theorems about the operations on representatives, e.g. the associative law:

$$x + (y + z) = (x + y) + z$$

The procedure first constructs a type of equivalence classes. The characteristic predicate required to select the set $\{R \; x\}$ is formally:

$$\lambda C. \; \exists x. \; C = R \; x$$

Next, the appropriate operations on the new type are defined. For example $+$ gives rise to a new operator $+^*$ (we use the star consistently, but in fact the user specifies the name of the operator) on the equivalence relation as follows:

$$X +^* Y = R((\varepsilon x. \; x \in X) + (\varepsilon y. \; y \in Y))$$

In other words, pick, using the ε operator, arbitrary representatives of each equivalence class, operate on them and then take the equivalence class of the result. If arguments or result are not of the representing type, then we avoid picking representatives or applying R, respectively. For example an 'is positive' predicate would be elevated as follows:

$$ispos^* X = ispos(\varepsilon x. \; x \in X)$$

and the *addn* function mentioned above as

$$addn^* \; n \; X = R(addn \; n \; (\varepsilon x. \; x \in X))$$

Finally, the proof procedure tries to convert the theorems on representatives into theorems about the new type. This is the only part which requires R to be an equivalence relation and the operations to be well defined. In the case of the associative law, we get:

$$X +^* (Y +^* Z) = (X +^* Y) +^* Z$$

Our procedure does *not* work if we have to deal with a subset of the basic type. We could have generalized the procedure: one would also have to supply it with theorems expressing the closure of the operations with respect to this subset, and the other theorems may become conditional on the variables belonging to the subset. The extra complexity was not necessary for the task in hand; further, there are a few problems to be resolved. Consider a definition of multiplicative inverse for a field of fractions in an integral domain:

$$(x,y)^{-1} = (y,x)$$

Then one needs a condition $x \neq 0$ on the closure theorem. Dealing with things like this in a regular way seems quite awkward.

1.6 From $\mathbb{N}$ to $\mathbb{Q}^+$

This is reasonably straightforward, but we do have to deal with the problem that the natural numbers contain 0 when we would rather they did not. The solution chosen was to use (x, y) to represent $(x+1)/(y+1)$.

The use of $y+1$ avoids zero denominators without using subsets, which would defeat the equivalence class procedure. Using $x+1$ avoids including zero in the rational semiring, which is what we want, and also makes the proofs more symmetrical and regular. The only drawback is that the definitions of the operations are somewhat more complicated. For example the addition of (x_1, y_1) and (x_2, y_2) is defined as

$$((x_1+1)(y_2+1)+(x_2+1)(y_1+1)-1, (y_1+1)(y_2+1)-1)$$

The apparent extra difficulty of the proofs can be overcome by a simple tactic which manages to eliminate a lot of the above complexity.

1.7 From $\mathbb{Q}^+$ to $\mathbb{R}^+$

This is the most difficult part of the whole procedure. Although we are dealing with semirings, the Dedekind cuts procedure is essentially identical to the full case, but includes none of the complicated case splits. We define the operations on cuts as follows (taking some liberties with the HOL notation):

- $sup\ S = \bigcup S$
- $X + Y = \{x + y \mid x \in X \ \wedge \ y \in Y\}$
- $XY = \{xy \mid x \in X \ \wedge \ y \in Y\}$
- $X^{-1} = \{w \mid \exists d < 1.\ \forall x \in X.\ wx < d\}$

Only the last of these is unobvious: the more natural definition is

$$X^{-1} = \{w \mid \forall x \in X.\ wx < 1\}$$

However this would mean that unless a cut denoted an irrational real, its inverse would not be a cut, for example:

$$\{x \in \mathbb{Q}^+ \mid x < 1\}^{-1} = \{x \in \mathbb{Q}^+ \mid x \leq 1\}$$

The construction then consists of proving closure, i.e. that when applied to cuts, all the operations yield a cut (including the trivial instance of proving that the set representing real 1 is a cut), and that all the required axioms hold. These proofs, while mostly routine, are sometimes quite long, so it is not possible to discuss them all here. We sketch only the proof of the axiom

$$\forall X.\ X^{-1}X = 1$$

which is a fairly representative example. In the above and what follows, we assume X denotes a cut, without stating it explicitly; in fact in the HOL proofs X becomes *cut* X, where *cut* is the type bijection from the real number type to the set of rational cuts. This means that no explicit set constraint is necessary since *cut* X is *always* a cut. Firstly we need the following sequence of lemmas.

Lemma 1

$$\forall X\ x\ y.\ x \in X \wedge y \notin X \Longrightarrow x < y$$

This follows easily from cut property 3.

Lemma 2

$$\forall X\ x\ y.\ x \notin X \wedge x < y \Longrightarrow y \notin X$$

Also a straightforward consequence of cut property 3.

Lemma 3 This states that we can get arbitrarily close to the "top" of a cut.

$$\forall X\ e.\ \exists x.\ x \in X \wedge x + e \notin X$$

To prove this, choose any $x_0 \in X$, and $x_1 \notin X$ (this is possible by virtue of cut properties 1 and 2). Then consider the sequence of rationals

$$\{x_0 + \mathit{addn}\ n\ e \mid n \in \mathbb{N}\}$$

It is a simple consequence of the Archimedean property that there is an n with

$$x_0 + \mathit{addn}\ n\ e > x_1$$

Then from Lemma 2 and the wellfoundedness of the naturals, we know there is a *least* n such that $x_0 + \mathit{addn}\ n\ e \notin X$, say k. Further, $k \neq 0$ because we know $x_0 \in X$. Consequently

$$x = x_0 + \mathit{addn}\ (k-1)\ e$$

has the required property. (Strictly, *addn* is not defined at 0 because $\mathbb{Q}^+$ has no zero, but the reasoning is essentially the same as the above.)

Lemma 4 This is just a multiplicative rather than additive version of Lemma 3.

$$\forall X\ u.\ u > 1 \Longrightarrow \exists x.\ x \in X \wedge ux \notin X$$

To prove this, choose any $x_0 \in X$, possible by cut property 1. If $ux_0 \notin X$, then we are finished. Otherwise let

$$e = x_0(u - 1)$$

so by Lemma 3, we can pick an x with

$$x \in X \wedge x + e \notin X$$

But now we have

$$x_0 + x_0(u - 1) \in X \wedge x + x_0(u - 1) \notin X$$

Consequently, using Lemmas 1 and 2, together with simple properties of $<$, we deduce that $x_0 < x$, and therefore $x + x(u - 1) \notin X$, as required.

Main theorem

$$\forall X.\ X^{-1}X = 1$$

Translated into cuts, we want to show

$$\{wx \mid x \in X \wedge \exists d < 1.\ \forall y \in X.\ wy < d\} = \{z \mid z < 1\}$$

Expressed formally in HOL, this means establishing the following logical equivalence

$$\forall z.\ (\exists w\ x.(z = wx) \wedge x \in X \wedge (\exists d < 1.\ \forall y \in X.\ wy < d)) = z < 1$$

This reduces to two implications. A little thought will show that the left to right implication is straightforward. For the other, suppose $z < 1$. Then it is a simple property of the rationals that we can find a d such that $z < d$ and $d < 1$ (for example $d = (z + 1)/2$). Therefore $z^{-1}d > 1$. By Lemma 4, we can choose an x such that

$$x \in X \wedge z^{-1}dx \notin X$$

Now let $w = zx^{-1}$. Then w, x and d are going to be witnesses for the correspondingly named existentially quantified variables. It remains only to prove

$$\forall y \in X.\ wy < d$$

But if $y \in X$, then by Lemma 1, $y < z^{-1}dx$; and $z = wx$, so we have $y < w^{-1}x^{-1}dx$, i.e. $wy < d$, as required.

1.8 From $\mathbb{R}^+$ to $\mathbb{R}$

Most of this is the same as the construction of $\mathbb{Z}$ and $\mathbb{Q}$ from their half-counterparts. The extra theorem that has to be transferred across is the supremum property. The supremum property for the positive reals states that

> Every nonempty set of positive reals which is bounded above has a supremum.

The first step is to transfer this result to the type of reals. Although not vacuous (formally, the positive reals are a completely different type), this is straightforward because the type bijections define an isomorphism between the *type* of positive reals and the positive elements of the type of reals. The theorem now becomes

> Every nonempty set of real numbers which is bounded above, and all of whose elements are strictly positive, has a supremum.

We generalize this in two stages. Firstly it is simple to prove the following strengthening:

> Every nonempty set of real numbers which is bounded above, and which contains at least one strictly positive element, has a supremum.

(The property 'nonempty' is actually superfluous here, but we keep it in for regularity.) This follows because l is a supremum of the whole set if and only if it is a supremum of the strictly positive elements of it, since any positive number is greater than any negative number.

Finally we prove the lemma that for any d, positive or negative, l is a supremum of S if and only if $l + d$ is of $\{x + d \mid x \in S\}$. Now this can be used to reduce the case of any nonempty bounded set to the above, by choosing a d to 'translate' it such that it has at least one strictly positive element. We now have the full result:

> Every nonempty set of real numbers which is bounded above has a supremum.

1.9 Interface issues

It is desirable to use the normal arithmetic symbols like $+$ both for the real numbers and the natural numbers (not to mention the integers, the rationals, and other uses such as binary summation in process calculi). At present, HOL does not have a mechanism to support overloading of constant names, but it is hoped to add something in the future. At present the interface map feature is used to allow the user to swap easily between using $+$ etc. for different purposes, but it is not possible to use the same symbols for different operations in the same term. Furthermore, a different notation is required for real number constants. At present the ampersand is used as an interface map for the 'inclusion' function $\iota : \mathbb{N} \to \mathbb{R}$, so the real constants can be written &0, &1 etc. This is reminiscent of programming languages like C and Standard ML where the floating-point constants are distinguished syntactically from integer ones by being written 1.0 or 1e1 etc.

2 Building on the real number axioms

To make the theory useful as a library, we have built a fair amount of theory on top of the basic construction. This includes a large number of algebraic lemmas, and enough mathematical analysis to define and prove the main properties of the transcendental functions like *exp* and *sin*. We will outline the main results.

2.1 Topology and metric spaces

Recall that a metric on a set S is a function $\rho : S \times S \to \mathbf{R}$ with the following two properties:

1. $\forall x\ y.\ \rho(x, y) = 0 \equiv x = y$

2. $\forall x\ y\ z.\ \rho(y, z) \leq \rho(x, y) + \rho(x, z)$

The latter is usually called the *triangle law*; in the usual metric on $\mathbf{R}^2$ it states that one side of a triangle is no greater than the sum of the other two sides. From the above two properties, it follows that a metric is both nonnegative ($\rho(x, y) \geq 0$) and symmetric($\rho(x, y) = \rho(y, x)$).

A metric is a generalization of the notion of distance between points; in subsequent developments we invariably use the usual metric on $\mathbf{R}$, namely

$$\rho(x, y) = |x - y|$$

but many of the theorems we prove are true in a more general framework. (In fact some are proved in the even more general structure of a topological space.) Since on *any* set we can define the *discrete* metric:

$$\rho(x, y) = (x = y) \to 0 \mid 1$$

we can define a type operator $(\alpha)metric$. (HOL insists that all types be inhabited.)

2.2 Convergence nets

Nets generalize the notions of sequences and pointwise limits, and allow us to avoid proving twice many theorems like

$$x \to l_x \wedge y \to l_y \implies (\lambda n.\ x_n + y_n) \to (l_x + l_y)$$

Nets are simply functions out of a set with a directed partial order, i.e. one where for any x, y there is a z with $z \geq x$ and $z \geq y$. In the case of sequences, the order is just the usual order on the natural numbers. For more details see [13].

2.3 Sequences and series

This consists in part of specializing net theorems, but then some theorems specifically about sequences are added. These include:

- A bounded and monotonic sequence converges. Suppose the sequence is increasing, the other case being analogous. Consider the set $\{x_n | n \in \mathbb{N}\}$. This must have a supremum l such that for any $\epsilon > 0$, there exists an N with $|x_N - l| < \epsilon$. But because the sequence is increasing, this means that $\forall n \geq N.\ l - \epsilon < x_n \leq l$, so the sequence in fact converges to l.

- Every sequence has a monotonic subsequence. Call n a *terrace point* if we have $\forall m > n.\ x_m \leq x_n$. If there are infinitely many such terrace points, we can just form a decreasing sequence by successively picking them. If on the other hand there are only finitely many terrace points, then suppose N is the last one (or $N = 0$ if there are none). Now for *any* $n > N$, there is an m with $x_m > x_n$ (otherwise n would be a terrace point). Hence we can choose a (strictly) increasing subsequence by repeatedly making such choices.

- Every Cauchy sequence converges. A Cauchy sequence is bounded because for any $\epsilon > 0$, say $\epsilon = 1$, we can find an N such that $\forall n \geq N,\ m \geq N.\ |x_m - x_n| < \epsilon$, so

$$max(x_0, \ldots, x_{N-1}, x_N + \epsilon)$$

  is an upper bound. Hence we can find a subsequence which is both bounded and monotonic, and hence convergent. But now because of the Cauchy criterion, the limit of the subsequence is in fact a limit for the sequence itself.

Then we move on to infinite series, defined as limits of a finite sums. Additional theorems proved are mainly tests for convergence such as the comparison test and ratio test.

2.4 Limits, continuity and differentiability

Part of this is specializing the net theorems again; next we define continuity and differentiability, proving the various combining theorems such as the derivative of a product and the chain rule. A valuable function we define here is a conversion which will automatically apply the chain rule and any particular differentiation theorems it is given in order to prove theorems about differentiating arbitrary expressions. We move on to prove various classical theorems, leading up to Taylor's Theorem:

- A function continuous on a closed interval is bounded. Suppose the function is f, and the interval is $\lambda x.\ a \leq x \leq b$. Consider the set

$$\{x | a \leq x \leq b + 1 \wedge \exists K.\ \forall x'.\ a \leq x' \leq x \Longrightarrow |f(x')| < K\}$$

This is nonempty (it contains a) and bounded above (by $b+1$). It therefore has a supremum, s. It is clear that $s > b$, proving the theorem, because if $s \leq b$, continuity means we can pick some $\epsilon > 0$ so that for any x' with $x - \epsilon \leq x' \leq s + \epsilon$, f is bounded, contradicting the fact that s is a supremum. (We assume two-sided continuity at a and b, although this is not necessary.)

- A function continuous on a closed interval attains its supremum and infimum. The following slick proof is taken from [2]. Suppose f does not attain its supremum M. Then the function defined by $\lambda x.\ (M - f(x))^{-1}$ is continuous on the interval (a previous theorem about continuity assures us of this because the denominator is never zero), and therefore *it* is bounded, by K say, which must be strictly positive. But this means that we have $M - f(x) \geq K^{-1}$, which is a contradiction because M is a *least* upper bound.

- Rolle's theorem: if f is continuous for $a \leq x \leq b$ and differentiable for $a < x < b$, and in addition $f(a) = f(b)$, then there is some $a < x < b$ with $f'(x) = 0$. We know that f attains its bounds, and in fact its derivative must be zero there, otherwise it would exceed its bounds on one side or the other.

- The Mean Value Theorem states that if f is continuous for $a \leq x \leq b$ and differentiable for $a < x < b$, then there is some $a < x < b$ with $f(b) - f(a) = (b - a)f'(x)$. A proof is easy by applying Rolle's theorem to the function:

$$\lambda x.\ f(x) - (f(b) - f(a))x/(b - a)$$

- A function whose derivative is zero on an interval is constant on that interval. This is an immediate corollary of the Mean Value Theorem.

2.5 Power series

Here we bring together the theories of infinite series and differentiability, proving a few results about power series, in particular that they are characterized by a 'circle of convergence' and within it we can differentiate them term-by-term. This latter result was in fact the most difficult proof in the whole undertaking. Had we been developing analysis for its own sake, we would have proved some general results about uniform convergence. As it is, we prove the result by direct manipulation of the definition of derivative, following Theorem 10.2 in [3]. The theorem requires both the first and second formal derivative series to converge within the radius of convergence. This does in fact follow in general, but we did not need to prove it because the set of power series we are concerned with differentiate to 'each other', so we already have convergence theorems.

We also prove Taylor's theorem in its full infinite series form. In fact it is no longer used in the subsequent development, but could be useful for giving error bounds when truncating infinite series.

2.6 The transcendental functions

The functions *exp*, *sin* and *cos* are defined by their power series expansions (we do not need Taylor's theorem to do this):

$$\begin{aligned} exp(x) &= 1 + x + x^2/2! + x^3/3! + \dots \\ sin(x) &= x - x^3/3! + x^5/5! - x^7/7! + \dots \\ cos(x) &= 1 - x^2/2! + x^4/4! - x^6/6! + \dots \end{aligned}$$

We show using the ratio test that the series for *exp* converges, and by the comparison test that the other two do. Now by our theorem about differentiating infinite series, we can show that the derivative of *sin* is *cos*, and so on. Furthermore, a few properties like $cos(0) = 1$ are more or less immediate from the series. Now we are in a position to prove more interesting properties. As an example, to show that $exp(x + y) = exp(x)exp(y)$, consider the function

$$\lambda x.\ exp(x + y)exp(x)^{-1}$$

Our automatic conversion, with a little manual simplification, shows that this has a derivative which is 0 everywhere. Consequently, by our previous theorem, it is constant everywhere. But at $x = 0$ it is just $exp(y)$, so the result follows.

The addition formulas for *sin* and *cos* can also be proved in this way, but we need the slight trick of combining the two as follows to make the procedure work:

$$\begin{aligned} \lambda x.\ & (sin(x + y) - (sin(x)cos(y) + cos(x)sin(y)))^2 + \\ & (cos(x + y) - (cos(x)cos(y) - sin(x)sin(y)))^2 \end{aligned}$$

By the same method as above we can prove that this function is always zero, and the addition formulas follow easily. Other function like *ln* and *asn* are defined as the inverses of *exp* and *sin*, and their basic properties are derived. Periodicity of the trigonometric functions follows from the addition formulas and the fact that there is a least $x > 0$ with $cos(x) = 0$. This latter fact is proved by observing that $cos(0) > 0$ and $cos(2) < 0$. Since $sin(x)$ is positive for $0 < x < 2$, *cos* is strictly decreasing there, so by the Intermediate Value Theorem there must be a unique zero. (These proofs involve some fiddly manipulations of the first few terms of the series for *sin* and *cos*.) The zero is of course $\pi/2$, and this serves as our definition of π.

3 Applications

There seem to be several promising areas of application, which have only been partially investigated so far.

3.1 Verification of floating-point hardware

This seems an ideal area for theorem-proving; it is hard to see how one could verify by model-checking a circuit to calculate logarithms, for example. We have already done a verification of a toy floating-point square root circuit. (By toy we mean that it uses a simple floating point format rather than the full IEEE [10] standard with special cases and denormalized numbers. Also our circuit is probably inefficient compared with a commercial design.) It is hoped in the future to do verifications of more realistic circuits and/or circuits for more complicated functions like *sin*.

3.2 Numerical work

It would be quite easy to program HOL to produce mathematical tables with high assurance (for human consumption or insertion of constants into hardware or software). After all, this is what Babbage designed his Difference Engine to do! More generally, there are many areas of application in the error analysis of numerical methods [18].

3.3 Computer Algebra

Computer algebra systems are widely used by applied mathematicians and others. In view of their complexity it seems likely that they include bugs, or consciously implement rules of a theoretically dubious nature. It would be an interesting project to build a computer algebra system which works by proof on top of a programmable theorem-prover like HOL which offers a fairly high degree of security (no complex decision procedures built in). We have taken a step towards this goal with our expression-differentiation conversion; this is rather a trivial computer algebra system, but it represents a start!

3.4 Hybrid Systems

Various formalisms for dealing with hybrid systems have been proposed [15]. Mixing the discrete and the continuous may well require use of the real numbers.

4 Conclusion and related work

As far as we are aware, the only previous construction of the classical reals in a computer theorem prover was by Jutting [12], who translated Landau's book [14] into Automath.

The definition of the constructive reals has been done in NuPRL, with a proof of their completeness, i.e. that every Cauchy sequence converges [5]. Much of the construction has also been done in the LEGO prover, as well as some work on completing a general metric space [11].

The full construction described here, from $\mathbb{N}$ to $\mathbb{R}$, took about two weeks, but it would have taken much longer without careful selection of strategy. In particular, it seems that a quotient procedure tends to be much easier that picking canonical elements. Previous constructions of the integers from the naturals have been made by others using canonical representations, and their greater complexity seems to bear out this point.

The additional work on mathematical analysis took several months, on and off. Analytical proofs tend to have quite a lot of minor details which need to be filled in, particularly tedious bits of arithmetic reasoning. In the near future, derived decision procedures will become available which will greatly ease this sort of task without compromising the security of the system.

The fact that the HOL system can be programmed easily (the presence of ML rather than just an ad-hoc macro language, together with the simplicity of the underlying term structures) is a major advantage. Implementing procedures like the quotient types function and the differentiation conversion would otherwise be very difficult. Furthermore, even though some proofs are long and tedious, one can always get there with a little patience because of the system's great flexibility.

Here are some indicators of the 'size' of the proof. The complete theories described here generated 167608 primitive inferences, and took 92 minutes to build on a 48Mb SPARCserver. The total ML source is 10080 lines, including comments. The parts leading just to the real number "axioms" generated 49017 primitive inferences, took 14 minutes to build, and consisted of 2098 lines of ML.

The nature of the underlying logic has some impact on the formalization. We have already discussed the effect of total functions on the division operation. They also mean that certain traditional notations are less useful; for example we cannot infer from $lim(x_n) = l$ that the sequence (x_n) actually tends to a limit; the *lim* function is *always* defined. Instead, we tend to use relational notations like $x \to l$ in preference. In most cases this is no handicap; indeed it is often clearer.

If analysis were to be taken further, some extensions to the logic would be convenient. For example, it is difficult to reason in a clean way about arbitrary n-ary Cartesian products without some simple form of dependent types. Furthermore, subtypes would allow more transparent embedding of one number system in another. It is perhaps difficult to know how difficult some parts of mathematics are to formalize without actually trying. Devices like adjoining infinities to the real line are easily waffled over, but perhaps not so easily formalized.

Formalization itself can be clarifying. One is forced to be less sloppy about things like variable binding (for example, what does $f'(x) = g(x)$ mean?). To avoid proving almost identical theorems twice, we saw the need for some common framework for limiting processes before we were actually aware that such frameworks (nets and filters) already existed. This constitutes an example of how abstraction can be driven by mundane considerations of economy, rather than of beauty.

Applications have only just got off the ground, but it seems to be fertile territory for computer-aided verification by theorem proving.

5 Acknowledgements

I would like to thank Mike Gordon for his help and encouragement, including his patient debugging when I was learning HOL, and for drawing my attention to Kelley's discussion of convergence nets [13]. Thomas Forster and Larry Paulson gave me some useful advice about constructing the reals. Everyone in the group at Cambridge has contributed to making such a pleasant and stimulating work environment, and many of them, too numerous to mention here, have helped my work in one way or another. I am specially grateful to Richard Boulton, Jim Grundy, Monica Nesi and Eike Ritter, who read an early version of this paper and made many helpful comments. Thanks are also due to the Science and Engineering Research Council for financial support.

References

[1] N. G. de Bruijn, *Defining Reals Without the Use of Rationals*, Indagationes Mathematicae, vol. 38, no. 2, 1976.

[2] J. C. Burkill, *A First Course in Mathematical Analysis*, Cambridge 1962.

[3] J. C. Burkill, H. Burkill, *A Second Course in Mathematical Analysis*, Cambridge 1970.

[4] C. W. Burrill, *Foundations of Real Numbers*, McGraw-Hill 1967.

[5] J. Chirimar, D. J. Howe, *Implementing Constructive Real Analysis*, preprint 1992.

[6] L. W. Cohen, G. Ehrlich, *Structure of the real number system*, Van Nostrand 1963.

[7] J. H. Conway, *On Numbers and Games*, Academic Press 1976.

[8] F. Faltin, N. Metropolis, B. Ross, G.-C. Rota, *The Real Numbers as a Wreath Product*, Advances in Mathematics, vol. 16, pp. 278-304, 1975.

[9] J. S. Golan, *The Theory of Semirings with Applications in Mathematics and Computer Science*, Longman 1992.

[10] IEEE, *Standard for Binary Floating Point Arithmetic*, ANSI/IEEE Standard 754-1985.

[11] C. Jones, *Completing the Rationals and Metric Spaces in LEGO*, Proceedings of the 2nd Workshop on Logical Frameworks, Edinburgh 1991.

[12] L. S. Jutting, *Checking Landau's "Grundlagen" in the Automath System*, PhD thesis, Eindhoven University of Technology 1977.

[13] J. L. Kelley, *General Topology*, Van Nostrand 1955.

[14] E. Landau, *Foundations of analysis*, Chelsea 1951.

[15] A. Pnueli, *From Timed To Hybrid Systems*, preprint 1992.

[16] R. R. Stoll, *Set theory and logic*, Dover 1979.

[17] H. A. Thurston, *The number system*, Blackie 1956.

[18] P. J . L. Wallis, *Improving floating-point programming*, Wiley 1990.

Higher Order Logic Theorem Proving and its Applications (A-20)
L.J.M. Claesen and M.J.C. Gordon (Editors)
Elsevier Science Publishers B.V. (North-Holland)

Modelling Generic Hardware Structures by Abstract Datatypes

Klaus Schneider, Ramayya Kumar and Thomas Kropf

Institute of Computer Design and Fault Tolerance, University of Karlsruhe
(Prof. Dr. D. Schmid) P.O. Box 6980, W-7500 Karlsruhe, Germany,
e-mail:{schneide,kumar,kropf}@ira.uka.de

Abstract

Although formal hardware specifications using natural numbers result in clear and succint descriptions, the proofs using them are not easy to automate. In this paper, an alternative means of specifying simple cascadable hardware using predefined operators is presented. These operators give the user a feeling that natural numbers are being used for specification and at the same time lend themselves to easy automation. Such operators work on n-bit bitvectors and their validation theorems, i.e. their correspondence to operations on natural numbers within the n-bit range, are also given. An example circuit is used to illustrate the use of these operators for specification and verification.

Keywords: Hardware Verification; Bit-vector theory; Arithmetic and Logic Circuits

1 Introduction

In digital circuit implementations natural numbers are always represented by bitvectors. If one wants to prove the correctness of such a circuit relative to a specification which uses the peano-axiomatized natural numbers, the disparity in the representations makes the automation of such proofs difficult. In this paper we present an abstract datatype for natural numbers, represented as bitvectors, which lends itself to easy automation. The advantage of this form of representation is that it is decidable and can therefore be automated. Furthermore, the semantical clarity of the operators are also retained.

In our approach we use bitvectors with a variable length n as opposed to the 'wordn' [HOL91] library implementations, which use an arbitrary but instantiated length of the bitvector. This allows us to define single n-bit operators for each function instead of defining operators for each specific bitvector. The 'wordn' library is more suited for interactive proofs where a set of possibly automatically derived theorems are used. On the other hand using the abstract data types as defined in this paper, most of the process can be automated from within MEPHISTO, a specialized hardware verification environment embedded in HOL [ScKK91b], [KuKS91b]. MEPHISTO provides a set of tactics for semi-automatically breaking up the proof goals into smaller subgoals and is augmented with a first-order prover called FAUST [ScKK91a], [KuKS91a], [ScKK92a], [ScKK92b], which is invoked to automatically prove the subgoals. In the context of specifications using the abstract data types, the proofs are performed using induction over n and the subgoals are automatically solved within MEPHISTO. When the recursive implementation has been proven correct with respect to a specification, the circuit implementation with a specific bitwidth is also generated automatically.

This paper is organized as follows: the next section deals with the implementation and validation of the abstract data types for natural numbers. Section 3 illustrates the use of the pre-defined operators and the tactics within MEPHISTO for specification, verification and generation of specific implementations. The paper is concluded with a summary and comments about our future work.

2 The Abstract Datatype

2.1 Representing Natural Numbers using Bitvectors

Natural numbers are represented by bitvectors which are in turn represented by pairs (n, f), where $n \in \mathbf{N}$ and $f : \mathbf{N} \rightarrow bool$. n denotes the length of the bitvector and the function f [1] determines the values of the bits, e.g. the bitvector $(b_n, \ldots, b_0)$ is represented by a pair (n, f) such that $f(i) = b_i$ for $i \in \{0, \ldots, n\}$. Of course there are infinitely many pairs for each bitvector, so we cannot define any isomorphism between them. However, the following equivalence relation can be defined :

$$(n, f) \approx (m, g) :\Leftrightarrow [m = n] \wedge \forall i.\, i \in \{0, \ldots, m\}\, [f(i) = g(i)]$$

In other words, two pairs (n, g) and (n, f) are identical if and only if the functions f and g are equal for all numbers less than or equal to n. Let $\mathcal{BV}_n$ denote the set of all pairs (n, f), and $\mathcal{BV}_n^{\approx}$ denote the set of all equivalence classes of $\approx$. The notion of equivalence classes can then be used in validating the operators which are defined over such bitvectors.

2.2 Validation of the Operators on Bitvectors

We define the following function for conversion between bitvectors and $\mathbf{N}$.

$$VAL\ n\ f := \textstyle\sum_{i=0}^{(n-1)} \Omega(f(i)) * 2^i$$

$$\text{where } \Omega : bool \rightarrow \mathbf{N} := \begin{cases} 1 & : if\ f(i) = T \\ 0 & : if\ f(i) = F \end{cases}$$

[1]when sequential specifications are to be described then $f : \mathbf{N} \rightarrow \mathbf{N} \rightarrow bool$, i.e. $f\ t : \mathbf{N} \rightarrow bool$, where t represents time

VAL assigns each element of $\mathcal{BV}_n^{\approx}$ to one element of $Z_n := \{0,\ldots,2^{n-1}\}$. Suppose, we have an operator $\pi : \mathcal{BV}_n^{\approx} \rightarrow \mathcal{BV}_n^{\approx}$, which should represent a function $\pi_V : \mathbf{N} \rightarrow \mathbf{N}$, then we validate the operator by proving the following theorem:

$$\vdash \forall n.\forall f.\pi_V(VAL\ n\ f) = VAL\ \pi(n,f)$$

Figure 1 illustrates the homomorphism used for the validation of the unary operator π.

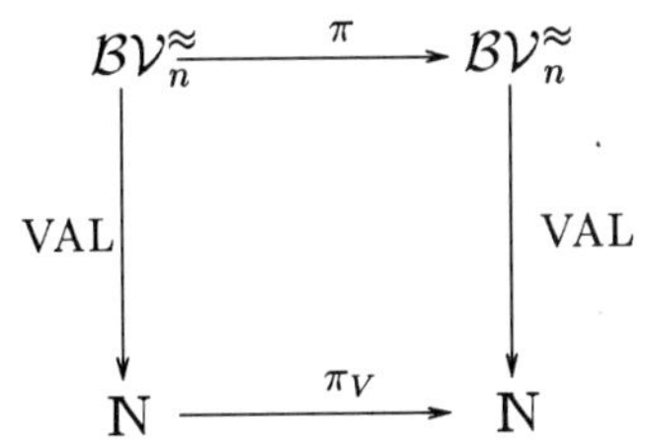

Figure 1: Validation of ADT-Operators

The validation theorem for a binary operator $\pi' : \mathcal{BV}_n^{\approx} \times \mathcal{BV}_n^{\approx} \rightarrow \mathcal{BV}_n^{\approx}$ would have the following form, (given that, $\pi'_V : \mathbf{N} \times \mathbf{N} \rightarrow \mathbf{N}$) —

$$\vdash \forall n.\forall f.\forall g.\ \pi'_V((VAL\ n\ f),(VAL\ n\ g)) = VAL\ \pi'((n,f),(n,g))$$

2.3 Predefined Operators

We now give a short description of the operators that we have defined within MEPHISTO and state the corresponding validation theorems. These validation theorems have been manually proved, once and for all, in HOL.

ALLZERO $(n, ALLZERO)$ represents the zero-valued n-bit vector.
Validation Theorem :$VAL\ n\ ALLZERO = 0$

SCARRY n f is true when a carry bit is generated by incrementing the pair (n, f).
Validation Theorem: $SCARRY\ n\ f = ((VAL\ n\ f) + 1 = 2^{n+1})$

SUCC n f g is true iff (n, g) is the successor of (n, f) modulo 2^{n+1}.
Validation Theorem : $SUCC\ n\ f\ g = VAL\ n\ g = ((VAL\ n\ f) + 1)\ MOD 2^{n+1}$

PCARRY n f g c is true when a carry bit is generated while adding (n, f) and (n, g) with the input carry bit c.
Validation Theorem :
$PCARRY\ n\ f\ g\ c = ([VAL\ n\ f + VAL\ n\ g + BV$[2] $c]DIV2^{n+1} = 1)$

PLUS n f g h c is true iff (n, h) is the sum of (n, f), (n, g) and the carry bit c modulo 2^{n+1}.
Validation Theorem:

[2]a function from bool → $\mathbf{N}$ for converting a bit to 1 or 0.

$$PLUS\ n\ f\ g\ h\ c = \\ VAL\ n\ h = (VAL\ n\ f + VAL\ n\ g + BV\ c) MOD 2^{n+1}$$

ADDER n f g h c is true iff $(n+1, h)$ is the sum of (n, f) and (n, g) the carry bit c.
Validation Theorem :

$$ADDER\ n\ f\ g\ h\ c = \\ (VAL(SUC\ n)h = VAL\ n\ f + VAL\ n\ g + BV\ c)$$

LS n f g is true iff (n, f) is less than (n, g).
Validation Theorem : $LS\ n\ f\ g = (VAL\ n\ f) < (VAL\ n\ g)$

EQQ n f g is true iff (n, f) and (n, g) belong to the same equivalence class.
Validation Theorem: $EQQ\ n\ f\ g = (VAL\ n\ f) = (VAL\ n\ g)$

COMPL n f g is true iff (n, g) is the one's complement of (n, f).
Validation Theorem: $COMPL\ n\ f\ g = (VAL\ n\ f + VAL\ n\ g + 1 = 2^{n+1})$

L_SHIFT n f g is true iff (n, g) is the left shifted value of (n, f).
Validation Theorem:$L\_SHIFT(SUCn)f\ g = (VAL(SUC\ n)g) = 2 * (VAL\ n\ f)$

R_SHIFT_HELP n f g is true iff $\forall i. 0 \leq i < n \rightarrow g(i) = f(i+1)$
Validation Theorem:

$$R\_SHIFT\_HELP(SUC\ n)f\ g = \\ (VAL\ n\ g) = ((VAL(SUC\ n)f) DIV 2)$$

R_SHIFT n f g is true iff $[\forall i. 0 \leq i < n \rightarrow g(i) = f(i+1)] \wedge g(n) = F$
Validation Theorem:$R\_SHIFT\ n\ f\ g = (VAL\ n\ g) = ((VAL\ n\ f) DIV 2)$

In the next section, we shall illustrate the use of these operators via an example.

3 Using the ADT

The definition of the pre-defined operators could also be interpreted as a hardware implementation of it, because they involve functions on bitvectors. Verifying another circuit for the same function now becomes much easier, since both the specification and the implementation are formulated at the same level of abstraction. The proof itself waters down to a mere comparison of the two 'circuits'. Specifications which are given in 'natural' form can be easily translated in a formula using the ADT.

For example consider the following specification for a comparator having two inputs I_1 and I_2 and two outputs *grtout* and *lessout*:

$$(grtout :\leftrightarrow I_1 > I_2) \wedge (lessout :\leftrightarrow I_1 < I_2)$$

Using the operators defined above and compiler construction techniques, the specification can be automatically translated into:

$$\begin{array}{l} GrtLss\_N\_SPEC\ n\ in1\ in2\ grtout\ lessout\ := \\ \qquad (grtout \leftrightarrow LS\ n\ in_2\ in_1) \\ \qquad (lessout \leftrightarrow LS\ n\ in_1\ in_2) \end{array}$$

Starting from this specification we illustrate the interactions with MEPHISTO using HOL-sessions given in numbered boxes. The specification in HOL syntax is given below:

```
new_definition                                                    1
  (`GrtLss_N_SPEC`,
   "GrtLss_N_SPEC n in1 in2 grtout lessout =
      (grtout  = LS n in2 in1)  /\ (lessout = LS n in1 in2)");;
```

The circuit implementing this function corresponds to a cascade of the 1-bit comparators which may correspond to the following basic circuit (GrtLss_1_IMP):

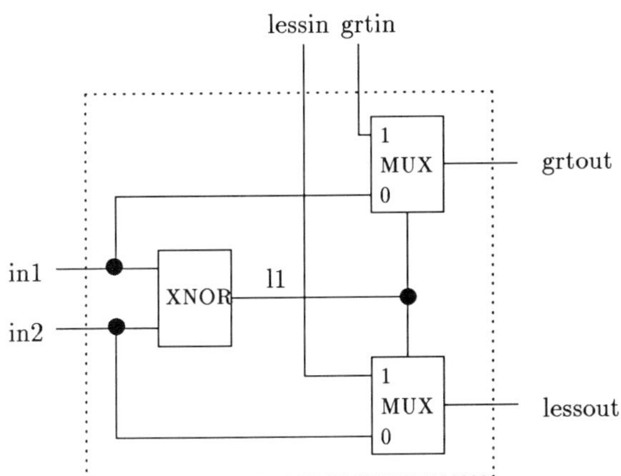

Figure 2: Implementation of 1-bit comparator (GrtLss_1_IMP)

The corresponding formal description of GrtLss_1_IMP is automatically derived by MEPHISTO and is given as follows:

```
new_definition                                                    2
   (`GrtLss_1_IMP`,
    "GrtLss_1_IMP grtin lessin in1 in2 grtout lessout =
       ? l1.
              XNOR(in1,in2,l1) /\
              MUX(l1,grtin,in1,grtout) /\
              MUX(l1,lessin,in2,lessout)");;
```

The 1-bit module is now used to recursively define an $n+1$-bit implementation (GrtLss_N_IMP) as shown in figure 3. This implementation, uses an n-bit blackbox and is interconnected to the 1-bit module.

Using the prim_rec_definition this structure can be formally coded as:

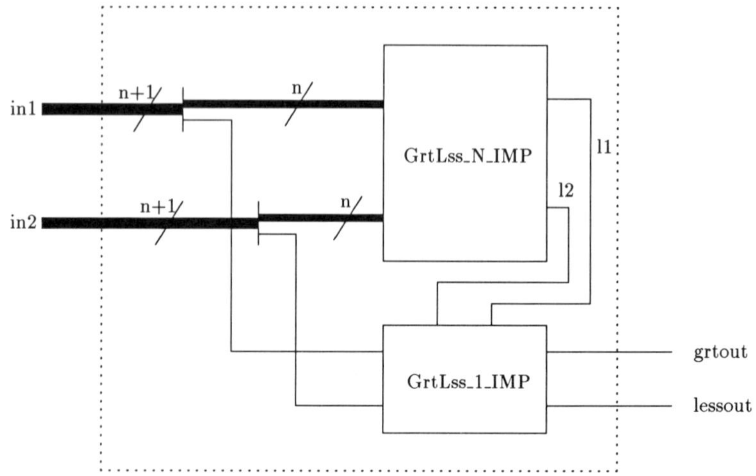

Figure 3: Recursive implementation scheme for the comparator (`GrtLss_N_IMP`)

```
new_prim_rec_definition                                                    3
    (`GrtLss_N_IMP`,
     "(GrtLss_N_IMP 0 in1 in2 grtout lessout =
        GrtLss_1_IMP F F (in1 0) (in2 0) grtout lessout) /\
      (GrtLss_N_IMP (SUC n) in1 in2 grtout lessout =
        ?l1 l2.
            GrtLss_N_IMP n in1 in2 l1 l2 /\
            GrtLss_1_IMP l1 l2 (in1 (SUC n)) (in2 (SUC n)) grtout lessout)");;
```

The goal to be proven, i.e. the equivalence between the recursive structure and the specification, can now be set:

```
#set_goal                                                                  4
    ([],
     "!n in1 in2 grtout lessout.
       GrtLss_N_IMP n in1 in2 grtout lessout =
       GrtLss_N_SPEC n in1 in2 grtout lessout");;
```

The first step in proving this goal is to apply induction and to rewrite with the definitions of the n-bit specification and implementation and the 1-bit implementation. This yields two subgoals as shown below:

5

```
#e(INDUCT_TAC THEN PURE_ASM_REWRITE_TAC defs);;
OK..
2 subgoals
"!in1 in2 grtout lessout.
  (?l1 l2.
    ((l1 = LS n in2 in1) /\ (l2 = LS n in1 in2)) /\
    (?l1'.
      (l1' = (in1(SUC n) = in2(SUC n))) /\
      (grtout = (l1' => l1 | in1(SUC n))) /\
      (lessout = (l1' => l2 | in2(SUC n))))) =
  (grtout =
   ((in2(SUC n) = in1(SUC n)) =>
    LS n in2 in1 |
    (~in2(SUC n) /\ in1(SUC n)))) /\
  (lessout =
   ((in1(SUC n) = in2(SUC n)) =>
    LS n in1 in2 |
    (~in1(SUC n) /\ in2(SUC n))))"
    [ "!in1 in2 grtout lessout.
        GrtLss_N_IMP n in1 in2 grtout lessout =
        GrtLss_N_SPEC n in1 in2 grtout lessout" ]

"!in1 in2 grtout lessout.
  (?l1.
    (l1 = (in1 0 = in2 0)) /\
    (grtout = (l1 => F | in1 0)) /\
    (lessout = (l1 => F | in2 0))) =
  (grtout = ~in2 0 /\ in1 0) /\ (lessout = ~in1 0 /\ in2 0)"

() : void
Run time: 3.3s
Intermediate theorems generated: 1276
```

Applying the tactic SIMPLIFY_TAC within MEPHISTO for removing the internal line l1, yields a simplified goal which can then be automatically solved by the first-order, automatic prover, FAUST[3].

6

```
#e(SIMPLIFY_TAC);;
OK..
"(grtout = ~(in1 0 = in2 0) /\ (~(in1 0 = in2 0) ==> in1 0)) /\
 (lessout = ~(in1 0 = in2 0) /\ (~(in1 0 = in2 0) ==> in2 0)) =
 (grtout = ~in2 0 /\ in1 0) /\ (lessout = ~in1 0 /\ in2 0)"

() : void
Run time: 0.6s
Intermediate theorems generated: 257
```

[3]The output churned out of HOL will be abbreviated to cut down the size of the paper

```
#e(PROP_FAUST_TAC);;                                                    7
OK..
goal proved
|- (grtout = ~(in1 0 = in2 0) /\ (~(in1 0 = in2 0) ==> in1 0)) /\

     .......................

() : void
Run time: 1.0s
Intermediate theorems generated: 14
```

Having solved the base case of the induction, the step case is also solved using the same principle. An application of SIMPLIFY_TAC, removes the internal lines and the simplified goal can then be solved using PROP_FAUST_TAC.

```
#e(SIMPLIFY_TAC);;                                                      8
OK..
"(grtout =
   ((in1(SUC n) = in2(SUC n)) ==> LS n in2 in1) /\
   (~(in1(SUC n) = in2(SUC n)) ==> in1(SUC n))) /\
  (lessout =
   ((in1(SUC n) = in2(SUC n)) ==> LS n in1 in2) /\
   (~(in1(SUC n) = in2(SUC n)) ==> in2(SUC n))) =
  (grtout =
   ((in2(SUC n) = in1(SUC n)) ==> LS n in2 in1) /\
   (~(in2(SUC n) = in1(SUC n)) ==> ~in2(SUC n) /\ in1(SUC n))) /\
  (lessout =
   ((in1(SUC n) = in2(SUC n)) ==> LS n in1 in2) /\
   (~(in1(SUC n) = in2(SUC n)) ==> ~in1(SUC n) /\ in2(SUC n)))"
    [ "!in1 in2 grtout lessout.
        GrtLss_N_IMP n in1 in2 grtout lessout =
        GrtLss_N_SPEC n in1 in2 grtout lessout" ]

() : void
Run time: 1.5s
Intermediate theorems generated: 571

#e(POP_ASSUM (\x. PROP_FAUST_TAC));;
OK..
goal proved
    ..............................
|- !n in1 in2 grtout lessout.
    GrtLss_N_IMP n in1 in2 grtout lessout =
    GrtLss_N_SPEC n in1 in2 grtout lessout

Previous subproof:
goal proved
() : void
Run time: 3.3s
Intermediate theorems generated: 26
```

This correctness theorem is then stored for future use.

```
                                                                    9
#save_top_thm 'GrtLss_N_CORRECT';;
|- !n in1 in2 grtout lessout.
    GrtLss_N_IMP n in1 in2 grtout lessout =
    GrtLss_N_SPEC n in1 in2 grtout lessout
Run time: 0.1s
```

Having proven an n-bit comparator, we can now generate an implementation corresponding to a specific bitwidth, using the function called WRAP_OUT. This function takes in the name of the generalized n-bit component and the number of the bits, and creates a new definitional constant for the specific bitwidth and also returns an appropriate correctness theorem. For example, instantiating the n-bit comparator to a 4-bit comparator, yields a circuit as shown in figure 4. The HOL session contains the theorem that has been proved using the new constant GrtLss_4_IMP, which has been introduced. It is to be noted that, the n-bit correctness theorem, the specification and the implementation are accessed by the function WRAP_OUT, in order to generate the specialized implementation and the correctness theorem. We are building an interface between MEPHISTO and a commercial circuit design framework CADENCE, which will then generate a netlist corresponding to the specialized implementation and introduce the module within the user's design library.

```
#WRAP_OUT 'GrtLss_N' 4;;                                            10
|- GrtLss_4_IMP in1 in2 grtout lessout =
   GrtLss_N_SPEC 3 in1 in2 grtout lessout
Run time: 1.8s
Intermediate theorems generated: 359
```

The use of the generalized recursive implementation followed by a specialization process, makes it possible to generate large formally verified circuits within acceptable times as shown below:

```
#WRAP_OUT 'GrtLss_N' 64;;                                           11
|- GrtLss_64_IMP in1 in2 grtout lessout =
   GrtLss_N_SPEC 63 in1 in2 grtout lessout
Run time: 70.8s
Intermediate theorems generated: 19199

#WRAP_OUT 'GrtLss_N' 256;;
;;; GC
;;; GC
;;; GC
;;; GC
|- GrtLss_256_IMP in1 in2 grtout lessout =
   GrtLss_N_SPEC 255 in1 in2 grtout lessout
Run time: 1471.4s
Garbage collection time: 20.4s
Intermediate theorems generated: 273023
```

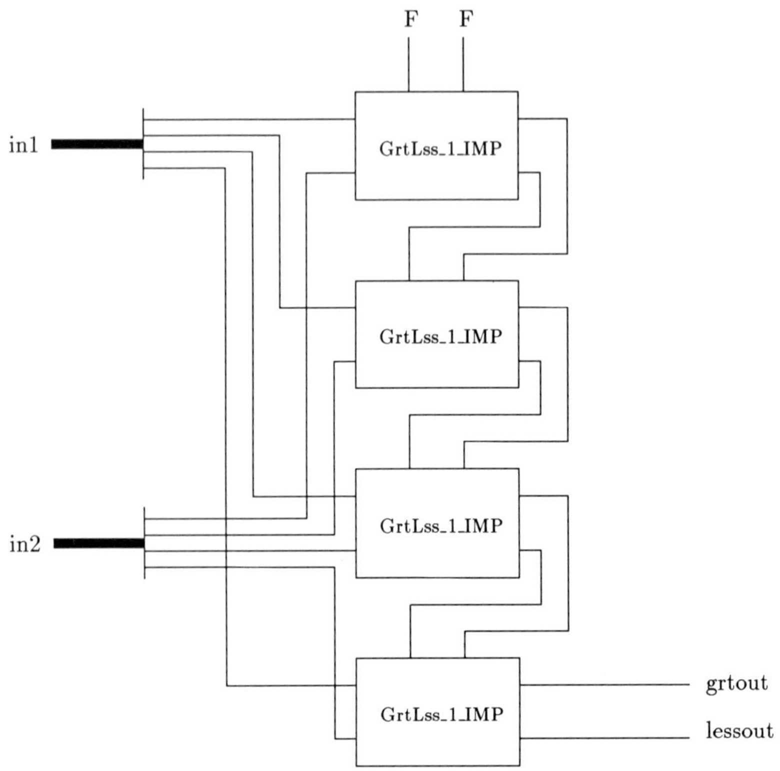

Figure 4: Implementation of `GrtLss_4_IMP`

4 Summary and future work

We have defined and validated several arithmetical operations on bitvectors represented by elements of $\mathcal{BV}_n^{\approx}$ such as addition, complementation, etc. The length of the bitvectors is an extra parameter for the operation, so a single operation can deal with arbitrary lengths. Such representations are cut out for use within MEPHISTO. The disadvantage of the representation is that a single bit of a vector cannot be changed, since this changes the whole function which is the second component of the representation pair. Thus, changing single bits means changing the whole pair.

Up to now we have only dealt with one dimensional digital circuits, but the concepts stated above also carry over to multi-dimensional circuits. In our future work we shall look at circuits which are realized using more complex interconnections, such as trees, arrays, etc.

References

[HOL91] Public Domain Distribution of HOL, *wordn Library in eval*

[KuKS91a] R. Kumar, T. Kropf, K. Schneider: *Integrating a First-Order Automatic Prover in the HOL Environment*; Proc. 1991 International Workshop on the HOL Theorem Proving System and its Applications, IEEE Press, pp. 170- 176, 1991.

[KuKS91b] R. Kumar, T. Kropf, K. Schneider: *First Steps Towards Automating Hardware Proofs in HOL*; Proc. 1991 International Workshop on the HOL Theorem Proving System and its Applications, IEEE Press, pp. 190-193, 1991.

[ScKK91a] Schneider K., Kumar R., Kropf T., *Structuring Hardware Proofs: First steps towards Automation in a Higher-Order Environment*, in Proc. VLSI '91, Edinburgh, pp. 3a.4.1-3a.4.10., North-Holland

[ScKK91b] K. Schneider, R. Kumar, T. Kropf: *Automating most parts of hardware proofs in HOL*; Proc. Workshop on Computer Aided Verification, K.G. Larsen, A. Skou (eds.), Aalborg, 1991, pp. 365-375, Lecture Notes in Computer Science, No. 575

[ScKK92a] Schneider K., Kumar R., Kropf T., *The FAUST-prover*, in Proc. of Conf. on Automated Deduction, CADE-11, Albany, New York, 1992

[ScKK92b] Schneider K., Kumar R., Kropf T., *Efficient Representation and Computation of Tableaux Proofs*; in Proc. of IFIP TC10/WG10.2 Workshop on Higher Order Logic Theorem Proving and its Applications, Leuven, Belgium, 21-24 Sept. 1992.

Higher Order Logic Theorem Proving and its Applications (A-20)
L.J.M. Claesen and M.J.C. Gordon (Editors)
Elsevier Science Publishers B.V. (North-Holland)

A Methodology for Reusable Hardware Proofs

Mark Aagaard and Miriam Leeser*

School of Electrical Engineering
Cornell University, Ithaca NY

Abstract

We present a methodology for reusing proofs of hardware components. Our methodology supports both bottom up and top down verification styles and allows multiple implementations of the same specification. We provide mechanical assistance for our methodology in the Nuprl proof development system. Our method exploits Nuprl's rich type theory to encode the specification of a module in the type of the module. This allows us to elegantly describe parameterized hardware modules. We illustrate our approach with several examples of circuit verification.

Keyword Codes: B.5.1, I.2.3
Keywords: Register Transfer Level Implementation, Design Aids; Deduction and Theorem Proving

1 Introduction

Using theorem provers to verify hardware has several advantages over ad hoc methods. These include an increased confidence in the correctness of the resulting designs, data independent analysis and more precise specification of behavior. Among the significant achievements in this area are the proofs of two microprocessors: the Viper chip with the HOL system [Coh88, Gor85a], and the FM8501 design with the Boyer-Moore system [BM88, Hun86]. Despite such successes, the application of theorem proving to hardware verification has been growing slowly. The main reasons for this are that proving hardware can be very tedious and that verification engineers require expertise in both hardware and mathematical logic.

The number of qualified verification engineers are few. Unfortunately, they frequently spend their time redoing proofs already done by others and reverifying designs with small

*Mark Aagaard is supported by a fellowship from Digital Equipment Corporation. This research was supported in part by the National Science Foundation under Award No. MIP-9100516.

changes in the implementation. While sharing proofs and proving each circuit once have always been the goals of theorem based proof methods, little support has been provided to facilitate these goals.

We present a methodology for specifying and verifying circuits that supports both bottom up and top down verification. We use the Nuprl proof development system [C+86], and exploit Nuprl's rich type theory to encode the specification of a module in the type of the module. Researchers in verification have been advocating the use of design parameters in hardware verification for quite some time. For example, it is common to parameterize modules such as adders and multiplexers by the width of the input and output ports and then to verify that the parameterized design works correctly for all port widths. In the past, design parameters have usually been limited to structural variables such as bit-widths of signals, size of memory, etc [Joy89]. Hanna has parameterized proofs of hardware modules by behavioral parameters such as radices of numbers and mathematical functions [HDL90]. We have extended the structural parameters just mentioned to include parameterizing by the implementations of submodules. In verifying a parameterized module the theorem which is proved is that the module will be correct for all choices of design parameters and for all correct implementations of the submodules.

This mechanism provides support for proving that the module is correct for any correct implementation of the submodules. Hence verification engineers can verify a module before verifying any or all of its submodules. In addition, they can change the implementation of a submodule without affecting the proof of the modules which use it. Our methodology also provides support for associating several different implementations of a submodule with the same specification.

Nuprl is a tactic oriented theorem prover developed at Cornell University. It is a descendent of the LCF project at Edinburgh [GMW79] and is based on a sequent version of Martin-Löf's constructive type theory [ML82], a rich and expressive type theory. The system has a graphical user interface which allows the user to walk over and manipulate the proof tree. Nuprl was developed for automated reasoning in many domains and has integers as a built-in type, as well as decision procedures to support arithmetic reasoning. Our experience has shown that these features make Nuprl a good choice for hardware reasoning. Display forms allow the user to define symbols so that proofs appear on screen exactly as they would on paper. Constructive type theory, and in particular, dependent types, allow for a natural specification style. We make heavy use of dependent types in our methodology for reusing proofs.

The research presented here is most similar to work done with the HOL theorem prover [Gor85a]. HOL differs from Nuprl in several ways. HOL's logic is classical and is based on a much simpler set of types than Nuprl. HOL was developed explicitly for hardware verification; Nuprl was developed as a general purpose reasoning system. Our approach follows the philosophy of similar proofs done in HOL. The difference is that our methodology provides support for sharing and reusing proofs.

Other researchers are investigating the use of dependent types for reasoning about hardware. Hanna [HDL90] is developing a system called VERITAS+. Suk [Suk91] is using Isabelle [PN90] to define such a logic. Both these systems take a similar approach to that used in Nuprl. We have an advantage in that Nuprl is much more developed than either of these systems, with large tactic libraries, well developed decision procedures, and a sophisticated user interface incorporating a proof editor and definition facility.

The rest of the paper is organized as follows. Section 2 introduces Nuprl and the use of dependent types. In Section 3 we describe our methodology, and in Section 4 we illustrate our methodology with examples. We conclude with an analysis of the methodology and its implementation.

2 Background

Nuprl's logic is a descendent of Martin-Löf's constructive type theory [ML82]. Nuprl supports a rich set of built-in types. Primitive types include *integer, atom*, and *void.* Type constructors include *list, union, function, product, quotient* and *subset.* Integer induction and list induction are built in. Types are stratified in an unbounded hierarchy of universes. U_1 is the first universe and contains all small types, including integers, lists, pairs, disjoint union, function space, equality (e. g. $a{=}b$ *in int*), and first order propositions. U_i and all elements of U_i are in U_{i+1}. This concept of universes allows the Nuprl user to quantify over types in a very natural manner. Quantification over types is not currently supported by HOL.

Nuprl's logic is higher-order. The logic is based on a *propositions-as-types* correspondence: a proposition is true if and only if the type associated with that proposition is inhabited. In other words, a proposition P is a type whose elements are proofs of P.

A proof in Nuprl may be thought of as a tree. Associated with each node of the tree is a sequent and, if the node is not a leaf, a proof rule. A *sequent* is a number of hypotheses and a goal, and can be written in the form:

$x_1{:}H_1,\ x_2{:}H_2,\ \ldots,\ x_n{:}H_n \vdash P$

Here the H_i are hypotheses and P is the conclusion. A sequent is true if the conclusion follows from the hypotheses. In constructive logic, this means that given members x_i of the types H_i we can construct a member of the type P.

Nuprl's proofs are developed in a top-down fashion. The root of the tree is the goal to be proved. The user applies inference rules which refine the goal into subgoals by which the truth of the goal may be established. The children of a node are uniquely determined by the sequent and rule of that node. (1) is the fragment of a proof tree where the rule *&-Intro* is applied. The rule refines the goal P & Q into two subgoals, one for proving P, and the other for proving Q.

...
$\vdash P$ & Q
BY *&-Intro*
$\vdash P$
$\vdash Q$ (1)

Inference rules in Nuprl may either be primitive rules or tactics written in the ML programming language. Nuprl tactics are similar to HOL tactics: given a sequent, they apply primitive inference rules and other tactics to the proof tree.

Reasoning about hardware frequently requires manipulating Booleans and bit vectors as well as arithmetic reasoning. We define Booleans as a subset of the integers:

$\{x\text{: } int \mid x{=}0 \text{ } in \text{ } int \text{ } \vee x{=}1 \text{ } in \text{ } int\}$

In Nuprl, integers, lists and pairs are built in to the system. This, combined with arithmetic decision procedures, allows for efficient arithmetic reasoning. Using Nuprl's evaluation facility, we can easily do case analysis over finite types such as Booleans.

A great deal of the expressive power in constructive type theory is due to dependent types. Dependent types allow for natural and general specifications [HDL90]. Nuprl provides several dependent types as primitives, including dependent function, dependent product, and subset types. Dependent product types are a generalization of Cartesian products where the type of the second element of the pair is dependent on the value of the first element. If the pair $<x,y>$ has dependent product type $z{:}X \times Y$, then x is of type X and y is of type $Y[x/z]$, which is read as *"Y with x substituted for z."*

A dependent function type, written $x{:}X \rightarrow Y$, is the type of functions from type X to type Y. Occurrences of the variable x which is considered to be of type X, are bound in type Y. For example, if f is a dependent function with type $x{:}X \rightarrow Y$ and t is a term with type X, then $f(t)$ has type $Y[t/x]$. If t has subset type $\{x{:}X|P\}$ where X is a type and P is a proposition possibly with a free variable x, then x has type X and $P[t/x]$ is a true proposition.

We can illustrate the use of dependent types with a definition of the *mod* function for division of non-negative integers. Without dependent types, *mod* would have type $N \rightarrow N+ \rightarrow N$ in Nuprl, where N is the type of non-negative integers, and $N+$ is the type of positive integers. However, since we know that the result of *mod* is less than its second input, we may wish to encode this in the type. We can express *mod* using the dependent function and subset types as: $N \rightarrow m{:}N+ \rightarrow \{n{:}N \mid n < m\}$

Unlike in HOL, type checking in Nuprl is in general undecidable. In practice, type checking which does not involve complicated uses of dependent types can almost always be handled automatically by tactics available in Nuprl.

3 Hardware Verification with Dependent Types

Our approach to hardware verification is based on that developed by the HOL group at Cambridge, but has been extended to exploit Nuprl's rich type theory. As in HOL and with previous work in Nuprl, hardware components are modeled as relations on their inputs and outputs and are either considered primitive or are built from other, simpler hardware components. The most significant difference between previous work and our method is that we use the type of a hardware module to describe both the externally visible structure of the module (the input and output ports) and the behavioral specification of the module. For example, we can use this technique to ensure that any module which is of type *adder_type* is a correct implementation of an adder. Previously, types have only been used to describe the externally visible structure of modules [HDL90].

Based upon this notion of using types to capture both structure and behavior, we have developed a methodology of describing and verifying digital hardware which provides an improved mechanism for supporting hierarchal designs and reusable hardware modules. This advantage comes at the cost of an increased number of library objects required to describe a hardware module and additional steps required at the beginning of a proof of correctness for a hardware module. We have addressed these issues by providing users with functions which automatically create the required library objects and we have developed tactics to handle the preliminary work in verifying a hardware module. In addition, ongoing research at Cornell is aimed at creating an improved mechanism for structuring libraries.

In hardware verification it is common practice to parameterize modules by structural variables such as the number of bits in a word and the size of memory. We parameterize modules by the implementations of submodules as well as by structural parameters. In verifying a parameterized module we prove that the module will be correct for all correct implementations of the submodules and for all choices of design variables. When we wish to create a fully instantiated design, we choose implementations for each of the submodules and values for each of the design variables. We can automatically verify that this instantiated module is a correct implementation because we have proved that the parameterized implementation is correct for all instantiations of the submodules and design variables. This technique could also be used in other proof systems, such as HOL, which have a higher order logic but do not have Nuprl's rich type theory. What makes Nuprl so well suited for doing verification in this style is that we can use the type theory to make the correctness theorems very concise and very readable.

In this paper we model time with a global clock and signals are either individual wires or busses. In general, a signal is a function from time to values of some type and a bus is a function from an index (a positive integer) to a signal. A *bit signal* is a function from time to Boolean values and a *bit bus* is a function from an index to time to Boolean values. The types for *time*, *vector*, *signal*, and *bus* are given below. All are functions from parameters to types in U_1, the first universe in Nuprl's hierarchy of universes. α is any type in U_1, and N is the type of non-negative integers. The types of vectors and busses are parameterized by the type to return (α) and the width of the vector or bus; the type of a signal is parameterized by the type to return. The implementations of these types rely upon dependent product, dependent function and subset types. A significant advantage of using these definitions is that we can enforce design rules through type checking. Because the widths of busses and ports are parts of the types, connecting a sixteen bit bus to a thirty-two bit port will not typecheck.

$time$	N
$\alpha\ vector(n)$	$\{0..n\text{-}1\} \to \alpha$
$\alpha\ signal$	$time \to \alpha$
$\alpha\ bus(n)$	$\{0..n\text{-}1\} \to time \to \alpha$

To create a hardware module in Nuprl, we first define the specification of the component, create one or more parameterized implementations of the component, prove that the parameterized implementations satisfy the specification, and finally instantiate the parameterized implementations to create fully defined implementations. Parameterized implementations are verified to be correct for all values of design parameters. Thus, once a parameterized implementation has been verified, design variables may be instantiated and the final implementation is verified automatically.

We can describe the type of n-bit wide adders as shown in Definitions 1 and 2. Definition 1, the specification for an adder, describes the correct behavior for an adder module. It says that the concatenation of c_{out} and sum is equal to the sum of a, b and c_{in}.

We use the function *cat* to concatenate a signal onto a bus and the function *bitVec2num* to convert a bit vector into a number. In Definition 2 the first part (before |) says that an adder is a module which has two n bit wide inputs (a and b), a one bit wide input (c_{in}), an n bit wide output (sum) and a one bit output (c_{out}). The second part says that the module satisfies the behavioral specification of an adder.

Definition 1 *adder_spec*

$$\lambda n\ (a,\ b,\ c_{in})\ (c_{out},\ sum).$$
$$bitVec2num(n{+}1)(cat\ c_{out}\ sum) =$$
$$bitVec2num(n)(a) + bitVec2num(n)(b) + bit\_val(c_{in})$$

Definition 2 *adder_type*

$$\{adder\colon (n{:}int \rightarrow$$
$$(bit\ bus(n) \times bit\ bus(n) \times bit\ signal) \rightarrow$$
$$(bit\ signal \times bit\ bus(n)) \rightarrow U_1)$$
$$|$$
$$\forall n : N{+}.$$
$$\forall a,\ b,\ sum : bit\ bus(n).$$
$$\forall c_{in},\ c_{out} : bit\ signal$$
$$adder(n)(a,b,c_{in})(c_{out},sum) \Rightarrow adder\_spec(n)(a,b,c_{in})(c_{out},sum)$$
$$\}$$

We can describe an implementation of an n-bit adder as a parameterized module which takes implementations of submodules as parameters. For example, if we want to create a ripple carry adder the submodule would be a single-bit full-adder.

A correct implementation of an adder has type *adder_type*. Our parameterized n-bit ripple-carry adder is a function which takes an implementation of a full adder and a positive integer n and returns a module whose type is *adder_type*. Specifically, the type of the parameterized module is: $full\_add\_type \rightarrow N{+} \rightarrow adder\_type$. Once we have verified that our parameterized module has the correct type, we can create a full implementation of the ripple carry adder by instantiating the design variables with an implementation of a full adder and a positive integer for the width of the design. This instantiated implementation can be automatically proved correct using existing tactics in Nuprl.

This design style allows us to easily modify designs by changing the implementations of the submodules, without requiring any additional verification effort. For example, there are many possible implementations of adders: ripple carry, carry-save, carry-propagate, etc. We can prove that each one is a correct implementation of an adder by proving that each one has type *adder_type*. Later on, we may design a module, such as a fixed point multiplier which uses an adder as a submodule.

After creating the parameterized module for the multiplier, we can choose which of the various implementations of the adder is most appropriate. For example, if we choose a ripple carry adder implementation named *add_ripple*, we could instantiate our parameterized module as: *mult(add_ripple)*. This implementation is verified automatically, because we have proved that the parameterized module is correct for all correct implementations of the submodule.

If we later wish to change our adder from a ripple carry implementation to a carry propagate adder, we simply create a new instantiation of the parameterized module: *mult(add_carry_prop)*. Again, we do not need to reverify any theorems by hand, because of our use of the parameterized implementation of the multiplier cell.

This style of design and verification easily scales to larger designs: when we build a module which contains a multiplier, we will use our specification of a multiplier in the parameterized module. We will then be able to choose whichever implementation of a multiplier we believe is most appropriate.

4 Examples

Our design methodology supports both top-down and bottom-up design styles as well as a mixture of the two styles. In the top-down style, a designer defines the implementation of a module in terms of specifications of submodules, which may not have been implemented yet. In the bottom-up style, a designer implements and verifies submodules before creating modules which will use the submodules. The examples in this section illustrate how we build hierarchal designs using both bottom-up and top-down design styles.

The process of designing and verifying a hardware module begins with the user providing a specification and parameterized implementation to an ML function. This function then creates several objects in the Nuprl library. These objects include the specification of the object, which is stated as a relation over the input and output signals; the type of the object; and a parameterized implementation of the object. All of the objects except the parameterized implementation are automatically verified; the user's task is to prove that the parameterized implementation satisfies the specification. The proof obligations appearing in the verification of the parameterized implementation are essentially the same as the obligations appearing in a conventional *implementation* $\Rightarrow$ *specification* proof.

module_*spec*	Specification for the module
module_*type*	Type of the module
module_*type_thm*	Proof showing that everything which has the type satisfies the specification
module_*pmod*	Parameterized module implementation
module_*mod*	Instantiated module implementation

Figure 1: Library objects for hardware module

The Nuprl user works within a library. A complete hardware module requires the library objects shown in Figure 1. We have written ML functions which prompt the user for information, automatically create the objects in the library and either completely prove or begin to prove the theorems. This is similar to the approach that Melham has taken in providing HOL users utilities for defining types [Mel90]. The only theorem which has non-trivial proof content is the parameterized module. All other theorems are proved automatically.

The goal of a proof of a parameterized module is to show that there exists a circuit which satisfies the specification. The proofs begin by defining an implementation of the parameterized module, exposing internal wires and then simplifying with propositional reasoning. The proofs are completed using general mathematical reasoning and previously proven lemmas.

To illustrate how we verify modules, we present two examples. In the first example, we progress in a bottom up style: we begin by creating an inverter, then build a buffer using two inverters. In the second example, we verify a counter using a top down style of design.

4.1 Verifying a Simple Buffer

As our first example, we present a simple buffer. This buffer is made up of two inverters cascaded together, as shown in Figure 3. The purpose of this example is to demonstrate our techniques; it is not representative of the level of designs that we plan to concentrate on. The inverters are primitive components in this example. To create the inverter we invoke the ML function *create_hardware_primitive* (Figure 2) with the name (*inv*), the position in the library where the module is to appear, the input and output port types and names and the specification for the inverter. The ML function then creates the library objects defining the specification and implementation of the inverter. Because we have decided to treat the inverter as a primitive component, the user is not asked to provide an implementation. Also, there is no parameterized module and the specification and implementation in the library are identical.

```
create_hw_primitive
  % name          :  tok        %   `inv`
  % position      :  tok        %   `after and_mod`
  % input types   :  term list  %   ["bit"]
  % input vars    :  tok list   %   [`i`]
  % output types  :  term list  %   ["bit"]
  % output vars   :  tok list   %   [`o`]

  % spec          :  term       %   "o=¬i"
;;
```

Figure 2: ML function to create hardware primitive

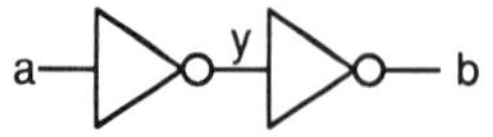

Figure 3: Implementation of a buffer

Nuprl libraries can contain ML objects which are executable ML code. We use this feature to provide a mechanism for prompting the user for the parameters to provide the function *create_hardware_primitive*. The ML object for calling the function is shown in Figure 2. The character "%" denotes a comment in ML; the text inside the comments are the prompts for the user. The user fills in the values for the parameters after the prompt. Once all of the values have been filled in, the user executes the ML object, which causes the library objects to be created.

For the inverter, the specification (Definition 3) says that the output is the logical negation of the input. The type of the inverter (Definition 4) says that the module has one input and one output and the input and output satisfy the specification for an inverter. These two theorems are automatically created and verified by the function *create_hw_primitive.*

Definition 3 *inv_spec*

$\lambda i\ o.\ o = \neg(i)$

Definition 4 *inv_type*

$\{\ inv{:}(bit \rightarrow\ bit \rightarrow\ U_1)\ |$
$\quad \forall i,o{:}bit.\ inv(i)(o)\ \Rightarrow inv\_spec\ (i)(o)\ \}$

The typing theorem (Theorem 1) is proved automatically and says that if a module is of type inverter then its input and output satisfy the specification of an inverter. This theorem is used to create a rewrite rule which may be used to replace an instantiation of an inverter with the specification of an inverter. This rewrite rule is useful in proving the correctness of modules which include inverters as submodules.

Theorem 1 *inv_type_thm*

$\vdash \forall inv{:}inv\_type\ .$
$\quad \forall i,o{:}bit.$
$\qquad inv(i)(o)\ \Rightarrow o{=}\neg(i)$

Because the inverter is a primitive component, there is no need for a parameterized module. The implementation of the module (Definition 5) is the same as the specification, thus the correctness can be proved automatically.

Definition 5 *inv_mod*

$\lambda i\ o.\ o = \neg(i)$

Now that we have built an inverter, we can create a buffer. To specify a buffer built from two inverters, we will use the ML function *create_hardware_module*, which is similar to *create_hardware_primitive*, except that we provide the design variables (submodules) and an implementation in addition to the other parameters. The resulting entries in the Nuprl library are shown in Definitions 6 – 9 and Theorem 2.

One of the potential disadvantages to our methodology is the large number of library objects created to describe a single hardware module. We are dealing with this in two ways: first, as just shown, we provide functions which automatically create the library objects and prove the theorems with trivial proof content for the user; second, in the near future Nuprl will have an improved library mechanism which will make it easier to deal with large structured libraries. Finally, only one of the library objects created has significant proof content, all others are dealt with completely automatically.

Definition 6 *buf_spec*

$\lambda i\ o.\ o = i$

Definition 7 *buf_type*

$\{\ buf{:}(bit\rightarrow\ bit\rightarrow\ U_1)\ |\ \forall i,o{:}bit.\ buf(i)(o)\ \Rightarrow buf\_spec\ (i)(o)\}$

Theorem 2 *buf_type_thm*

$\vdash \forall buf : buf\_type\ .$
$\quad \forall i,\ o : bit\ .$
$\qquad buf(i)(o)\ \Rightarrow o = i$

Definition 8 *buf_pmod*

$\lambda inv.\ \lambda i\ o.\ \exists x{:}bit.\ inv(i)(x)\ \&\ inv(x)(o)$

Definition 9 *buf_mod*

buf_pmod(inv_mod)

The buffer is verified by proving that the parameterized implementation is correct. That is, a correct implementation of a buffer may be constructed by taking a correct implementation of an inverter and putting two copies of the inverter in series.

The complete proof of the parameterized buffer module in Nuprl is shown in Figure 4. This is the only proof that the user must do in order to verify the buffer. The top line, beginning with $\vdash$, shows the goal to be proved. The user types in all lines beginning with **BY**. The other lines of the proof are generated by Nuprl. The lines beginning with **BY** are the rules applied to refine the proof. Underneath each rule are the unproven subgoals generated by applying the rule. Above each subgoal is its hypothesis list which is numbered and displayed vertically. In the interest of brevity, we do not repeat hypotheses which are carried forward from earlier steps in the proof. In general, several subgoals may be generated by each rule, and the proof has the form of a tree. In this proof, only one unproven subgoal is generated at each step. A proof is complete when a rule generates no unproven subgoals.

$\vdash$ *inv_type* $\rightarrow$ *buf_type*

BY *DefineImplementation*
'$\lambda inv.\ \lambda i\ o.$
$\exists x{:}bit.$
$inv(i)(x)\ \&inv(x)(o)$'

---

1. *inv:inv_type*
2. *i:bit*
3. *o:bit*
4. $\exists x{:}bit.\ inv(i)(x)\ \&inv(x)(o)$

$\vdash$ *buf_spec* $(i)(o)$

BY *HWStartTac*

---

4. *x:bit*
5. $x{=}\neg(i)$
6. $o{=}\neg(x)$

$\vdash$ $o{=}i$

BY *HWChain*

---

5. $o{=}\neg(\neg(i)\ )$

$\vdash$ $o{=}i$

BY *BoolSimp*

Figure 4 Proof of buffer

The proof of the buffer is completed in four steps. In the first step the implementation is defined; this step is done automatically by the ML function which also creates the library objects. In the second step we use the tactic *HWStartTac*, which exposes the internal wire x, replaces the inverter submodules with their specifications. and unfolds the specification of the buffer. Replacing the inverter submodules with their specifications is done using a rewrite rule based on Theorem 1. The third step uses the tactic *HWChain*, which does forward chaining and equality reasoning. In this case it substitutes hypothesis five into hypothesis six. So far, we have not done any work which is specific to the buffer; in fact, we can write a single tactic encompasses the first three steps of this proof.

The purpose of the tactics used so far have been to set up the proof to the point where we are able to begin reasoning about why two inverters in series constitute a correct implementation of a buffer. In this simple example, we can complete the proof by using a tactic which performs Boolean simplification.

As can be seen in this example, the use of dependent types does not restrict the reasoning power of the user or force the user to satisfy additional proof obligations. In fact, the reasoning shown here is the same that would be required in a conventional proof of this circuit. By using our methodology, users gain the advantages of hierarchal descriptions with re-usable modules but do not accrue any additional obligations or restrictions in their proof style.

4.2 Verifying a Counter

The buffer example in the previous section illustrated a bottom up design style. We now contrast that with an example showing how our methodology supports a top down design style. We also demonstrate our ability to reason about sequential hardware, bit vectors, and feedback. Our example is a resettable, sequential counter. Other presentations of proofs of this circuit appear in two earlier papers [Gor85b, Lee92]. The first paper describes a proof that was done in HOL and the second paper describes a proof that was done in Nuprl before we developed this new methodology.

Because we are using a top down design style, we will first create the specification for the counter, and then decide how to implement it. The specification for the counter (Definition 10) says that it outputs a zero on the next clock tick if *reset* is true on the current clock tick. Otherwise it outputs the previous output value plus one modulo 2^n. The implementation is shown in Figure 5.

We now must design the specifications of the three submodules: the incrementer, the register, and the init module. The incrementer increments the input by one modulo 2^n (Definition 12) and the register delays its input by one clock cycle (Definition 13). The init module (Definition 14) takes a one bit input signal *reset* and an n bit wide input signal a. If *reset* is L, then the output signal b is equal to a, otherwise all the bits in b are cleared. In the specification *zvec* is a function which given a positive integer n, returns a vector of length n whose elements are all zero. Having written a specification for each of the submodules, we can build a parameterized implementation for the counter (Definition 15).

The next step is to verify our parameterized module. Note that we have not yet implemented any of the submodules for the counter. After we have verified the parameterized

Definition 10 *counter_spec*

λn *reset out.*
  $\forall t$*: time.*
    bitVec2num(n,out(t+1))=
      if reset(t) then 0 *else (bitVec2num(n,out(t))+1) mod* 2^n

Definition 11 *counter_type*

{*counter : (n:N+* $\rightarrow$ *bit signal* $\rightarrow$ *bit bus(n))* |
  $\forall n$ *: N+.*
    $\forall$*reset : bit signal.*
      $\forall$*out : bit bus(n).*
        counter n reset out $\Rightarrow$ *counter_spec n reset out*
}

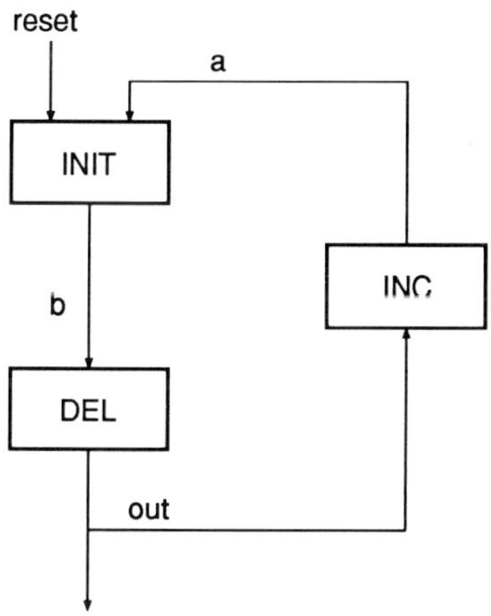

Figure 5: Implementation of Counter

Definition 12 *inc_spec*

$\lambda n\ a\ b.$
$\quad \forall t : time\ .\ bitVec2num(n,b(t)) = (bitVec2num(n,a(t))) + 1)\ mod\ 2^n$

Definition 13 *reg_spec*

$\lambda a\ b.$
$\quad \forall t : time.\ b(t+1) = a(t)$

Definition 14 *init_spec*

$\lambda n\ reset\ a\ b.$
$\quad \forall t : time\ .$
$\quad\quad b(t) =$
$\quad\quad\quad if\ reset(t)\ then\ zvec(n)\ else\ a(t)$

Definition 15 *parameterized module for counter*

$\lambda init\ reg\ inc\ .$
$\quad \lambda n\ reset\ out\ .$
$\quad\quad \exists a,\ b{:}bit\ bus(n).$
$\quad\quad\quad init\ n\ (reset,\ a)\ b\ \&$
$\quad\quad\quad reg\ n\ b\ out\ \&$
$\quad\quad\quad inc\ n\ out\ a$

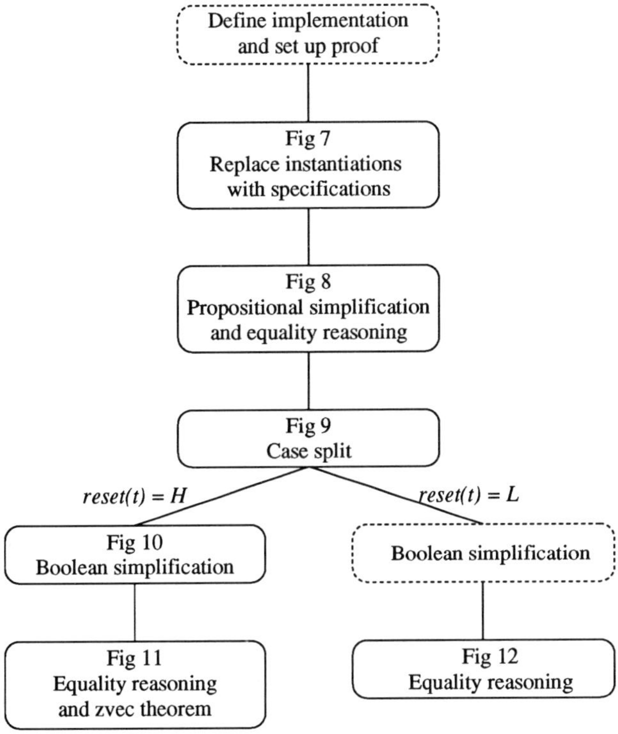

Figure 6: Proof tree for counter

module for the counter we will know that the specifications are correct. Once we know this we can implement the submodules without fear that we will have to later go back and change the implementation because of design errors in the counter. This is the real advantage of the top-down style of design; we never have to implement a module until we are sure that the specification for the module is correct. Unfortunately, the disadvantage is that in designing a parameterized module we may end up writing a specification for a submodule that we can't implement. For example, we may under-estimate the number of clock cycles required to calculate a result.

The proof tree for the counter is shown in Figure 6. Nodes in the proof tree are steps in the proof and refer to figures; dashed nodes do not have corresponding figures. The captions of the nodes correspond to the captions of the figures and describe the reasoning that is performed to get to the next step in the proof. As with the proof of the buffer, the proof of the counter begins by defining the implementation, exposing existentially quantified wires and unfolding the definition of the specification for the counter. Once the proof has been set up, we are left with the proof as shown in Figure 7. Hypotheses one through nine just contain variable declarations and typing information, so they are not shown in the proof.

10. *reg*(*n*)(*b*)(*out*)

11. *inc*(*n*)(*out*(*t*))(*a*(*t*))

12. *init*(*n*)(*b*(*t*),*reset*(*t*))(*b*(*t*))

⊢ $bitVec2num(n)(out(t+1))=$
  $\text{if } reset(t) \text{ then } 0 \text{ else } (bitVec2num(n)(out(t))+1) \bmod 2^n$

Figure 7 Replace instantiations with specifications

10. $\forall t{:}time.\ out(t+1) = b(t)$

11. $\forall t{:}time.\ bitVec2num(n)(a(t))=(bitVec2num(n)(out(t))+1) \bmod 2^n$

12. $\forall t{:}time.\ b(t) = \text{if } reset(t) \text{ then } zvec(n) \text{ else } a(t)$

⊢ $bitVec2num(n)(out(t+1))=$
  $\text{if } reset(t) \text{ then } 0 \text{ else } (bitVec2num(n)(out(t))+1) \bmod 2^n$

Figure 8 Instantiate t in hypotheses 10, 11 and 12 and equality reasoning

Applying the rewrite rules to replace instantiations of the modules with their specifications results in Figure 8. Instantiating the universally quantified variable t in hypotheses

10. $out(t{+}1) = \mathit{if}\ reset(t)\ \mathit{then}\ zvec(n)\ \mathit{else}\ a(t)$

11. $bitVec2num(n)(a(t)) = (bitVec2num(n)(out(t))+1)\ mod\ 2^n$

$\vdash$ $bitVec2num(n)(out(t{+}1)) =$
$\quad \mathit{if}\ reset(t)\ \mathit{then}\ 0\ \mathit{else}\ (bitVec2num(n)(out(t))+1)\ mod\ 2^n$

Figure 9 Case split on *reset*(t)

ten, eleven and twelve; then substituting the term $b(t)$ from hypothesis twelve into hypothesis ten leads to Figure 9.

We now perform a case split on the value of *reset*(t). This sets up two subgoals, one in which *reset*(t) is equal to H and one in which it is equal to L. Figure 10 shows the case when *reset*(t) is H and Figure 12 shows the case where *reset*(t) is L after some simplification has been done.

10. $out(t{+}1) = \mathit{if}\ H\ \mathit{then}\ zvec(n)\ \mathit{else}\ a(t)$

11. $bitVec2num(n)(a(t)) = (bitVec2num(n)(out(t))+1)\ mod\ 2^n$

$\vdash$ $bitVec2num(n)out(t{+}1)) =$
$\quad \mathit{if}\ H\ \mathit{then}\ 0\ \mathit{else}\ (bitVec2num(n)out(t))+1)\ mod\ 2^n$

Figure 10 Boolean simplification on first side of case split (*reset*(t) $= H$)

For the case where *reset*(t) is H, we can simplify the *if-then-else* statements to give us Figure 11. We can now substitute the term *out*(t+1) in hypothesis ten into the conclusion and apply Theorem 3 to complete this side of the proof.

10. $out(t{+}1) = zvec(n)$

$\vdash$ $bitVec2num(n)(out(t{+}1)) = 0$

Figure 11 Equality reasoning and apply *zvec* Theorem to first side of case split (*reset*(t) $= H$)

For the case where *reset*(t) is L, we simplify the *if-then-else* statements and are left with Figure 12. We can now complete the proof by substituting hypothesis ten into the conclusion.

Theorem 3 *zvec*

$\vdash \forall n{:}N{+}.\ bitVec2num(n)(zvec(n)){=}0$

10. $out(t{+}1) = a(t)$

11. $bitVec2num(n)(a(t)){=}(bitVec2num(n)(out(t)){+}1)\ mod\ 2^n$

$\vdash\ bitVec2num(n)(out(t{+}1)) = bitVec2num(n)(out(t)){+}1\ mod\ 2^n$

Figure 12 Equality reasoning on second side of case split $(reset(t){=}L)$ after Boolean simplification

5 Discussion

We have demonstrated our methodology for describing and verifying hardware with three examples: the adder illustrated our ability to build hierarchal designs and reuse the proofs of modules; the buffer was done in a bottom up style and included the use of primitive components; and the counter was done in a top down style and showed that we can easily reason about sequential circuits, vectors, and feedback. In this section we provide more analysis of our methodology.

Hardware design is naturally hierarchal: modules at one level become submodules at the next level. Previous methods have handled this hierarchy in two different ways. In the first method, the *specifications* of submodules are used in the implementation of a module; in the second method, the *implementations* of submodules are used in the implementation of a module.

In our methodology, we solve this question of whether to use implementations or specifications of submodules through the use of parameterized modules. Using this style we gain the power of using abstract specifications, rather than concrete implementations of submodules in the verification. And by forcing parameterized modules to be instantiated with correct implementations we get a very high degree of assurance that there are no inconsistencies introduced between levels of the design hierarchy. Nuprl is particularly well suited for this application because of its higher order logic and rich type theory. The use of higher order logic and dependent types allows us to write concise specifications and definitions for the parameterized modules. Other proof systems, such as HOL, which have higher order logics but simpler type theories, can also support this style of design. But it will be more cumbersome and less elegant.

In addition to increasing engineers' ability to reuse verified hardware modules, we have provided a number of tools and specialized tactics to assist users in interacting with the proof system. We have written functions which prompt the user for required information, automatically create the required Nuprl objects for primitive and parameterized modules, and then verify the trivial proof objects. The designer is left with only the verification of the parameterized module. The specialized tactics hide the low level details of the proof process and allow the user to verify modules in a more natural style.

It is doubtful that the day will come when all designs are fully verified using theorem proving based methods. Instead, verification engineers will use a variety of different tools and methods to verify large systems. In this scenario, design and verification engineers will need to be able to share information between themselves and between a large number of tools. Our methodology is well suited to such an environment, because in the verification effort we manipulate specifications, not implementations. For example a module may be comprised of a combinational logic submodule, a pipelined arithmetic unit, and a register file. Each of these submodules may require different design and verification techniques. The combinational logic may be implemented using a verified logic synthesis tool such as *Pbs* [AL91], the arithmetic pipeline may break down into more primitive submodules whose composition is verified within the proof system, and the register file may be hand designed and extensively simulated using a circuit analysis tool.

Requiring that the complete design be done inside the proof system will be too restrictive. But there needs to be a framework within which the verification and design effort can proceed. This is one of the roles of the proof system. Using our methodology, the specification for the combinational logic circuit could be taken from the proof system and used by the logic synthesis tool to create an implementation of the circuit. The pipeline design group could be given the specification for the pipeline and proceed on their own, keeping up with requirement changes through modifications to their specification. The circuit designers working on the register file would also be provided with a specification which they would then work with on their own.

The key to this process is that the verification within the proof system is done using *specifications* and not *implementations.* If we used the implementations of the combinational logic circuit, pipeline and register file, then all of the designs would have to be described within the proof system and the verification of one level of the design could not proceed until all lower modules had been designed and verified. By using specifications, the verification of a module can proceed as soon as its specification and the specifications of the submodules are complete.

Acknowledgements

This work owes a great debt to Mike Gordon and the research done in hardware verification at Cambridge University. In addition we would like to thank Robert Constable, Doug Howe, Paul Jackson and the rest of the Nuprl group for many helpful discussions and comments on earlier versions of this paper.

References

[AL91] M. Aagaard and M. Leeser. The implementation and proof of a boolean simplification system. In Geraint Jones and Mary Sheeran, editors, *Designing Correct Circuits, Oxford 1990.* Springer-Verlag, 1991.

[BM88] R. S. Boyer and J. S. Moore. *A Computational Logic Handbook.* Academic Press, 1988. Volume 23 of Perspectives in Computing.

[C+86] R. L. Constable et al. *Implementing Mathematics with the Nuprl Proof Development System.* Prentice Hall, 1986.

[Coh88] A. Cohn. A proof of correctness of the Viper microprocessor: The first level. In *VLSI Specification, Verification, and Synthesis*, pages 27–71. Kluwer Academic Publishers, 1988.

[GMW79] M. Gordon, R. Milner, and C. Wadsworth. *Edinburgh LCF: A Mechanized Logic of Computation*, volume 78. Springer Verlag, 1979. Lecture Notes in Computer Science.

[Gor85a] M. J. C. Gordon. HOL: A machine oriented formulation of higher order logic. Technical Report 68, Cambridge University Computer Laboratory, 1985.

[Gor85b] Mike Gordon. Hardware verification by formal proof. Technical Report 74, University of Cambridge Computer Laboratory, 1985.

[HDL90] F. K. Hanna, N. Daeche, and M. Longley. Specification and verification using dependent types. *IEEE Transactions on Software Engineering*, 16(9):949–964, September 1990.

[Hun86] W. A. Hunt, Jr. *FM8501: A Verified Microprocessor.* PhD thesis, Institute for Computing Science, The University of Texas at Austin, 1986.

[Joy89] Jeffrey Joyce. *Multi-Level Verification of Microprocessor-Based Systems.* PhD thesis, Cambridge University Computer Laboratory, 1989.

[Lee92] Miriam E. Leeser. Using Nuprl for the verification and synthesis of hardware. In C. A. R. Hoare and M. J. C. Gordon, editors, *Mechanized Reasoning and Hardware Design.* Prentice-Hall International Series on Computer Science, 1992.

[Mel90] Thomas Melham. *Formalizing Abstraction Mechanisms for Hardware Verification in Higher Order Logic.* PhD thesis, Cambridge University Computer Laboratory, August 1990. Technical Report No. 201.

[ML82] Per Martin-Löf. Constructive mathematics and computer programming. In *Sixth International Congress for Logic, Methodology, and Philosophy*, pages 153–175. North-Holland, 1982.

[PN90] L. Paulson and T. Nipkow. Isabelle tutorial and user's manual. Technical Report 189, University of Cambridge Computer Laboratory, January 1990.

[Suk91] Danny Suk. Hardware synthesis in constructive type theory. In Geraint Jones and Mary Sheeran, editors, *Designing Correct Circuits, Oxford 1990.* Springer-Verlag, 1991.

Higher Order Logic Theorem Proving and its Applications (A-20)
L.J.M. Claesen and M.J.C. Gordon (Editors)
Elsevier Science Publishers B.V. (North-Holland)

Abstract Theories in HOL

Phillip J. Windley

University of Idaho Laboratory for Applied Logic, Moscow ID 83843

Abstract

Abstract theories are widely used in mathematics because the provide a convenient way to reason about classes of structures. Abstract theories provide proofs about abstract structures which can then be used to reason about specific instances of those structures. Abstract theories are useful for specifying and verifying computer systems because they provide structure and guidance in the specification and verification processes and provide a convenient method of theorem reuse. This report describes and documents a package for using abstract theories in the HOL theorem proving system.

Keyword Codes: F.4.1; F.4.3
Keywords: Mathematical Logic; Formal Languages

1 Introduction

A theory is a set of types, definitions, constants, axioms and parent theories. Logics are extended by defining new theories. An abstract theory is parameterized so that some of the types and constants in the theory are undefined inside the theory except for their syntax and an algebraic specification of their semantics. Group theory provides an example of an abstract theory: the multiplication operator is undefined except for its syntax (a binary operator on type `:group`) and a semantics given by the axioms of group theory.

Abstract theories are useful because they provide proofs about abstract structures which can then be used to reason about specific instances of those structures. In groups, for example, after showing that addition over the integers satisfies the axioms of group theory, we can use the theorems from group theory to reason about addition on the integers.

This report describes the use of abstract theories in the HOL theorem prover [Gor88]. The current version of the abstract theory package has evolved greatly since the original report was published [Win90].

2 Abstract Theories

There are two key components of an abstract theory: (1) the *abstract representation* and (2) the *theory obligations*. The *abstract representation* is a set of abstract objects and a set of abstract operations. The operations are unspecified; that is, we don't know (inside the theory) what the objects and operations mean. Their meaning is specified through the theory obligations: a set of predicates that define relationships among members of the abstract representation. The abstract theory describes a model. Any structure with objects and operations that satisfy the predicates is a homomorphism of that model.

The theory obligations axiomatize the theory. Using the obligations as axioms, we prove theorems of interest about the abstract objects and operations. The goal is to use the abstract theory to reason about specific objects by instantiating the abstract theory with a concrete representation which has been shown to meet the obligations. The instantiation specializes the abstract theorems, producing a set of theorems about the concrete representation. The concrete representation is an instance of the abstract theory and represents a *member* of the class of abstract objects that it describes.

HOL, the verification environment used in the research reported here, does not explicitly support abstract theories; however, HOL's metalanguage, ML, combined with higher–order logic, provides a framework sufficient for implementing abstract theories. Several specification and verification systems, such as OBJ and EHDM, offer explicit support for abstract theories. We briefly describe them for comparison.

OBJ. OBJ is a specification and programming language developed by Joseph Goguen *et al.* that has most recently been described in [GW88]. OBJ is widely known and the semantics of its theories and views match our use of abstract theories. OBJ is based on a many–sorted (or typed) algebraic semantics and supports parameterized specification and programming [Gog84]. OBJ has three kinds of entities:

1. **Objects**, which are concrete modules that encapsulate executable code,

2. **Theories**, which are parameterized modules that correspond to abstract theories as used in this report, and

3. **Views**, which bind objects and theories to parameters in another theory.

Objects are said to contain executable code because the expressions in an object module give the initial algebraic semantics of the sorts and operations being defined. The fact that their semantics is initial implies that they describe just one model (up to isomorphism). Theories, on the other hand, are said to have a "loose" semantics since they define a variety of models. A loose semantics describes a class of objects; any member of that class will satisfy the theory.

A view is *not* an instantiation. Instantiation is done using a special command, **make**, after the view has been established. A view can be seen as a mapping of the operators and objects from one module onto a theory, as well as a declaration of intent that the module meets the obligations set forth in the equations of the theory module. OBJ does not require that the user prove that the obligations are met—a simple declaration is sufficient. Of course, if the view is not proper, then the OBJ program will not operate as intended.

EHDM. EHDM is a specification and verification system developed by SRI International [EHD88]. The language of EHDM is based on first–order predicate logic, but includes some elements of higher–order logic as well. For example, variables can range over functions, functions can return other functions, and functions can appear in quantifications. Parameterized modules are an important part of the EHDM language where they are used to organize specifications. Modules can be parameterized with types, constants, and functions. The module parameters can have constraints placed on them that must be met before the module can be instantiated.

In EHDM, a parameterized module is called a *generic module* and an instantiation is called a *module instance*. EHDM module declarations give the uninterpreted types, constants, and functions over which the module is parameterized. This declaration is analogous to our abstract representation.

The module body contains (among other things) an **ASSUMING** clause that gives the properties of the module parameters. The formulae in the **ASSUMING** clause are analogous to our theory obligations.

The module can also contain declarations of concrete types, constants, and functions that define the theory associated with the module and proofs of theorems about the abstract operations in the theory. These proofs may rely on the formulae in the **ASSUMING** clause.

3 Using the Abstract Theory Package

This section briefly describes the major functions in the abstract theory package. The following section provides an example of its use. Complete descriptions of all the commands are given in the user's manual that accompanies the package. For an example of how abstract theories can be used in computer system specification and verification see [Win91, Win92].

Before beginning an abstract theory, the ML file `abs_theory` must be loaded. This defines the commands in the abstract package and modifies some of the standard HOL commands to support its operation.

One declares a new abstract theory in the same way that one declares a standard theory, using `new_theory`. One is free to use any of the standard HOL commands for manipulating a draft theory in their usual manner. For example, definitions are made in the usual way using `new_definition`.

3.1 Abstract Representations

The abstract representation describes the abstract objects and operators in the abstract theory. The abstract theory package defines `new_abstract_representation` for declaring the abstract representation. The function is applied to a string representing the name of the abstract object and a list of string–type pairs. The first member of the pair gives the name of the abstract operator and the second member of the pair is the its type. Any number of abstract operators can be defined in an abstract representation. One can use `new_abstract_representation` more than once in a single theory defining more than one abstract object.

The system does not require that abstract objects be specifically declared. We represent abstract objects as type variables in HOL (denoted by a prepended asterisk). Since HOL does not require that type variables be declared, we are free to use them wherever we wish. The declaration of abstract objects is implicit, being the set of type variables occurring in the abstract representation.

The result of declaring a new abstract representation is a theorem that can be saved to use with `abs_type_info` to retrieve the type of the abstract object when this is difficult to otherwise discern.

When one defines a constant in the abstract theory, by convention, the first argument to the constant will be a variable with the same type as the abstract representation. This variable must, in turn be the first argument to any of the abstract constants from the abstract representation used in the definition. Later, during instantiation, the definition will be applied to a concrete representation and the instantiation functions will replace the abstract constants with the appropriate concrete constants in the instantiation.

3.2 Theory Obligations

The theory obligations are declared using the ML function `new_theory_obligations`. The function takes a string–term pair as its sole argument. The pair represents a theory obligation, giving the name and the predicate defining the theory obligation. The pair can be thought of as an axiom defining the semantics of the abstract objects.

The predicate is usually a conjunction of obligations. These obligations will be available for use in the draft theory. The system will automatically add them to the assumption list when the HOL commands for declaring abstract goals and proving abstract theorems, such as `set_abs_goal` and `prove_abs_thm`, are used. The HOL command `close_theory` closes the current draft and flushes the theory obligations.

3.3 Instantiating Theories

One makes use of an abstract theory by instantiating it. The first step is to make the abstract theory a parent of the draft theory using the ML function `new_abstract_parent`.

HOL theories differentiate between definitions and theorems and so there is an ML function for instantiating each. Instantiating any abstract object requires that we create a concrete representation. Concrete representations are created by applying the name of the abstract object to the concrete objects that are to be used for the instantiation.

Abstract Definitions. Abstract definitions are definitions which use the abstract objects from the abstract representation. By convention, the first variable in an abstract definition is the representation variable and has the same type as the abstract representation. Creating a concrete definition from an abstract one requires two steps:

1. Make an auxiliary definition that uses the abstract definition and applies it to a concrete representation.

2. Use the ML function `instantiate_abstract_definition` to produce a concrete instance of the abstract definition from the auxiliary definition.

The result of this instantiation is a theorem that defines a concrete instance of the abstract definition and makes no reference to the abstract definition.

Abstract Theorems. In drafting an abstract theory, one normally proves theorems about the abstract representation using the theory obligations as axioms. When the abstract theory is used, we instantiate the theorems in it so that the theory obligations are discharged and the new concrete theorems stand on their own.

The ML function `instantiate_abstract_theorem` instantiates one abstract theorem. The function takes four arguments:

1. The name of the abstract theory where theorem reside.
2. The name of the abstract theorem to instantiate.
3. A list of term pairs that instantiate variables with concrete representations. The first term in the pair is the variable to instantiate and the second is the concrete representation.
4. A list of theorems that satisfy the theory obligations in the subject abstract theorem. These theorems discharge the antecedents of the abstract theorems.

The new theorem resulting from the instantiation is *not* automatically saved in the current theory, but must be explicitly saved using `save_thm`.

4 Example: Group Theory

This section demonstrates the major features of the abstract theory package. We begin by defining an abstract theory for *monoids*. The following section uses the theory of monoids to create a theory of *groups*. The last section instantiates the theory of groups using exclusive–or as the group operator.

4.1 Defining Monoids

A monoid is an algebra with a binary operator and an identity element. The operator is associative and operating on the identity element with any object in the algebra, x yields that same object, x.

We begin the the session by loading the abstract theory package and entering draft mode:

```
loadf `abs_theory`;;

new_theory `monoid_def`;;
```

The abstract representation is declared using `new_abstract_representation`. We declare the binary operator, `op`, and an identity element, `e`:

```
let MONOID = new_abstract_representation `monoid`
   [
    (`op`,":* -> * -> *")
   ;
    (`e`,":*")
   ];;
```

The abstract representation gives only the name and type of the operators.

A declaration of an abstract representation will almost always be followed immediately by a declaration of the theory obligation since the abstract representation introduces the structure of the abstract entity, but not the semantics. The theory obligations declare the semantics (or axioms) for the abstract entity. The theory obligations are contained in a predicate that states the required properties of op and e:

```
new_theory_obligations
   (`IS_MONOID`,
    "∀m:(*)monoid . IS_MONOID m =
        (∀x:* .op m x (e m) = x) ∧
        (∀x:* .op m (e m) x = x) ∧
        (∀ x y z:*. (op m x (op m y z)) = (op m (op m x y) z))");;
```

Note that when op and e are used in a term, they always take as their first argument a variable with the type of the abstract object. While this is an implementational requirement, one can think of it as identifying op and e as abstract operators and giving the specific abstract object of which they are part. This last point is important since some theorems may use more than one abstract object and the abstract operators are differentiated by their argument (as shown in the example below).

We will begin with an interactive proof that the identity element is unique. We place the goal on the goal stack in the usual manner:

```
g
"∀ (m:(*)monoid) (f:*) .
        (∀(a:*) .(op m a f = a) ∧ (op m f a = a)) ⇒  (f = (e m))";;
```

HOL responds with the goal. Note that the theory obligations are automatically placed on the assumption list.

```
"∀f. (∀a. (op m a f = a) ∧ (op m f a = a)) ⇒  (f = e m)"
    [ "∀x y z. op m x(op m y z) = op m(op m x y)z" ]
    [ "∀x. op m(e m)x = x" ]
    [ "∀x. op m x(e m) = x" ]
```

We begin by stripping the universally quantified variables and the antecedent of the implication:

```
e(
REPEAT STRIP_TAC
);;

OK..
"f = e m"
    [ "∀x y z. op m x(op m y z) = op m(op m x y)z" ]
    [ "∀x. op m(e m)x = x" ]
    [ "∀x. op m x(e m) = x" ]
    [ "∀a. (op m a f = a) ∧ (op m f a = a)" ]
```

We can specialize the antecedent that we just placed on the assumption list with the identity element. Either conjunct can be used to solve the goal, we'll use the first.

```
e(
ASSUME_TAC (
  (CONV_RULE SYM_CONV) (
  CONJUNCT1 (
  SPEC "e (m:(*)monoid)" (
  ASSUME "∀a. (op m a f = a) ∧ (op m f a = a)"))))
);;

OK..
"f = e m"
    [ "∀x y z. op m x(op m y z) = op m(op m x y)z" ]
    [ "∀x. op m(e m)x = x" ]
    [ "∀x. op m x(e m) = x" ]
    [ "∀a. (op m a f = a) ∧ (op m f a = a)" ]
    [ "e m = op m(e m)f" ]
```

The result of the last step can be substituted into the goal:

```
e(
SUBST1_TAC ASSUME "e m = op m(e m)f"
);;

OK..
"f = op m(e m)f"
    [ "∀x y z. op m x(op m y z) = op m(op m x y)z" ]
    [ "∀x. op m(e m)x = x" ]
    [ "∀x. op m x(e m) = x" ]
    [ "∀a. (op m a f = a) ∧ (op m f a = a)" ]
    [ "e m = op m(e m)f" ]
```

The result can be rewritten with the assumptions to solve the goal:

```
e(
ASM_REWRITE_TAC []
);;

OK..
goal proved
. ⊢ f = op m(e m)f
.. ⊢ f = e m
.. ⊢ f = e m
. ⊢ ∀f. (∀a. (op m a f = a) ∧ (op m f a = a)) ⇒  (f = e m)

Previous subproof:
goal proved
```

The interactive proof can be packaged up as a proof script:

```
let IDENTITY_UNIQUE = ABS_TAC_PROOF
   (([],"∀ (m:(*)monoid) (f:*) .
       (∀(a:*) .(op m a f = a) ∧ (op m f a = a)) ⇒  (f = (e m))"),
    REPEAT GEN_TAC
    THEN STRIP_GOAL_THEN (⊢ .
        SUBST1_TAC (
        SYM_RULE (
        CONJUNCT1 (
        SPEC "e (m:(*)monoid)" thm))))
    THEN ASM_REWRITE_TAC []
   );;


IDENTITY_UNIQUE =
. ⊢ ∀f. (∀a. (op m a f = a) ∧ (op m f a = a)) ⇒  (f = e m)
```

The function ABS_TAC_PROOF is like TAC_PROOF except that when used on a goal with an abstract variable, the theory obligations are added to the assumption list before the proof is started. If there is more than one abstract variable the theory obligations are added to the assumption list for each. The fact that the theory obligations are needed to establish the theorem and are on the assumption list obviously influences the style of proof. An example in the next section demonstrates how to explicitly declare the obligations.

The next theorem gives an example that uses two abstract objects. The theorem proves that if the operations for two monoids are the same then their identity elements must be the same as well.

```
let OP_DETERMINES_IDENTITY = ABS_TAC_PROOF
   (([],
     "∀ m1 (m2:(*)monoid) . (op m1 = (op m2)) ⇒  (e m1 = (e m2))"),
    REPEAT STRIP_TAC
    THEN let t1 = ASSUME "∀x:*. op m1 (e m1) x = x" in
         SUBST_TAC (map SYM_RULE [SPEC "e m2:*" t1])
    THEN let t2 = ASSUME "∀x:*. op m2 x (e m2) = x" in
         SUBST_TAC (map SYM_RULE [SPEC "e m1:*" t2])
    THEN ASM_REWRITE_TAC []
  );;


OP_DETERMINES_IDENTITY = .. ⊢ (op m1 = op m2) ⇒  (e m1 = e m2)
```

In this example m1 and m2 are two distinct abstract objects. The term (op m1) represents the binary operation of monoid m1 and the term (op m2) represents the binary operation of monoid m2. When there is more than one abstract object in a goal, the abstract theory package places theory obligations on the assumption list for each. Note that the style of proof in this example is more goal oriented than the last example even though, as in the last example, the goal is proven using the assumptions.

4.2 Defining Groups

A group is a monoid extended with an inverse operator defined such that when the binary operator is applied to any element and its inverse, the identity element is the result.

To define a theory of groups, we take care of the usual front matter and also declare monoids_def to be an abstract parent since we are going to establish some properties of groups from our theory of monoids.

```
loadf `abs_theory`;;

new_theory `group_def`;;

new_abstract_parent `monoid_def`;;
```

The call to new_theory clears the global theory obligation list. new_abstract_parent puts the theory obligations from monoid_def on the theory obligation list.

The next step is to define the abstract representation:

```
let GROUP = new_abstract_representation `group`
   [
    (`fn`,":* -> * -> *")
   ;
    (`id`,":*")
   ;
    (`inv`,":* -> *")
   ];;
```

Note that we do not simply define the abstract representation for groups as an extension to the representation for monoids. Without subtypes, the implementation and use of extensions becomes too unwieldy. Also, since HOL does not allow operator overloading, we had to give the operator and identity element for groups different names from the operator and identity element for monoids.

The theory obligations are those for monoids extended with two additional facts for the inverse function:

```
new_theory_obligations
    ('IS_GROUP',
     "∀g:(*)group . IS_GROUP g =
        (∀x:* .fn g x (id g) = x) ∧
        (∀x:* .fn g (id g) x = x) ∧
        (∀x:* .fn g x (inv g x) = (id g)) ∧
        (∀x:* .fn g (inv g x) x = (id g)) ∧
        (∀ x y z:*. (fn g x (fn g y z)) = (fn g (fn g x y) z))");;
```

Again, because HOL does not have a notion of subtypes, we do not define the theory obligations for groups in terms of the theory obligations for monoids.

The connection between groups and monoids is established by the following trivial theorem:

```
let GROUP_EXTENDS_MONOID = ABS_TAC_PROOF
   (([],
     "∀ g:(*)group . IS_MONOID(monoid (fn g) (id g))"),
    EXPAND_THOBS_TAC 'monoid_def'
    THEN ASM_REWRITE_TAC []
   );;

GROUP_EXTENDS_MONOID = ... ⊢ IS_MONOID(monoid(fn g)(id g))
```

The theorem states that the operator and identity element for groups can be used in place of the operator and identity element for monoids. The tactic EXPAND_THOBS_TAC is used to expand the theory obligations for monoids as the first step of the proof. The result is an abstract theorem since it relies on the theory obligations of group_def.

We can make use of the fact that a group is a monoid by instantiating the theorem that the identity is unique from monoid_def:

```
let IDENTITY_UNIQUE =
    instantiate_abstract_theorem 'monoid_def' 'IDENTITY_UNIQUE'
        ["m","monoid (fn (g:(*)group)) (id (g:(*)group))"]
        [GROUP_EXTENDS_MONOID];;

IDENTITY_UNIQUE =
... ⊢ ∀f. (∀a. (fn g a f = a) ∧ (fn g f a = a)) ⇒  (f = id g)
```

We instantiate the abstract representation variable "m" with a concrete (from the standpoint of monoid_def) representation made from the operator and identity element of group_def. The theorem that a group extends a monoid is used to discharge the theory obligations for monoids. Note that the instantiated theorem must be explicitly saved on the current theory.

The resulting theorem is really a theory about groups, not monoids, as can be seen from the following expression which checks the type of the abstract representation variable "g" in IDENTITY_UNIQUE:

```
(type_of o hd o frees o concl o DISCH_ALL) IDENTITY_UNIQUE;;

":(*)group" : type
```

In addition to instantiating theorems from monoid_def, we can prove theorems about groups directly. For example, we can prove that left cancellation holds for groups:

```
let LEFT_CANCELLATION = prove_abs_thm
   (`LEFT_CANCELLATION`,
     "∀ (g:(*)group) (x y a:*) .
      ((fn g) a x = ((fn g) a y)) ⇒  (x = y)",
    REPEAT STRIP_TAC
    THEN ACCEPT_TAC (
      let t1 = (ASSUME "∀x y z. fn g x(fn g y z) = fn g(fn g x y)z") and
          t2 = (ASSUME "∀x. fn g(inv g x)x = id g") and
          t3 = (ASSUME "∀x. fn g(id g) x  = x") and
          t4 = (ASSUME "fn g a x = fn g a y") in
      SYM_RULE (
      REWRITE_RULE [t1;t2;t3] (
      REWRITE_RULE [t2;t3;t4] (
      ISPECL ["(inv g a)";"a";"x"] t1))))
   );;

LEFT_CANCELLATION = ... ⊢ ∀x y a. (fn g a x = fn g a y) ⇒  (x = y)
```

In this case, we used prove_abs_thm, so the the resulting theorem is explicitly saved in the current theory.

Because some people may be uncomfortable with implicit theory obligations and their automatic inclusion in the assumptions, the abstract theory package supports explicit theory obligations as well. For example, consider the following proof that the inverse function is reversible (that is that the inverse operator applied twice is the identity function):

```
let INVERSE_INVERSE_LEMMA = prove_thm
   (`INVERSE_INVERSE_LEMMA`,
    "∀(g:(*)group) . IS_GROUP(g) ⇒  ∀a .(((inv g) ((inv g) a)) = a)",
    STRIP_THOBS_TAC
    THEN GEN_TAC
    THEN ACCEPT_TAC (
        let t1 = ASSUME "∀x. fn g x(inv g x) = id g" and
            t2 = ASSUME "∀x. fn g (inv g x)x = id g" in
        let LC_LEMMA =
              ISPECL ["inv g (inv g a)";"a";"inv g a"] LEFT_CANCELLATION in
        MATCH_MP LC_LEMMA
                 (TRANS (ISPEC "(inv g) a" t1)
                        (SYM_RULE (ISPEC "a" t2))))
   );;

INVERSE_INVERSE_LEMMA = ⊢ ∀g. IS_GROUP g ⇒  (∀a. inv g(inv g a) = a)
```

In this case, we use the standard prove_thm function and so the theory obligations were not automatically added to the assumption list. The theory obligation are given explicitly as the antecedent to an implication. The tactic STRIP_THOBS_TAC is a convenient way to strip explicit theory obligations from the goal and expand them on the assumption list.

We can use a different proof style using the theorem continuation STRIP_THOBS_THEN for the same theorem:

```
let ALTERNATE_INVERSE_INVERSE_LEMMA = TAC_PROOF
   (([],
    "∀(g:(*)group)  . IS_GROUP(g) ⇒  ∀ a .(((inv g) ((inv g) a)) = a)"),
    STRIP_THOBS_THEN (λ thm .
        let thl = CONJUNCTS thm in
        MAP_EVERY ASSUME_TAC thl THEN
        GEN_TAC THEN
        MATCH_MP_TAC
         (ISPECL ["inv g (inv g a)";"a";"inv g a"] LEFT_CANCELLATION) THEN
        REWRITE_TAC thl)
  );;

ALTERNATE_INVERSE_INVERSE_LEMMA =
   ⊢ ∀g. IS_GROUP g ⇒  (∀a. inv g(inv g a) = a)
```

The only caveat to using STRIP_THOBS_THEN is that the accompanying theorem–tactic must put the theory obligations on the assumption list if any other abstract theories are to be used. Since we use LEFT_CANCELLATION, the tactic MAP_EVERY ASSUME_TAC thl is used to add the theory obligations to the assumption list.

4.3 Using Groups

We can instantiate the theory of groups using exclusive–or as the operator, false as the identity element, and the identity function as the inverse operator. As usual, we load the abstract theory package, enter draft mode, and declare groups_def an abstract parent.

```
loadf `abs_theory`;;

new_theory `example`;;

load_library `taut`;;

new_abstract_parent `group_def`;;
```

Before we can instantiate the theorems of group_def we must show that our proposed instantiation discharges the theory obligations:

```
let GROUP_THOBS = TAC_PROOF
   (([],"IS_GROUP(group (λx y.¬(x=y)) F I)"),
    EXPAND_THOBS_TAC `group_def`
    THEN BETA_TAC
    THEN REWRITE_TAC [I_THM]
    THEN TAUT_TAC
   );;
```

The term (group (λ x y. ¬(x=y)) F I) creates the concrete representation.

Using instantiate_abstract_theorem, we can instantiate the theorem of group theory for our concrete representation. The abstract variable "g" is instantiated with our concrete representation.

```
instantiate_abstract_theorem `group_def` `IDENTITY_UNIQUE`
    ["g","group (λx y.¬(x=y)) F I"]
    [GROUP_THOBS];;

⊢ ∀f. (∀a. (¬(a = f) = a) ∧ (¬(f = a) = a)) ⇒  ¬f

instantiate_abstract_theorem `group_def` `LEFT_CANCELLATION`
    ["g","group (λx y.¬(x=y)) F I"]
    [GROUP_THOBS];;

⊢ ∀x y a. (¬(a = x) = ¬(a = y)) ⇒  (x = y)

instantiate_abstract_theorem `group_def` `INVERSE_INVERSE_LEMMA`
    ["g","group (λx y.¬(x=y)) F I"]
    [GROUP_THOBS];;

⊢ ∀a. I(I a) = a
```

The instantiation is done in the same way regardless of whether the theory obligations for the abstract theorems were explicitly stated or not.

5 Conclusions

Abstract theories are widely used in mathematics because the provide a convenient way of reasoning classes of structures. Abstract theories are useful for specifying and verifying computer systems because they provide structure and guidance in the specification and verification processes and provide a convenient method of theorem reuse. The package we have described shows how abstract theories can be easily used in HOL.

References

[EHD88] SRI International Computer Science Laboratory. *EHDM Specification and Verification System: User's Guide,* Version 4.1, 1988.

[Gog84] J. Goguen. Parameterized programming. *IEEE Transactions on Software Engineering,* SE-10(5):528–543, September 1984.

[Gor88] Michael J.C. Gordon. HOL: A proof generating system for higher-order logic. In G. Birtwhistle and P.A Subrahmanyam, editors, *VLSI Specification, Verification, and Synthesis.* Kluwer Academic Press, 1988.

[GW88] J. Goguen and T. Winkler. Introducing OBJ3. Technical Report SRI-CSL-88-9, SRI International, August 1988.

[Win90] Phillip J. Windley. A poor man's implementation of abstract theories. Technical Report CSE-90-06, University of California, Davis, Division of Computer Science, 1990.

[Win91] Phillip J. Windley. Abstract hardware. In *Proceedings of the ACM/SIGDA International Workshop in Formal Methods in VLSI Design*, January 1991.

[Win92] Phillip J. Windley. A theory of generic interpreters. Technical Report LAL-92-06, University of Idaho, Laboratory for Applied Logic, April 1992.

Higher Order Logic Theorem Proving and its Applications (A-20)
L.J.M. Claesen and M.J.C. Gordon (Editors)
Elsevier Science Publishers B.V. (North-Holland)

Machine Abstraction in Microprocessor Specification

Michael McAllister

Integrated Systems Design Laboratory, Department of Computer Science, University of British Columbia, Vancouver, British Columbia, Canada, V6T 1Z2.

Abstract

Microprocessor verification to date has primarily focused on single microprocessors in isolation, yet families of microprocessors exist which share a common base instruction set. We propose a "machine abstraction" method for specifying the behaviour of a microprocessor in a manner which encourages portability of specifications to other microprocessors in the same family. Areas of an assembly language which are restricted by hardware constraints rather than language semantics are grouped within a "machine abstraction" and are supplied to a specification of the language as parameters. We hope such "machine abstraction" will simplify the verification of families of microprocessors. This paper extends earlier work on generic approaches by Joyce, Windley, and Herbert.

Keyword Codes: B.7.1;F.3.2;I.2.3
Keywords: Integrated Circuits, Types and Design Styles; Logics and meanings of Programs, Semantics of Programming Languages; Deduction and Theorem Proving.

1 Introduction

Microprocessor verification has primarily focused on a single microprocessor and its individual assembly language. However microprocessors are being constructed in families which share the syntax, semantics, and op-code format of a base instruction set. We propose a microprocessor specification format based on "machine abstraction" which separates assembly language semantics from restrictions imposed on the language by a particular microprocessor. The separation parallels a technique of generic specification in the HOL [G+85] community for hardware verification [Joy89] [Win90]. While generic specification distinguishes between microprocessors specification and primitive operations within the specification, "machine abstraction" creates an abstraction of components within a microprocessor and describes the microprocessor in terms of component behaviour rather than the component implementation. The machine abstraction can also be applied in conjunction with incremental verification techniques [Her91] for families of large microprocessors.

Machine abstractions were devised as a solution for sharing specifications while creating a programming-level specification of the Motorola 68000 microprocessor. We wanted to develop a specification which would describe several processors from one processor family. Parameterization of the hardware differences between the processors seemed to be a natural course and later developed into "machine abstractions."

In this paper we will review the basic structure of generic specification for hardware verification and touch on operational semantics. We then combine these two concepts into a machine abstraction technique for microprocessor specification. An example of machine abstraction for a single instruction is presented. Advantages and disadvantages of the technique are discussed and are followed by some possible areas of application.

2 A Separation of Concerns

Early microprocessor specifications focused on obtaining an accurate and unambiguous definition of a microprocessor's behaviour. The battle developed on two fronts: first to devise true semantics for the microprocessor's assembly language and second to represent the microprocessor's hardware structure. As researchers became familiar with specification techniques their focus shifted to reducing the inclusion of irrelevant information in specifications. A generic specification technique was devised in the HOL community to address this problem. Now we are in a position to confront a different concern: shared specifications among processors within a processor family (or specification "re-use in the large"). We can draw on our experience in both generic specification and operational semantics to address this issue.

Generic specification encourages us to isolate non-essential information into functional parameters for the specification. The specification and some verification can then be accomplished without ever instantiating these parameters – we rely on the parameters' semantics and their consistent usage in both the specification and hardware description. Operational semantics reminds us that a specification can be based on a more malleable abstraction of some physical or mathematical system as long as the semantics of the abstraction is well-defined.

We begin by presenting an overview of generic specification and operational semantics. The combination of these ideas is developed in the final subsection.

2.1 Generic Specification

Generic specification in the hardware verification arena originated as a microprocessor specification technique for masking specification details which are not essential in verification. It was first used to verify the Tamarack-3 microprocessor [Joy89] and then appeared in microprocessor verifications of the Viper [Cur] and AVM-1 [Win90].

Generic specification represents a separation of concerns within a proof. Some operations may be extremely basic and yet be distracting or time consuming to prove. For instance, in a microprocessor specification we may wish to add numbers together frequently. These numbers may have different (bit) sizes or different signs. Moreover, foreknowledge may indicate that all additions for the microprocessor will be performed by the arithmetic logic unit (ALU). Consequently, we define some "increment" operation to use in the microprocessor specification and define the ALU behaviour in terms of this same "increment" operator. The microprocessor proof proceeds without requiring any further details and ensures a consistent use of "increment". A secondary verification proves the correct "increment" implementation by the ALU. This also allows the ALU details to be separate from the microprocessor details and facilitates hardware interchangeability.

In a second case, some operations may not be relevant to the microprocessor specification. Microprocessors generally do not care how the external memory is structured beyond the "access protocol" at their processor pins. Generic specification can provide "store" and "fetch" primitives as a memory abstraction for external memory. Proofs then assume that the "store" and "fetch" operations behave in accordance with the access protocol. In this case the generic specification operators separate the microprocessor specification from the environment in which it is embedded.

2.2 Operational Semantics

Classical operational semantics define the effect of instructions for a language by describing them in terms of a different (and usually less complex or less expressive) language whose semantics are well-defined. These semantics provide a formal description of a language while maintaining an abstract view of the mechanisms through with the instructions are performed and is a key aspect of our "machine abstraction" technique. In this manner, proofs concerning a large or complex language are reduced to proofs in a much simpler language. A more recent variation called Structured Operational Semantics is described in [Plo81].

At a very coarse level we can draw an analogy from a compilation algorithm to classical operational semantics. A compiler takes instructions in, for example, C and converts them into some assembly code which performs an "equivalent" operation. The expressive C commands are thus expressed as a larger sequence of simpler assembly commands. The analogy is not complete since we are making assumptions about the semantics of the compiler, compilation algorithm, and assembly language but the intuitive transformation is similar to classical operational semantics.

3 Machine Abstraction

Machine abstraction is the recognition of an abstraction level within the specification hierarchy (see figure 1). In a flavour similar to generic specification, machine abstractions introduce a level above (and somewhat in parallel with) the generic specification level. The machine abstraction allows the specifier to focus on the structure of the assembly language rather than the hardware limitations imposed on the language.

From a hardware verification viewpoint, a machine abstraction allows multiple microprocessors to share programming-level specifications as long as the microprocessors share the same assembly language. The differences between the microprocessors are captured by the machine abstraction and a unified structure is presented to the assembly language description level. Components of the assembly language specification which involve restrictions from hardware are isolated into specification parameters. The restrictions which arose while creating the Motorola 68000 specification include

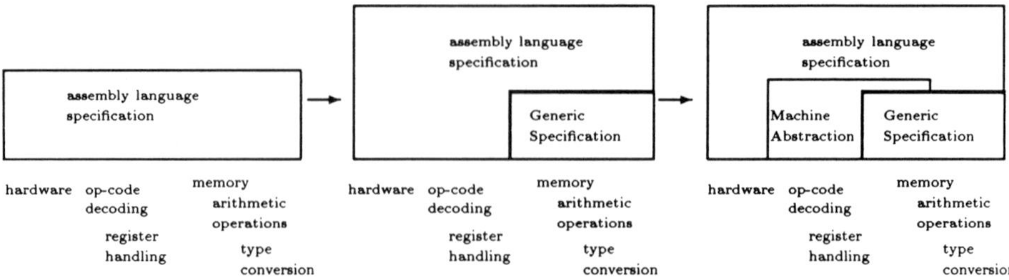

Figure 1: The evolution of the specification hierarchy

- the number or names of registers on a particular microprocessor
- the word sizes to which an instruction may or may not be applied
- the addressing schemes which are permissible for a particular instruction
- the instructions which are permissible in the various execution modes of the processor

For some of these restrictions such as the addressing modes the set of possible values is a characteristic of the assembly language while the allocation of the values to individual instructions is a characteristics of individual processors.

The machine abstraction can also lead to a simplified verification for families of microprocessors. If we can develop a proof at the programming-level for one microprocessor which is based on some assumed theorems for the abstract machine then showing that the instantiation of the machine abstraction for a particular microprocessor of the family satisfies these assumptions is equivalent to performing the same verification proof for the particular microprocessor.

Machine abstractions combined with generic specification parameters, can also play the role of a simple machine model on which the semantics of the assembly language are defined. The set of machine and generic operations is usually small and simple and its elements perform hardware-dependent tasks. Before the assembly language specification can be completed a precise meaning for each of these operations must be known. The semantics of the assembly language are then defined in terms of this smaller set of routines thus providing a semantic definition. The language semantics are independent of the hardware platform (subject to the restrictions on the abstraction functions) and can be studied independently of the microprocessors on which it is implemented. For instance, compiler algorithms which target families of processors could use a processor-independent specification of an assembly language as a compilation target.

4 An Example of Machine Abstraction

The idea of machine abstractions developed in the course of formally specifying a Motorola 68000 microprocessor in HOL. The 68000 microprocessor is the first of a series of microprocessors which share a common kernel instruction set. Machine abstractions arose as a solution to capturing the complexity of the Motorola 68000 family within one specification. Since we were trying to capture the complexity of the whole Motorola 68000 family, the specifications we developed are large and sometimes cryptic. Consequently, we have chosen to illustrate machine abstractions by modifying the assembly language of a simpler microprocessor which is well-known in the HOL community. Our example consists of modifying the assembly language format for the ADD instruction in the Tamarack-3 microprocessor.

Throughout the example we will assume the definition of some data types. The data type names should make their meaning self-evident. A syntactic definition of these types is presented in figure 2 and a description of each function follows:

- bool – a boolean T or F value
- memory – the memory external to the microprocessor
- wordn – a machine word of "n" bits.
- rep_typ – a list of functions which form the generic specification and include the functions
  - address – convert a wordn type to a memory address
  - fetch – retrieve a wordn from external memory at the supplied address
  - store – store a wordn to external memory at the supplied address
  - add – add two wordn numbers
  - inc – add 1 to a wordn number
  - operand – retrieve the operand from a wordn by treating it as an encoded instruction
  - zeroWord – a representation of 0 as a wordn
- regBank – a collection of data registers, each register having type wordn
- regNumber – a collection of identifiers which can be placed in a bijection with the contents of the regBank data type
- mach_typ – a list of functions which form the machine abstraction and include the functions
  - validAddressing – indicates if a pair of source and destination address modes are allowable for a particular instruction
  - getInstArgs – retrieve addressing information from an instruction (where the instruction is given by its program counter memory address)
  - putRegValue – modify the value of a register in a regBank
  - getRegValue – retrieve the value of a register from a RegBank

```
% data types we assume are previously defined:   %
%     memory, wordn, addr, regBank, regNumber    %

addressMode = Immediate | Absolute | RegisterDirect | RegisterIndirect
instruction = ADD | SUB | JRZ | JMP | LDA | STA | NOP | RFI`;;

rep_ty = :(wordn -> addr) #                           %function address%
        (memory # addr -> wordn) #                    %function fetch%
        (memory # addr # wordn -> memory) #           %function store%
        (wordn # wordn -> wordn) #                    %function add%
        (wordn -> wordn) #                            %function inc%
        (wordn -> wordn) #                            %function operand%
        wordn                                         %function zeroWord%

addressInfo = :addressMode # regNumber # wordn

machine_ty =
   :(rep_ty -> instruction # addressInfo # addressInfo -> bool) #
                                                  %function validAddressing%
   (rep_ty -> (memory # addr) -> (addressInfo # addressInfo)) #
                                                      %function getInstArgs%
   (rep_ty -> regBank # regNumber # wordn -> regBank) #
                                                      %function putRegValue%
   (rep_ty -> regBank # regNumber -> wordn)           %function getRegValue%

|- address (rep:rep_ty) = FST(rep)
|- fetch (rep:rep_ty) = FST(SND(rep))
|- store (rep:rep_ty) = FST(SND(SND(rep)))
|- add (rep:rep_ty) = FST(SND(SND(SND(rep))))
|- inc (rep:rep_ty) = FST(SND(SND(SND(SND(rep)))))
|- operand (rep:rep_ty) = FST(SND(SND(SND(SND(SND(rep))))))
|- zeroWord (rep:rep_ty) = SND(SND(SND(SND(SND(SND(rep))))))

|- validAddressing (mach:machine_ty) = FST(mach)
|- getInstArgs (mach:machine_ty) = FST(SND(mach))
|- putRegValue (mach:machine_ty) = FST(SND(SND(mach)))
|- getRegValue (mach:machine_ty) = SND(SND(SND(mach)))
```

Figure 2: Data type definitions for generic specification and machine abstraction.

The "state" of the Tamarack-3 consists of some external memory, a program counter, a single accumulator, a return address register, and an interrupt acknowledge flag. In this example we will be primarily concerned with the external memory, the program counter, and the accumulator. The ADD instruction on the Tamarack-3 takes one memory address as a parameter (absolute memory address), adds the word value located at the indicated address to the contents of the accumulator, and places the result of the addition in the accumulator. The actual addition is performed by an ALU within the microprocessor itself.

A simple description of the basic Tamarack-3 ADD instruction is shown in figure 3. The description of the instruction is very simple and straightforward. It accepts two processor states as input — one state before the instruction and one state after the instruction. The definition returns a boolean value of *True* if the designated *out* state follows from the designated *in* state by performing the ADD instruction.

For this example, we want to extend the Tamarack-3 assembly language to have more than one accumulator and to support address modes beyond absolute addressing. The "new and improved" language will have addressing modes of immediate data, absolute addressing, values from a register, and register indirection.

Our extension of the ADD instruction could be added to the description in figure 3 with special cases throughout the definition. However, our extension of the ADD instruction has not defined how many more accumulator registers exist nor has it defined which new addressing modes are actually applicable to the ADD instruction. These details are left to an embedding microprocessor so our specification of the ADD instruction should not restrict or specify any of these parameters.

A modified description of the ADD instruction is presented in figures 4 and 5. Two routines *getArgument* and *putArgument* are defined to retrieve and store (respectively) data in all the addressing modes and give all instructions access to every addressing mode. These routines are then complemented by the machine abstraction parameter *validAddressing* used in the semantics of ADD itself which restricts which of the addressing modes are applicable to ADD.

The semantics of the ADD instruction are divided into 3 parts. First, *Get2Args* is used to retrieve the arguments for the addition as well as the addressing modes for the arguments. The retrieval process may modify memory or register values if addressing modes such as pre-increment indirection is used so new copies of the memory and registers are also returned. The actual addition is performed in the next stage and then the result is stored according to the addressing mode by a call to *putArgument*. The final "truth" value of the ADD instruction is *True* only if the proposed *out* state matches the computed *out* state *and* all the "restriction" routines for the instruction are satisfied.

We must now address whether or not the new instruction description over-specifies or under-constrains the ADD instruction relative to its extended definition. First, notice how a common routines for retrieving and storing arguments gives the ADD instruction access to all new addressing modes, a condition which was not specified in the extended definition. To counter-balance this generality we use the machine abstraction routine *validAddressing* within the ADD description to ensure that only a restricted set of (unspecified) addressing modes are used by the instruction. An invalid addressing mode would result in the overall definition return value of *False* rather than *True* to indicate that the proposed *out* state

```
|-  ADD_SEM (rep:rep_ty)
      (omem:*memory,opc:*wordn,oacc:*wordn,ortn:*wordn,oiack:bool)
      (mem:*memory,pc:*wordn,acc:*wordn,rtn:*wordn,iack:bool) =

      let inst = (fetch rep) (mem, (address rep) pc) in
      let operand = (fetch rep) (mem, (address rep) inst) in

        (mem, (inc rep) pc, (add rep) (acc,operand), rtn, iack) =
        (omem, opc, oacc, ortn, oiack)
```

Figure 3: HOL specification of the Tamarack-3 ADD instruction

does not logically follow from the *in* state.

Second, the getArgument and putArgument routines access the set of accumulators through machine abstraction interface routines. These routines leave the number of accumulators unspecified until the machine abstraction is instantiated. So, in the uninstantiated version, we see that the semantics of the ADD instruction is independent of the number of registers available. The machine abstraction interface to the accumulators also takes advantage of strong data-typing to ensure a named register is always available. Typically we would want to define a bijection between the regBank and regNumber data types so any register which can be named by a regNumber will have a value in regBank which can be retrieved. Consequently the ADD instruction has access to an arbitrary number of accumulators as arguments and cannot request values from registers which do not exist.

This extended definition of the Tamarack-3 ADD instruction still includes the original definition of the Tamarack-3. Moreover, the original definition is easily retrieved from the description in figures 4 and 5 by instantiating the machine abstraction routines with the routines in figure 6.

4.1 Some Quick Comments on the Motorola 68000

As previously mentioned, the concept of machine abstraction was developed in the course of formally specifying the programming-level of the Motorola 68000. The initial goals of the programming-level specification included applying earlier specification techniques to a large microprocessor while maintaining a specification which could be easily adapted to other microprocessors within the same family. Software engineering techniques such as modularization and code re-use were absolutely necessary for developing a consistent, coherent, and comprehensive specification.

Machine abstraction arose near the first third marker of the specification effort, just as problems relating to scale were appearing. Rather than create custom procedures which handled addressing modes for each individual instruction we began to create routines which could handle all addressing modes and then applied restrictions on a per-instruction basis. Consequently, the addressing mode interpretation code became centralised (which

```
|- putArgument (mach:machine_ty) (rep:rep_ty)
  ((mode:addressMode, regNo:regNumber, data:wordn),
  mem:memory, accs:regBank, newValue:wordn) =

      (mode = Absolute) =>
        (store rep (mem, (address rep) (data), newValue), accs) |
      (mode = RegisterDirect) =>
        (mem, (putRegValue mach) rep (accs, regNo, newValue)) |
      (mode = RegisterIndirect) => ((store rep) (mem,
            (address rep) ((getRegValue mach) rep (accs, regNo)),
        newValue), accs) |
      (mem, accs)

|- getArgument (mach:machine_ty) (rep:rep_ty)
  ((mode:addressMode, regNo:regNumber, data:wordn),
  mem:memory, accs:regBank) =

      (mode = Absolute) => (fetch rep) (mem, (address rep) (data)) |
      (mode = RegisterDirect) =>  (getRegValue mach) rep (accs, regNo) |
      (mode = RegisterIndirect) => (fetch rep) (mem,
        (address rep) ((getRegValue mach) rep (accs, regNo))) |
      data

|- Get2Args (mach:machine_ty) (rep:rep_ty)
  (mem:memory, pc:wordn, accs:regBank) =

  let (sourceInfo, destInfo) =
      (getInstArgs mach) rep (mem, (address rep) pc) in
  let arg1data = getArgument mach rep (sourceInfo, mem, accs) in
  let arg2data = getArgument mach rep (destInfo, mem, accs) in

  (mem, accs, sourceInfo, destInfo, arg1data, arg2data)
```

Figure 4: HOL specification of an extended Tamarack-3 ADD instruction

```
|- ADD_SEM (mach:machine_ty) (rep:rep_ty)
  (omem:memory, opc:wordn, oaccs:regBank)
  (mem:memory, pc:wordn, accs:regBank) =

  let (newMem, newData, sourceInfo, destInfo, arg1data, arg2data) =
    Get2Args mach rep (mem, pc, accs) in

  let newValue = (add rep) (arg1data, arg2data) in
  let (finalMem, finalData) =
    putArgument mach rep (destInfo, newMem, newData, newValue) in

 (
  ((validAddressing mach) rep (ADD, sourceInfo, destInfo)) /\

  ((finalMem, (inc rep) pc, finalData) = (omem, opc, oaccs))
 )
```

Figure 5: HOL specification of an extended Tamarack-3 ADD instruction (continued)

we believe is also true for most hardware implementations) and changing addressing mode restrictions became more flexible. These techniques were then applied to sets of registers, data sizes on which instructions will operate, and the execution mode in which instructions could be executed.

We envisioned a "future" microprocessor in the 68000 family which may have larger word sizes, more registers than the current processors, or more execution modes than the current processors and were pleased to find that our model with its machine abstraction could handle these variations with no change to the programming-level specification. All the changes were isolated in the machine abstraction variable or the generic specification variable.

As of the writing of this paper, the programming-level specification of the Motorola 68000 includes approximately 80 percent of the user mode and supervisor mode instructions.

5 Advantages

The benefits of machine abstractions arise in three domains. First, machine abstractions introduce a degree of portability to programming-level specifications. Areas of the specification which are constrained by the hardware implementation are isolated and the behaviour of these hardware dependencies rather than their implementation characterise the proof. As these "parameters" vary between microprocessors their impact on a common proof is restricted by the constraints placed on the proof's parameters. Also, as a specification is migrated from one microprocessor to another, only machine abstraction instantiations need to be created (in HOL) before the specification is usable for the new

```
define_type `regNumber_DEF` `regNumber = accumulator`;;
new_type_abbrev(`regBank`, ":wordn");;

|- tamValidAddressing (rep:rep_ty) (instructName:instruction,
(sourceMode:addressMode, sourceReg:regNumber, sourceAddr:wordn),
(destMode:addressMode, destReg:regNumber, destAddr:wordn)) =

    (((instructName = ADD) \/
     (instructName = SUB) ) =>
         ((sourceMode = Absolute) /\ (destMode = RegisterDirect)) |
    ((instructName = JRZ) \/
     (instructName = JMP) \/
     (instructName = LDA) \/
     (instructName = STA) ) => (sourceMode = Absolute) |
    T)


|- tamGetInstArgs (rep:rep_ty) (mem:memory, pc:addr) =
    ((Absolute,accumulator,(operand rep)((fetch rep) (mem, pc))),
        (RegisterDirect,accumulator,(zeroWord rep)))

|- tamPutRegValue (rep:rep_ty)
    (accum:regBank, regName:regNumber, data:wordn) =
        data

|- tamGetRegValue (rep:rep_ty)
    (accum:regBank, regName:regNumber) =
        accum
```

Figure 6: Machine abstraction instantiation for the Tamarack-3 microprocessor

hardware platform.

Second, machine abstractions separate the structure of the assembly language from the restrictions made by the microprocessor implementation. For instance, the semantics of the assembly language may include a set of different execution modes in which the language can operate in addition to the set of instructions. However, the language may leave the decision of which instructions belong to which execution mode to be a function of the embedding processor. With a machine abstraction, the execution mode becomes a component of the microprocessor's state and the assignment between modes and instructions is handled with a restriction predicate similar to the addressing modes in the previous Tamarack-3 ADD example.

Finally, machine abstractions encourage the prover to develop verification theorems for the assembly language rather than for an individual microprocessor. Theorems for the language can then be applied to each microprocessor of a family which shares a base instruction set. The programming-level verification for the microprocessors becomes an instantiation of the machine abstraction and proving the instantiations satisfy the assumptions of the theorems for the assembly language. This strategy will hopefully reduce the verification effort for a family of microprocessors once one of its members has been verified.

6 Disadvantages

Disadvantages of machine abstraction arise in both the specification description of microprocessors and the definition of the abstractions themselves. Microprocessor specifications become more verbose when machine abstractions are included since the specification is trying to capture concepts and complexities which normally appear in a restricted form for a single microprocessor. Machine abstractions also introduce general concepts into specifications which must be accompanied by corresponding "restriction" routines which focus the specification to one microprocessor instance. Incorporating these generalizations into a specification and creating the "restriction" routines require additional work and forethought on the specifier's part.

The choice of components from a microprocessor which are included into a machine abstraction is neither clear nor unambiguous. The selection requires some knowledge of the hardware configuration for other processors in the same family or some vision of the direction in which the processor family is heading. Even with guidelines from the family, we must still disentangle assumptions of the assembly language from the language semantics and from the restrictions of a single microprocessor.

In closing, the concepts of machine abstraction has not been drawn through any verification yet. The usefulness of the technique will depend on the quantity of complexity versus benefits it yields during a full hardware verification.

7 Application Areas

Machine abstractions were developed with two purposes in mind: first and foremost, to permit the verification of a family of processors and second, to provide an operational-like semantics for assembly languages. In the verification aspect we can use the specification of a family to boot-strap the specification and verification of a new microprocessor for the family, hopefully prior to its fabrication. Moreover, machine abstractions with the possible addition of one extra abstraction level, may allow specifications to be shared between families of microprocessors since many basic operations are common to most assembly languages.

In the domain of language semantics, machine abstractions can assist the verification of machine-coded algorithms which are intended for any one of the processors from a family. Boyer and Yu have already demonstrated the feasibility of proving machine-coded algorithms for microprocessors as complex as the Motorola 68020 [BY92].

Many microprocessor families require that existing software for the family be forward compatible to a new processor. In this respect, Boyer and Yu's work on the Motorola 68020 apply equally well to the 68030 and 68040 microprocessors but no conclusions can be made about the 68000 or 68010 microprocessors of the same family.

We hope that machine abstractions will permit similar machine-code proofs at the assembly language level which facilitate proofs of compatibility with prior generations of microprocessors (such as a Motorola 68020 program works equally well, under certain restrictions, on a Motorola 68000.)

8 Conclusions

Machine abstractions can be seen as a natural extension of generic specification as we escalate specifications from single processors to families of processors. In such instances, masking hardware differences within the family provides a unified "abstract machine" view from which an assembly language can be described. This unified view can also be used as a basis for defining the semantics of the assembly language. The programming-level specification (for verification) can be re-used as a study base for the language itself.

The concepts of machine abstraction was developed as we created a programming-level specification in HOL for the Motorola 68000 microprocessor. Within this context, machine abstraction relies heavily on the "higher order" capabilities of HOL. In particular, the machine abstractions we use are functions which are passed as parameters to other functions. Two other aspects of HOL which are crucial for machine abstraction are strong type-checking and enumerated types. The enumerated types permit us to define index sets for other sets (such as sets of registers) which are both inclusive and exhaustive. The strong type-checking ensure a consistent usage of functions throughout the specification and often identified errors in our initial specification attempts.

Areas which we hope to investigate include re-using specifications between families of processors where the semantics of their assembly languages overlap, developing verification strategies which apply to a machine abstracted specification, instantiating these general proofs for individual processors of a family, and using the programming-level specifications to verify the execution of assembly language algorithms.

9 Acknowledgements

The author was supported by a post-graduate scholarship from the National Sciences and Engineering Research Council of Canada while this work was developed. He extends his thanks to Stephan Mueller for his invaluable assistance with the programming-level specification of the Motorola 68000 microprocessor and to various members of the Integrated Systems Design group at the University of British Columbia for many discussions and comments.

References

[BY92] B. Boyer and Y. Yu. A formal specification of some user mode instructions for the motorola 68020. Technical Report TR-92-04, University of Texas, Austin, Texas 78712-1188, February 1992.

[Cur] P. Curzon. Generic specification of the viper microprocessor. Unpublished notes.

[G+85] Michael J. C. Gordon et al. *The HOL System Description.* Cambridge Research Centre, SRI International, 1985.

[Her91] John M.J. Herbert. Incremental design and formal verification of microcoded microprocessors. Technical report, SRI International, 1991.

[Joy89] Jeffrey J. Joyce. Multi-level verification of microprocessor-based systems. Ph.D. Thesis 195, Computer Laboratory, University of Cambridge, December 1989.

[Plo81] Gordon D. Plotkin. A structural approach to operational semantics. Technical report, Aarhus University, September 1981.

[Win90] P.J. Windley. The formal verrification of generic interpreters. Ph.D. Thesis, Research Report CS-90-01, University of California, Davis, 1990.

Chapter 4

Formalizing and Modelling of Automata

Higher Order Logic Theorem Proving and its Applications (A-20)
L.J.M. Claesen and M.J.C. Gordon (Editors)
Elsevier Science Publishers B.V. (North-Holland)

A Formal Theory of Simulations Between Infinite Automata

Paul Loewenstein

Sun Microsystems Inc., Mountain View, California, USA

Abstract

Automata are suitable for modelling a wide range of sequential and concurrent hardware. They can be used at many levels of abstraction, from top-level specifications to register transfer descriptions suitable for input to synthesis tools. This paper covers approaches for relating one automaton with another using simulations and tranformations on automata. The entire theory is mechanically derived in, and intended for use in a higher-order logic theorem prover. Because automaton-based models can be used at multiple abstraction levels, much of the formal verification of sequential and concurrent designs can be performed by composing and relating automata.

Keyword Codes: F.1.1; I.2.3
Keywords: Models of Computation; Deduction and Theorem Proving

1 Introduction

Design of sequential and especially concurrent systems poses severe challenges to the designer. In the absence of a rigorous proof of correctness, design errors are almost impossible to avoid.

One approach to the modelling of sequential and concurrent systems is to use automata. Automaton models can be used to abstract away the notion of an infinite time-line, and replace it with a notion of "now" and "next". This can simplify proofs considerably.

Much verification work is performed using mechanical theorem provers, such as HOL[7] and Boyer-Moore[3]. This results of this paper allow the integration of automata-theoretic reasoning into the HOL theorem prover. The theory is developed formally, by definition of automata using the built-in theories of HOL, and subsequent proof of theorems demonstrating relationships between automata.

The theory presented in this paper differs from much published material in that the cardinality of the state-space is not restricted. When deriving theories by formal proof one attempts to state as few preconditions as possible; for most theorems derived here, finiteness or countability of the state space is not required. Allowing infinite state-spaces allows the expression of liveness and fairness without denoting "final states" and special acceptance conditions as for Büchi, Rabin and Muller automata[4, 16, 15]. The unrestricted state-space cardinality causes some familiar "theorems" to no longer be theorems. For example, the subset construction does not preserve behaviour of infinite automata.

This paper is not concerned with decidability of particular problems[4, 16, 15]; it is more concerned with formally justifying soundness. In particular, the paper is concerned with relating the languages accepted by different automata.

To some extent this paper is overkill. For formal verification, all that is strictly required are the soundness theorems; however, some completeness theorems are also presented.

1.1 Related work

Early on we formally derive the use of *simulation relations*, a concept attributed to Milner[14].

Abadi and Lamport[1] demonstrate the existence of "refinement mappings" from an implementation modified with "history variables" and "prophecy variables" to a finitely non-deterministic specification. The results presented in this paper do not require these modifications to the implementation automaton.

The latter part of the paper owes much to the work of Klarlund and Schneider[8], which introduces a more versatile relation between countable-state automata. The results in this paper are not restricted to countable state spaces. More significantly, this paper extends the Klarlund-Schneider soundness results to include some specification automata with infinite non-determinism. It also extends the completeness results.

1.2 Presentation of Proofs

All theorems in this paper have been developed formally using the HOL theorem prover. One of the joys of mechanical theorem proving is that nobody need ever read a complete formal proof — not even the proof developer. The proofs are therefore presented in outline, in a very similar style typically used for informal proofs. For those readers who wish to examine the proof in more detail, both HOL and the input files are available by ftp.[1] By running the proofs interactively, line by line, the steps can be seen as the proof developer sees them. Each line, however, can correspond to hundreds of steps in a fully expanded formal proof.

1.3 Rest of Paper

The paper starts by introducing the HOL logic in sufficient detail for understanding the paper. The automaton models are then introduced. Simulation relations are then introduced, and their sufficiency for demonstrating that one automaton implements another is shown. The concepts of safety and liveness properties are then introduced, and it is shown that safety properties correspond to those properties definable by deterministic automata. The paper then defines three closely related constructions on automata; one of which is the familiar "subset" construction, which is not in general behaviour-preserving. These constructions are used to derive more widely applicable extensions to simulation relations. Some completeness results are also presented. A proof of the well-known König's lemma, which is key to some of the results, is deferred to an appendix.

2 The HOL logic

The logic used in this paper is the logic of the HOL theorem-prover ("Higher-Order Logic")[7, 6], typographically embellished to improve readability. HOL uses a very expressive logic, in which variables can range over functions as well as values (unlike first-order logic).

HOL logic reads very much like ordinary predicate logic with a few differences. Variables and constants can range over functions or values, depending on their type. Types are left implicit

[1] It is available with the HOL90 sources from fsa.cpsc.ucalgary.ca.

in this paper; they are explained in the text unless they are obvious from context. Function application is indicated by juxtaposition: $f\,x$ denotes function f applied to x. Often multiple argument functions are *curried*; $g\,x\,y$ denotes a function g applied to x, the result (again a function) is then applied to y. Informally one can consider g to be merely a multiple argument function. The (x, y) notation is used for pairing (cartesian product). When pairs are used as function arguments, as in $f(x, y)$, the notation appears conventional. Sometimes, (as in this paper), currying and pairing are mixed, as in $g(u, v)(x, y)$.

Proven theorems are indicated by the turnstile symbol $\vdash$. These have been derived using the mechanical theorem prover, HOL[7], and should be correct.[2] Defined constants are printed in sans serif typeface. Formally, definitions look just like theorems, and also carry the turnstile.

Sets are represented in HOL as *types*. In this paper we shall be concerned with booleans $\{\mathsf{T}, \mathsf{F}\}$, the natural numbers $\{0, 1, 2, \ldots\}$. We shall also be implicitly using the function type $(\rightarrow)$ and cartesian products. Not all types have to be explicit; arbitrary non-empty well-ordered sets can be represented using *type variables* $(\alpha, \beta, \gamma, \ldots)$. Because functions are primitive in HOL, subsets (of the set represented by a type) are represented by their characteristic function. $x \in A$ is then $A\,x$. $A \subseteq B$ is indicated by: $\forall x. A\,x \Rightarrow B\,x$. Although we shall occasionally refer to "sets" in the text, they will be denoted by types or predicates in HOL.

HOL functions are total. This leads to some unconventional definitions of usually partial functions, such as natural number subtraction; $a - b = 0$ if $a < b$.

Infinite sequences of values are denoted as functions from a natural number to the value at each instant. We have adopted some special conventions for describing sequences and automata. Variables denoting sequences are emboldened $(\boldsymbol{e})$, as opposed to other variables (e). Often sequences are sequences of tuples, and it is convenient to define a "zipping" operator $\circledcirc$ to join two sequences to form a sequence of tuples:

$$\vdash \forall t. (\boldsymbol{a} \circledcirc \boldsymbol{b})t = (\boldsymbol{a}\,t, \boldsymbol{b}\,t)$$

We shall also use finite sequences, which will usually be prefixes of infinite sequences. Rather than define the concept of a finite sequence directly, and the operators to concatenate finite and infinite sequences, we instead refer to a finite number of elements of an infinite sequence. This allows fewer definitions, simpler formulæ and less tedious reasoning.

Most defined constants are given alphabetic names; these are typeset in sans-serif, such as in Powerset.

3 The Automaton Model

For the definition of a (potentially) infinite automaton we shall initially use a traditional set-theoretic formulation, then introduce the equivalent HOL definition, and then define the acceptance conditions, both informally and in HOL.

Definition 1. A set-theoretic definition of an infinite automaton is a four-tuple (α, β, Q, N)[3] where:

- α is the set of externally visible states.

[2] In the unfortunate circumstance that any errors are found in the formal theorems, they would almost certainly have arisen during typographic transcription.

[3] I have broken with tradition and used minuscule greek characters to denote sets; this is for compatibility with the use of minuscule greek characters for HOL types.

- β is the set of internal, or hidden states.
- $Q \subseteq \alpha \times \beta$ is a set of initial states.
- N is a transition function from $\alpha \times \beta$ to $2^{\alpha\times\beta}$. This can also be considered a transition relation between the current state $s \in \alpha \times \beta$ and the next state $s' \in \alpha \times \beta$.

In HOL, the automaton four-tuple becomes a pair:

$$(Q{:}(\alpha{\times}\beta){\rightarrow}bool, N{:}(\alpha{\times}\beta){\rightarrow}(\alpha{\times}\beta){\rightarrow}bool)$$

The sets of visible and internal states have been denoted in HOL by *type variables*. We have also chosen to represent N in curried form.

The automaton can also be viewed as a state-transition digraph, with a node for each element of $\alpha \times \beta$, with edges denoting allowed transitions. The α component can be viewed as a *label* on each state. It is the sequence of labels which is visible to the outside world; the β component exists to distinguish between states with the same label.

Definition 2. A sequence $e_0, e_1, \ldots$ is an *infinite trace* or *trace* of an automaton (Q, N) if there exists a state sequence, $s_0, s_1, \ldots$ such that:

- $(e_0, s_0) \in Q$ and
- For all n, $(e_{n+1}, s_{n+1}) \in N(e_n, s_n)$.

In HOL this can be defined as a predicate Trace:

$$\vdash \forall Q N e.\, \mathsf{Trace}(Q, N)e = (\exists s.Q(e\,0, s\,0) \wedge (\forall t.N(e\,t, s\,t)(e(\mathsf{Suc}\,t), s(\mathsf{Suc}\,t))))$$

The predicate $\mathsf{Trace}(Q, N)$ can be viewed in several equivalent ways:

- The set of traces of the automaton.
- The *language* accepted by the automaton.
- Describing a *property* of a sequence.
- Defining the externally visible *behaviour* of the automaton.

The term "property" is used in section 5, which deals with safety and liveness. In hardware design, one typically assembles some components, each with its own behaviour, and then one derives the behaviour of the ensemble. For this paper "behaviour" and "property" are therefore synonymous, and for the case of automata, these two terms are also synonymous with "trace set".

3.1 Choice of automaton model for hardware modelling

The automaton as defined in this paper is a *labelled state automaton* rather than the more conventional *labelled transition automaton*. We define an automaton in this way for several reasons:

- Because the transition relation contains both the current and next labels, it allows the straightforward definition of stuttering automata (section 3.5) which are driven by externally visible transitions, rather than by the instantaneous values.

- Labelled transition automata can be easily represented as labelled state automata; the reverse operation is less straightforward.
- Because both the externally visible label and the internal state are treated similarly, the operation of hiding (transferring a label component to internal state) is clean and straightforward.

3.2 Universality of infinite automata

Lemma 1. *Infinite automata can specify any behaviour:*

$$\vdash \forall P.\exists Q N.P = \mathsf{Trace}(Q, N)$$

Proof. By construction of Q and N from P. The internal state space of the automaton ranges over infinite sequences; by selecting $Q(e, s) = P\, s$ the automaton initially selects some arbitrary sequence satisfying P. By choosing

$$N(e, s)(e', s') = (e = s\, 0) \wedge (\forall t.s'\, t = s(\mathsf{Suc}\, t))$$

the trace is then equal to that sequence. □

This "universal construction" is useless for the purposes of relating one automaton to another; we shall, however, be referring to lemma 1 to demonstrate some properties of constructions on automata in section 7.

3.3 Prefix and limit traces

Definition 3. A *prefix trace* is a finite sequence that is compatible with the initial state and transition relation:

$$\begin{aligned}&\vdash \forall n Q N e.\, \mathsf{Prefix\text{-}Trace}\, n(Q, N)e = \\ &\quad (\exists s.\, Q(e\, 0, s\, 0) \wedge (\forall t.t < n \Rightarrow N(e\, t, s\, t)(e(\mathsf{Suc}\, t), s(\mathsf{Suc}\, t))))\end{aligned}$$

where we have represented a finite prefix of length n using the first n elements of an infinite sequence s.

Definition 4. A *limit trace* is an infinite sequence of which every finite prefix is a prefix trace:

$$\vdash \forall Q N e.\, \mathsf{Limit\text{-}Trace}(Q, N)e = (\forall n.\, \mathsf{Prefix\text{-}Trace}\, n(Q, N)e)$$

The language defined by Trace in definition 2 will be referred to as the *infinite behaviour* when necessary to distinguish it from the prefix and prefix-limit behaviours.

Lemma 2. *The infinite traces are limit traces:*

$$\vdash \forall Q N e.\, \mathsf{Trace}(Q, N)e \Rightarrow \mathsf{Limit\text{-}Trace}(Q, N)e$$

Proof. All prefixes of any sequence accepted by $\mathsf{Trace}(Q, N)$ satisfy the initial conditions and transition relation. They are therefore accepted by $\mathsf{Limit\text{-}Trace}(Q, N)$. □

The converse does not in general hold, as will be seen later in section 7.

3.4 Deterministic automata

Definition 5. An automaton which has a unique initial internal state for every initial visible state, and for each state, has a unique internal next state for every visible next state, is *deterministic*:

$$\vdash \forall Q N.\, \mathsf{Deterministic}(Q, N) =$$
$$(\forall e s s'. Q(e, s) \wedge Q(e, s') \Rightarrow (s = s')) \wedge (\forall e s e' s' s''. N(e, s)(e', s') \wedge N(e, s)(e', s'') \Rightarrow (s' = s''))$$

As will be seen in section 6, deterministic automata are less expressive than non-deterministic automata.

Theorem 1. *For deterministic automata, every limit trace is an infinite trace:*

$$\vdash \forall Q N.\, \mathsf{Deterministic}(Q, N) \Rightarrow (\forall \mathbf{e}.\, \mathsf{Limit\text{-}Trace}(Q, N)\mathbf{e} \Rightarrow \mathsf{Trace}(Q, N)\mathbf{e})$$

Proof. For any given limit trace $\mathbf{e}$, we have an infinite array of prefix traces, each defining a unique finite sequence of states $\mathbf{s}_n\, 0, \mathbf{s}_n\, 0, \ldots, \mathbf{s}_n\, n$.

$$\begin{array}{lllll} \mathbf{s}_0\, 0 & & & & \\ \mathbf{s}_1\, 0 & \mathbf{s}_1\, 1 & & & \\ \mathbf{s}_2\, 0 & \mathbf{s}_2\, 1 & \mathbf{s}_2\, 2 & & \\ & \vdots & & \ddots & \\ \mathbf{s}_n\, 0 & \mathbf{s}_n\, 1 & \mathbf{s}_n\, 2 & \ldots & \mathbf{s}_n\, n \\ & \vdots & & & \ddots \end{array}$$

Because the automaton is deterministic, the initial states and next states are unique, so, if $t \leq m$ and $t \leq n$, then $\mathbf{s}_m\, t = \mathbf{s}_n\, t$. The $\mathbf{s}_n\, t$ in each column t are equal to each other. Because each row satisfies the initial condition Q, and the transition relation N, and the values in the columns are the same, the diagonal $\mathbf{s}_0\, 0, \mathbf{s}_1\, 1, \mathbf{s}_2\, 2, \ldots$ also satisfies the initial condition and the transition relation. The diagonal is also an infinite sequence and therefore the $\mathbf{e}$ accepted by $\mathsf{Limit\text{-}Trace}(Q, N)$ is also accepted by $\mathsf{Trace}(Q, N)$. □

Definition 6. A state is *reachable* if it is either an initial state, or can be reached from the initial state via a finite number of transitions:

$$\vdash \forall Q N e' s'.\, \mathsf{Reachable}(Q, N)(e', s') = (\exists n \mathbf{e} \mathbf{s}.$$
$$(\mathbf{e}\, n = e') \wedge (\mathbf{s}\, n = s') \wedge Q(\mathbf{e}\, 0, \mathbf{s}\, 0) \wedge (\forall t. t < n \Rightarrow N(\mathbf{e}\, t, \mathbf{s}\, t)(\mathbf{e}(\mathsf{Suc}\, t), \mathbf{s}(\mathsf{Suc}\, t))))$$

Definition 7. An automaton has *no dead states* if there is an infinite path out of every reachable state:

$$\vdash \forall Q N.\, \mathsf{No\text{-}Dead}(Q, N) = (\forall e' s'.\, \mathsf{Reachable}(Q, N)(e', s') \Rightarrow$$
$$(\exists \mathbf{e} \mathbf{s}.(\mathbf{e}\, 0 = e') \wedge (\mathbf{s}\, 0 = s') \wedge (\forall t. N(\mathbf{e}\, t, \mathbf{s}\, t)(\mathbf{e}(\mathsf{Suc}\, t), \mathbf{s}(\mathsf{Suc}\, t)))))$$

3.5 Stuttering automata

Definition 8. A *stuttering automaton* is an automaton in which every reachable state has a self loop:

$$\forall Q N.\, \mathsf{Stuttering}(Q, N) = (\forall e s.\, \mathsf{Reachable}(Q, N) \Rightarrow N(e, s)(e, s))$$

The trace set of a stuttering automaton is closed under stuttering (repetition of trace values). This class of automata is particularly useful for modelling asynchronous systems[11, 12]. Because no component is forced to advance its state at any particular time, "lock-step" composition can be used for asynchronous composition. Liveness properties, if any, would normally be specified separately[1].[4] Clearly, stuttering automata have no dead states.

3.6 Finitely non-deterministic automata

We start with a formal definition of *finiteness* of a predicate (set):

Definition 9. A predicate (set) P is *finite* if there exists a bound b, and a function f which maps from natural numbers less than b to every value $x \in P$:

$$\vdash \forall P. \mathsf{Finite}\, P = (\exists b f. \forall x. P\, x \Rightarrow (\exists n. n < b \wedge (x = f\, n)))$$

Definition 10. A *finitely non-deterministic* automaton has a finite number of possible initial states for every initial visible state, and a finite number of next states for every current state and visible next states:

$$\vdash \forall Q N. \mathsf{Fin\text{-}Non\text{-}Det}(Q, N) = (\forall e. \mathsf{Finite}(\lambda s. Q(e, s))) \wedge (\forall e e' s. \mathsf{Finite}(\lambda s'. N(e, s)(e', s'))$$

4 Simulations between Automata

One of the most common requirements in hardware verification is to demonstrate that the behaviour of an implementation is contained in a specification. If both the implementation and specification are expressed as automata, then we can often use a *simulation relation* to demonstrate that one automaton implements another.

4.1 Relating two automata

One automaton *implements* another if every trace of the first is an trace of the second. A powerful method for showing that this relation holds is to find a *simulation relation* between them.

Definition 11. One automaton (implementation state graph) *simulates* another automaton (specification state graph) if there exists a simulation relation between the implementation and specification graphs. A simulation relation is a three-way relation $R\, e\, s_1\, s_2$, where (e, s_1) is a state of the implementation graph and (e, s_2) is a state of the specification graph (the visible components must be the same in both states). It must satisfy two properties: first, every initial state of the implementation is related by R to some initial state of the specification; and, second, whenever $R\, e\, s_1\, s_2$ and $N_1(e, s_1)(e', s_1')$ both hold, there exists some specification state (e', s_2') such that $N_2(e, s_2)(e', s_2')$ and $R\, e'\, s_1'\, s_2'$ both hold:

$$\begin{aligned}
&\vdash \forall Q_1 N_1 Q_2 N_2. (Q_1, N_1)\, \mathsf{Simulates}(Q_2, N_2) = (\exists R. \\
&\quad (\forall e s_1. Q_1(e, s_1) \Rightarrow (\exists s_2. Q_2(e, s_2) \wedge R\, e\, s_1\, s_2)) \wedge \\
&\quad (\forall e e' s_1 s_1' s_2. R\, e\, s_1\, s_2 \wedge N_1(e, s_1)(e', s_1') \Rightarrow (\exists s_2'. R\, e'\, s_1'\, s_2' \wedge N_2(e, s_2)(e', s_2'))))
\end{aligned}$$

[4] As shown in section 5, stuttering automata can express liveness properties.

Theorem 2. *To show that one automaton is an implementation of another, it is sufficient to produce a simulation relation from the first to the second:*

$$\vdash \forall Q_1 Q_2 N_1 N_2.(Q_1, N_1)\,\mathsf{Simulates}(Q_2, N_2) \Rightarrow (\forall \boldsymbol{e}.\,\mathsf{Trace}(Q_1, N_1)\boldsymbol{e} \Rightarrow \mathsf{Trace}(Q_2, N_2)\boldsymbol{e})$$

Proof. It is sufficient to construct, for a given $\boldsymbol{e}$, a state sequence $\boldsymbol{s}_2$ of automaton 2 from a state sequence $\boldsymbol{s}_1$ of automaton 1, where $\forall t.R(\boldsymbol{e}\,t)(\boldsymbol{s}_1\,t)(\boldsymbol{s}_2\,t)$. We pick some $\boldsymbol{s}_2\,0$ related to $\boldsymbol{s}_1\,0$ using:

$$\forall e s_1.Q_1(e, s_1) \Rightarrow (\exists s_2.Q_2(e, s_2) \wedge R\,e\,s_1\,s_2)$$

From $\boldsymbol{s}_1\,t$,$\boldsymbol{s}_1(\mathsf{Suc}\,t)$ and $\boldsymbol{s}_2\,t$, we pick some $\boldsymbol{s}_2(\mathsf{Suc}\,t)$ using:

$$\forall e e' s_1 s_1' s_2.R\,e\,s_1\,s_2 \wedge N_1(e, s_1)(e', s_1') \Rightarrow (\exists s_2'.R\,e'\,s_1'\,s_2' \wedge N_2(e, s_2)(e', s_2'))$$

The core of the formal proof is induction on time. The proof is complicated by having to find an existential witness for $\boldsymbol{s}_2$, before performing the induction. This involves the skolemisation of the existentially quantified s_2' to provide a function with which to recursively define $\boldsymbol{s}_2$. Also, before performing the induction, the goal is strengthened with $\forall t.R(\boldsymbol{e}\,t)(\boldsymbol{s}_1\,t)(\boldsymbol{s}_2\,t)$. □

Theorem 2 is the mainstay of much sequential and concurrent verification, including serial-parallel multipliers[10], single and multi-processor caches[11, 12]. Informally, the simulation relation R indicates how the implementation state models the specification state.

Lemma 3. *To show that every prefix trace of automaton 1 is also a prefix trace of automaton 2, it is also sufficient to supply a simulation relation:*

$$\begin{aligned}&\vdash \forall Q_1 Q_2 N_1 N_2.(Q_1, N_1)\,\mathsf{Simulates}(Q_2, N_2) \Rightarrow\\ &\quad(\forall \boldsymbol{e} n.\,\mathsf{Prefix\text{-}Trace}\,n(Q_1, N_1)\boldsymbol{e} \Rightarrow \mathsf{Prefix\text{-}Trace}\,n(Q_2, N_2)\boldsymbol{e})\end{aligned}$$

Proof. One can proceed along similar lines to the proof of theorem 2. However it is more elegant to perform induction over n, the length of the prefix.[5] This avoids any need to skolemise, since the induction can be performed before any existential witness for $\boldsymbol{s}_2$ need be found. The rest of the proof is along similar lines to that of theorem 2. □

Lemma 4. *To show that every limit trace of automaton 1 is also a limit trace of automaton 2, it is also sufficient to supply a simulation relation:*

$$\begin{aligned}&\vdash \forall Q_1 Q_2 N_1 N_2.(Q_1, N_1)\,\mathsf{Simulates}(Q_2, N_2) \Rightarrow\\ &\quad(\forall \boldsymbol{e}.\,\mathsf{Limit\text{-}Trace}(Q_1, N_1)\boldsymbol{e} \Rightarrow \mathsf{Limit\text{-}Trace}(Q_2, N_2)\boldsymbol{e})\end{aligned}$$

Proof. Trivially follows from lemma 3. □

4.2 Completeness of simulation

As will be seen in the examples (section 9.3), one automaton can implement another without its simulating the other. However, simulation of a determinitic automaton by an automaton with no dead states is complete.

Theorem 3. *If an automaton with no dead states implements a deterministic automaton then one also simulates the other.*

$$\begin{aligned}&\vdash \forall Q_1 Q_2 N_1 N_2.\,\mathsf{No\text{-}Dead}(Q_1, N_1) \wedge \mathsf{Deterministic}(Q_2, N_2) \Rightarrow\\ &\quad(\forall \boldsymbol{e}.\,\mathsf{Trace}(Q_1, N_1)\boldsymbol{e} \Rightarrow \mathsf{Trace}(Q_2, N_2)\boldsymbol{e}) \Rightarrow (Q_1, N_1)\,\mathsf{Simulates}(Q_2, N_2)\end{aligned}$$

Proof. We choose an $R\,e\,s_1\,s_2$ specifying that if (e, s_1) is reachable via a finite prefix of some $\boldsymbol{e}$, then (e, s_2) must be reachable via the same prefix. The no dead state condition ensures that there is an infinite extension of that prefix which is accepted by automaton 1. Because automaton 1 implements automaton 2, then the same extension must be accepted by automaton 2. The initial state component of definition 11 is satisfied straightforwardly. We then proceed to the next-state component, where we are given that $R\,e\,s_1\,s_2$ holds. We then use a similar argument to find an s_2' which satisfies $R\,e'\,s_1'\,s_2'$. It remains to show that $N(e, s_2)(e', s_2')$ also holds. The key to this is to use definition 5 to show that the $\boldsymbol{s}_2\,n$ is unique for a given n-prefix of $\boldsymbol{e}$, so we can use the same state sequence as we used for satisfying $R\,e'\,s_1'\,s_2'$. □

5 Safety Properties

To save space we start by defining the following operator:

Definition 12. $\mathsf{Safe}\,P$ defines the infinite sequences of which all finite prefixes have infinite extensions satisfying P:

$$\vdash \forall P \boldsymbol{w}.\, \mathsf{Safe}\,P\,\boldsymbol{w} = (\forall n. \exists \boldsymbol{v}.(\forall t. t < n \Rightarrow (\boldsymbol{v}t = \boldsymbol{w}t)) \wedge P\,\boldsymbol{v}))$$

Informally, a Safety property on an infinite sequence states that "something bad *never* happens". By contrast, a liveness property states that "something good *eventually* happens". As we shall see later, many classes of automata can express only safety properties.

Alpern and Schneider[2] provide a precise notion of a safety property:

Definition 13. A property P is a *safety property* if and only if every infinite sequence of events $\boldsymbol{w} \notin P$ has some finite prefix $\boldsymbol{v}$ such that no infinite extension of $\boldsymbol{v}$ exists which makes P hold.

The formal version is based on the contrapositive of the above definition.

$$\vdash \forall P.\, \mathsf{Safety}\,P = (\forall \boldsymbol{w}.\, \mathsf{Safe}\,P\boldsymbol{w} \Rightarrow P\,\boldsymbol{w})$$

I shall state the following two lemmas without their straightforward proofs:

Lemma 5. $\mathsf{Safe}\,P$ *is a safety property:*

$$\vdash \forall P.\, \mathsf{Safety}(\mathsf{Safe}\,P)$$

Lemma 6. $\mathsf{Safe}\,P$ *is the smallest safety property containing* P*:*

$$\vdash \forall PQ.\, \mathsf{Safety}\,Q \wedge (\forall \boldsymbol{w}.P\,\boldsymbol{w} \Rightarrow Q\,\boldsymbol{w}) \Rightarrow (\forall \boldsymbol{w}.\, \mathsf{Safe}\,P\,\boldsymbol{w} \Rightarrow Q\,\boldsymbol{w})$$

The next lemma is illustrative, and is needed for later rigour:

Lemma 7. *Some properties are not safety properties:*

$$\vdash \exists P. \neg\, \mathsf{Safety}\,P$$

Proof. Let P be $\lambda \boldsymbol{w}.\exists t.\boldsymbol{w}\,t$. Then $\lambda t.\,\mathsf{F}$ is not in P. However all finite prefixes of $\lambda t.\,\mathsf{F}$ can be extended to be in P by having $\boldsymbol{w}\,t$ hold for some t greater than the length of the prefix. □

6 Automata and Safety Properties

Theorem 4. *Prefix-limit automata express safety properties:*

$$\vdash \forall Q N.\, \mathsf{Safety}(\mathsf{Limit\text{-}Trace}(Q, N)))$$

Proof. Because Limit-Trace defines a behaviour over prefixes, then for each prefix length n', we can, by choosing n' for the n in definition 13, show that the property satisfies definition 13. □

Lemma 8. *Deterministic automata express safety properties:*

$$\vdash \forall Q N.\, \mathsf{Deterministic}(Q, N) \Rightarrow \mathsf{Safety}(\mathsf{Trace}(Q, N)))$$

Proof. By theorems 1 and 4. □

Theorem 5. *Deterministic automata are less expressive than non-deterministic automata.*

Proof. Non-deterministic automata can express any property by lemma 1. Deterministic automata can only express safety properties by lemma 8. Since not all properties are safety properties (lemma 7), deterministic automata are less expressive than non-deterministic automata. □

Theorem 6. *Safety properties are precisely those definable by deterministic automata:*

$$\vdash \forall P.\, \mathsf{Safety}\, P = (\exists Q N.\, \mathsf{Deterministic}(Q, N) \wedge (P = \mathsf{Trace}(Q, N)))$$

Proof. [5] Every deterministic automaton expresses a safety property by lemma 8, so it remains to construct a deterministic automaton satisfying safety property P. The initial state is the empty prefix. The nth state is the n-element prefix of $\boldsymbol{e}$. It advances to a next state (extends the prefix) provided there is an infinite extension of the extended prefix that satisfies P. Provided P is a safety property, by definition 13, we shall eventually reach a point at which any $\boldsymbol{e} \notin P$ is rejected. Conversely, any $\boldsymbol{e} \in P$ is accepted. This is clearly a deterministic automaton whose behaviour is P. □

Lemma 9. *Prefix-limit behaviour of an automaton does not in general describe all safety properties of the infinite behaviour:*

$$\vdash \exists Q N. \neg(\forall \boldsymbol{e}.\, \mathsf{Limit\text{-}Trace}(Q, N)\boldsymbol{e} \Rightarrow \mathsf{Safe}(\mathsf{Trace}(Q, N))\boldsymbol{e})$$

Proof. Let the internal state range over the natural numbers. The visible state can be taken from any non-empty set. Let the witness for Q be $\lambda(e, s).\,\mathsf{T}$ and the witness for N be $\lambda(e, s)(e', s').s = \mathsf{Suc}\, s'$. $\mathsf{Limit\text{-}Trace}(Q, N)$ accepts all sequences $\boldsymbol{e}$, whereas $\mathsf{Trace}(Q, N)$ rejects all sequences. □

The above example demonstrates an extreme pathology. The automaton has infinite non-determinism; it also has a dead state. The resulting "eventual contradiction" (the internal state cannot decrement past 0) is then used to reject all infinite traces. However, for any finite prefix we can find an initial value of s so that the contradiction is not reached.

One way to avoid this situation is to use automata with no dead states:

---

[5]This proof is derived from a proof in [8].

Lemma 10. *For automata with no dead states, the prefix-limit behaviour preserves the safety properties of the infinite behaviour:*

$$\vdash \forall Q N.\, \mathsf{No\text{-}Dead}(Q, N) \Rightarrow (\forall e.\, \mathsf{Limit\text{-}Trace}(Q, N)e \Rightarrow \mathsf{Safe}(\mathsf{Trace}(Q, N))e)$$

Proof. Because the automaton has no dead states, every prefix satisfying the initial state and transition relation has an infinite extension which also satisfies the transition relation, and is therefore an infinite trace. This is the condition prescribed by Safe. □

6.1 Liveness and stuttering

Because stuttering automata have a self-loop on every state, they can "stutter forever". One might think that they are incapable of expressing liveness properties. This is not true:

Theorem 7. *Stuttering automata can express liveness properties.*

$$\vdash \exists Q N.\, \mathsf{Stuttering}(Q, N) \wedge \neg\, \mathsf{Safety}(\mathsf{Trace}(Q, N))$$

Proof. Let the internal state s range over natural numbers, and the visible state space be some set of more than one element. Let $Q(e, s) = \mathsf{T}$ and $N(e, s)(e', s') = ((e' = e) \wedge (s' = s)) \vee (s = \mathsf{Suc}\, s')$. Whenever there is a transition in the trace sequence, the internal state is forced to decrement. Since it cannot decrement forever (it will reach zero), there cannot be an infinite number of transitions. However, any prefix can be extended by infinite stuttering to be a trace of the automaton. It is therefore expressing a liveness (non-safety) property. □

7 Constructions on Automata

This section covers three closely related constructions on automata. The first is the familiar subset construction; the second, for lack of a better name, I shall call the powerset construction. In general, these two constructions do not preserve behaviour. The third, which does preserve behaviour, is a restricted form of the powerset construction.

7.1 The subset construction

This is the familiar construction on finite automata, extended to infinite automata.

Definition 14. The state-space of the subset automaton is the powerset of the state-space of the original defining automaton. The initial state of the subset automaton is the set of possible initial states of the defining automaton. The transition relation defines the set of all possible next states of the defining automaton. This set is constrained to be non-empty. Formally the definition of the subset construction is:

$$\begin{aligned}
&\vdash \forall Q N.\, \mathsf{Subset}(Q, N) = \\
&\quad (\lambda(e, x).(\forall s.x\; s = Q(e, s)) \wedge (\exists s.x\; s)), \\
&\quad (\lambda(e, x)(e', x').(\forall s'.x'\; s' = (\exists s.x s \wedge N(e, s)(e', s'))) \wedge (\exists s'.x'\; s'))
\end{aligned}$$

Lemma 11. *The subset automaton is deterministic:*

$$\vdash \forall Q N.\, \mathsf{Deterministic}(\mathsf{Subset}(Q, N))$$

Proof. Follows directly from the uniqueness of the sets defined by the subset construction. □

Theorem 8. *The subset construction does not in general preserve behaviour.*

Proof. By theorem 5, since deterministic automata are less expressive than non-deterministic automata, behaviour cannot always be preserved. □

Definition 15. The *powerset construction* is a relaxed version of the subset construction. Its state-space is the same as that of the subset automaton. The initial state of the powerset automaton is some set of possible initial states of the defining automaton. Because the initial state not unique, ranging over all subsets of the initial state predicate of the defining automaton, the powerset construction is not in general deterministic. The transition relation is also relaxed, defining some set of states that is compatible with some non-empty subset of the current state.

Formally the powerset construction resembles the subset construction with the equalities replaced with implications:

$$
\begin{aligned}
&\vdash \forall Q N.\ \mathsf{Powerset}(Q,N) = \\
&\quad (\lambda(e,x).(\forall s.x\ s \Rightarrow Q(e,s)) \wedge (\exists s.x\ s)), \\
&\quad (\lambda(e,x)(e',x').(\forall s'.x'\ s' \Rightarrow (\exists s.x\ s \wedge N(e,s)(e',s'))) \wedge (\exists s'.x'\ s'))
\end{aligned}
$$

7.2 The behaviour of subset and powerset automata

In this section we shall demonstrate equivalence of the following behaviours by a five-way circular implication:

$$
\begin{array}{ccc}
 & \mathsf{Limit\text{-}Trace}(Q,N) & \\
\nearrow & & \searrow \\
\mathsf{Limit\text{-}Trace}(\mathsf{Powerset}(Q,N)) & & \mathsf{Limit\text{-}Trace}(\mathsf{Subset}(Q,N)) \\
\uparrow & & \downarrow \\
\mathsf{Trace}(\mathsf{Powerset}(Q,N)) & \leftarrow & \mathsf{Trace}(\mathsf{Subset}(Q,N))
\end{array}
$$

Lemma 12.

$$\vdash \forall Q N e n.\ \mathsf{Prefix\text{-}Trace}\, n(\mathsf{Powerset}(Q,N))e \Rightarrow \mathsf{Prefix\text{-}Trace}\, n(Q,N)e$$

Proof. The proof proceeds by induction on n. For each internal state prefix of the powerset automaton, we need to find a corresponding prefix of the defining automaton. Before performing the induction we strengthen the goal to state that every element of the powerset automaton's nth state defines an n element state prefix of the defining automaton's state. Clearly, every element of the initial state of the powerset automaton is an initial state of the defining automaton. This covers the case of $n = 0$. For the inductive case we find a suitable $n+1$-prefix by picking some element of the $n+1$th state of the powerset automaton. By definition 15, we know this exists, and it is related by the transition relation to some element of the nth state. Since each element of the nth state defines a valid n-prefix, the element we have chosen defines an $n+1$-prefix. Also, by definition 15, we can show that every element of the $n+1$th state defines a valid $n+1$-prefix. □

Lemma 13.

$$\vdash \forall Q N e.\ \mathsf{Limit\text{-}Trace}(\mathsf{Powerset}(Q,N))e \Rightarrow \mathsf{Limit\text{-}Trace}(Q,N)e$$

Proof. Follows trivially from lemma 12. □

Lemma 14.

$$\vdash \forall Q N e.\, \mathsf{Limit\text{-}Trace}(Q,N)e \Rightarrow \mathsf{Limit\text{-}Trace}(\mathsf{Subset}(Q,N))e$$

Proof. It follows directly by theorem 4. The simulation relation between s, the state of (Q, N), and x, the state of $\mathsf{Subset}(Q, N)$, is $x\ s$ (or in set notation: $s \in x$). □

Lemma 15.

$$\vdash \forall Q N e.\, \mathsf{Limit\text{-}Trace}(\mathsf{Subset}(Q,N))e \Rightarrow \mathsf{Trace}(\mathsf{Subset}(Q,N))e$$

Proof. Follows directly from lemma 11 and theorem 1. □

Lemma 16.

$$\vdash \forall Q N e.\, \mathsf{Trace}(\mathsf{Subset}(Q,N))e \Rightarrow \mathsf{Trace}(\mathsf{Powerset}(Q,N))e$$

Proof. The proof is straighforward because the powerset construction is a relaxed version of the subset construction. □

Lemma 17.

$$\vdash \forall Q N e.\, \mathsf{Trace}(\mathsf{Powerset}(Q,N))e \Rightarrow \mathsf{Limit\text{-}Trace}(\mathsf{Powerset}(Q,N))e$$

Proof. By lemma 2. □

Theorem 9. *The five behaviours denoted by* $\mathsf{Limit\text{-}Trace}(\mathsf{Powerset}(Q,N))$, $\mathsf{Limit\text{-}Trace}(Q,N)$, $\mathsf{Limit\text{-}Trace}(\mathsf{Subset}(Q,N))$, $\mathsf{Trace}(\mathsf{Subset}(Q,N))$ *and* $\mathsf{Trace}(\mathsf{Powerset}(Q,N))$ *are all equivalent.*

Proof. By transitivity of implication and lemmas 13–17. □

Lemma 18. *An automaton implements its own subset construction.*

$$\vdash \forall Q N e.\, \mathsf{Trace}(Q,N)e \Rightarrow \mathsf{Trace}(\mathsf{Subset}(Q,N))e$$

Proof. By lemma 2 and theorem 9. □

7.3 A behaviour-preserving construction

Definition 16. The *finite powerset construction* is a powerset construction, where the state of the powerset automaton is also constrained to be finite:

$$\begin{aligned}&\vdash \forall Q N.\, \mathsf{Finite\text{-}Powerset}(Q,N) = \\ &\quad (\lambda(e,x).(\forall s.x\ s \Rightarrow Q(e,s)) \wedge (\exists s.x\ s) \wedge \mathsf{Finite}\ x), \\ &\quad (\lambda(e,x)(e',x').(\forall s'.x'\ s' \Rightarrow (\exists s.xs \wedge N(e,s)(e',s'))) \wedge (\exists s'.x'\ s') \wedge \mathsf{Finite}\ x')\end{aligned}$$

Before demonstrating that the finite powerset construction preserves behaviour, we shall need[9]:

König's Lemma. *Let $E_1, E_2, E_3, \ldots$ be a countable sequence of non-empty finite sets, and let R be a relation such that each element x_{n+1} of each set E_{n+1} corresponds with at least one element x_n of E_n by the relation R, which we write as $x_n \, R \, x_{n+1}$ $(n = 1, 2, 3, \ldots)$. Then we can choose an element a_n from each set E_n such that for the infinite sequence $a_1, a_2, a_3, \ldots$, we have $a_n \, R \, a_{n+1}$ $(n = 1, 2, 3, \ldots)$.*[6]

$$\begin{aligned}&\vdash \forall E R. \\ &\quad (\forall n.\, \mathsf{Finite}(E\, n) \wedge (\exists x. E\, n\, x) \wedge (\forall x'. E(\mathsf{Suc}\, n) x' \Rightarrow (\exists x. E\, n\, x \wedge R\, x\, x'))) \Rightarrow \\ &\quad (\exists a. \forall n. E\, n (a\, n) \wedge R(a\, n)(a(\mathsf{Suc}\, n)))\end{aligned}$$

Lack of space prevents the inclusion of a proof of König's lemma.

Theorem 10. *The finite powerset automaton has the same behaviour as the original defining automaton:*

$$\vdash \forall Q N e.\, \mathsf{Trace}(\mathsf{Finite\text{-}Powerset}(Q, N))e = \mathsf{Trace}(Q, N)e$$

Proof. Every trace of the defining automaton is also a trace of the finite powerset automaton:

For a given trace $\boldsymbol{e}$ and state sequence $\boldsymbol{s}$ we choose an infinite sequence of singleton sets $\boldsymbol{x}$ such that $\forall t.(\boldsymbol{x}\, t)(\boldsymbol{s}\, t)$ (or $\boldsymbol{s}\, t \in \boldsymbol{x}\, t$). This is clearly a valid state sequence of the finite powerset automaton.

Every trace accepted by the finite powerset automaton is also accepted by the defining automaton:

For any trace $\boldsymbol{e}$ accepted by the finite powerset automaton we demonstrate the existence of an infinite sequence of states of the original automaton from the infinite sequence of states (finite sets) of the finite powerset automaton. This follows directly by König's lemma. □

Theorem 11. *The behaviour of finitely non-deterministic automata is preserved by the subset (or powerset) construction:*

$$\vdash \forall Q N.\, \mathsf{Fin\text{-}Non\text{-}Det}(Q, N) \Rightarrow (\mathsf{Trace}(\mathsf{Subset}(Q, N)) \Rightarrow \mathsf{Trace}(Q, N))$$

Proof. Because the set of initial states is finite, and the set of possible next states for any reachable current state is finite, the set of possible states is always finite. The subset automaton's state is therefore always a finite set. Because the powerset construction is a relaxed form of the subset construction, the subset construction implements the finite powerset construction, which by theorem 10, also implements the defining automaton. □

8 Extending the use of Simulation

Despite the restriction of deterministic specification on the completeness theorem 3, for most engineering applications one can usually find a simulation relation even for non-deterministic specifications.

However, when the specification makes a non-deterministic choice *before* the corresponding choice is made by the implementation, implementation automaton may not simulate the specification automaton. A simple example can be seen in section 9. We can overcome this problem by relating the implementation to the powerset construction of the specification rather than to the specification directly. If the specification's behaviour is preserved by the powerset construction, then this will suffice for demonstrating that one automaton implements the other.

[6] Translated directly from [9] (in French).

Theorem 12. *Demonstrating that automaton 1 simulates the powerset construction of automaton 2 is sufficient to show that automaton 1 implements automaton 2 provided that automaton 2 is finitely non-deterministic:*[7]

$$\vdash \forall Q_1 Q_2 N_1 N_2.\, \mathsf{Fin\text{-}Non\text{-}Det}(Q_2, N_2) \Rightarrow$$
$$(Q_1, N_1)\,\mathsf{Simulates}(\mathsf{Powerset}(Q_2, N_2)) \Rightarrow (\forall e.\, \mathsf{Trace}(Q_1, N_1)e \Rightarrow \mathsf{Trace}(Q_2, N_2)e)$$

Proof. By theorems 2, 11 and 9. □

Theorem 13. *If automaton 1 has no dead states, and automaton 1 implements automaton 2, then automaton 1 also simulates the powerset construction of automaton 2:*[8]

$$\vdash \forall Q_1 Q_2 N_1 N_2.\, \mathsf{No\text{-}Dead}(Q_1, N_1) \Rightarrow$$
$$(\forall e.\, \mathsf{Trace}(Q_1, N_1)e \Rightarrow \mathsf{Trace}(Q_2, N_2)e) \Rightarrow (Q_1, N_1)\,\mathsf{Simulates}(\mathsf{Powerset}(Q_2, N_2))$$

Proof. By lemma 18 automaton 1 implements the subset construction of automaton 2. Because the subset construction is deterministic (lemma 11), and automaton 1 has no dead states, automaton 1 must also simulate the subset construction of automaton 2. Because the powerset construction is a relaxed version of the subset construction, automaton 1 also simulates the powerset construction of automaton 2. □

8.1 Further remarks on the use of the powerset construction

The requirement that the specification be finitely non-deterministic is often stronger than necessary. We can use any condition sufficient for the preservation of the specification's behaviour by the powerset construction. One example is given by the following theorem:

Theorem 14. *If an automaton has no dead states and its behaviour is a safety property, then the powerset construction preserves that behaviour:*

$$\vdash \forall Q N.\, \mathsf{Safety}(\mathsf{Trace}(Q, N)) \Rightarrow \mathsf{No\text{-}Dead}(Q, N) \Rightarrow$$
$$(\forall e.\, \mathsf{Trace}(\mathsf{Powerset}(Q, N))e \Rightarrow \mathsf{Trace}(Q, N)e)$$

Proof. By theorem 9 and lemma 10 we have

$$\forall e.\, \mathsf{Trace}(\mathsf{Powerset}(Q, N))e \Rightarrow \mathsf{Safe}(\mathsf{Trace}(Q, N))e$$

By definition 13 we also have that $\mathsf{Safe}(\mathsf{Trace}(Q, N))e \Rightarrow \mathsf{Trace}(Q, N)e$. □

8.2 Using the finite powerset construction

Because the finite powerset construction preserves behaviour, we can use it to demonstrate conformance to a specification automaton, even if that automaton's behaviour is not preserved by the powerset construction. In practice, this is applicable when the non-deterministic choices made by the specification before the corresponding choices are made by the implementation, are all finite. The specification may still also make infinitely non-deterministic choices, but these must not be made before corresponding choices have been made by the implementation.

[7] This is the primary soundness result of [8].

[8] This is the primary completeness result of [8], but without the requirement of countable state spaces and finitely non-deterministic specifications.

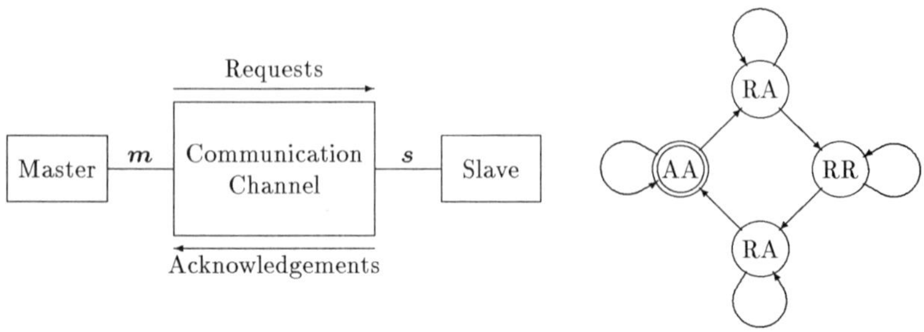

Figure 1: Request-acknowledge communication channel, and its state-transition graph schema

Theorem 15. *To show that one automaton implements another, it is sufficient to demonstrate the first automaton simulates the finite powerset construction of the specification:*

$$\vdash \forall Q_1 N_1 Q_2 N_2.\ (Q_1, N_1)\ \mathsf{Simulates}(\mathsf{Finite\text{-}Powerset}(Q_2, N_2)) \Rightarrow (\forall e.\ \mathsf{Trace}(Q_1, N_1)e \Rightarrow \mathsf{Trace}(Q_2, N_2)e)$$

Proof. From theorems 2 and 10. □

9 Examples

This section presents three examples. The first is a stuttering automaton model of a communication channel. The second demonstrates how to show by simulation relation that two channels in series implement a single channel. The third example demonstrates the incompleteness of simulation as a means of showing that one automaton implements another.

9.1 A request-acknowledge communication channel

This is an example of a system modelled by a stuttering automaton, driven by signal transitions rather than by signal values. It is therefore naturally represented as a labelled-state transition graph. It models a request-acknowledge communication channel with arbitrary delay for both requests and acknowledgements (figure 1). The values transmitted range over an arbitrary alphabet α, which is assumed to be divided into two disjoint sets, one contaim request values, and one containing acknowledge values. Predicate A defines the acknowledge subset; the remaining elements are assumed to denote requests.

The nodes in the state-transition graph schema in figure 1 denote sets of states rather than individual states. The double-circle node denotes the initial states. The self-loops denote self loops on every element of the set, and the transitions between different nodes denote transitions from elements of one set to another, with the condition that there is no transition from one request value to another, or one acknowledge value to another on either port. There are two nodes with the same label; internal state is required to distinguish between the forward propagation af requests and the backward propagation of acknowledgements:

$$
\begin{array}{l}
\vdash \forall \boldsymbol{A m s}.\, \mathsf{MS}\, \boldsymbol{A}\, \boldsymbol{m}\, \boldsymbol{s} = \mathsf{Trace} \\
\quad ((\lambda((m,s),b).A\, m \wedge (m = s)), \\
\quad\; (\lambda((m,s),b)((m',s'),b') = \\
\quad\;\; (A\, m \wedge A\, s \Rightarrow (s' = s) \wedge ((m' = m) \vee \neg A\, m' \wedge \neg b')) \wedge \\
\quad\;\; (\neg A\, m \wedge A\, s \wedge \neg b \Rightarrow (m' = m) \wedge ((s' = s) \wedge \neg b' \vee (s' = m))) \wedge \\
\quad\;\; (\neg A\, m \wedge \neg A\, s \Rightarrow (m' = m) \wedge ((s' = s) \vee A\, s' \wedge b')) \wedge \\
\quad\;\; (\neg A\, m \wedge A\, s \wedge b \Rightarrow (s' = s) \wedge ((m' = m) \wedge b' \vee (m' = s))))) \\
\quad (\boldsymbol{m} \circledcirc \boldsymbol{s})
\end{array}
$$

9.2 Example simulation relation

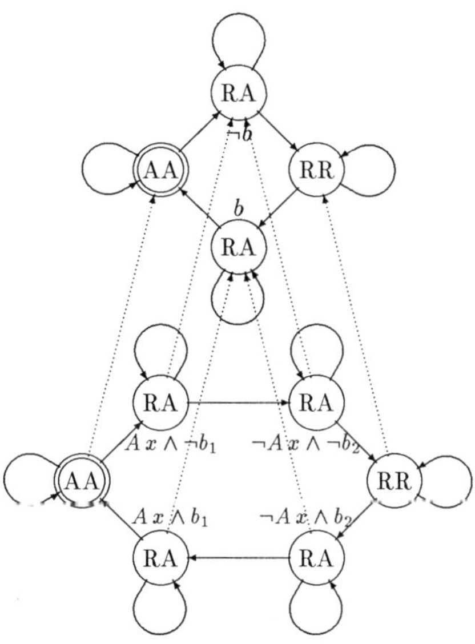

Label	Visible State
AA	$A\, m \wedge A\, s$
RA	$\neg A\, m \wedge A\, s$
RR	$\neg A\, m \wedge \neg A\, s$

Figure 2: State-transition graph schema of two communication channels in series and simulation relation with single communication channel

Figure 2 shows the state transition graph schema of two communication channels (figure 1) in series. The common port $\boldsymbol{x}$ becomes a hidden component of the composed automaton's state space. The two internal bits b are renamed b_1 and b_2 in the composed automaton. The RA states are annotated with their corresponding internal state predicates. Also shown as dotted arrows is a simulation relation between that state graph and a single communication channel state graph. This simulation relation is sufficient to show that two communication channels in series implements a single communication channel. By visual inspection it can be seen that the simulation relation conforms with definition 11.

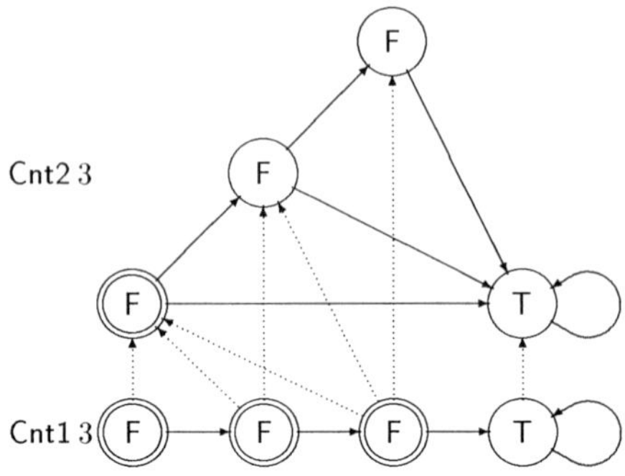

Figure 3: Two equivalent automata

9.3 Example of incompleteness of simulation

Figure 3 shows two state-transition graphs of two equivalent automata. The automata have boolean visible state. Initially the visible state is F, but it becomes T within 3 time units after initialisation. There are three possible traces: FTTTTT..., FFTTTT... and FFFTTT.... Both automata have a natural number as internal state. The visible state is T if and only if the internal state is 0. **Cnt1** non-deterministically chooses some initial state s where $s \neq 0$ and $s \leq 3$. It then decrements s until it reaches 0. **Cnt2**, however initialises s to 3, but later chooses between either setting s to 0 or decrementing it.

The HOL definitions of the automata have a parameter n instead of the fixed value 3:

$$\mathsf{Cnt1}\,n = ((\lambda(e,s).s \leq n \wedge 0 < s \wedge \neg e), (\lambda(e,s)(e',s').(e = (s = 0)) \wedge (s' = s - 1))$$

$$\mathsf{Cnt2}\,n = ((\lambda(e,s).(s = n) \wedge 0 < s \wedge \neg e), (\lambda(e,s)(e',s').(e = (s = 0)) \wedge ((s' = s - 1) \vee (s' = 0))$$

That **Cnt1** implements **Cnt2** can be shown by simulation relation. The dotted arrows in figure 3 show this simulation relation. By inspection it can be seen that it has the properties of definition 11. To show that **Cnt2** implements **Cnt1** is less straightforward. Turning the dotted arrows to point backwards does not define a simulation relation from **Cnt2** to **Cnt1**. It is, however, possible to interpret the backward arrows as a relation between states of **Cnt2** and sets of states of **Cnt1**. These sets of states also correspond to the powerset construction[9] of **Cnt1**. We can therefore use the powerset construction of **Cnt1** to demonstrate that **Cnt2** implements **Cnt1**.

This approach works for any n. However, if we hide (existentially quantify) n, then the automata express a non-safety property. Any prefix consisting entirely of Fs can be extended to be a valid behaviour; however the infinite sequence of F is not included in the behaviour of the automata.

This hiding of n can be expressed by including n as a component of the internal state, and adding the term $n' = n$ as a conjunct of the transition relation. Because the resulting automata express a non-safety property, the powerset construction can no longer be used, because it does not preserve behaviour (it would accept the infinite sequence of Fs). However, in this case, both automata are making the infinitely non-deterministic choice (selecting n) at initialisation time.

[9] For this example the construction turns out to also be a subset construction; Cnt2 is isomorphic to the subset construction of Cnt1.

We can therefore form a finite powerset construction, which has the same form as the original powerset construction on the s component of the state, and also includes the constraint that both automata select the same n.

10 Conclusion

For relating different automata I have presented three approaches to using simulation relations:

- Directly between the implementation and specification automata.
- Between the implementation and the powerset construction of the specification.
- Between the implementation and the finite powerset construction of the specification.

The first is always sound. Despite the rather restricted completeness theorem (theorem 3), in practice it usually succeeds. It fails, however, when the specification makes non-deterministic choices before the implementation makes the corresponding choice.

The second is sound when the powerset construction preserves the behaviour of the specification, and is complete when the implementation has no dead states. This approach is therefore limited to specifications which describe safety properties. The behaviour of both finitely deterministic automata, and automata with no dead states and which specify safety properties, is preserved by the powerset construction.

The third is always sound. It introduces extra proof obligations, but allows specifications whose behaviour is not preserved by the powerset construction. It also allows the specification to make finitely non-deterministic choices before corresponding choices are made by the implementation.

It is possible to effectively use simulation relations to verify non-safety properties. However, because this involves infinitely non-deterministic specifications, these non-deterministic choices must be made in the implementation before or at the same time as in the specification. One approach to ensure this is to use a style of specification which makes all infinitely non-deterministic choices at initialisation time. It may, however, be more practical to specify liveness properties separately using temporal logic.

This paper presents the first mechanically derived theory of infinite automata. Although much of the material has been presented before in different forms, a significant proportion of the material is new.

10.1 Further work

One problem not addressed in this paper is "forced stuttering". In figure 2 the lower automaton implements the upper automaton. However, the upper does not implement the lower because the lower has a minimum delay of two time units for the propagation of requests and acknowledgements, whereas the upper has a minimum of one time unit. When using a discrete time model for asynchronous systems, the "unit" of time should not be considered fixed, but just representing some arbitrary time interval. When interpreted this way, any number of time units with no change of visible state should be implementable with a single time unit.

One approach to this would be to introduce the concept of "stuttering closure" of an automaton. This can be used to introduce extra transitions to avoid forced stuttering. This closure can be defined with the inductive definition package which is now included in HOL[13].

References

[1] Martín Abadi and Leslie Lamport. The existence of refinement mappings. *Theoretical Computer Science*, 82(2):253–284, May 1991.

[2] Bowern Alpern and Fred B. Schneider. Recognising safety and liveness. Technical Report TR 86-727, Department of Computer Science, Cornell University, 1986.

[3] Robert S. Boyer and J. Strother Moore. *A Computational Logic*. Academic Press, 1979.

[4] J. R. Büchi. On a decision method in restricted second-order arithmetic. In *Proceedings International Congress on Logic, Methodology and Philosophy of Science, 1960*, pages 1–12. Stanford University Press, 1962.

[5] Mike Fourman. Avoiding use of the axiom of choice. Private Communication, 1990.

[6] Mike Gordon. HOL: A machine oriented formulation of higher-order logic. Technical Report 68, University of Cambridge Computer Laboratory, 1985.

[7] Mike Gordon. HOL: A proof generating system for higher-order logic. In G. Birtwistle and P. A. Subrahmanyam, editors, *VLSI Specification, Verification and Synthesis.* Kluwer Academic Publishers, 1988.

[8] Nils Klarlund and Fred B. Schneider. Verifying safety properties using non-deterministic infinite-state automata. Technical Report TR 89-1037, Cornell University Computer Science Department, 1989.

[9] Dénes König. Sur les correspondances multivoques des ensembles. *Fundamenta Mathematicae*, 8:114–134, 1926.

[10] Paul Loewenstein. The formal verification of state-machines using higher-order logic. In *IEEE International Conference on Computer Design*, 1989.

[11] Paul Loewenstein and David Dill. Formal verification of cache systems using refinement relations. In *IEEE International Conference on Computer Design*, pages 228–233, 1990.

[12] Paul Loewenstein and David L. Dill. Verification of a multiprocessor cache protocol using simulation relations and higher-order logic. In E. M. Clarke and R. P. Kurshan, editors, *Computer-Aided Verification '90*, pages 75–84. American Mathematical Society, 1991.

[13] T. F. Melham. Inductive relation definitions in HOL. In *International Workshop on the HOL theorem proving system and its applications.* IEEE Computer Society Press, 1991.

[14] R. Milner. An algebraic definition of simulation between programs. In *Proceedings 2nd International Joint Conference on Artificial Intelligence*, 1971.

[15] D. E. Muller. Infinite sequences and finite machines. In *Switching circuit theory and logical design: Proceedings Fourth Annual Symposium*, pages 3–16, New York, 1963. Institute of Electrical and Electronic Engineers.

[16] M. O. Rabin. Decidability of second-order theories and automata on infinite trees. *Transactions American Mathematical Society*, 141:1–35, 1969.

Higher Order Logic Theorem Proving and its Applications (A-20)
L.J.M. Claesen and M.J.C. Gordon (Editors)
Elsevier Science Publishers B.V. (North-Holland)

A Comparison between Statecharts and State Transition Assertions

Nancy Day

Integrated Systems Design Laboratory, Department of Computer Science, University of British Columbia, Vancouver, BC, V6T 1Z2, Canada.

Abstract

This paper compares statecharts, a specification formalism for reactive systems, to state transition assertions, a verification method for hard real-time systems. While these two methods are used for different tasks and they take different points of view in describing a system, it is useful to compare them to determine what is necessary in a formal specification notation for real-time systems. In this paper, we conclude with a list of issues that need to be resolved when integrating formal verification with a specification notation. The future goal of this work is to provide a more readable front-end specification formalism which can be used for verification. The purpose of doing a formal verification of specifications is to check for correctness early in the system development process and discover errors which can prove costly in later stages. If a more readable notation like statecharts is embedded in the theorem-prover, HOL (Higher Order Logic), it would provide the tools necessary to do mechanized verification.

Keyword Codes: D.2.1;D.2.4;F.3.1
Keywords: Software Engineering, Requirements/Specifications; Program Verification; Logics, Specifying and Verifying and Reasoning about Programs

1 Introduction

Discovering errors early in the design and implementation of any system, reduces the cost of development and increases one's confidence in the reliability of the end result. Top-down system development begins with a very high-level, abstract description of the system, taking into account its environment and the results it is expected to produce. More and more details are added as the process moves closer to a final product. Each level can be viewed operationally as a way of moving from one moment in execution to the next or it can be viewed as a specification for what the next, more detailed level must produce at the end of a step or series of steps. Phases in this development process are commonly refered to as specification, design, and implementation.

Although formal specification and verification methods show promise for reducing or eliminating errors[13], the difficulty in learning these techniques makes industry reluctant to use them except possibly as an afterthought[13]. Perhaps before using the system, it is "verified" using some method, but this is separate from the steps of the development process. However, work in industry has already demonstrated the advantages of using formal specification notations early in the development process[13]. If we take the second point of view mentioned above, that one level of description is a specification for the next level, then there is room to use formal verification techniques to ensure that a lower level description correctly implements a higher level one. It would also be possible to show overall system properties, such as safety requirements, continue to hold as more implementation details are added.

The general purpose theorem prover HOL (Higher Order Logic)[5] makes it possible to machine-check proofs of this form. Previous work in verifying that level by level descriptions correctly interpret a more abstract level has been done for hardware[14][17], and is currently being completed for the complete "stack" of a system including the hardware, compiler, and software (SAFEMOS at SRI International). One of the advantages of using a theorem-prover is that a compositional approach is possible which is well-suited to viewing parts of the system at different levels of abstraction and reusing completed proofs.

This paper looks at two different approaches to system specification: statecharts (D. Harel [7]), and state transition assertions(STAs) (M. Gordon [4]). These two formalisms were chosen because they approach the problem from the two different points of view that we would like to see integrated. Statecharts have an elegant visual notation for describing the operation of a system and are supported by the commercial tool STATEMATE [10]. STAs are designed for use in formally verifying systems and expressing limitations on the system's implementation.

At the end of this comparison, we are able to conclude that five issues which need to be considered when looking for a real-time specification formalism to be used for verification are:

- Does it express operation or assertions?
- Can it handle complete system description at any level?
- Does it have a formal model of time?
- Can it integrate data and control descriptions?
- Is it a visual notation?

The first section of this paper defines the type of system these specification techniques are used for. Then a brief description of each method is given. An example described in each notation is presented in Section 5 to provide a basis for comparing them. Finally, we present the list of issues outlined above.

2 Terminology

Before beginning to look at methods for describing real-time reactive systems, we should define precisely the meanings of "reactive" and "real-time".

In Harel's original paper on statecharts[7], he defines reactive systems as being event driven, meaning they have to respond to both internal and external stimuli. In a later paper[10], he states that it is necessary to "specify the relationship between inputs and outputs over time". Gordon places the emphasis on the timing requirements in saying hard real-time systems "are required to meet explicit timing constraints, such as responding to an input within 100 milliseconds of a change" [4].

For our purposes, real-time systems are a type of reactive system where the response time to an event is an essential part of the description of a system. It is important to note that the word "system" is meant to describe the complete set-up including software, computing hardware, and any specialized peripherals.

3 Statecharts

Statecharts build on the ideas of finite state machines (FSMs), extending the notation to overcome some of its limitations. The system is described as being in one of a number of possible states, and it changes from state to state as a result of events which occur.

Since statecharts are a type of higraph, they are a *visual formalism.* Higraphs are a notation which combines the ideas of graphs and Venn diagrams [9]. The nodes (or blobs) in the graph represent states the system can be in. The edges indicate transitions between states. This movement is considered to happen instantaneously. Please note that in the following description, transition labels such as **T1** have no semantic meaning and are only used to reference certain transitions.

With reference to the statechart of Figure 1, the main features of statecharts are the following:

- *Concurrency:* The Cartesian product of states is represented by orthogonal components of a state. In the example, states **A** and **D** are orthogonal components which make up state **Y**. They are separated by a dashed line through state **Y**. (Also note that the labelling of the state **Y** is in a small box outside of its contour.) This means that for the system to be in state **Y**, it must be in state **A** *and* in state **D**. This notation prevents the explosion of states in a FSM resulting from the Cartesian product of state diagrams.

  Synchronization between the various substates the system is in simultaneously can be accomplished using events, conditions, and actions which are described below.

- *Hierarchy:* A state can be decomposed into substates (often called OR states) indicating more of the internal workings of a state. A state which is not refined is called a *basic state.* In the example, state **A** is decomposed (or refined) into the basic states **B** and **C**. This means that when the system is in state **A**, it is in either state **B** *or* state **C** (exclusive or). (**B** is called a descendant of **A** and **A** is an ancestor of **B**.) This hierarchy of states allows a system to be initially specified at a higher level of abstraction, and later decomposed, as the design process progresses.

- *Transitions:* Transitions are labelled with the notation $e[c]/a$. The occurrence of event e enables the transition (meaning it can be taken) providing the condition c is

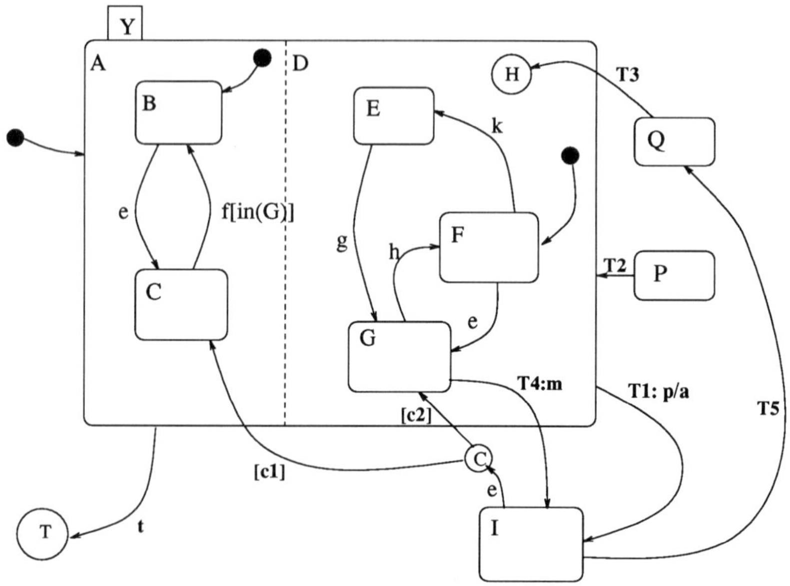

Figure 1: An Example of a Statechart

true. (If no condition is given then "true" is assumed.) A transition can be enabled if it originates in any descendant or ancestor of a state the system is currently in. If the transition is taken then the actions a are carried out. These actions can be to start and stop data processes (called *activities*), to generate events, such as internal events for synchronization, or to change the values of variables or conditions. For example, if the system is currently in the basic states (**B**,**F**) and the event **p** occurs, then transition **T1** is enabled. All parts of the transition label are optional. If neither an event nor a condition is given then the transition is always enabled.

- *Broadcast communication*: This type of communication is implicit in the operation of statecharts. This means that every external and internal event which occurs can be seen in all parts of the system.

- *Default states:* The small arrow originating from a filled-in circle pointing at state **B** in Figure 1 gives a default state for state **A**. It means that whenever a transition terminates at the outside boundary of a non-basic state the default arrow will be followed to enter a basic state. The transition labelled **T2** originating at state **P** is an example of this. Following this transition will cause the system to follow default entries for state **Y**, which in turn means default entries for both states **A** and **D**, leaving the system in state **B** and **F**.

- *History states:* A history connector, marked with an **H** can also be the destination of a transition. It dynamically represents the substate which this state was in at the time it was last exited. For example, if the transition labelled **T4** from state **G** triggered by event **m** was taken to leave state **D** and the system then followed **T5** to arrive in state **Q**, and **T3** terminating at the history connector, to return to **D**, it would re-enter state **G**. If there is no history (ie the system has never been in this state before or the history has been cleared) then a transition from the history connector is followed if it exists or else the default transition is taken.

- *Terminal connectors:* Entering a terminal connector, labelled by **T** in a circle, stops all processing in the system.
- *Transition connectors:* These connectors are a way of decomposing transitions into smaller parts. For example the three transitions joined by the **C** connector in the diagram really represent two transitions from state **I**, one going to **G** and labelled **e[c2]**, and the other going to state **C**, labelled **e[c1]**.

3.1 The Semantics of Statecharts

While statecharts have the advantage of using pictures to portray a great deal of information about the system being specified, there are instances in which it is not clear from the diagram what is intended by the specification. These problems have been noted and discussed by Harel and others[6][12]. Harel's description of the semantics [6] differs from those used in STATEMATE somewhat. In particular, some operators which he describes for checking the state of the system in between "steps" (see the third section below) are not implemented. In this section, we characterize some of these subtleties and describe the semantics which STATEMATE uses in simulating the specification.

A great deal of work has been completed recently on incorporating timing models into statecharts[11] [15]. Since STATEMATE was developed before this work, it is based on the original form of statecharts. Because our future goal is to provide a connection between a commercial CASE tool and formal verification through the theorem-prover HOL, we have decided, at this time, to consider only the semantics of statecharts in STATEMATE.

3.1.1 Non-determinism

A non-deterministic situation exists when an event enables two or more possible transitions leaving the same state. Even if the transitions have the same target, there could be different actions associated with each one. The simulator of STATEMATE prompts the user to choose which transition to follow in a case like this.

3.1.2 Structural Non-determinism

Statecharts are very good at graphically describing a hierarchy of states. This allows for refinement of specifications, or the opposite, abstraction. This is accomplished using OR states. However, transitions can leave any state boundary and it is possible the same event will trigger a transition from both a state and one of its ancestor states. An example is given in Figure 2 where the event **e** triggers the transition **T1**, leaving state **B** as well as one from its parent, state **A**, labelled **T2**.

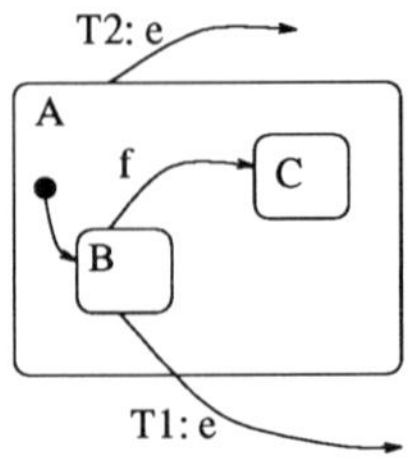

Figure 2: Structural Ambiguity

In STATEMATE, priority is given to transitions leaving states higher up in the hierarchy. In the preceding example, the transition **T2** would be taken.

3.1.3 Timing

There is no inherent model of timing associated with statecharts other than the movement between states by following transitions. STATEMATE offers both an asychronous and a synchronous model of time. The first stage in the algorithm for both models is to determine all the enabled compound transitions. A compound transition(CT) is the connection of all the transitions necessary to go from one basic state to another (through default and history connectors, etc). The trigger for a CT is the conjunction of the triggers for its component transitions. A maximal, non-conflicting set of enabled transitions for the system is then determined by randomly picking one when two or more transitions leave the same state.

In the synchronous model, time is incremented just before the set of CTs to be executed is determined and then the system performs these transitions in a random order. Performing a transition consists of doing the actions for exiting one state, entering the new state and then the actions for the transition itself. The execution of the complete set of CTs is called a "step" and the results of actions and any internal events generated are not available until the next step.

In the asynchronous model, the system repeatedly determines and executes all enabled CTs without incrementing time or considering any new external events generated until there are no more transitions enabled. This is called a "super step". This is the model Harel describes in his paper[6] but there is some question as to whether external events persist for the length of the entire super step or whether only internal events are relevant in later steps of a super step. The examples in Harel's paper indicate that the external events persist for the whole time. The author has yet to experiment with this in STATEMATE.

Another model of time which could have been used is to determine the set of enabled transitions after each execution of a compound transition. This way, internally generated events could have an effect on the next step (where step means executing only one CT). However, this would create more opportunities for race conditions.

These three possible models raise questions about how much time a transition should take. (Are they really instantaneous?) Also, when should the system react to external events?

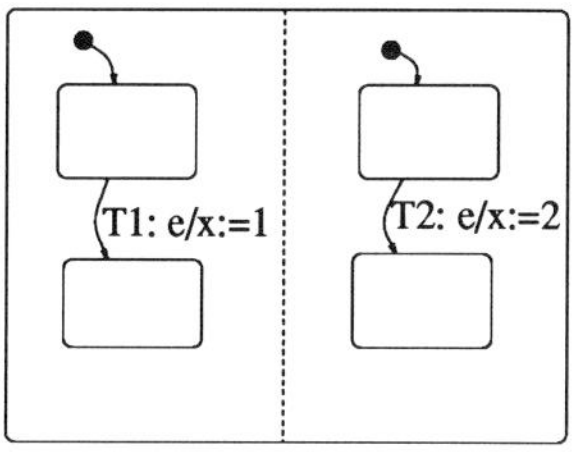

Figure 3: Race conditions

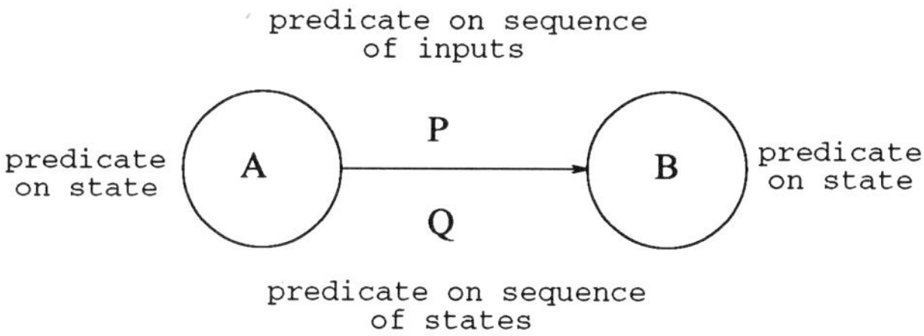

Figure 4: State Transition Assertion

3.1.4 Race Conditions

Race conditions occur when two transitions are enabled at the same time but if executed in different orders sequentially will have different results. An example is given in Figure 3 when the transitions **T1** and **T2** are both enabled and the last one to finish execution will determine the value of **x**. STATEMATE chooses randomly which transition to follow first although this effect is repeatable.

4 State Transition Assertions

State transition assertions are logical statements of constraints on the sequence of state transitions of a machine. They use the semi-graphical notation described in Figure 4 (from Figure 4 in [4]).

The notation means that if the system is ever in a state satisfying the predicate **A** and the next sequence of inputs satisfies **P**, then the system will arrive in a state satisfying **B**, having gone through a series of states satisfying **Q**. The STA must hold true everywhere in the system.

STAs are formulated about the problem at the specification level. An STA can also describe the semantics of a single machine instruction of a simple microprocessor where the transition only takes one step to execute (=1).

For example, for a typical jump instruction, JMP n would be described by:

$$< pc, stk, mem > \xrightarrow[=1]{[T]} < n, stk, mem >$$

where *pc*, *stk*, and *mem* mean program counter, stack, and memory respectively. The predicate on the sequence of inputs is [T] (true) since this operation will hold true for any input.

Every transition (even ones composed of several steps) must take at least one time unit. Using rules in the axiomatic style, Gordon is able to compose lower level STAs to see if the higher level ones hold. The following is an example of one of these rules describing a form of transitivity:

$$\frac{A \xrightarrow[\leq \delta_1 \wedge [q_1]]{P_1} B \qquad B \xrightarrow[\leq \delta_2 \wedge [q_2]]{P_2} C}{A \xrightarrow[\leq (\delta_1 + \delta_2) \wedge ([q_1] \vee [q_2])]{P_1 \wedge P_2} C}$$

The timing restrictions on all these phases are given by the notation $\leq \delta$ or $= \delta$ in the STA to indicate the number of time steps or transitions taken between states.

One of the most important features of Gordon's work is his use of axiomatic semantics to compose STAs to produce higher level statements about the system. It also bridges the gap between the specification and the implementation in the machine code. This makes the timing more exact.

> Verification of the system implementation is accomplished by translating the low level machine code into primitive STAs and showing that these STAs correspond to the STAs of the specification [1].

This method could be used as a bottom-up approach to verification. Difficulties arise in attempting to use it as a top-down approach because of the strong constraints placed on inputs by the predicate **P**, which must hold in all transitions decomposed from a higher level STA.

5 An Example

To provide a basis for comparing statecharts and STAs, it is useful to look at a simple example. The device OP given by Figure 5 and its corresponding description are from Gordon's paper [4].

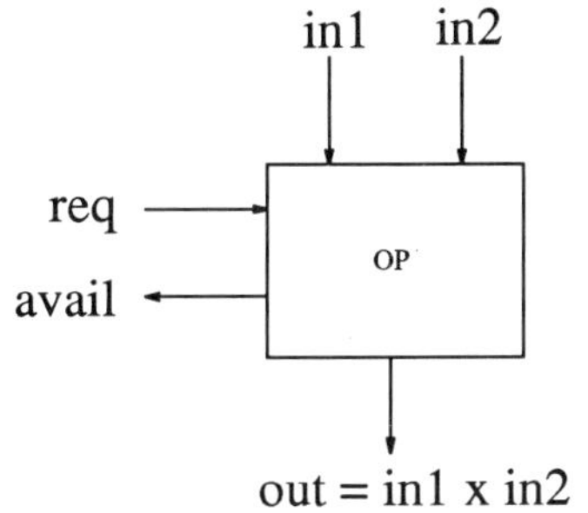

Figure 5: The binary device OP

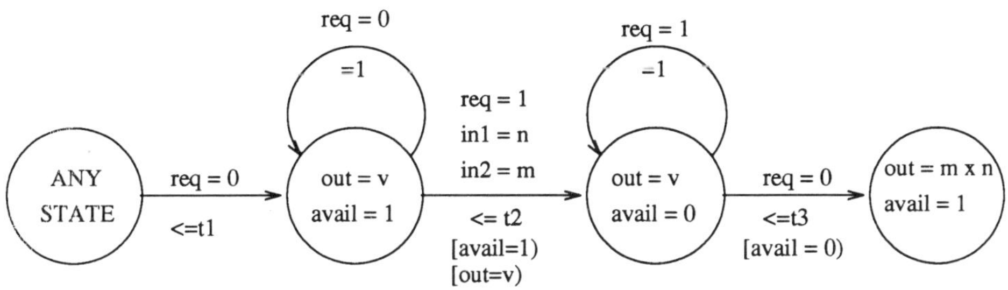

Figure 6: A STA specification of OP

OP is a device which calculates the operation **x**, given two operands **in1** and **in2**, and produces output **out**. (The **x** can be any binary operation.) Two signals **req** and **avail** control its execution. **Req** is provided by the user (or "client") and **avail** is driven by OP.

Whenever **avail** is 1, the client can initiate a request to perform the operation by setting **req** to 1. The client must keep its inputs stable and **req** at 1 until **avail** is set to 0. If the client continues to keep **req** at 1 then the output remains stable. After the client drops **req** to 0, the new output is available when **avail** goes to 1. The output then stays stable while **req** is 0.

Figure 6 gives five STAs which must hold in the implementation of the device. The left-most one says that from any state, if the **req** signal remains low, then the system must move into a state where **avail** is high in less than **t1** time units. The two main other timing restrictions are that the system receives the inputs and therefore moves into a state where **avail** is low in less than **t2** units after **req** goes high and that once **req** goes low again, the system will produce valid output and raise **avail** in less than **t3** time units. (The notation **[p)** means **p** will hold true for all states in the transition except the last one. The first state is included in the sequence of states which must satisfy the predicate below the line in an STA.)

In attempting to express the behaviour of this device in a statechart it is more intuitive

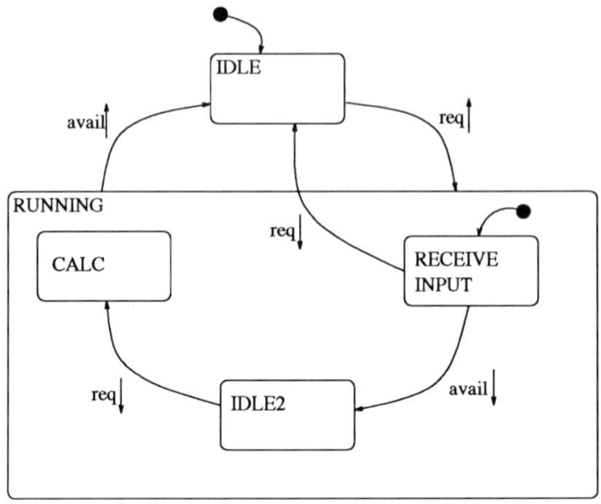

Figure 7: A statechart specification of OP

to think of the events which cause transitions to happen. Therefore we look at events like **req**↑ when the **req** signal goes high. Figure 7 is a statechart specification for OP. (There are many statecharts which could be used to describe the execution of this device.)

This second specification contains less information about the state of the system at each point since no predicates are attached to the states and we can only give them meaningful names. To associate these states with the operation of the device, some invariants on the control signals which would hold in these states are:

	avail	**req**
IDLE	1	0
RECEIVE INPUT	1	1
IDLE2	0	1
CALC	0	0

Time is passing and possibly work is being done (except for in the IDLE states) within the state and the transitions take no time.

We also need to distinguish between internal and external events. The **req** signal is controlled by the client but the changes to **avail** must be internally generated. This information would appear in a more detailed statechart where the states RECEIVE INPUT and CALC were decomposed further. For example, the last action in RECEIVE INPUT would be to generate the internal event **avail**↓. The constraints on how long the system can take to perform these tasks are also not expressed.

The set of STAs gives no indication of what the system should do if the client lowers **req** before the system has raised **avail** when reading in the inputs. Procedurally, it makes sense, to return to the IDLE state since the output has not yet been changed and in doing this, we satisfy the initial STA which must hold from any state.

The statechart highlights one constraint given by the STAs which is not immediately apparent from their visual representation. In the model of operation which the statechart presents, we can see that in attempting to check the first STA, which applies to "any state", it is only relevant in the three states of RUNNING because in IDLE the output is already valid and **avail** is high. From RECEIVE INPUT, when **req** goes low, we return to IDLE because the inputs have not yet been read in. This transition must take less than **t1** time units to satisfy the first STA. From IDLE2, when **req** goes low, the system will attempt to calculate the output going through CALC. This calculation phase is bounded in the last STA by time **t3**, but it must also be limited by **t1** since both assertions apply to the same path through the statechart in the execution of OP. Therefore **t3** must be less than or equal to **t1**. (The first STA has wider applicability than the last one therefore **t1** could be larger.)

6 Differences Between STAs and Statecharts

Many of the differences between statecharts and STAs are already apparent from the example. This section will briefly summarize these:

Events vs Conditions In statecharts, transitions can be triggered by both events and conditions. STAs rely entirely on conditions. Conditions are appropriate if, for example, as in hardware, an input is always available on an input line, but in other cases it may be more natural to speak of events, like a button being pushed. It is also possible to show inputs persisting over time by using events when they change their values. In the state RECEIVE INPUT of the example, the **req** signal must stay high otherwise the transition triggered by **req**↓ would be taken.

Visual Notation The actual arrangement of states in a statechart has meaning, whereas the ideas of concurrency and modularity are not expressed visually with STAs. The example presented above shows how it is possible to group the states RECEIVE INPUT, IDLE2, and CALC in the state RUNNING.

State Conditions STAs allow the specifier to give a meaning to a state in terms of predicates which must hold, rather than just a mnemonic name.

Synchronization Between Components In statecharts, it is possible to synchronize concurrent components using events, conditions and actions. While concurrency is not really represented in STAs, the only tools available for synchronization are predicates on the variables.

Passage of Time In statecharts, time passes, and possibly work is done, while the system is in a state, since the transitions are instantaneous. In STAs, the state describes a moment in time, and time passes and work happens on transitions. Although one STA could be true at several moments in time. These are dual notations in this sense and each developer will have a different preference on the notation used for the passage of time.

Incrementing Time As stated earlier, statecharts have no particular model of time associated with them, whereas, STAs have a very clear synchronous model. STAs also can have predicates on the amount of time a transition can take. In statecharts, it is possible to check how long the system has been in a particular state using conditions. It is also possible to schedule events to happen in the future.

7 Issues for Connecting Real-time Specification and Verification Techniques

The purpose of this study was to look at how a real-time specification technique differs from a method designed for verification with the aim of connecting a readable specification formalism with verification. The results which can be presented at this time are a list of issues, drawn from looking at both techniques, which need to be resolved in a formal specification method which includes verification as an integral part.

It should be first noted though that different types of systems will have different requirements of the specification technique. In describing a general notation and method, it may become less useful for certain problems. However, the usefulness of having a common standard which allows for comparison and integration between components of a project or projects themselves may outweigh this disadvantage.

Following is a list of these issues and a brief discussion of each. This list is not intended to be exhaustive and the usual requirements for a specification technique like readability, expressibility and support tools are not examined here. (See [2][16] for a report on these.) This list is based on items particularly concerning an approach which integrates verification into the design process for real-time systems.

Does it express operation or assertions? The fundamental difference between the two notations is the point of view from which they approach the system. STAs are limitations on the implementation's operation whereas statecharts are high level programs for the operation of the system. Timing conditions are a good example of this. A STA on the number of steps in a transition is a constraint on how much time the system can take to complete the task described by the transition. A statechart condition checking how long the system has been in a state which triggers a transition means the system should follow that path when the condition becomes true, regardless of whether it has finished the task assigned in that state.

Can it handle complete system description at any level? Rather than speak of either "hardware" or "software" specification, real-time systems are usually a combination of the two working together. From this point of view, the method should be equally applicable to both, and, as the specification becomes more detailed, the developer can decide which parts will be implemented in either hardware or software. Ideally, the process should begin with a very high-level description which can be decomposed and have more details added to it using the same notation, right down to the gate level for hardware, and the microcode level for software. It should also be possible, to move in the other direction and compose parts for reusability or to look at higher levels of abstraction. This is also the direction verification would take after a level has been given more details.

Does it have a formal model of time? A major difference in these two methods is the way time passes. Statecharts are good at specifying required reactions to events, whereas STAs give a more precise idea of how many time units will pass. In real-time systems, both these elements are essential. It should be possible to state that the time it takes for a system to react to an event will satisfy a predicate on time. We might also want to say that the time it takes a particular process to complete its execution is limited by a certain time. Notice, that these statements do not say "within" a limited time since there may be cases requiring a minimum amount of time before the system should react. The ideas of timed transition systems [11] and timed statecharts [15] will be explored to see if these modifications of statecharts provide a sufficiently clear model of time which can be used to describe the operation of the system at all levels.

Can it integrate data and control descriptions? This is an area where neither of the methods score particularly highly. Statecharts can really only describe control of the system with a limited ability for assignment to variables. The tool STATEMATE completely separates control and data descriptions, although data process can be associated with being in a particular state. STAs integrate the two aspects of the system by having predicates on the state which are really predicates on the variables at the current time in the system. Any kind of large data description would become unmanageable in this notation. The argument against STATEMATE's division of interests is that the line between what is control information and what is data is not clear in most systems and each developer would make a different division.

Is it a visual notation? To make specification and accompanying verification more amenable to industry, a specification technique should use a meaningful, compact, visual notation in which modularity and concurrency can be represented, and which it is possible to animate.

8 Future Work

8.1 A Specification Formalism

Looking beyond statecharts and STATEMATE in particular, examining these two notations has provided many ideas for what needs to be in a formal specification notation used for verification. A formalism satisfying these conditions will probably combine ideas from both statecharts and STAs. In particular, this notation should be able to express:

- Hierarchy of states
- Concurrency of operation

These first two points should be represented visually.

- Operation so the specifications would be executable (this includes having default states and history states or some form of memory, although these are really just a notational convenience). A benefit of taking the operational approach is that more automatic techniques such as symbolic simulation or model checking can be applied to the specification to test conditions, particularly timing conditions.
- Events and conditions to trigger transitions
- Communication - ideally both broadcast and some form of local communication or the ability to hide internal events in a module. Local communication would make it possible to combine specifications, worrying only about the interface between the two, rather than also having to consider if the same names are used for local events or variables.
- Timing conditions and constraints
- Data and control specifications
- Axiomatic proof rules for composition

It should be possible to develop a notation which includes all these ideas. Given the two possible ways of indicating where time passes, the notation should be able to be used in either form (ie where time passes within a state or where time passes on a transition). The two notations would be complete duals and have the same semantic meaning.

8.2 Embedding the Formalism within HOL

The semantics of this formalism can be developed using HOL as a research tool to check their preciseness, and to check certain properties to ensure that the semantics give the correct meaning. Later it can be used to verify mechanically that lower level descriptions imply the same behaviour as a more abstract view of the system.

The ability to quantify over functions in higher order logic makes it possible to use functions in the predicates constraining the system, like those used in STAs. The type of hierarchy of states used in statecharts can also be represented using higher order functions. This could be particularly useful in data process specifications. It would also be possible to take advantage of inductive proofs in cases where there are multiple instances of a type of object in the system. Another benefit to using a theorem prover is to link this type of "control" specification with more detailed data specifications. Finally, given that we have chosen to use an operational system description, it will be within the theorem-prover that we can look at constraints which must apply at more than one state within the system, such as the predicate on "any state" given in the STA of the example.

9 Summary

The main goal of this work is to integrate a more readable specification language with formal verification. Accomplishing this will make it possible for verification to start earlier in the system development process. We present in this paper a study of two existing notations from the two areas which we are trying to connect, to determine what some of the issues are in working towards this main goal. We showed that statecharts offer the advantages of readability which we desire and the availability of a commercial tool to support the notation, however, their model of operation and timing is different from STAs which have the primary purpose of verification.

We can conclude by saying that several issues which need to be examined when working towards a specification formalism for verification are:

- Does it express operation or assertions?
- Can it handle complete system description at any level?
- Does it have a formal model of time?
- Can it integrate data and control descriptions?
- Is it a visual notation?

Combining a readable front-end specification formalism with a theorem- prover and possibly other verification tools will make formal methods more appealing to industry.

10 Acknowledgements

This work was completed while I was funded by a National Science and Engineering Research Council Post Graduate Scholarship and was written up while working at SRI International, Cambridge. I am indebted to the Integrated Systems Design Laboratory at UBC, the Hardware Verification Group at Cambridge, and my colleagues at SRI for interesting discussions where many of these ideas originated.

References

[1] Victor A. Carreno. Specification of real time reactive systems using state transition assertions. a research proposal. Computer Laboratory, Cambridge, UK.

[2] Paul C. Clements, Carolyn E. Gasarch, and Ralph D. Jeffords. Evaluation criteria for real-time specification languages. Technical Report NRL Memorandum Report 6935, Naval Research Laboratory, February 1992.

[3] Derek Coleman, Fiona Hayes, and Stephan Bear. Introducing objectcharts or how to use statecharts in object-oriented design. *IEEE Transactions on Software Engineering*, 18(1):9–18, January 1992.

[4] Mike Gordon. A formal method for hard real-time programming. Computer Laboratory, Cambridge, UK.

[5] Mike Gordon. A proof generating system for higher-order logic. Technical Report No. 103, University of Cambridge Computer Lab, January 1987.

[6] D. Harel, A. Pnueli, J.P. Schmidt, and R. Sherman. On the formal semantics of statecharts. In *Proceedings of the 2nd IEEE Symposium on Logic in Computer Science*, pages 54–64, Ithaca, New York, June 1987.

[7] David Harel. Statecharts: A visual formalism for complex systems. *Science of Computing*, 8:231–274, 1987.

[8] David Harel. On visual formalisms. *Communications of the ACM*, 31(5):514–530, May 1988.

[9] David Harel. Biting the silver bullet. *IEEE Computer*, 25(1):8–20, January 1992.

[10] David Harel and et al H. Lachover. Statemate: A working environment for the development of complex reactive systems. *IEEE Transactions on Software Engineering*, 16(4):403–414, April 1990.

[11] Thomas A. Henzinger, Zohar Manna, and Amir Pnueli. Timed transition systems. Technical Report TR 92-1263, Department of Computer Science, Cornell University, Jaunary 1992.

[12] i-Logix Inc., Burlington, MA. *The Semantics of Statecharts*, January 1991.

[13] Victoria Stavridou Jonathon Bowen. Safety-critical systems, formal methods and standards. Technical Report No. PRG-TR-5-92, Oxford University Computing Laboratory, 1992.

[14] Jeffrey Joyce. *Multi-Level Verification of Microprocessor Based Systems.* PhD thesis, University of Cambridge Computer Laboratory, 1989.

[15] Y. Kesten and A. Pnueli. Timed and hybrid statecharts and their textual representation. Weizmann Institute of Science.

[16] Place, Good, and Tudball. Survey of formal specification techniques for reactive systems. Technical Report CMU/SEI 90-TR-5, Carnegie Mellon University/Software Engineering Institute, May 1990.

[17] P.J. Windley. *The Formal Verification of Generic Interpreters.* PhD thesis, University of California, Davis, 1990.

Higher Order Logic Theorem Proving and its Applications (A-20)
L.J.M. Claesen and M.J.C. Gordon (Editors)
Elsevier Science Publishers B.V. (North-Holland)

An Embedding of Timed Transition Systems in HOL

Rachel Cardell-Oliver, Roger Hale and John Herbert

Cambridge Computer Science Research Centre, SRI International, Suite 23 Millers Yard, Mill Lane, Cambridge CB2 1RQ, England

Abstract

The theory of timed transition systems provides a formal framework for specifying and reasoning about real-time systems. In this paper, we report on work in progress to mechanize the published theory of timed transition systems using the HOL theorem prover.

Keyword Codes: F.3.1
Keywords: Specifying and Verifying and Reasoning about Programs.

1 Introduction

In the theory of timed transition systems (TTSs) a system is described by a set of state transitions with associated time constraints [HMP91b]. Different specification languages may be defined in terms of this general model. In particular, we have used a real-time temporal logic (RTTL) for specifying requirements and a graphical notation for specifying system designs. A semantics for each of these languages can be given in terms of TTSs and these common semantic interpretations enable formal proofs that one specification satisfies another. Figure 1 shows the hierarchy of semantic interpretations.

We have defined a model of TTSs in HOL [SRI91]. Within this model we have verified the proof rules for TTSs which were proposed by Henzinger *et al* [HMP91a], thus demonstrating the soundness of the rules with respect to the (HOL) semantics. We are now using these rules on verification examples; further details will be reported in the full paper.

Section 2 contains a review of the main ideas of TTSs by reference to a particular example. The example is specified in a graphical specification language. We explain the underlying TTS informally and propose some real-time requirements. Sections 3, 4 and 5 present an embedding in HOL of the the main components of the theory of TTSs and show how the example is represented in this theory. Finally, in section 6, the verification of TTSs is discussed with reference to the example of section 2.

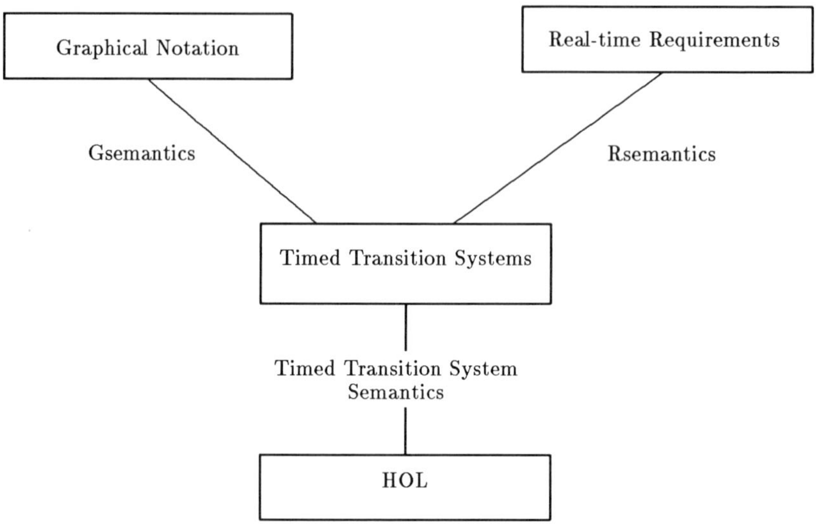

Figure 1: Hierarchy of models

2 Example: A Traffic Light Controller

The example chosen to illustrate the theory of TTSs is a simple pedestrian traffic light controller (taken from [HMP91b]).

System Description

The controller changes a pedestrian light from red to green and green to red in response to requests from pedestrians. A pedestrian may request the lights at any time. On detecting a request, the controller cancels the request and sets the pedestrian light to green. The light remains green for at least δ seconds and eventually becomes red again, whereupon the controller waits indefinitely for the next request. If any further requests are received whilst the light is green then it remains green for a further δ seconds. Thus, it is possible for pedestrians to keep the light green indefinitely.

The behaviour of the environment (pedestrians) is captured by the diagram in figure 2, and the pedestrian light controller is specified by the diagram in figure 3. The behaviour of the whole system is given by the concurrent composition of these two behaviours (papraméterised by the delay, δ).

$$\mathtt{LIGHT}_{\mathrm{TTD}}(\delta) \quad \stackrel{\mathrm{def}}{=} \quad (\mathtt{init}_{\mathrm{TTD}})[\mathtt{E} \parallel \mathtt{L}(\delta)]$$

where $\mathtt{init}_{\mathrm{TTD}}$ is a condition on the initial state of the system, E is the Timed Transition Diagram (TTD) representing the environment, and L is the TTD representing the lights. An informal interpretation of the diagrams follows.

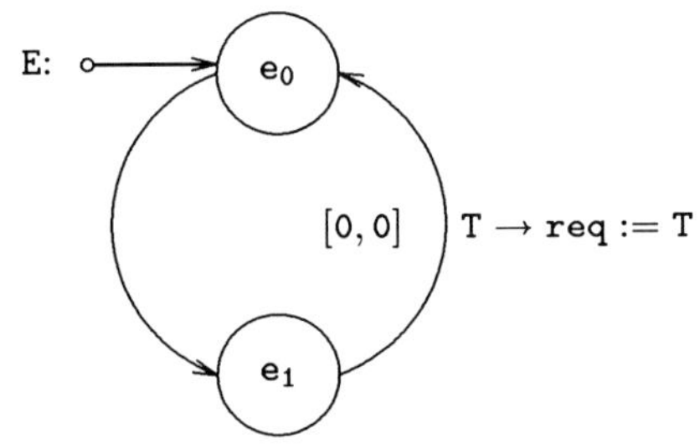

Figure 2: Environment

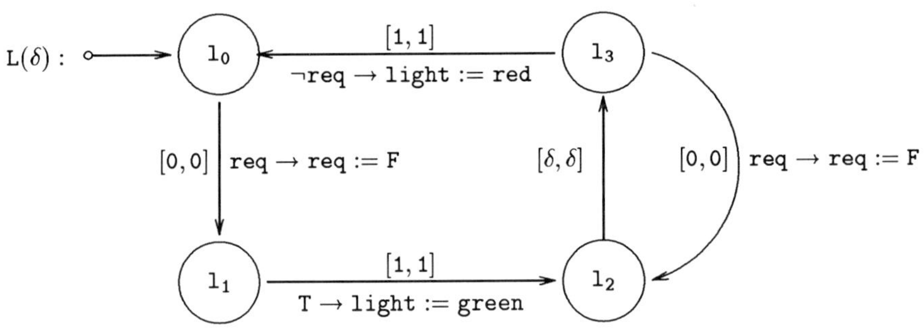

Figure 3: Traffic Light Controller

Locations $\mathtt{e_i}$ and $\mathtt{l_i}$ are names for states of the processes E and L, respectively, and each labelled arrow represents a timed transition. A transition changes the state of the system. A state usually represents a class of variable to value mappings. For example, $\mathtt{l_2}$ represents states in which the value of light is green and the value of req may be either true (T) or false (F). Initially, the environment is in state $\mathtt{e_0}$, the controller is in state $\mathtt{l_0}$, req is false and the light is red.

A timed transition has three parts: an enabling condition, an action and a timing constraint. A transition is written $c \rightarrow a$. When the enabling condition, c, is true the transition is said to be enabled. The action, a, specifies the value of variables in the state reached by performing the transition. For example, the action light := green on the transition between states $\mathtt{l_1}$ and $\mathtt{l_2}$ means that in $\mathtt{l_2}$ the value of the variable light is green. It is assumed that variables remain unchanged across the transition unless explicitly changed by an action. With each transition is associated a timing constraint, $[l, u]$, comprising a lower bound l and an upper bound u. The transition can only be taken after it has been continuously enabled for l time units and must be taken once it has been enabled for u time units.

A TTS consists of a set of timed transitions together with an initial condition which must be satisfied by the first state of the system. Thereafter, the system progresses from one state to the next by taking any transition that is enabled and has been waiting for at least its lower bound. That is, concurrency is modelled by the interleaving of transitions from all the processes in a system, in this case the controller and environment. There may, of course, be more than one transition that can be taken, in which case one is chosen non-deterministically.

Communication between processes is via state variables which are global to the system. A model encompassing shared and private variables may be considered in future work. It is also possible to define synchronous and asynchronous message passing between processes [HMP91b] and these primitives could also be mechanised in HOL.

System Requirements

Consider the following three real-time requirements for the traffic light:

- Whenever a request is registered (a pedestrian is waiting) the light must become green within 1 second.
- Having become green the light must remain green for at least $\delta + 1$ seconds.
- The light must be red within $\delta + 2$ seconds if no request is registered.

In Section 3 it is shown how these properties can be conveniently expressed using real-time temporal logic (RTTL).

A requirement is satisfied by a particular TTS if all allowable behaviours of the system satisfy that requirement. There are certain constraints on behaviours besides those that are dictated by the transitions themselves. These are detailed in section 4 below, but one is worth noting here: to avoid ambiguity about the times of state changes, time and state may not change together.

A behaviour which meets the constraints is called a computation. A possible computation of the traffic light controller satisfying these restrictions is (π_P denotes the location in process P):

#	π_E	π_L	req	light	t
0	e_0	l_0	F	red	0
1	e_0	l_0	F	red	0
2	e_0	l_0	F	red	1
3	e_1	l_0	T	red	1
4	e_1	l_1	F	red	1
5	e_1	l_1	F	red	2
6	e_1	l_2	F	green	2
⋮	⋮	⋮	⋮	⋮	⋮

Requirements are interpreted over computations. For the computation above it can be seen that whenever `req` is true the light becomes green within a time interval of length 1: `req` is only true in state 3 at time 1, the `light` = `green` in state 6 at time 2 and state 6 occurs within time 1 of state 3, as required.

The following sections present a formalisation of the components of this theory — RTTL, TTSs and TTDs — in HOL.

3 Real-Time Temporal Logic

RTTL is a metric temporal logic in which operators are bounded by intervals of time during which they apply. They may refer to past as well as future states. The following temporal operators are used.

- $\bigcirc p$, meaning "next p".
- $\Diamond_I p$, meaning "eventually p during I".
- $\Box_I p$, meaning "always p during I".
- $p \mathsf{U}_I q$, meaning "p unless q during I".

Past time variants of the above operators are permitted, but only $\ominus p$, meaning "previously p", is needed in the following.

A time interval I may be any set of time points, but we are chiefly interested in intervals of the forms:

$$[l,u] \quad [l,u) \quad (l,u] \quad (l,u)$$

For example, the formula $\Box_{[0,\infty)}p$ means that p is true throughout the infinite interval $[0,\infty)$. This is equivalent, under a conventional model of time, to the regular untimed operator, $\Box p$.

The operators of RTTL are defined over infinite sequences of *situations*. A situation has a state component and a time component. The state at position n in a situation sequence σ is given by $\mathtt{State}(\sigma,n)$ and the time is given by $\mathtt{Time}(\sigma,n)$.[1] States are mappings from variables to values; times are non-negative numbers (actually natural numbers in this work, but they could just as well be non-negative reals using an embedding of the reals in HOL [Har]).

Variables, expressions and equality

The value of a variable, written $\hat{x}$, is determined by the state on which it is evaluated.

$$\hat{x}(\sigma,n) \stackrel{\text{def}}{=} \mathtt{State}(\sigma,n)x$$

Constant expressions, such as $0,1,2,\ldots$, are simply lifted forms of the corresponding constants in underlying theories. The same syntax can be used for both forms without confusion. Overloading the equality operator, we define

$$(e_1 = e_2)(\sigma,n) \stackrel{\text{def}}{=} e_1(\sigma,n) = e_2(\sigma,n)$$

For example, $\hat{x} = 0$ denotes that the value of variable x is 0.

[1]The time component of the situation at position n is actually the time *difference*, $\mathtt{Time}(\sigma,n) - \mathtt{Time}(\sigma,n-1)$. The time at position n, $\mathtt{Time}(\sigma,n)$, is the summation of these non-negative differences, thus ensuring that time is non-decreasing and is initially 0 (see below).

Boolean Operators

These are lifted forms of the usual first-order connectives. We shall use the same syntax to denote both the first-order and lifted forms. For example, the temporal negation, $\neg p$, is true at position n of a sequence σ iff p is not true at that position.

$$(\neg p)(\sigma, n) \stackrel{\text{def}}{=} \neg(p(\sigma, n))$$

Other operators are defined similarly.

Next

The formula $\bigcirc p$ is true at position n of a sequence iff p is true at position $n+1$.

$$(\bigcirc p)(\sigma, n) \stackrel{\text{def}}{=} p(\sigma, n+1)$$

The definition works for arbitrary temporal expressions.

Previous

The formula $\ominus p$ is true at position n of a sequence iff n is greater than 0 and p is true at position $n-1$.

$$(\ominus p)(\sigma, n) \stackrel{\text{def}}{=} (0 < n) \wedge p(\sigma, n-1)$$

A similar definition can be made for arbitrary temporal expressions, but leaving the value undefined at position 0.

Eventually

The formula $\Diamond_I p$ is true at position n of a situation sequence σ iff p is true at some $n+n'$ such that the time difference between positions n and $n+n'$ is within I.

$$(\Diamond_I p)(\sigma, n) \stackrel{\text{def}}{=} \exists n' \bullet (\mathtt{Time}(\sigma, n+n') - \mathtt{Time}(\sigma, n)) \in I \ \wedge \ p(\sigma, n+n')$$

The untimed version is true for arbitrary time differences: $\Diamond p \stackrel{\text{def}}{=} \Diamond_{[0,\infty)]} p$.

Always

The formula $\Box_I p$ is true at position n of a situation sequence σ iff p is true at every $n+n'$ for which the time difference between positions n and $n+n'$ is within I.

$$(\Box_I p)(\sigma, n) \stackrel{\text{def}}{=} \forall n' \bullet (\mathtt{Time}(\sigma, n+n') - \mathtt{Time}(\sigma, n)) \in I \ \Rightarrow \ p(\sigma, n+n')$$

Again, the untimed version is true for arbitrary time differences: $\Box p \stackrel{\text{def}}{=} \Box_{[0,\infty)} p$.

Unless

The formula $p \; \mathsf{U}_I \; q$ is true at position n of a situation sequence σ iff either p is always true or q is eventually true within I and p is true at every position until then.

$$(p \; \mathsf{U}_I \; q) \;\stackrel{\text{def}}{=}\; (\Box p) \;\vee\; (p \, \mathcal{U}_I \; q)$$

where

$$\begin{aligned}(p \, \mathcal{U}_I \; q)(\sigma, n) \;\stackrel{\text{def}}{=}\; & \exists n' \bullet \\ & \quad (\mathtt{Time}(\sigma, n+n') - \mathtt{Time}(\sigma, n)) \in I \;\wedge \\ & \quad q(\sigma, n+n') \;\wedge \\ & \quad \forall n'' \bullet n'' < n' \;\Rightarrow\; p(\sigma, n+n'')\end{aligned}$$

The untimed version is true for arbitrary time differences: $p \; \mathsf{U} \; q \stackrel{\text{def}}{=} p \; \mathsf{U}_{[0,\infty)} \; q$, as before.

Example

Real-time temporal logic can be used to capture the requirements of the traffic light controller which were given informally in Section 2.

- Whenever a request is registered (a pedestrian is waiting) the light must become green within 1 second.

$$\Box((\widehat{\mathtt{req}} = \mathtt{T}) \;\Rightarrow\; \Diamond_{[0,1]}(\widehat{\mathtt{light}} = \mathtt{green}))$$

- Having become green the light must remain green for at least $\delta + 1$ seconds.

$$\Box(((\widehat{\mathtt{light}} = \mathtt{green}) \;\wedge\; \ominus(\widehat{\mathtt{light}} = \mathtt{red})) \;\Rightarrow\; \Box_{[0,\delta+1]}(\widehat{\mathtt{light}} = \mathtt{green}))$$

- The light must be red within $\delta + 2$ seconds if no request is registered. That is, if request is false then within $\delta + 2$ seconds either request has become true or the light is red.

$$\Box((\widehat{\mathtt{req}} = \mathtt{F}) \;\Rightarrow\; \Diamond_{[0,\delta+2]} \; (\widehat{\mathtt{light}} = \mathtt{red} \;\vee\; \widehat{\mathtt{req}} = \mathtt{T}))$$

How to formalise, and ultimately verify, these requirements for a particular TTS is discussed in the following sections.

4 Timed Transition Systems

A TTS is represented in HOL as a pair, $(\theta, \mathcal{T})$, comprising an initial condition, θ, and a set of timed transitions, $\mathcal{T}$.

Transitions

Each timed transition is a triple (τ, l, u), comprising:

- State transition τ, which is a set of state pairs. If (s, s') is in τ, it is possible to go from s to s' by taking τ; that is, s is a state in which τ is enabled and s' is a possible successor of s after performing the action associated with τ.

- Minimum delay l. Transition τ must be continuously enabled for l time units before it can be taken.
- Maximum delay u. Transition τ cannot be be continuously enabled for longer than u without being taken.

It is always possible to take the idle transition, $(\mathtt{Idle}, 0, \infty)$, where

$$\mathtt{Idle} \stackrel{\text{def}}{=} \{(s, s') \mid s' = s\}$$

This transition is always enabled; it leaves the state unchanged.

Formally, a transition, τ, is enabled at position n in a sequence if there is a possible successor state in τ; it is taken if the state at position $n+1$ is a successor in τ of the state at n.

$$\begin{aligned} \mathtt{Enabled}\,\tau(\sigma, n) &\stackrel{\text{def}}{=} \exists s' \bullet (\mathtt{State}(\sigma, n), s') \in \tau \\ \mathtt{Taken}\,\tau(\sigma, n) &\stackrel{\text{def}}{=} (\mathtt{State}(\sigma, n), \mathtt{State}(\sigma, n{+}1)) \in \tau \end{aligned}$$

The time for which a transition has been continuously enabled but not taken is the time for which it has been *waiting*.

$$\mathtt{Waiting}\,\tau \stackrel{\text{def}}{=} \mathtt{Age}((\mathtt{Enabled}\,\tau) \wedge \neg\ominus(\mathtt{Taken}\,\tau))$$

where the age of a formula, i.e. the time for which it has been continuously true, has the primitve recursive definition:

$$\begin{aligned} (\mathtt{Age}\,p)(\sigma, 0) &\stackrel{\text{def}}{=} 0 \\ (\mathtt{Age}\,p)(\sigma, n{+}1) &\stackrel{\text{def}}{=} p(\sigma, n{+}1) \rightarrow (\mathtt{Age}\,p(\sigma, n) + \Delta\mathtt{Time}(\sigma, n)) \mid 0 \end{aligned}$$

where

$$\Delta\mathtt{Time} \stackrel{\text{def}}{=} (\bigcirc\mathtt{Time}) - \mathtt{Time}$$

Computations

The allowable behaviours of a TTS, its computations, can now be defined formally. A computation of a TTS, $(\theta, \mathcal{T})$, with initial condition θ and set of timed transitions $\mathcal{T}$, is an infinite sequence of situations satisfying the following conditions.

- The initial condition, θ, must be true in the first state.
- Time behaves reasonably; that is, it is initially 0, never decreases and does not converge.

$$\mathtt{Time} = 0 \;\wedge\; \Box(\mathtt{Time} \le \bigcirc\mathtt{Time}) \;\wedge\; \forall t \bullet \Diamond(t \le \mathtt{Time})$$

- Time and state do not change together.

$$\Box((\mathtt{State} = \bigcirc\mathtt{State}) \vee (\mathtt{Time} = \bigcirc\mathtt{Time}))$$

- At each step, some transition from $\mathcal{T}^*$ (where $\mathcal{T}^* \stackrel{\text{def}}{=} \mathcal{T} \cup \{(\mathtt{Idle}, 0, \infty)\}$) which has been enabled for at least its lower bound must be taken.

$$\Box(\exists(\tau, l, u) \in \mathcal{T}^* \bullet \mathtt{Taken}(\tau) \;\wedge\; l \le \mathtt{Waiting}(\tau))$$

- A transition that has been continuously enabled must be taken before its upper time limit expires.

$$\forall(\tau, l, u) \in \mathcal{T}^* \bullet \Box(\texttt{Waiting}(\tau) \leq u)$$

We define $\texttt{Computation}(\theta, \mathcal{T})$ to be true of a sequence of situations when these conditions hold.

Properties of Computations

The formulae of RTTL are defined over infinite sequences of situations, and thus they can be interpreted over the computations of a TTS. The relation **Sat** is defined between TTSs and formulae of real-time temporal logic. A particular TTS satisfies a temporal requirement if the temporal formula is true for *all* of the computations of that TTS. Formally,

$$(\theta, \mathcal{T}) \ \texttt{Sat}\ p \ \stackrel{\text{def}}{=} \ \forall\sigma.\ \texttt{Computation}(\theta, \mathcal{T})\sigma \ \Rightarrow \ p(\sigma, 0)$$

For example, we want to show that the TTS corresponding to the pedestrian light system satisfies each of the three requirements given in section 3.

5 Timed Transition Diagrams

In Section 2 TTDs were used to specify the behaviour of a traffic light controller. In this section we show how such designs can be specified formally using our HOL embedding of TTSs.

TTD Representation

TTD transitions are characterised by a process name, an initial location, a final location, an enabling condition, an action and upper and lower time bounds. For example, the diagram fragment

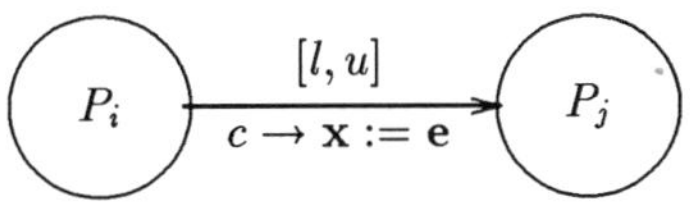

in a process P means that if P is in location P_i and c is true for at least l time steps then the transition may be taken. The effect of the transition is to change P's current location, given by π_P, to P_j and update process variables according to the action $\mathbf{x} := \mathbf{e}$. The transition must be taken before it has been continuously enabled for more than u time steps.

If there is no time interval specified for a TTD transition, it is assumed to be $[0, \infty)$. If no enabling condition is given, the transition is assumed to be always enabled, and if no action is given then the values of the variables in the new state P_j remain the same as those in state P_i.

TTD transitions are represented in HOL by a structured type, which can be thought of as a record. There are fields for, respectively, the process identifier, initial and final locations, an enabling condition, an action and the lower and upper time bounds. For example, the TTD transition from $\texttt{l}_0$ to $\texttt{l}_1$ in the traffic light

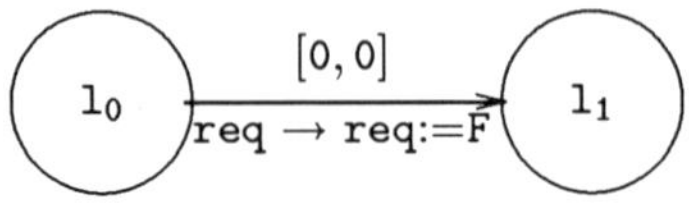

is represented in our HOL embedding by the term,[a]

$$\texttt{TRANS L } \texttt{l}_0\ \texttt{l}_1\ \{\widehat{\texttt{req}} = \texttt{T}\}\ \{(\texttt{req}, \texttt{F})\}\ \texttt{0 0}$$

The enabling condition is a set of conditions that must be true for the transition to be enabled, which, by default, is the empty set. An action is represented by a set of variable, expression pairs. For example, the pair $(\texttt{req}, \texttt{F})$ denotes that the transition assigns the value F to the variable req. Again, the default is the empty set which means that no process variables are updated by the transition.

In the shared variable communication model there is also a frame assumption; i.e. all variables, other than location variables, not explicitly listed in the action set remain unchanged by the transition. For the moment, we assume all variable names are global to the system.

Recall that a system specified using TTDs comprises an initial condition together with a set of concurrent processes. A corresponding syntactic structure may be defined in HOL, with each TTD may be represented as a set of transitions of the above form.

TTD to TTS Mapping

Let us now give a formal definition of the mapping from TTD transitions to TTS transitions. Recall that a TTS transition is a triple, (τ, l, u), comprising a set of state pairs, τ, a lower bound, l and an upper bound u. The TTS transition corresponding to the TTD transition

$$\texttt{TRANS}\ P\ P_i\ P_j\ C\ A\ l\ u$$

is the set of state pairs (σ_1, σ_2) for which

- all the conditions in the set $C \cup \{\widehat{\pi_P} = P_i\}$ are true in σ_1,
- all the assignment actions in the set $A \cup \{(\pi_P, P_j)\}$ are achieved in state σ_2,
- all variables not explicitly assigned have the same values in σ_2 as in σ_1.

[a]We are overloading the notation $\hat{\texttt{x}}$. Here, it denotes the value of a variable on a state rather than its value on a situation sequence.

This set is defined formally, as follows, denoting by $\mathtt{mapfst}\ S$ the set $\{x \mid \exists y \bullet (x,y) \in S\}$ of all first elements of pairs in S,

$$\begin{array}{lcl}\mathtt{Tr}\ P\ P_i\ P_j\ C\ A & \stackrel{\text{def}}{=} & \{(\sigma_1,\sigma_2) \mid \\ & & \quad \forall c \in (C \cup \{\widehat{\pi_P} = P_i\}) \bullet c(\sigma_1) \ \wedge \\ & & \quad \forall (x,e) \in (A \cup \{(\pi_P, P_j)\}) \bullet \sigma_2(x) = e(\sigma_1) \ \wedge \\ & & \quad \forall x \notin (\mathtt{mapfst}\ A) \cup \{\pi_P\} \bullet \sigma_2(x) = \sigma_1(x)\}\end{array}$$

The time bounds, l and u, are the same for the TTS transition as for the TTD transition. Thus, the meaning of a TTD transition is given by

$$\mathtt{MkTrans}\ (\mathtt{TRANS}\ P\ P_i\ P_j\ C\ A\ l\ u) \stackrel{\text{def}}{=} ((\mathtt{Tr}\ P\ P_i\ P_j\ C\ A), l, u)$$

and the meaning of a system specified using TTDs is a TTS, $(\theta, \mathcal{T})$, where the initial condition, θ, is the same as for the TTD and the set of timed transitions, $\mathcal{T}$, is obtained by mapping MkTrans over the concurrent composition of TTDs.

Example

Recall that the behaviour of the pedestrian light system described in Section 2 was specified by

$$\mathtt{LIGHT}_{\mathrm{TTD}}(\delta) \stackrel{\text{def}}{=} (\mathtt{init}_{\mathrm{TTD}})[\mathtt{E} \parallel \mathtt{L}(\delta)]$$

for the TTDs E and L. The corresponding TTS, say $\mathtt{LIGHT}_{\mathrm{TTS}}$, comprises an initial condition and the set of timed transitions corresponding to the TTD transitions.

The initial condition, $\mathtt{init}_{\mathrm{TTS}}$, is a formula that must be true on the initial state.[3]. It is the same as $\mathtt{init}_{\mathrm{TTD}}$, i.e.

$$\mathtt{init}_{\mathrm{TTS}} \stackrel{\text{def}}{=} \widehat{\pi_{\mathtt{E}}} = \mathtt{e}_0 \ \wedge \ \widehat{\pi_{\mathtt{L}}} = \mathtt{l}_0 \ \wedge \ \widehat{\mathtt{req}} = \mathtt{F} \ \wedge \ \widehat{\mathtt{light}} = \mathtt{red}$$

The whole system is then given by the TTS:

$$\begin{array}{lcl}\mathtt{LIGHT}_{\mathrm{TTS}}(\delta) & = & (\mathtt{init}_{\mathrm{TTS}}, \\ & & \{\mathtt{MkTrans}(\mathtt{TRANS\ L\ l_0\ l_1}\ \{\widehat{\mathtt{req}} = \mathtt{T}\}\ \{(\mathtt{req},\mathtt{F})\}\ 0\ 0), \\ & & \ \mathtt{MkTrans}(\mathtt{TRANS\ L\ l_1\ l_2}\ \{\}\ \{(\mathtt{light},\mathtt{green})\}\ 1\ 1), \\ & & \ \mathtt{MkTrans}(\mathtt{TRANS\ L\ l_2\ l_3}\ \{\}\ \{\}\ \delta\ \delta), \\ & & \ \mathtt{MkTrans}(\mathtt{TRANS\ L\ l_3\ l_2}\ \{\widehat{\mathtt{req}} = \mathtt{T}\}\ \{(\mathtt{req},\mathtt{F})\}\ 0\ 0), \\ & & \ \mathtt{MkTrans}(\mathtt{TRANS\ L\ l_3\ l_0}\ \{\widehat{\mathtt{req}} = \mathtt{F}\}\ \{(\mathtt{light},\mathtt{red})\}\ 1\ 1), \\ & & \ \mathtt{MkTrans}(\mathtt{TRANS\ E\ e_0\ e_1}\ \{\}\ \{\}\ 0\ \infty), \\ & & \ \mathtt{MkTrans}(\mathtt{TRANS\ E\ e_1\ e_0}\ \{\}\ \{(\mathtt{req},\mathtt{T})\}\ 0\ 0)\})\end{array}$$

Further work is needed to capture the syntax and semantics of TTDs more conveniently in HOL. Eventually, users should be able to enter specifications using a graphical tool and the translation to the underlying TTS, and any checking that the TTD is well formed, should be performed automatically by HOL in the background.

---

[3]Note that our interpretation differs slightly from Henzinger's in which all processes start in an "undefined" location, $\perp$, and then proceed to the "initial" location at any time. This assumption can be specified explicitly in our interpretation but it is not assumed implicitly.

6 Verification

In order to prove that the traffic light design satisfies a temporal requirement p we must prove that the TTS interpretation, $\texttt{LIGHT}_{\texttt{TTS}}(\delta)$, satisfies p. That is, we must prove the theorem,

$$\vdash \ \forall\delta \bullet (\texttt{LIGHT}_{\texttt{TTS}}(\delta)) \ \texttt{Sat} \ p$$

Recall that a TTS satisfies a requirement, p, if p is true in all situation sequences which are computations of the TTS.

Proof Rules

A set of proof rules, which is complete for showing that a particular class of TTSs satisfies a given class of RTTL formulae, has been proposed by Henzinger *et al* [HMP91a]. The class of RTTL requirements captured are bounded response and bounded invariance properties. Bounded response properties are temporal formulae of the form

$$p \ \Rightarrow \ \Diamond_{[0,u]}q$$

Bounded invariance properties are temporal formulae of the form

$$p \ \Rightarrow \ \Box_{[0,l]}q \quad \text{or, more generally,} \quad p \ \Rightarrow \ q \ \mathsf{U}_{[l,\infty)} \ r$$

There are certain restrictions on the class of TTSs to which the rules apply. For instance, every loop within a TTD must have at least two edges and the enabling conditions for two transitions from the same location must be mutually exclusive[4]. The TTS, $\texttt{LIGHT}_{\texttt{TTS}}(\delta)$, and the temporal requirements given in Section 3 meet these conditions.

At the core of Henzinger's approach are two step rules, which give the maximum time until a given transition is taken and the minimum time which must elapse before the transition can be taken. RTTL rules for transitivity, disjunction and so on, may then be invoked to combine properties of a chain of transitions. There are also induction rules for proving response times of TTDs which contain terminating loops. All these rules have been verified in HOL with respect to the TTS definitions of computations. Thus, we have proved mechanically that the rules are sound with respect to TTS semantics.

The *upper bound rule*, SS_UB, is used to prove bounded response properties of the form

$$(p \ \Rightarrow \ \Diamond_{[0,u]}q)$$

for a particular timed transition, (τ, l, u). If τ is enabled from the time p is true until it is taken and taking τ makes q true, then q must become true within an interval $[0, u]$ of p being true. Note that $\Diamond_{[0,u]}q$ is still satisfied if q becomes true before τ is taken. This rule depends on finding an invariant, φ, that is true from when p becomes true until q becomes true.

SS_UB: For a transition (τ, l, u) belonging to a TTS, $(\theta, \mathcal{T})$, the following four premises must be satisfied.

[4]Recent work by Henzinger relaxes this requirement

1. There is an invariant, φ, such that the truth of p implies that either φ or q is true.
2. The invariant ensures that τ is enabled.
3. If φ is true, every transition except τ either preserves φ or makes q true.
4. Taking τ when φ is true makes q is true.

If these four premises are satisfied then we may conclude that q is always true within u time units of p being true. This rule is formalised as follows:

SS_UB (for $(\tau, l, u) \in \mathcal{T}$):

$$
\begin{array}{lll}
(1) & (\theta, \mathcal{T}) \ \texttt{Sat} & \Box(p \Rightarrow (\varphi \vee q)) \\
(2) & (\theta, \mathcal{T}) \ \texttt{Sat} & \Box(\varphi \Rightarrow \texttt{Enabled}\ \tau) \\
(3) & (\theta, \mathcal{T}) \ \texttt{Sat} & \Box(\forall \tau' \in (\mathcal{T}^* - \{\tau\}) \bullet \varphi \wedge \texttt{Taken}\ \tau' \Rightarrow \bigcirc(\varphi \vee q)) \\
(4) & (\theta, \mathcal{T}) \ \texttt{Sat} & \Box(\varphi \wedge \texttt{Taken}\ \tau \Rightarrow \bigcirc q) \\
\hline
\vdash & (\theta, \mathcal{T}) \ \texttt{Sat} & \Box(p \Rightarrow \Diamond_{[0,u]} q)
\end{array}
$$

For TTDs, a special instantiation of the upper bound rule is useful. For a transition

$$\tau \ = \ \texttt{MkTrans}(\texttt{TRANS}\ P\ P_i\ P_j\ C\ A\ l\ u)$$

choose:

$p = \texttt{Enabled}\ \tau$ $\qquad$ ($\widehat{\pi_P} = P_i$ and all conditions in C satisfied)
$\varphi = p$
$q = \texttt{After}\ \tau$ $\qquad$ (where $\texttt{After}\ \tau \ = \ (\widehat{\pi_P} = P_j)$)

For this choice of p, φ and q, all but the third of the four premises for the upper bound rule are always satisfied. Thus, for the TTS, $(\theta, \mathcal{T})$, representing a TTD system we have the rule, TTD_UB:

TTD_UB (for $(\tau, l, u) = \texttt{MkTrans}(\texttt{TRANS}\ P\ P_i\ P_j\ C\ A\ l\ u)$ and $(\tau, l, u) \in \mathcal{T}$):

$$
\begin{array}{lll}
(1) & (\theta, \mathcal{T}) \ \texttt{Sat} & \Box(\forall \tau' \in (\mathcal{T}^* - \{\tau\}) \bullet \\
 & & \quad (\texttt{Enabled}\ \tau \wedge \texttt{Taken}\ \tau') \Rightarrow \bigcirc(\texttt{Enabled}\ \tau \vee \texttt{After}\ \tau)) \\
\hline
\vdash & (\theta, \mathcal{T}) \ \texttt{Sat} & \Box(\texttt{Enabled}\ \tau \Rightarrow \Diamond_{[0,u]}(\texttt{After}\ \tau))
\end{array}
$$

In the HOL mechanisation of this rule, the premise can be proved automatically, provided that certain conditions are met. For a transition τ, these conditions are:

- if τ is enabled, no other transition of the same process, except `Idle`, is enabled.
- τ must not be disabled by a transition of any other process.

The first of these conditions dictates that transitions starting from the same location are mutually exclusive.

A number of standard rules and theorems of RTTL are also required. In the verification example presented below we use the following, all of which have been mechanically verified in HOL:

`TRANS_UB` expresses the transitivity of the diamond operator.

$$\begin{array}{lll} & \multicolumn{2}{c}{\texttt{TRANS\_UB}:} \\ (1) & (\theta,\mathcal{T}) & \texttt{Sat}\ \Box(p \Rightarrow \Diamond_{[0,u_1]}\, q) \\ (2) & (\theta,\mathcal{T}) & \texttt{Sat}\ \Box(q \Rightarrow \Diamond_{[0,u_2]}\, r) \\ \hline \vdash & (\theta,\mathcal{T}) & \texttt{Sat}\ \Box(p \Rightarrow \Diamond_{[0,u_1+u_2]}\, r) \end{array}$$

`WITHIN_0_lemma` expresses the fact that if p is true now then it also is true within an interval of length 0.

$$\vdash\ (\theta,\mathcal{T})\ \texttt{Sat}\ \Box(p\ \Rightarrow\ \Diamond_{[0,0]}\ p)$$

`DIAMOND_LESS_lemma` states that if p is true within an interval of length u_1 then it is also true within any longer interval of length $u_1 + u_2$.

$$\vdash\ (\theta,\mathcal{T})\ \texttt{Sat}\ \Box(\Diamond_{[0,u_1]}\ p\ \Rightarrow\ \Diamond_{[0,u_1+u_2]}\ p)$$

Example

These theorems and rules are now used to prove that $\texttt{LIGHT}_{\texttt{TTS}}(\delta)$ satisfies the requirement that if request is true the light becomes green within 1 time step.

We have also used the upper bound rule to prove that if request is initially false then within $\delta+2$ time steps either request will have become true or the light controller will have entered location $\texttt{l}_0$ and thus the light will be red. To prove that once the light becomes green it remains green for at least $\delta + 1$ time steps we used a variant of Henzinger's lower bound rule. Both of these proofs have been performed in HOL, but we do not present the details here.

Required to prove:

$$\texttt{LIGHT}_{\texttt{TTS}}(\delta)\ \ \texttt{Sat}\ \ \Box(\widehat{\texttt{req}} = \texttt{T}\ \Rightarrow\ \Diamond_{[0,1]}\widehat{\texttt{light}} = \texttt{green})$$

Proof:

We can prove by induction on the states of all computations of the traffic light system that,

$$(0)\quad \texttt{LIGHT}_{\texttt{TTS}}(\delta)\ \ \texttt{Sat}\ \ \Box((\widehat{\texttt{light}} = \texttt{green})\ =\ (\widehat{\pi_{\texttt{L}}} = \texttt{l}_2\ \lor\ \widehat{\pi_{\texttt{L}}} = \texttt{l}_3))$$

Thus, to prove the original goal it is sufficient to prove the following,

$$\texttt{LIGHT}_{\texttt{TTS}}(\delta)\ \ \texttt{Sat}\ \ \Box(\widehat{\texttt{req}} = \texttt{T}\ \Rightarrow\ \Diamond_{[0,1]}\widehat{\pi_{\texttt{L}}} = \texttt{l}_2)$$

Our proof shows that no matter which state of the traffic light controller we begin in, if request is initially true then within a time interval of length 1 the light process will be in location $\texttt{l}_2$.

From the upper bound rule TTD_UB derive consequences (1), (2), (3).

(1) $\mathtt{LIGHT_{TTS}}(\delta)$ Sat $\Box(\widehat{\pi_L} = \mathtt{l}_0 \ \wedge \ \widehat{\mathtt{req}} = \mathtt{T} \ \Rightarrow \ \Diamond_{[0,0]}\widehat{\pi_L} = \mathtt{l}_1)$

(2) $\mathtt{LIGHT_{TTS}}(\delta)$ Sat $\Box(\widehat{\pi_L} = \mathtt{l}_1 \ \Rightarrow \ \Diamond_{[0,1]}\widehat{\pi_L} = \mathtt{l}_2)$

(3) $\mathtt{LIGHT_{TTS}}(\delta)$ Sat $\Box(\widehat{\pi_L} = \mathtt{l}_3 \ \wedge \ \widehat{\mathtt{req}} = \mathtt{T} \ \Rightarrow \ \Diamond_{[0,0]}\widehat{\pi_L} = \mathtt{l}_2)$

From the the transitivity rule TRANS_UB, (1) and (2) derive

(4) $\mathtt{LIGHT_{TTS}}(\delta)$ Sat $\Box(\widehat{\pi_L} = \mathtt{l}_0 \ \wedge \ \widehat{\mathtt{req}} = \mathtt{T} \ \Rightarrow \ \Diamond_{[0,1]}\widehat{\pi_L} = \mathtt{l}_2)$

By elementary logic and (2) we have,

(5) $\mathtt{LIGHT_{TTS}}(\delta)$ Sat $\Box(\widehat{\pi_L} = \mathtt{l}_1 \ \wedge \ \widehat{\mathtt{req}} = \mathtt{T} \ \Rightarrow \ \Diamond_{[0,1]}\widehat{\pi_L} = \mathtt{l}_2)$

Similarly, from WITHIN_0_lemma and DIAMOND_LESS_lemma deduce,

(6) $\mathtt{LIGHT_{TTS}}(\delta)$ Sat $\Box(\widehat{\pi_L} = \mathtt{l}_2 \ \wedge \ \widehat{\mathtt{req}} = \mathtt{T} \ \Rightarrow \ \Diamond_{[0,1]}\widehat{\pi_L} = \mathtt{l}_2)$

Also, DIAMOND_LESS_lemma and (3) gives,

(7) $\mathtt{LIGHT_{TTS}}(\delta)$ Sat $\Box(\widehat{\pi_L} = \mathtt{l}_3 \ \wedge \ \widehat{\mathtt{req}} = \mathtt{T} \ \Rightarrow \ \Diamond_{[0,1]}\widehat{\pi_L} = \mathtt{l}_2)$

Combining (4), (5), (6) and (7) using elementary logic gives,

(8) $\mathtt{LIGHT_{TTS}}(\delta)$ Sat

$$\Box((\widehat{\pi_L} = \mathtt{l}_0 \ \vee \ \widehat{\pi_L} = \mathtt{l}_1 \ \vee \ \widehat{\pi_L} = \mathtt{l}_2 \ \vee \ \widehat{\pi_L} = \mathtt{l}_3) \ \wedge \ \widehat{\mathtt{req}} = \mathtt{T} \ \Rightarrow \ \Diamond_{[0,1]}\widehat{\pi_L} = \mathtt{l}_2)$$

Finally, we can prove by mathematical induction on situation sequences and the definition of a computation of a TTS that the traffic light process is always in one of its locations $\mathtt{l}_0$, $\mathtt{l}_1$, $\mathtt{l}_2$ or $\mathtt{l}_3$.

(9) $\mathtt{LIGHT_{TTS}}(\delta)$ Sat $\Box(\widehat{\pi_L} = \mathtt{l}_0 \ \vee \ \widehat{\pi_L} = \mathtt{l}_1 \ \vee \ \widehat{\pi_L} = \mathtt{l}_2 \ \vee \ \widehat{\pi_L} = \mathtt{l}_3)$

From (8) and (9) derive,

(10) $\mathtt{LIGHT_{TTS}}(\delta)$ Sat $\Box(\widehat{\mathtt{req}} = \mathtt{T} \ \Rightarrow \ \Diamond_{[0,1]}\widehat{\pi_L} = \mathtt{l}_2)$

and by (0) this implies the original goal,

$$\mathtt{LIGHT_{TTS}}(\delta) \ \mathtt{Sat} \ \Box(\widehat{\mathtt{req}} = \mathtt{T} \ \Rightarrow \ \Diamond_{[0,1]}\widehat{\mathtt{light}} = \mathtt{green})$$

as required which completes the proof.

End proof.

In the HOL proof of this theorem there are steps corresponding to each application of the rules TRANS_UB and TTD_UB. Much of the proof effort for checking each rule application has been automated in HOL using ML tactics, but with further work more automation should be possible. Our aim is to automate as much as possible of the mundane checking of details, to enable a user to concentrate on the high level proof steps, which require insight.

References

[Har] J. R. Harrison. Constructing the real numbers in HOL. In *this volume.*

[HMP91a] T. A. Henzinger, Z. Manna, and A. Pnueli. Temporal proof methodologies for real-time systems. In *Proceedings of the 18th Symposium on Principles of Programming Languages.* ACM Press, 1991.

[HMP91b] T. A. Henzinger, Z. Manna, and A. Pnueli. Timed transition systems. In *Proceedings of the 1991 REX Workshop*, November 1991.

[SRI91] SRI International and DSTO Australia. *The HOL System.* Cambridge Computer Science Research Center, SRI International, revised version, 1991. Four volumes.

Higher Order Logic Theorem Proving and its Applications (A-20)
L.J.M. Claesen and M.J.C. Gordon (Editors)
Elsevier Science Publishers B.V. (North-Holland)

Formalizing a Modal Logic for CCS in the HOL Theorem Prover

Monica Nesi

University of Cambridge, Computer Laboratory, New Museums Site, Pembroke Street, Cambridge CB2 3QG, England

Abstract

The operational semantics of concurrent processes is given via labelled transition systems, and logical properties of processes can be expressed by means of modal/temporal logics, which are interpreted on transition systems as well. A recently developed package for inductive definitions in the HOL theorem prover allows one to formalize the transition relation for processes and to prove many interesting theorems about it. Based on this formalization, we consider the mechanization in HOL of (a slight extension of) Hennessy-Milner logic for a subset of Milner's CCS, and show how modal properties can be mechanically verified in the resulting framework.

Keyword Codes: D.3.1, F.3.1, I.2.3
Keywords: Programming Languages, Formal Definitions and Theory; Specifying and Verifying and Reasoning about Programs; Deduction and Theorem Proving

1 Introduction

In this paper we use higher order logic and the general purpose theorem prover HOL [8, 9] to formalize a modal logic for Milner's CCS [17]. This study is a step in the development of an interactive verification environment for supporting reasoning about process algebra specifications. In previous works [6, 19] verification strategies for behavioural equivalences between CCS specifications have been mechanized. These include strategies that exhibit different degrees of user interaction, depending on the subsets of CCS under consideration [6], and proofs of correctness by mathematical induction for parameterized specifications [19]. The approach is based on the algebraic nature of the CCS language and on the axiomatic characterization of behavioural semantics. The formal theory for a specific behavioural semantics, namely observational congruence [17], has been embedded in the HOL logic, and the resulting formalization supports verification strategies based on mechanized formal proof.

At the initial stage of this work our interest was in higher level axiomatic strategies. The decision was taken to skip the formalization of the operational semantics and of observational congruence and the derivation of the algebraic laws for such a behavioural semantics. The laws were just directly asserted in the HOL logic. This gap in the formalization has now been filled: operational semantics and observational congruence have been defined in HOL, and the algebraic laws for observational congruence have been derived by formal proof [20].

The aim is to build a verification system based on theorem proving which is logically sound, i.e. it is built up by following a purely definitional approach, in order to avoid introducing inconsistencies in the logic being used. In the HOL system, this can be achieved by means of primitive definition mechanisms for introducing new entities in a sound way, and by deriving all other kinds of definitions by formal proof.

All components of the formal theory for process algebras, such as the labelled transitions, the operational and axiomatic characterizations of behavioural semantics and modal/temporal logics, can be formalized in a unified framework, and users can define their own verification strategies. In this paper we reason about process algebra specifications by verifying that processes have or lack some given modal properties. We address *pure* CCS (no value passing) and a slight extension of Hennessy-Milner logic, in which modalities are indexed by a set of actions instead of a single action [10, 21]. The formal theory for pure CCS and for the modal logic is embedded in higher order logic, and the resulting formalization is the basis for property checking.

In what follows, the subset of CCS and the modal logic under consideration are first introduced. We then describe how the syntactic definitions, the operational semantics and the process logic can be formalized in HOL. Next, we show how modal properties can be checked by presenting the HOL proofs for two simple recursive processes like a vending machine and a counter. Finally, we discuss related work and possible extensions to the described approach.

2 CCS

In this section we recall only the essential information about CCS (*Calculus of Communicating Systems*) and refer the reader to [17] for more details about the calculus.

2.1 Syntax and Operational Semantics

Pure CCS is a subset of the language which does not involve value passing and consists of the inactive process **nil**, and the following operations on processes: *prefix* (.), *summation* (+), *restriction* (\), *relabelling* ([ ]), *parallel composition* (|) and *recursion* (**rec**). The syntax of pure CCS expressions, ranged over by $E, F, E', E1, \ldots$ is as follows:

$$E ::= \mathbf{nil} \mid X \mid u.E \mid E+E \mid E \setminus L \mid E[f] \mid E|E \mid \mathbf{rec}\, X.\, E$$

where L is a subset of visible actions, called labels (ranged over by l), u ranges over actions, which are either labels or the invisible action τ, and X ranges over process variables (which will be bound in recursive definitions). Labels consist of names and co-names where, for any name a, the corresponding co-name is written $\overline{a}$. This complement operation has the property that $\overline{\overline{l}} = l$. A relabelling function f is a function from labels to labels such that relabelling co-names has the property that $f(\bar{l}) = \overline{f(l)}$. A relabelling function f is then extended to actions by defining $f(\tau) = \tau$.

The expression **nil** represents a process which cannot perform any action. The process $u.E$ can only perform the action u and then behaves like E. The process $E_1 + E_2$ behaves like either E_1 or E_2. The process $E \setminus L$ behaves like E but cannot perform an action u if either u or $\overline{u}$ is in L. The actions of $E\,[f]$ are renamings of those of the process E via the relabelling function f. The process $E_1 \,|\, E_2$ can perform the actions of E_1 and E_2 in parallel; moreover, the processes E_1 and E_2 can synchronize through the action τ whenever they are able to perform complementary actions. The expression $\mathbf{rec}\, X.\ E$ denotes a recursive process.

The operational semantics of the above CCS operators is given via a transitional semantics based on *labelled interleaving transitions* $\xrightarrow{u}$ over CCS expressions. The transition relation $E \xrightarrow{u} F$ is inductively defined by the following rules:

$$\text{PREFIX:} \quad \frac{}{u\,.E \xrightarrow{u} E}$$

$$\text{SUM1:} \quad \frac{E \xrightarrow{u} E'}{E{+}F \xrightarrow{u} E'} \qquad\qquad \text{SUM2:} \quad \frac{E \xrightarrow{u} E'}{F{+}E \xrightarrow{u} E'}$$

$$\text{RESTR:} \quad \frac{E \xrightarrow{u} E'}{E\backslash L \xrightarrow{u} E'\backslash L} \quad u, \overline{u} \notin L \qquad\qquad \text{RELAB:} \quad \frac{E \xrightarrow{u} E'}{E\,[f] \xrightarrow{f(u)} E'\,[f]}$$

$$\text{PAR1:} \quad \frac{E \xrightarrow{u} E'}{E|F \xrightarrow{u} E'|F} \qquad\qquad \text{PAR2:} \quad \frac{E \xrightarrow{u} E'}{F|E \xrightarrow{u} F|E'}$$

$$\text{PAR3:} \quad \frac{E \xrightarrow{l} E' \quad F \xrightarrow{\bar{l}} F'}{E|F \xrightarrow{\tau} E'|F'}$$

$$\text{REC:} \quad \frac{E\{\mathbf{rec}\, X.\, E/X\} \xrightarrow{u} E'}{\mathbf{rec}\, X.\, E \xrightarrow{u} E'}$$

where the notation $E\{\mathbf{rec}\, X.\ E/X\}$ denotes the substitution of $\mathbf{rec}\, X.\ E$ for all free occurrences of X in the expression E.

2.2 A Modal Logic

The modal logic under consideration is a slight extension of Hennessy-Milner logic in the sense that modalities are indexed by a set of actions instead of a single action [10, 21]. Its formulas are defined by the following abstract syntax:

$$\Phi \ ::= \ \mathtt{tt} \ \mid \ \neg\Phi \ \mid \ \Phi \wedge \Phi \ \mid \ [A]\Phi$$

where A ranges over sets of actions. A formula is either the constant true formula $\mathtt{tt}$, or a negated formula $\neg\Phi$, or a conjunction of formulas $\Phi_1 \wedge \Phi_2$, or a modalized formula $[A]\Phi$, where the modal operator $[A]$ is sometimes referred to as *box*. The meaning of the first three formulas is familiar, while the modalized formula $[A]\Phi$ means that Φ holds after every execution of any action in A.

For any formula Φ of the logic we define when a process E has (or "satisfies") the property Φ. We write $E \models \Phi$ to mean E satisfies Φ, and $E \not\models \Phi$ to mean E fails to have the property Φ. The *satisfaction* relation $\models$ is inductively defined on the structure of formulas:

$$
\begin{array}{lll}
E \models \mathtt{tt} & & \\
E \models \neg\Phi & \textit{iff} & E \not\models \Phi \\
E \models \Phi_1 \wedge \Phi_2 & \textit{iff} & E \models \Phi_1 \textit{ and } E \models \Phi_2 \\
E \models [A]\Phi & \textit{iff} & \forall E'. \forall u \in A. \textit{ if } E \xrightarrow{u} E' \textit{ then } E' \models \Phi
\end{array}
$$

Every process has the property tt. A process has the property $\neg\Phi$ when it fails to satisfy the property Φ, and it has the property $\Phi_1 \wedge \Phi_2$ when it has both properties Φ_1 and Φ_2. Finally, a process satisfies $[A]\Phi$ if after every execution of any action in A all the resulting processes have the property Φ.

Derived operators, including the dual $\langle A\rangle$ (sometimes referred to as *diamond*) of $[A]$, are defined as follows:

$$
\begin{array}{lll}
\mathtt{ff} & \stackrel{\text{def}}{=} & \neg\mathtt{tt} \\
\Phi_1 \vee \Phi_2 & \stackrel{\text{def}}{=} & \neg(\neg\Phi_1 \wedge \neg\Phi_2) \\
\langle A\rangle\Phi & \stackrel{\text{def}}{=} & \neg[A]\neg\Phi
\end{array}
$$

The intended meaning of the diamond operator $\langle A\rangle$ is the following:

$$E \models \langle A\rangle\Phi \quad \textit{iff} \quad \exists E'. \exists u \in A.\ E \xrightarrow{u} E' \textit{ and } E' \models \Phi$$

Properties such as *capacity* and *necessity* can be expressed within this logic. The modal formula $\langle A\rangle\mathtt{tt}$ expresses a capacity to perform an action in A, since:

$$E \models \langle A\rangle\mathtt{tt} \quad \textit{iff} \quad \exists E'. \exists u \in A.\ E \xrightarrow{u} E'$$

and $\langle A\rangle\mathtt{ff}$ expresses an inability to perform any action in A. Using the notation in which, given a set of actions *Act*, $[-]$ stands for $[Act]$ and $[-u]$ for $[Act - \{u\}]$ (and analogously for the diamond operator), the property that a process E must perform a given action a (necessity) can be expressed as follows:

$$E \models \langle -\rangle\mathtt{tt} \wedge [-a]\mathtt{ff}$$

where the formula $\langle -\rangle\mathtt{tt}$ states that some action can be performed, and $[-a]\mathtt{ff}$ expresses that every action but a is impossible.

3 Mechanization of CCS in HOL

3.1 The Syntax

The first step in the HOL formalization of the CCS syntax is the mechanization of labels and actions using the derived principle for recursive type definitions [14]. This principle allows one to define arbitrary concrete recursive types in terms of their constructors. The input to this definition mechanism is a specification of the syntax of the operators written in terms of existing types and recursive calls to the type being defined. The syntactic types *label* and *action* can be defined as follows:

$$\textit{label} = \mathsf{name}\ \textit{string} \mid \mathsf{coname}\ \textit{string}$$

$$action = \mathsf{tau} \mid \mathsf{label}\ label$$

where name, coname, tau and label are distinct constructors. Given the above specifications, the HOL system automatically derives a theorem of higher order logic for each type being defined, which characterizes the type in a complete and abstract way. These theorems assert the admissibility of defining functions over the types *label* and *action* by primitive recursion. Structural induction theorems for both types are provided as well.

The notion of complement can be defined by a function over the type *label* as follows:

$$(\forall s.\ \mathsf{Compl}\,(\mathsf{name}\ s) = \mathsf{coname}\ s)\ \wedge\ (\forall s.\ \mathsf{Compl}\,(\mathsf{coname}\ s) = \mathsf{name}\ s)$$

and then extended to actions with the following definition:

$$\forall l.\ \mathsf{Compl\_Act}\,(\mathsf{label}\ l) = \mathsf{label}\,(\mathsf{Compl}\ l)$$

Using case analysis on the type *label*, we can then prove that $\bar{\bar{l}} = l$ for all l, thus obtaining the following theorem:

$$\vdash\ \forall l.\ \mathsf{Compl}\,(\mathsf{Compl}\ l)\ =\ l$$

The type *relabelling* for relabelling functions is defined as the set of functions of type $action \rightarrow action$ such that relabelling respects complements and τ is renamed as τ.

The type *CCS* of CCS expressions can then be defined by means of the above derived HOL rule for recursive type definitions as follows:

$$\begin{aligned} CCS = {} & \mathsf{nil} \mid \\ & \mathsf{var}\ string \mid \\ & \mathsf{prefix}\ action\ CCS \mid \\ & \mathsf{sum}\ CCS\ CCS \mid \\ & \mathsf{restr}\ CCS\ (label)set \mid \\ & \mathsf{relab}\ CCS\ relabelling \mid \\ & \mathsf{par}\ CCS\ CCS \mid \\ & \mathsf{rec}\ string\ CCS \end{aligned}$$

where nil, var, prefix, sum, restr, relab, par and rec are distinct constructors. A theorem of higher order logic which completely characterizes the type *CCS* is automatically derived, as for the types *label* and *action*, and this theorem is the basis for reasoning about the type *CCS*.

3.2 The Operational Semantics

The next step in our formalization is the definition of the labelled transition relation which gives the operational meaning of the CCS operators. This relation can be formalized in HOL using the derived principle for inductively defined relations [15]. The transition relation $E \xrightarrow{u} E'$ is represented by $\mathsf{Trans}\ E\ u\ E'$, where the relation

$$\mathsf{Trans} : CCS \rightarrow action \rightarrow CCS \rightarrow bool$$

is defined as the intersection of all relations that satisfy the rules of the operational semantics. The mechanism for inductive definitions proves that this intersection is closed under the transition rules and is the least such relation. Proving that Trans satisfies the transition rules results in the following list of theorems, which state the labelled transition rules given at the end of Section 2.1:

PREFIX: $\vdash \forall u\, E.\ \mathsf{Trans}\,(\mathsf{prefix}\ u\ E)\ u\ E$

SUM1: $\vdash \forall E\ u\ E1.\ \mathsf{Trans}\ E\ u\ E1\ \supset (\forall E'.\ \mathsf{Trans}\,(\mathsf{sum}\ E\ E')\ u\ E1)$

SUM2: $\vdash \forall E\ u\ E1.\ \mathsf{Trans}\ E\ u\ E1\ \supset (\forall E'.\ \mathsf{Trans}\,(\mathsf{sum}\ E'\ E)\ u\ E1)$

RESTR: $\vdash \forall E\ u\ E'\ L.\ (\exists l.\ \mathsf{Trans}\ E\ u\ E' \wedge ((u = \tau) \vee ((u = \mathsf{label}\ l) \wedge (l \notin L) \wedge (\mathsf{Compl}\ l \notin L)))) \supset \mathsf{Trans}\,(\mathsf{restr}\ E\ L)\ u\ (\mathsf{restr}\ E'\ L)$

RELAB: $\vdash \forall E\ u\ E'.\ \mathsf{Trans}\ E\ u\ E' \supset (\forall f.\ \mathsf{Trans}\,(\mathsf{relab}\ E\ f)\ (\mathsf{Apply\_Relab}\ f\ u)\ (\mathsf{relab}\ E'\ f))$

PAR1: $\vdash \forall E\ u\ E1.\ \mathsf{Trans}\ E\ u\ E1\ \supset (\forall E'.\ \mathsf{Trans}\,(\mathsf{par}\ E\ E')\ u\ (\mathsf{par}\ E1\ E'))$

PAR2: $\vdash \forall E\ u\ E1.\ \mathsf{Trans}\ E\ u\ E1\ \supset (\forall E'.\ \mathsf{Trans}\,(\mathsf{par}\ E'\ E)\ u\ (\mathsf{par}\ E'\ E1))$

PAR3: $\vdash \forall E\ E1\ E'\ E2.\ (\exists l.\ \mathsf{Trans}\ E\ (\mathsf{label}\ l)\ E1 \wedge \mathsf{Trans}\ E'\ (\mathsf{label}\,(\mathsf{Compl}\ l))\ E2) \supset \mathsf{Trans}\,(\mathsf{par}\ E\ E')\ \tau\ (\mathsf{par}\ E1\ E2)$

REC: $\vdash \forall E\ X\ u\ E1.\ \mathsf{Trans}\,(\mathsf{CCS\_Subst}\ E\ (\mathsf{rec}\ X\ E)\ X)\ u\ E1 \supset \mathsf{Trans}\,(\mathsf{rec}\ X\ E)\ u\ E1$

where Apply_Relab $f\ u$ performs the renaming of the action u via the relabelling function f, and the function CCS_Subst implements the substitution $E\{\mathsf{rec}\,X.E/X\}$ of $\mathsf{rec}\,X.E$ for all free occurrences of X in E. Such a function can be defined in HOL through a primitive recursive definition over the type *CCS*.[1]

Proving that Trans is the least relation closed under the transition rules results in a "rule induction" theorem, from which a tactic is generated for proofs by induction over the structure of the derivations defined by the transition rules. The inductive definition package provides other tactics for supporting goal directed proofs about the relation Trans. A tactic that reduces a goal which matches the conclusion of a transition rule can be defined for each of the theorems corresponding to the transition rules. For example, a tactic `SUM1_TAC` is generated from the above rule SUM1 such that, given a list of assumptions Γ, a goal $\Gamma\ ?\!\!\vdash \mathsf{Trans}\,(\mathsf{sum}\ E1\ E2)\ u\ E$ is reduced to the subgoal $\Gamma\ ?\!\!\vdash \mathsf{Trans}\ E1\ u\ E$.

Finally, a theorem for performing exhaustive case analysis over the inductively defined relation is provided. This means that if there is a transition Trans $E\ u\ E'$, this can only happen if one of the cases given by the transition rules holds. From this theorem many other useful theorems about the relation Trans can be derived. Some of them will be used in Section 4. They are presented below:

Trans_Rec: $\vdash \forall X\ E\ u\ E'.\ \mathsf{Trans}\,(\mathsf{rec}\ X\ E)\ u\ E' = \mathsf{Trans}\,(\mathsf{CCS\_Subst}\ E\ (\mathsf{rec}\ X\ E)\ X)\ u\ E'$

[1] Note that, for the time being, CCS_Subst works under the assumption that variables bound in recursive processes are distinct, in order to avoid capture of free variables.

Trans_Sum: $\vdash \forall E\, E'\, u\, E''.$
$$\mathsf{Trans}\,(\mathsf{sum}\; E\; E')\; u\; E'' \supset \mathsf{Trans}\; E\; u\; E'' \lor \mathsf{Trans}\; E'\; u\; E''$$

Prefix_cases: $\vdash \forall u\, E\, u'\, E'.\ \mathsf{Trans}\,(\mathsf{prefix}\; u\; E)\; u'\; E' \supset (u = u') \land (E = E')$

All theorems and tactics provided by the inductive definition package greatly aid the task of proving properties of the transition relation and of deriving the algebraic laws for behavioural semantics [20].

3.3 The Modal Logic

The syntax of the modal logic can be formalized in HOL by defining a concrete data type *eHML* of formulas of the extended Hennessy-Milner logic, using the derived HOL rule for recursive type definition as follows:

$$\begin{aligned} \mathit{eHML} = {} & \mathtt{tt} \mid \\ & \mathsf{neg}\; \mathit{eHML} \mid \\ & \mathsf{conj}\; \mathit{eHML}\; \mathit{eHML} \mid \\ & \mathsf{box}\; (\mathit{action})\mathit{set}\; \mathit{eHML} \end{aligned}$$

where tt, neg, conj, and box are distinct constructors.

The satisfaction relation Sat : $\mathit{CCS} \to \mathit{eHML} \to \mathit{bool}$ can be defined using the derived principle for the definition of primitive recursion functions over the type *eHML*, thus obtaining the following list of theorems:

Sat_tt: $\vdash \forall E.\ \mathsf{Sat}\; E\; \mathtt{tt} = T$

Sat_neg: $\vdash \forall E\, Fm.\ \mathsf{Sat}\; E\,(\mathsf{neg}\; Fm) = \sim \mathsf{Sat}\; E\; Fm$

Sat_conj: $\vdash \forall E\, Fm\, Fm'.\ \mathsf{Sat}\; E\,(\mathsf{conj}\; Fm\; Fm') = \mathsf{Sat}\; E\; Fm \land \mathsf{Sat}\; E\; Fm'$

Sat_box: $\vdash \forall E\, A\, Fm.$
$$\mathsf{Sat}\; E\,(\mathsf{box}\; A\; Fm) = (\forall E'\, u.\ u \in A \land \mathsf{Trans}\; E\; u\; E' \supset \mathsf{Sat}\; E'\; Fm)$$

The derived operators of the modal logic can then be defined through basic HOL definition mechanisms:

$$\mathtt{ff} = \mathsf{neg}\,\mathtt{tt}$$
$$\forall Fm\, Fm'.\ \mathsf{disj}\; Fm\; Fm' = \mathsf{neg}\,(\mathsf{conj}\,(\mathsf{neg}\; Fm)\,(\mathsf{neg}\; Fm'))$$
$$\forall A\, Fm.\ \mathsf{dmd}\; A\; Fm = \mathsf{neg}\,(\mathsf{box}\; A\,(\mathsf{neg}\; Fm))$$

The theorems for the relation Sat can be easily proved by rewriting with the definitions of these operators and the above satisfaction rules for the basic operators:

Sat_ff: $\vdash \forall E.\ \mathsf{Sat}\; E\; \mathtt{ff} = F$

Sat_disj: $\vdash \forall E\, Fm\, Fm'.\ \mathsf{Sat}\; E\,(\mathsf{disj}\; Fm\; Fm') = \mathsf{Sat}\; E\; Fm \lor \mathsf{Sat}\; E\; Fm'$

Sat_dmd: $\vdash \forall E\, A\, Fm.$
$$\mathsf{Sat}\; E\,(\mathsf{dmd}\; A\; Fm) = (\exists E'\, u.\ u \in A \land \mathsf{Trans}\; E\; u\; E' \land \mathsf{Sat}\; E'\; Fm)$$

A tactic that reduces a goal which matches the structure of formulas can still be obtained for each of the cases for the satisfaction relation. For example, a tactic `Sat_conj_TAC` is generated from the above theorem Sat_conj such that a goal $\Gamma\ ?\!\!\vdash \mathsf{Sat}\; E\,(\mathsf{conj}\; Fm\; Fm')$ is reduced to two subgoals $\Gamma\ ?\!\!\vdash \mathsf{Sat}\; E\; Fm$ and $\Gamma\ ?\!\!\vdash \mathsf{Sat}\; E\; Fm'$.

4 Proving Modal Properties of CCS Processes in HOL

In this section we show how modal properties of CCS processes can be checked in our HOL-CCS enviroment by means of two simple examples, both taken from [21].

To help readability the HOL transcripts are edited to show proper logical symbols instead of their ASCII representations, and the parsing and pretty-printing facilities in the HOL system are extended to accept input and print output almost identical to the notation normally associated with CCS.[2]

4.1 A Vending Machine

Let a simple vending machine be defined by the following CCS recursive process:

$$V \;\equiv\; \mathbf{rec}\, X.\, 2p.\, big.\, collect.\, X + 1p.\, little.\, collect.\, X$$

which can be defined in HOL using a basic definition mechanism as follows:

V = rec ‘X‘ (‘2p‘.‘big‘.‘collect‘.(var ‘X‘) + ‘1p‘.‘little‘.‘collect‘.(var ‘X‘))

Several properties can be proved of this vending machine. For example, we can show that a button (*big* or *little*) cannot be depressed before money ($2p$ or $1p$) is inserted into the machine. This property can be expressed in Hennessy-Milner logic by the formula $[\{big, little\}]\,\texttt{ff}$.

Although the HOL proof of $V \models [\{big, little\}]\,\texttt{ff}$ is fairly simple, we show some of its steps in what follows. In fact, the tactic which solves this goal will be also used for proving some lemmas in Section 4.2. Thus, this tactic can be defined parametrically and invoked with appropriate arguments to solve a class of goals related to checking that a process has a given modal property.

Given the goal `([], "Sat V ([{‘big‘, ‘little‘}] ff)")` in HOL, we start the proof by rewriting with the definition of V and with the following theorem Inab_Act, which shows that the formula $\langle A\rangle\texttt{ff}$ expresses an inability to perform any action in A:

$$\text{Inab\_Act: } \vdash\ \forall E\, A.\, \mathsf{Sat}\, E\,(\langle A\rangle\, \texttt{ff}) = \sim(\exists E'\, u.\, u \in A \ \wedge\ \mathsf{Trans}\, E\, u\, E')$$

We then unfold the recursive expression by rewriting with the theorem Trans_Rec (Section 3.2), apply the substitution of processes with CCS_Subst, and finally fold back the obtained expression using the definition of V:

```
#e (REWRITE_TAC [V; Inab_Act] THEN PURE_ONCE_REWRITE_TAC [Trans_Rec] THEN
    REWRITE_TAC [CCS_Subst] THEN PURE_ONCE_REWRITE_TAC [SYM V]);;
OK..
"~(∃E' u.
    u ∈ {‘big‘,‘little‘} ∧
    Trans (‘2p‘.‘big‘.‘collect‘.V + ‘1p‘.‘little‘.‘collect‘.V) u E')"

() : void
```

[2]Modulo ASCII syntax, e.g. $\overline{a}$ is written $-a$ and τ is written tau.

Stripping quantified variables and moving antecedents into the assumptions of the goal results in the following:

```
#e (REPEAT STRIP_TAC);;
OK..
"F"
    [ "u ∈ {‘big‘,‘little‘}" ]
    [ "Trans (‘2p‘.‘big‘.‘collect‘.V + ‘1p‘.‘little‘.‘collect‘.V) u E’" ]

() : void
```

By applying the HOL resolution with the implicational theorem Trans_Sum (Section 3.2), we derive new assumptions and get two subgoals to prove:

```
#e (IMP_RES_TAC Trans_Sum);;
OK..
2 subgoals
"F"
    [ "u ∈ {‘big‘,‘little‘}" ]
    [ "Trans (‘2p‘.‘big‘.‘collect‘.V + ‘1p‘.‘little‘.‘collect‘.V) u E’" ]
    [ "Trans (‘1p‘.little‘.‘collect‘.V) u E’" ]

"F"
    [ "u ∈ {‘big‘,‘little‘}" ]
    [ "Trans (‘2p‘.‘big‘.‘collect‘.V + ‘1p‘.‘little‘.‘collect‘.V) u E’" ]
    [ "Trans (‘2p‘.‘big‘.‘collect‘.V) u E’" ]

() : void
```

In order to prove these subgoals, we have to show that the assumptions are inconsistent. Both subgoals are proved with the following tactic: new assumptions are derived by applying the HOL resolution using the theorem Prefix_cases (Section 3.2), and then assumptions are manipulated in such a way that a contradiction is derived. For the first subgoal this means the following:

```
#e (IMP_RES_TAC Prefix_cases);;
OK..
"F"
    [ "u ∈ {‘big‘,‘little‘}" ]
    [ "Trans (‘2p‘.‘big‘.‘collect‘.V + ‘1p‘.‘little‘.‘collect‘.V) u E’" ]
    [ "Trans (‘2p‘.‘big‘.‘collect‘.V) u E’" ]
    [ "‘2p‘ = u" ]
    [ "‘big‘.‘collect‘.V = E’" ]

() : void
```

By rewriting the assumption u ∈ {‘big‘, ‘little‘} with the newly-derived ‘2p‘ = u, the new assumption ‘2p‘ ∈ {‘big‘, ‘little‘} is obtained:

```
#e (ASSUME_TAC (ONCE_REWRITE_RULE [SYM (ASSUME "'2p' = u")]
               (ASSUME "u ∈ {'big','little'}")));;
OK..
"F"
    [ "u ∈ {'big','little'}" ]
    [ "Trans ('2p'.'big'.'collect'.V + '1p'.'little'.'collect'.V) u E'" ]
    [ "Trans ('2p'.'big'.'collect'.V) u E'" ]
    [ "'2p' = u" ]
    [ "'big'.'collect'.V = E'" ]
    [ "'2p' ∈ {'big','little'}" ]

() : void
```

By applying the HOL resolution using a pre-defined conversion `Action_IN_CONV` for deciding membership in a set of actions, a contradiction is derived, thus proving the first subgoal. The second subgoal can be proved in a similar way, thus solving the initial goal.

A tactic `MODAL_TAC` can then be defined to apply the above steps which solve the two subgoals:

```
#let MODAL_TAC as1 as2 =
     IMP_RES_TAC Prefix_cases THEN
     ASSUME_TAC (ONCE_REWRITE_RULE [SYM (ASSUME as1)] (ASSUME as2)) THEN
     IMP_RES_TAC (Action_IN_CONV (fst (dest_eq as1)) (snd (dest_comb as2)));;
MODAL_TAC = - : (term -> term -> tactic)
```

For example, the first subgoal above can be solved by invoking `MODAL_TAC` with the assumptions *as*1 and *as*2 given by `'2p' = u` and `u ∈ {'big','little'}`, respectively.

Many other and more complex properties of the vending machine can be checked similarly. Moreover, modal properties of parameterized specifications can be naturally checked using proofs by induction.

4.2 A Counter

Let us consider a (parameterized) specification of a simple counter and of a modal formula, as described in [21], and verify that such a specification of the counter has the given modal property.

The behaviour of a counter can be specified in terms of a parameter $m \in \mathbb{N}$, which represents the "state" of the counter, i.e. how many items are currently being counted. Let the following family of processes $Counter_m$ define the behaviour of the counter:

$$
\begin{aligned}
Counter_0 &\equiv up\,.\,Counter_1 + around\,.\,Counter_0 \\
Counter_{m+1} &\equiv up\,.\,Counter_{m+2} + down\,.\,Counter_m
\end{aligned}
$$

Whenever the counter is in the initial state of a counting process ($m = 0$), it can either count once by performing an action *up* and evolve to $Counter_1$, or execute an action *around* and still behave like the process $Counter_0$. Whenever the counter is in a state $m + 1$, i.e. something has already been counted, it can either perform a further counting action *up* and evolve to the counter in the state $m + 2$, or count down by executing an action *down* and move back to the process $Counter_m$.

We want to check that the above specification has the property that whatever goes up may come down in equal proportions. This property can be expressed in Hennessy-Milner logic by means of the following parameterized formula. Let $[\{up\}]^n \, \Phi$ be the formula Φ when $n = 0$ and $[\{up\}] \, [\{up\}]^{n-1} \, \Phi$ when $n > 0$, and similarly for $\langle\{down\}\rangle^n$. Consequently, our aim is to show that the following relation is satisfied for all $m, n \in \mathbb{N}$:

$$Counter_m \models [\{up\}]^n \, \langle\{down\}\rangle^n \, \mathtt{tt} \qquad (*)$$

This result can be proved by induction on the parameter n. The proof is sketched below in the way it has been formalized and carried out in HOL.

The specification $Counter_m$ can be defined in HOL by introducing a function Counter of type $num \rightarrow CCS$, and asserting its properties, thus obtaining the following definition:

$$\begin{array}{l} \forall m. \, (\mathsf{Counter}\, 0 = up. \, (\mathsf{Counter}\, (\mathsf{SUC}\, 0)) + around. \, (\mathsf{Counter}\, 0)) \wedge \\ \quad (\mathsf{Counter}\, (\mathsf{SUC}\, m) = up. \, (\mathsf{Counter}\, (\mathsf{SUC}\, (\mathsf{SUC} m))) + down. \, (\mathsf{Counter}\, m)) \end{array}$$

Note that this is actually an axiom. In order to formalize the family of processes $Counter_m$ according to a purely definitional approach, a different formulation of recursion is needed for dealing with systems of recursive equations [18, 11].

In order to define the parameterized formula $\Phi = [\{up\}]^n \, \langle\{down\}\rangle^n \, \mathtt{tt}$, the n-times application of a modal operator f of type $(action)set \rightarrow (eHML \rightarrow eHML)$ to a set of actions A and a modal formula Fm, has to be defined. This can be obtained by defining an operator Raise in a (primitive) recursive way as follows:

$$\begin{array}{lcl} (\forall \, f \, A \, Fm. \, \mathsf{Raise}\, f \, A \, 0 \, Fm & = & Fm) \wedge \\ (\forall \, f \, A \, n \, Fm. \, \mathsf{Raise}\, f \, A \, (\mathsf{SUC}\, n) \, Fm & = & f \, A \, (\mathsf{Raise}\, f \, A \, n \, Fm)) \end{array}$$

Our intuition in the proof of (*) is that the proof of the inductive case can be done in an simpler way by reducing it to showing (as a separate result) that for all n, a counter in any state m satisfies the above formula Φ for $n + 1$ if and only if it satisfies the same formula for n. This allows us to reduce the subgoal for the inductive case to the inductive hypothesis and prove (*) easily. Actually, we prove a stronger result since it holds for all formulas Fm, and not only for tt. The key result to be proved is thus the following:

$$\begin{array}{l} \forall n \, m \, Fm. \\ \quad Counter_m \models [\{up\}]^{n+1} \langle\{down\}\rangle^{n+1} Fm \;\; iff \;\; Counter_m \models [\{up\}]^n \langle\{down\}\rangle^n Fm \end{array}$$

which in our HOL formalization means proving the following lemma:

Key_Lemma:
$\vdash \; \forall n \, m \, Fm.$
$\quad \mathsf{Sat}\, (\mathsf{Counter}\, m) \, (\mathsf{Raise\ box}\, \{up\} \, (\mathsf{SUC}\, n) \, (\mathsf{Raise\ dmd}\, \{down\} \, (\mathsf{SUC}\, n) \, Fm)) \; =$
$\quad \mathsf{Sat}\, (\mathsf{Counter}\, m) \, (\mathsf{Raise\ box}\, \{up\} \, n \, (\mathsf{Raise\ dmd}\, \{down\} \, n \, Fm))$

Key_Lemma is itself proved by induction on n. The basis case means proving:

$$\forall m \, Fm. \; Counter_m \models [\{up\}] \, \langle\{down\}\rangle \, Fm \;\; iff \;\; Counter_m \models Fm$$

We prove this property separately and refer to it as Lemma3:

Lemma3: $\vdash \; \forall m \, Fm.$
$\quad \mathsf{Sat}\, (\mathsf{Counter}\, m) \, (\mathsf{box}\, \{up\} \, (\mathsf{dmd}\, \{down\} \, Fm)) = \mathsf{Sat}\, (\mathsf{Counter}\, m) \, Fm$

The proof is by cases on the value of m, and makes use of rewriting with the definitions of the counter and the satisfaction relation Sat, by applying the tactic MODAL_TAC (Section 4.1). Moreover, the following theorems about the counter are used in the proof of Lemma3:

$$\text{Lemma1: } \vdash\ \forall m\, Fm.\ \mathsf{Sat}\,(\mathsf{Counter}\, m)\,(\mathsf{box}\,\{up\}\ Fm) = \mathsf{Sat}\,(\mathsf{Counter}\,(\mathsf{SUC}\, m))\, Fm$$

$$\text{Lemma2: } \vdash\ \forall m\, Fm.\ \mathsf{Sat}\,(\mathsf{Counter}\,(\mathsf{SUC}\, m))\,(\mathsf{dmd}\,\{down\}\ Fm) = \mathsf{Sat}\,(\mathsf{Counter}\, m)\, Fm$$

Lemma1 and Lemma2 establish relations between the specifications of the counter with parameters m and $m+1$, and some modal formulas they satisfy. In particular, Lemma1 deals with formulas containing the box operator and shows that $Counter_m \models [\{up\}]\, Fm$ for all m and Fm, if and only if $Counter_{m+1} \models Fm$. Lemma2 considers formulas modalized with the diamond operator and proves that for all m and Fm, $Counter_{m+1} \models \langle\{down\}\rangle\, Fm$ if and only if $Counter_m \models Fm$. The tactic MODAL_TAC is still used to prove both Lemma1 and Lemma2.

As far as the inductive case of the proof of Key_Lemma is concerned, some manipulations of the modal formula are needed to transform it into a suitable form to which other rewritings can be applied. A transformation is given by the following theorem Raise_Perm expressing the equality $f^{n+1} = f(f^n) = f^n(f)$ for all n, f:

$$\text{Raise\_Perm: } \vdash\ \forall n\, f\, A\, Fm.\ \mathsf{Raise}\, f\ A\,(\mathsf{SUC}\, n)\ Fm\ =\ \mathsf{Raise}\, f\ A\, n\,(f\ A\, Fm)$$

Our HOL proof of Key_Lemma is the following, where Lemma3 solves the basis case of the induction, and the theorems Raise_Perm and Lemma1 allow us to manipulate the subgoal for the inductive case, in such a way that the inductive hypothesis can be applied and the theorem can be derived.

```
#let Key_Lemma =
      prove_thm
        ('Key_Lemma',
         "∀n m Fm.
           SAT (Counter m) (Raise box {'up'} (SUC n)
                                (Raise dmd {'down'} (SUC n) Fm)) =
           SAT (Counter m) (Raise box {'up'} n (Raise dmd {'down'} n Fm))",
         INDUCT_TAC THENL
         [REWRITE_TAC [Raise; Lemma3]
          ;
          REPEAT GEN_TAC THEN
          ONCE_REW_LHS_TAC [SPEC "box" (CONJUNCT2 Raise)] THEN
          PURE_ONCE_REWRITE_TAC [Lemma1] THEN
          ONCE_REW_LHS_TAC [SPECL ["SUC n"; "dmd"] Raise_Perm] THEN
          ONCE_ASM_REW_LHS_TAC [] THEN
          PURE_ONCE_REWRITE_TAC [SYM (SPEC_ALL Raise_Perm)] THEN
          PURE_ONCE_REWRITE_TAC [SYM (SPEC_ALL Lemma1)] THEN
          REWRITE_TAC [SPEC "box" (CONJUNCT2 Raise)] ]);;
Key_Lemma =
⊢ ∀n m Fm.
   SAT(Counter m)(Raise box{'up'}(SUC n)(Raise dmd{'down'}(SUC n)Fm)) =
   SAT(Counter m)(Raise box{'up'}n(Raise dmd{'down'}n Fm))
```

It often happens that only one side of an equation has to be rewritten with a list of theorems and/or assumptions of the goal. Some steps of the proof of Key_Lemma consist of the application of theorems and/or assumptions only on the left-hand side of the goal. For this purpose we have defined some rewriting tactics, such as `ONCE_REW_LHS_TAC` and `ONCE_ASM_REW_LHS_TAC` in the proof above, which rewrite the left-hand side of the goal only.

Finally, the main property (*) can be checked in HOL by induction on n, where the basis case is easily solved by rewriting with the definitions of the Raise operator and the satisfaction relation for `tt`, and the inductive case is solved by applying Key_Lemma and then rewriting with the inductive hypothesis.

5 Related Work and Conclusions

Several verification tools based on process algebras (Concurrency Workbench, Auto, TAV, Aldebaran, ... [22, 23, 24]) have been proposed for proving properties of concurrent systems. These tools work in the framework of CCS-like specifications and most of them resort to a finite state machine representation of processes. This internal representation is used to verify equivalences of processes and to show that a process satisfies a logical property by means of some reasonably efficient automatic algorithms. However, an automata based approach has the well-known problem of state explosion and the limitation that it can deal with only finite state specifications. Moreover, more general and powerful proof techniques are sometimes required, such as induction, contradiction, case analysis, etc., and it is often convenient to define proofs parametrically so that they can be used to deal with a class of processes and/or logical properties.

Recently, several investigations into verification environments based on the algebraic nature of the concurrency calculi have been carried out, which allow for a better understanding of the process algebra specifications one is trying to verify than the finite state machine approach. They include axiomatic tools in which the signature of a calculus and the laws for behavioural semantics are just entered and then used to construct proofs in specially designed proof tools [13, 24] or in general purpose theorem provers like LP, RRL and the Boyer-Moore theorem prover [12, 1]. Other work in this field includes the formalization in the HOL system of different CSP semantics [4, 5] and of Milner's π-calculus [16] following a purely definitional approach to using higher order logic. This means that only primitive definition mechanisms are used for introducing new entities in a sound way, and all other kinds of definitions, such as the laws for behavioural semantics, are derived by formal proof. To our knowledge, however, these tools have not yet addressed the issue of checking modal properties.

In this paper we have presented the formalization in HOL of some components of the CCS process algebra, i.e. its syntax and operational semantics, and a modal logic. We have shown how this representation can be actually used to check properties of processes. Other aspects of the formal theory for CCS, like behavioural semantics and their axiomatic characterization, have already been addressed and formalized in HOL, and strategies for verifying behavioural equivalences of processes have been mechanized [6, 19, 20]. Note that these axiomatic strategies have not been used when verifying the modal properties in the examples in Section 4. Those strategies provide tools for verifying that a certain behavioural relation exists between two specifications by manipulating them according to the algebraic laws that characterize the given behavioural relation. This is an axiomatic approach to verification, but the same relation can also be shown in an operational way, namely by checking if there exists such a relation, e.g. weak bisimulation, which contains the given specifications. Property checking is an operational kind of verification since modal/temporal logics for processes are interpreted on transition systems as described in Section 2.2. It can be useful to combine operational verification techniques with the axiomatic ones: for example, when proving that a given process has a certain modal property, the process can be transformed into an equivalent one for which it may be easier to show that the given modal property is satisfied. In fact, bisimulation equivalence preserves modal properties [21].

Extensions to the subset of CCS and the process logic can be embedded in the HOL system. For example, a more expressive *temporal* logic [21, 2] can be formalized in higher order logic, and proof tools, e.g. the *tableau system* (extended to deal with infinite state processes in [3]), can be soundly mechanized. The tableau system decision procedure has been implemented in some verification tools, e.g. the Concurrency Workbench [7]. On the other hand, such a technique is also naturally described as a goal directed proof system and, as such, is amenable to be formalized in a theorem proving system which provides goal directed proofs. We believe that this demonstrates further evidence that the formal theory for a process language can be embedded in a theorem proving system to provide an effective approach to the mechanical verification of concurrent systems.

Acknowledgements

I should like to thank Mike Gordon, Tom Melham and John Harrison for their advice on mechanization in HOL and for valuable comments on the presentation of this work. Thanks are also due to Consiglio Nazionale delle Ricerche (C.N.R., Italy) which has funded the research described in this paper.

References

[1] Aujla S. S., Fletcher M., 'The Boyer-Moore Theorem Prover and LOTOS', in *Formal Description Techniques*, Proceedings of *FORTE'88*, K. J. Turner (ed.), North-Holland, 1989.

[2] Bradfield J., Stirling C., 'Verifying Temporal Properties of Processes', in Proceedings of *Concur'90*, J. C. M. Baeten and J. W. Klop (eds.), Lecture Notes in Computer Science, Springer-Verlag, 1990, Vol. 458, pp. 115–125.

[3] Bradfield J., Stirling C., 'Local Model Checking for Infinite State Spaces', in Preliminary Proceedings of the *3rd Workshop on Computer Aided Verification*, Ålborg University, 1991, Technical Report IR-91-4, pp. 7–24.

[4] Camilleri A. J., 'Mechanizing CSP Trace Theory in Higher Order Logic', *IEEE Transactions on Software Engineering*, Special Issue on Formal Methods, N. G. Leveson (ed.), 1990, Vol. 16, No. 9, pp. 993–1004.

[5] Camilleri A. J., 'A Higher Order Logic Mechanization of the CSP Failure-Divergence Semantics', in Proceedings of the *4th Banff Higher Order Workshop*, 1990, G. Birtwistle (ed.), Workshops in Computing Series, Springer-Verlag, London, 1991, pp. 123–150.

[6] Camilleri A. J., Inverardi P., Nesi M., 'Combining Interaction and Automation in Process Algebra Verification', in Proceedings of *TAPSOFT'91*, S. Abramsky and T. S. E. Maibaum (eds.), Lecture Notes in Computer Science, Springer-Verlag, 1991, Vol. 494, pp. 283–296.

[7] Cleaveland R., Parrow J., Steffen B., 'The Concurrency Workbench', in [22], pp. 24–37.

[8] DSTO, The University of Cambridge, SRI International, 'The HOL System: DESCRIPTION', 1991.

[9] Gordon M. J. C., 'HOL—A Proof Generating System for Higher-Order Logic', in *VLSI Specification, Verification and Synthesis*, G. Birtwistle and P. Subrahmanyam (eds.), Kluwer Academic Publishers, 1988, pp. 73–128.

[10] Hennessy M., Milner R., 'Algebraic Laws for Nondeterminism and Concurrency', *Journal of ACM*, 1985, Vol. 32, No. 1, pp. 137–161.

[11] Inverardi P., Nesi M., 'Deciding Observational Congruence of Finite-State CCS Expressions by Rewriting', Technical Report B4-10, I.E.I.- C.N.R., Pisa, 1990, (revised version, September 1992).

[12] Kirkwood, C., Norrie, K., 'Some Experiments Using Term Rewriting Techniques for Concurrency', in *Formal Description Techniques III*, Proceedings of *FORTE'90*, J. Quemada, J. Mañas and E. Vazquez (eds.), North-Holland, 1991, pp. 527–530.

[13] Lin H., 'PAM: A Process Algebra Manipulator', in [24], pp. 136–146.

[14] Melham T. F., 'Automating Recursive Type Definitions in Higher Order Logic', in *Current Trends in Hardware Verification and Automated Theorem Proving*, G. Birtwistle and P. Subrahmanyam (eds.), Springer-Verlag, 1989, pp. 341–386.

[15] Melham T. F., 'A Package for Inductive Relation Definitions in HOL', in Proceedings of the *1991 International Workshop on the* HOL *Theorem Proving System and its Applications*, P. J. Windley, M. Archer, K. N. Levitt and J. J. Joyce (eds.), IEEE Computer Society Press, 1992, pp. 350–357.

[16] Melham T. F., 'A Mechanized Theory of the π-calculus in HOL', Technical Report No. 244, Computer Laboratory, University of Cambridge, January 1992.

[17] Milner R., *Communication and Concurrency*, Prentice Hall, 1989.

[18] Milner R., 'Operational and Algebraic Semantics of Concurrent Processes', in *Handbook of Theoretical Computer Science, Vol. B: Formal Models and Semantics*, J. van Leeuwen (ed.), North-Holland, 1990, pp. 1201–1242.

[19] Nesi M., 'Mechanizing a Proof by Induction of Process Algebra Specifications in Higher Order Logic', in [24], pp. 288–298.

[20] Nesi M., 'A Formalization of the CCS Process Algebra in Higher Order Logic', forthcoming, 1992.

[21] Stirling C., 'An Introduction to Modal and Temporal Logics for CCS', in Proceedings of the *joint UK/Japan Workshop on Concurrency*, Oxford, 1989, Lecture Notes in Computer Science, Springer-Verlag, Vol. 491, 1991, pp. 2–20.

[22] Proceedings of the *Workshop on Automatic Verification Methods for Finite State Systems*, Grenoble, 1989, J. Sifakis (ed.), Lecture Notes in Computer Science, Springer-Verlag, 1990, Vol. 407.

[23] Proceedings of the *2nd Workshop on Computer Aided Verification*, New Brunswick, New Jersey, 1990, E. M. Clarke and R. P. Kurshan (eds.), Lecture Notes in Computer Science, Springer-Verlag, 1991, Vol. 531.

[24] Proceedings of the *3rd Workshop on Computer Aided Verification*, Ålborg University, 1991, K. G. Larsen and A. Skou (eds.), Lecture Notes in Computer Science, Springer-Verlag, 1992, Vol. 575.

Higher Order Logic Theorem Proving and its Applications (A-20)
L.J.M. Claesen and M.J.C. Gordon (Editors)
Elsevier Science Publishers B.V. (North-Holland)

Modelling Non–Deterministic Systems in HOL

Jim Alves–Foss

Department of Computer Science, University of Idaho, Moscow, Idaho 83843, USA.
(jimaf@cs.uidaho.edu)

Abstract

When developing the specifications of computer system components, it is often necessary or desirable to include non–deterministic behavior into the specification. Unfortunately this non–determinism can cause difficulty when we try to verify properties of the specification. In this paper we present a method for modelling non–determinism in a way that allows the verifier to isolate it from other aspects of the system component.

Keyword Codes: F.3.1, K.6.5
Keywords: Specifying and Verifying and Reasoning about Programs; Security and Protection

1 Introduction

In safety and security critical applications it is vital that the computer system be correct. To ensure this correctness it is important to be able to detect flaws, incompleteness, and ambiguities in the software specification at an early stage in software development. Proving that the specification satisfies certain correctness properties is the first step in guaranteeing that the software will perform reliably. This early detection will also prevent costly redevelopment efforts that would occur if the flaws were left undiscovered until the testing and debugging phase.

When developing the specifications of computer system components, it is often necessary or desirable to include non–deterministic behavior in the specification. This non–determinism occurs either in the values of the outputs of the system, or the ordering of those outputs in relationship to the inputs. Sometimes we cannot accurately model the precise behavior of the system and must leave the non–determinism in the specification. At other times non–determinism can lead to more efficient implementations of the specification by providing the implementor with the ability to make some choices in the details of the system.

Unfortunately this non–determinism can cause difficulty when we try to verify properties of the specification. The predicates that define these properties may be insufficient to deal with the non–determinism. In this paper we present a method for modelling non–determinism in a way that allows the verifier to isolate it from other aspects of the system component. This approach depends on the type of non–determinism we are using but always involves modelling the system component as a composition of smaller more specialized components. To demonstrate our method we have modelled exemplary non–deterministic systems using the HOL theorem proving system [7].

The HOL theorem proving system permits us to write and verify system specifications using higher order logic. Specifications in this logic can thus contain functions as parameters and predicates can be quantified over these functions. This provides us the capability to create abstract specifications and generic predicates defining our reliability properties.

We have used this abstraction capability in our formal system model to simplify overall proof efforts through reuse of proofs on objects that are specific instances of abstractions. In Section 2 of this paper we present a brief description of the formal system model and security policy that we use in this work. In Section 3 we discuss how to model systems that are characterized by non–deterministic ordering. In Section 4 we discuss how to model systems that are characterized by non–deterministic output values.

2 System Model

The work presented in this paper involves the formal specification and verification of generic system components. This requires the development of the component specifications in a formal framework, the definition of certain reliability properties for the system, and the validation that these properties are held true for the system. Once we have these formal specifications we can then use them to develop the basic system. In this section we give a brief overview of the formal system model, the reliability property (in this case restrictiveness security policy) and our generic model of components. Complete HOL listings of these formal specifications can be found in [2, 1].

2.1 Event–System Model

We define system components using an event–system model. This model is derived from the one presented by McCullough in [9, 10], which is based on the processes (CCS, CSP) of Milner [12] and Hoare [8]. This model defines systems in terms of sequences of events where each event is either a communication event (input or output), or an internal event. McCullough's model of event–systems requires a 4–tuple representation consisting of the set of events, set of inputs, set of outputs, and set of valid traces for the system. A valid trace of the system is a sequence of events that is a possible event history for the system.

The theory of event–systems in the HOL system defines this 4–tuple representation and enforces relationships between the fields. The relationships ensure that the inputs and outputs of the event–system are disjoint subsets of the set of events and that the set of traces of the event–system are sequences of events.

2.2 Restrictiveness

Proving that a computer system is secure requires that one first formalize the security properties that the system must satisfy. There has been a large amount of research in the area of formalizing security policies, including [4, 5, 6, 9, 10]. In particular the *restrictiveness* security policy defined by McCullough [9, 10] is general enough to be useful in several different security–conscious environments. In the context of the event–system model, this policy defines security partitions in terms of user *views*. Each view consists of the collection of events that a user in this security partition can "see". In the simplest terms, an event–system is considered to be restrictive with respect to a view if the output events in the view are not a function of input events not in the view. The input events carry information into the event–system; the security policy defines how information leaves the event–system.

To prove that a system satisfies restrictiveness with respect to a view requires that, for every legal trace of the system, the user's view of possible future behavior of the system is a function of only the previous inputs in the user's view. This proof effort is generally difficult even for simple system specifications, and thus we want to reduce the proof effort as much as possible. To simplify this effort, McCullough designed restrictiveness to be a composable (hook–up) property [10]. As a result, if we decompose the system into a collection of very simple components, we need only show that each component is restrictive and that the components are composed correctly in order to prove that the system is restrictive. The *Hook–Up Theorem* states that such a composition is restrictive.

```
HOOK_UP_THEOREM
⊢ ∀es es1 es2 V.
   RESTRICT V es1 ∧
   RESTRICT V es2 ∧
   IS_HOOK_UP es es1 es2 ⟹
     RESTRICT V es
```

The justification for this approach to verification is that the combined proofs of the components of the system require less effort than the system proof. Any computer system can be designed as a collection of smaller subsystems, or of separate systems connected together. In either case we are concerned with the interaction of several concurrent components. A divide and conquer approach can be used on proofs of the reliability of these systems only if we can reliably combine the separate proofs to get a full system proof.

2.3 Generic Components

In [1, 3] we presented a method for the development of a collection of generic system components for use in system specification and verification. Each component is specified as an event–system and classified according to how they relate inputs to outputs. Each component is defined in terms of one of four classes:

- (1-1) Filters. The components map single inputs to 0 or 1 outputs.
- (1-Many) Generators. These components map single inputs to 1 or more outputs.
- (Many-1) Programmable Filters. These components map single inputs to 0 or 1 outputs based on a function of the previously processed inputs.

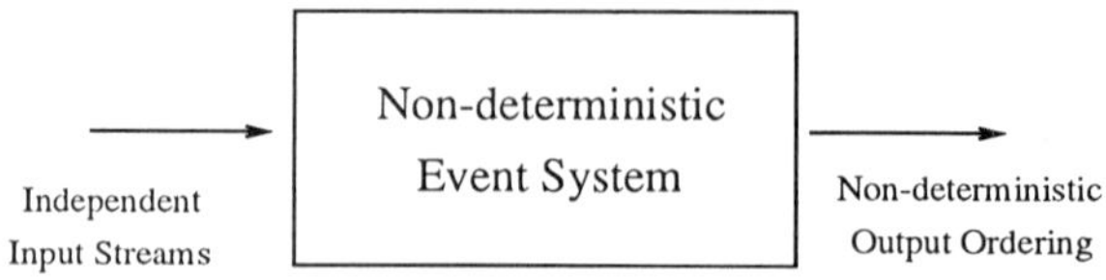

Figure 1: A system with non-deterministic output ordering.

- (Many-Many) Programmable Generators. These components map single inputs to 1 or more outputs based on a function of the previously processed inputs.

We demonstrated how we can prove that members of subclasses satisfy certain correctness properties (in this case the restrictiveness security property [9, 11]). The proof that a component is a member of the subclass is much easier than the actual proof of correctness, thus simplifying the overall correctness proof effort. To aid in the development of system components we presented a method for defining a set of specializations which satisfy the correctness properties. Developers can then instantiate the specializations to create specific instances of system components that are already verified to satisfy the correctness properties.

To maintain a level of abstraction, we use Windley's type abstraction package [15] for the HOL sytsem. This package permits us to define the system in terms of user provided parameters, encapsulating these parameters into a single system structure. All theorems and definitions are created in terms of this structure and can be automatically instantiated by the user.

As an initial phase of the method discussed above we limited the scope of system components to only those that are input–total (components that are always ready to receive new inputs) and deterministic. In this paper we extend the previous work to permit non–deterministic system components. Methods for modelling non–input total systems can be found in [13, 14].

3 Non–Deterministic Ordering

The first type non–deterministic systems we discuss is those in which non–determinism occurs in the ordering of the outputs as depicted in Figure 1. Usually in this case it is possible to find some partial ordering on the outputs, but not a total ordering.

For example we may notice that there are n independent streams of inputs from which the outputs are derived. The outputs derived from any one particular stream are in the same order as that streams inputs, but the interleaving of the streams is non–deterministic. Such a system is analogous to sequences of requests to a server by concurrently executing clients. The server will process the requests according to some internal policy which may not correspond to the relationship of the ordering between when clients sent requests but may depend on the ordering when the server recognized that it received the requests.

The problem we have with analyzing such a system is iterating over all possible interleavings between the independent streams. It is much easier to analyze behavior if the output ordering directly correlates with the input ordering. To provide this simplification

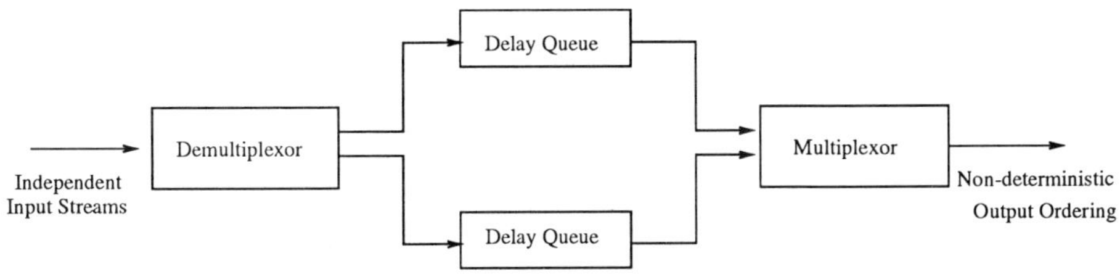

Figure 2: A model of a system with non–deterministic output ordering.

of the analysis of such a system it is necessary to separate out each of the input streams so that they may be processed independently. Since the precise ordering between streams is non–deterministic, we insert a delay queue onto each stream as depicted in Figure 2. The delay queue permits us to model the non–deterministic inter–stream ordering.

```
NOND_DEMUX_P =
⊢def ∀p es.
    NOND_DEMUX_TRANS_P p es =
    DEMUX_P es(EXT_PORT (abs_I_PORT p)) (DQ_IN_PORTS p)

NOND_DELAYQ_P =
⊢def ∀p es id.
    NOND_DELAYQ p es id =
    DELAYQ_P es(MSGS p)(PORT_ID_IN id)(PORT_ID_OUT id)

NOND_MUX_P =
⊢def ∀p es.
    NOND_MUX_P p es =
    MUX_P es (DQ_OUT_PORTS p) (EXT_PORT (abs_O_PORT p))
```

Each of the components of this decomposition are simply instantations of specialized simple filter components. We have three specializations, the demultiplexor, delay queues and the multiplexor. The generic specification of each of these devices has already been shown to be secure [3] and can thus be instantiated rather easily. The composition of the system is guaranteed to be secure by the *Hook–Up Theorem.*

```
SYSTEM_RESTRICT =
⊢ ∀v es lan_list.
    IS_LAN_LIST sys lan_list ∧
    IS_LIST_HOOK_UP es lan_list ∧
    v IN (PO_VIEW_SET sys) ⟹
    RESTRICT v es
```

4 Non–Deterministic Outputs

The other type of non–deterministic systems we encounter is those in which non–determinism occurs in the values of the outputs as depicted in Figure 3. This situation occurs when a

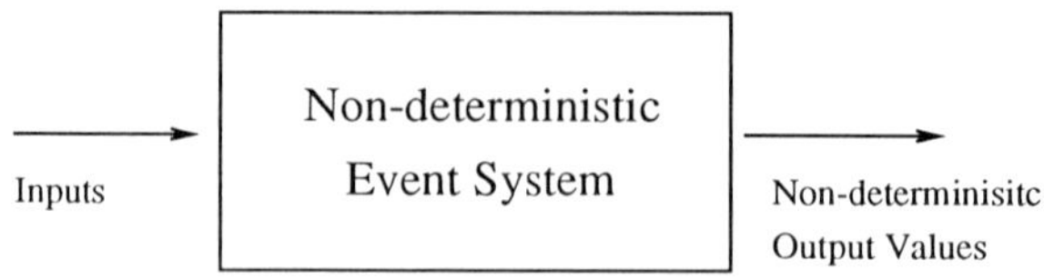

Figure 3: A device that generates outputs non–deterministically.

system is only partially specified, and that certain inputs are defined to generate DON'T CARE outputs. Systems in which the output is truly non–deterministic, or in which the output depends on some unspecified event also fall into this category.

When designing a computer system, it is often beneficial to leave some details of the system unspecified, either for the need to rapidly prototype a design, to create a simpler system abstraction, or because we really don't care what the output is for certain input values. When we partially specify a system, we allow the system implementor freedom in choosing more efficient implementation formats. A completely specified system that contains details that were chosen by the designer for no other reason than obtaining a complete specification can lock the system implementor into a less efficient format.

It is also sometimes necessary to specify a system who's behavior is actually non–deterministic, if for no other reason than because to deterministically specify the system requires the inclusion of details that would make the specification completely unwieldy. It is a system of this type that we will use as an example later in this section.

The problem we encounter with this type of system, is being able to apply our reliability predicates to the non–determinism, specifically when the predicate requires deteriministic knowledge about some of the output values of the system.

Although the specification defines the system to be non–deterministic, we can still separate out the non–deterministic portion of the specification and isolate it as a single separate component of the system. This single component can be used to identify the non–deterministic choice in a manner that is compatible with our reliability property. Using this identification the system can then deterministically generate the corresponding output.

As an example let us look at a device that receives input events and non–deterministically either retransmits them or upgrades their security classification. Since there is no information flow from previous inputs, this system is manifestly secure. Unfortunately, the HOL version of the restrictiveness security property is too strong to consider this system to be secure. To correct this we can either modify the system specification or the restrictiveness property. We chose the former.

The problem manifests itself in that we can not specify which output choice the system will make before the output is transmitted. Since this choice affects the security level of the output, it affects the user's view of possible future outputs. These views are the basis of the security property and therefore it is necessary to be able to deterministically analyze them.

Figure 4 depicts our solution to this problem. In this situtation we subdivide the system into two components: a non–deterministic choice identifier, and a deterministic output generator. The choice identifier takes an input message and attaches to it a value

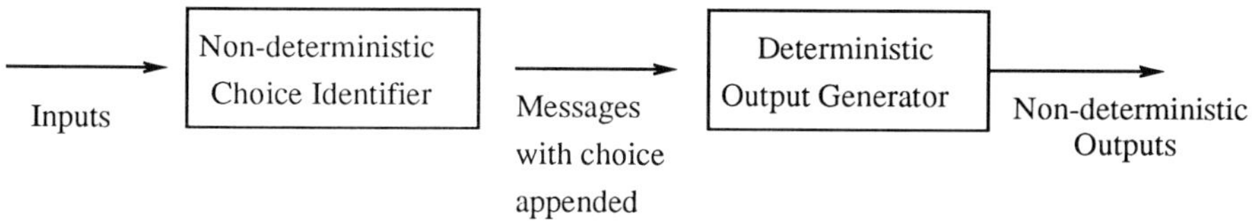

Figure 4: A decomposition of a device that generates outputs non–deterministically.

that identifies which output to generate, without modifying the security classification of the message. The output generator takes this modified message and generates the corresponding output. The external interface to the system is the same, we have just added auxillary internal functionality to simplify the analysis of the system.

```
NOND_TRANS_P =
⊢def ∀p es.
    NOND_TRANS_P p es =
    TRANS_P es(MSGS p)(NOND_TRANS_FN p)(abs_I_PORT p)(abs_TMP_PORT p)

NOND_OUT_P =
⊢def ∀p es.
    NOND_OUT_P p es =
    TRANS_P es(MSGS p)(NOND_OUT_FN p)(abs_TMP_PORT p)(abs_O_PORT p)

NOND_P =
⊢def ∀p es.
    NOND_P p es =
    (∃es1 es2.
      IS_HOOK_UP es es1 es2 ∧ NOND_TRANS_P p es1 ∧ NOND_OUT_P p es2)
```

According to our verification method [1, 3], we initially verify the security of each component separately and then are guaranteed that the composition is secure. We have previously shown that a simple tranformer component is secure as long as the component does not use information from previous inputs to generate outputs [1].

We can model the choice identifier as a simple transformer which simply appends a non–deterministic value to the input message and retransmits the pair. We can also model the output generator as a simple transformer which generates outputs based on the current input pair. The composition of these components is a straight–forward hook–up.

The secuirty verification has already been accomplished for us by in the verification of the generic transformer [1]. We simply reuse the transformer proofs and the *Hook–UP Theorem* to verify the security of the system.

```
NOND_TRANS_IS_RESTRICTIVE =
⊢ ∀p v es.
    VISIBLE_OUT_IMP_INP v es
      (TRANS_FN(abs_I_PORT p)(abs_TMP_PORT p)(NOND_TRANS_FN p)) ∧
    NOND_TRANS_P p es ⟹
    RESTRICT v es

NOND_OUT_IS_RESTRICTIVE =
⊢ ∀p v es.
    VISIBLE_OUT_IMP_INP v es
      (TRANS_FN(abs_TMP_PORT p)(abs_O_PORT p)(NOND_OUT_FN p)) ∧
    NOND_OUT_P p es ⟹
    RESTRICT v es

NOND_IS_RESTRICTIVE =
⊢ ∀p v es.
    NOND_P p es ∧
    v IN (PO_VIEW_SET p) ⟹
    RESTRICT v es
```

5 Conclusion

When developing the specifications of computer system components, it is often necessary or desirable to include non-deterministic behavior in the specification. This non-determinism occurs either in the values of the outputs of the system, or the ordering of those outputs in realationship to the inputs. Sometimes we can not accurately model the precise behvior of the system and must leave in the non-determinism. Other times non-determinism can lead to more efficient implementations of the specification by providing the implementor with the ability to make some choice in the details of the system.

Unfortunately this non-determinism can cause difficulty when we try to verify properties of the specification. The predicates that define these properites may be insufficient to deal with the non-determinism. In this paper we presented a method for modelling non-determinism in a way that allows the verifier to isolate it from other aspects of the system component. This model depends on the type of non-determinism we are using but always involves modelling the system component as a composition of smaller more specialized components. These specialized components isolate the non-determinism into special modules that define the result of the non-determinism without affecting the results of the reliability predicate. The rest of the system can then be defined as determinisitc system components, simplifying the sytsem specification and analysis. To demonstrate our method we modelled exemplary non-deterministic systems using the HOL theorem proving system [7].

Note that for both non-deterministic system the specification does differ from the original, only in the existence of the auxilary internal component specifications the messages between those component and the output generator component. The interface and behavior as seen by the outside world is unchanged.

Although we mentioned only non-deterministic components in this paper, the techniques we defined can also be used to model probabilistic systems. Specifically, systems with probabilistic output values can be modelled with an output identifier component as in the non–deterministic systems, except that the component is parameterized with a function that models the probabilistic output behavior of the system.

More work is needed in specifying and verifying these non–deterministic and probabilistic system components. We are currently exploring predicates which analyze the properties of these probabilistic components, especially when they are combined into a larger system. Such probabilisitic components are important in modelling data encryption systems and in modelling fault tolerant systems.

References

[1] J. Alves-Foss. *Mechanical Verification of Secure Distributed System Specifications.* PhD thesis, Department of Computer Science, University of California, Davis, 1991.

[2] J. Alves-Foss. Mechanical verification of secure distributed system specifications: The HOL proofs and definitions. Technical Report CSE-91-21, Department of Computer Science, University of California, Davis, July 1991.

[3] J. Alves-Foss and K. Levitt. Mechanical verification of secure distributed systems in higher order logic. In *1991 International Workshop on HOL Theroem Proving System and its Applications*, pages 263–278, 1991.

[4] D.E. Bell and L.J. LaPadula. Secure computer systems: Unified exposition and multics interpretation. Technical Report MTR-2997, The MITRE Corporation, Bedford, MA, July 1975.

[5] J.A. Goguen and J. Meseguer. Security policies and security models. In *Proc. IEEE Symposium on Security and Privacy*, pages 11–20, 1982.

[6] J.A. Goguen and J. Meseguer. Unwinding and inference control. In *Proc. IEEE Symposium on Security and Privacy*, pages 75–86, 1984.

[7] M. Gordon. A proof generating system for higher-order logic. Technical Report 103, University of Cambridge Computer Laboratory, January 1987.

[8] C.A.R. Hoare. *Communicating Sequential Processes.* Prentice Hall, London, 1985.

[9] D. McCullough. Specifications for multi-level security and a hook-up property. In *Proc. IEEE Symposium on Security and Privacy*, pages 161–166, 1987.

[10] D. McCullough. Foundations of ulysses: The theory of security. Technical Report RADC-TR-87-222, Odyssey Research Associates, Inc., July 1988.

[11] D. McCullough. Noninterference and the composability of security properties. In *Proc. IEEE Symposium on Security and Privacy*, pages 177–186, 1988.

[12] R.A. Milner. *Communicating and Concurrency.* Prentice Holl, New York, 1989.

[13] D. Rosenthal. Implementing a verification methodology for McCullough security. In *Proc. Computer Security Foundations Workshop*, pages 133–140. IEEE Computer Society Press, June 1989.

[14] D. Rosenthal. Security models for priority buffering and interrupt handling. In *Proc. Computer Security Foundations Workshop*, pages 91–97. IEEE Computer Society Press, June 1990.

[15] P.J. Windley. Abstract theories in hol. Technical Report LAL-92-07, Laboratory for Applied Logic, Department of Computer Science, University of Idaho, June 1992.

Chapter 5

Program Verification

Higher Order Logic Theorem Proving and its Applications (A-20)
L.J.M. Claesen and M.J.C. Gordon (Editors)
Elsevier Science Publishers B.V. (North-Holland)

Mechanising some Advanced Refinement Concepts

J. von Wright[1], J. Hekanaho[2], P. Luostarinen[2] and T. Långbacka[2]

[1]University of Cambridge Computer Laboratory, New Museums Site, Pembroke Street, Cambridge, CB2 3QG, England.

[2]Dept. of Computer Science, Åbo Akademi University, Lemminkäisenkatu 14, SF-20520 Turku, Finland.

Abstract

We describe how proof rules for three advanced refinement features are mechanically verified using the HOL theorem prover. These features are data refinement, backwards data refinement and superposition refinement of initialised loops. We also show how applications of these proof rules to actual program refinements can be checked using the HOL system, with the HOL system generating the verification conditions.

Keyword Codes: D.2.4; F.3.2; I.2.3
Keywords: Program Verification; Semantics of Programming Languages; Deduction and Theorem Proving

1 Introduction

Stepwise refinement is a methodology for developing programs from high-level program specifications into efficent implementations. The *refinement calculus* of Back [1, 2] is a formalisation of this approach, based on the weakest precondition calculus of Dijkstra [9].

Practical program derivation within the refinement calculus [5] has shown that refinement steps often tend to be very similar to each other. Rather than always inventing a refining statement and proving the correctness of the refinement, it is convenient to have access to a collection of program transformation rules whose correctness has been verified once and for all. Every step in the development process is then an application of such a rule.

General rules are easily formulated in the refinement calculus. The rules are verified within the calculus either by appealing to the definition of correct refinement or to some rule that has already been proved. Collections of rules can thus be designed in hierarchy: correctness proofs of higher-level rules appeal to rules on lower levels.

In this paper we describe a mechanisation in HOL of three advanced features of the refinement calculus. These features are *data refinement*, *backwards data refinement* and a special formulation of *superposition refinement*. For each of these concepts, we have formulated a proof rule and proved it in HOL. We also show by means of small examples how these proof rules can be used to verify actual program refinements using HOL.

As a basis we use previously reported work on mechanising a command language with a weakest precondition semantics [8, 14]. Our intention is to show how the HOL system can be used both to verify the correctness of general refinement rules and to verify applications of these rules.

Data refinement and superposition can be used in the transformational design of both software and hardware. Since there is a growing interest in using mechanised formal systems in such design, we think that it is important to investigate how a theorem prover can be used in proving data refinements and superposition.

We assume that the reader is familiar with the HOL system and its version of higher order logic (simple type theory), as described in the documentation of the HOL system [11].

When referring to HOL terms and interaction with the HOL system we use the syntax of HOL. In particular, we note that the scope of binders and quantifiers extends as far to the right as possible. To make formulas more readable, we often omit type information which can be inferred from the context. Also, we use the ordinary logical symbols (the boolean truth values are denoted `F` and `T`).

2 The refinement calculus

We shall give a brief presentation of the refinement calculus and its command language. A more detailed description and further justifications for the definitions can be found in, e.g., [6].

2.1 Predicate transformers

A *state space* Σ_u over the variables u is a mapping which assigns a value to every variable in u, for all lists u of (distinct) variables. We may assume that every variable x is associated with a value set T_x (the *type* of x).

The *predicate space* $Pred_u$ is the set of boolean functions (predicates) on Σ_u. We assume that the ordinary boolean operators (conjunction, negation, etc.) are lifted to $Pred_u$. The partial order $\leq$ on predicates is defined by lifting the implication ordering on the booleans. Thus $Pred_u$ is a complete boolean lattice. Its bottom and top elements are denoted *false* and *true*.

A *predicate transformer* of arity $u \leftarrow v$ is a monotonic function from $Pred_v$ to $Pred_u$. We let $Mtran_{u \leftarrow v}$ denote the space of such predicate transformers. We write $S : u \leftarrow v$ to indicate that the predicate transformer S has arity $u \leftarrow v$. The predicate transformer will be interpreted as a command executed in an initial state over u and intended to terminate in a final state over v, hence the reverse arrow notation.

By lifting we get lattice operations on $Mtran_{u \leftarrow v}$; we will use the partial order $(\leq)$, meet $(\wedge)$ and join $(\vee)$. Furthermore, we let $S; S'$ stand for the functional composition of the predicate transformers $S : u \leftarrow v$ and $S' : v \leftarrow w$.

A predicate transformer is *conjunctive* (*disjunctive*) if it distributes over nonempty conjunctions (disjunctions). Furthermore, a predicate transformer S is *strict* if $S(\mathit{false}) = \mathit{false}$ and *total* if $S(\mathit{true}) = \mathit{true}$. Finally, S is *continuous* if S distributes over ascending (countable) chains of predicates.

2.2 A language of commands

We use the *command lattice* introduced in [7]. We identify statements and predicate transformers, calling them *commands*. Thus we have three basic operations on commands

$$\begin{array}{cl} \bigwedge & \text{meet (demonic choice)} \\ \bigvee & \text{join (angelic choice)} \\ ; & \text{sequential composition} \end{array}$$

Commands are interpreted as weakest precondition predicate transformers; thus $S(Q)$ (which is traditionally written as $\mathrm{wp}(S, Q)$) holds in a state σ if execution of S from initial state σ is guaranteed to terminate in a final state where Q holds. This weakest precondition interpretation justifies the use of the names demonic choice, angelic choice and sequential composition for the operators meet, join and functional composition. For more details we refer to, e.g., [7].

We define two basic commands, called nondeterministic assignments. The *demonic assignment* is written $\langle \wedge x * y'.\ P \rangle$ while the *angelic assignment* is written $\langle \vee x * y'.\ P \rangle$. Here P is a predicate over the union of the initial state space x and a primed copy of the final state space y (the priming is necessary to distinguish variables that occur in both state spaces). Both the nondeterministic assignments choose (angelically and demonically, respectively) a final state which is related to the initial state by P. The definitions are as follows:

$$\begin{array}{rcl} \langle \wedge x * y'.\ P \rangle(Q) & = & (\forall y'.\ P \Rightarrow Q[y'/y]) \\ \langle \vee x * y'.\ P \rangle(Q) & = & (\exists y'.\ P \wedge Q[y'/y]) \end{array}$$

Both assignments have arity $x \leftarrow y$. Note that the demonic assignment is conjunctive and total while the angelic assignment is disjunctive and strict.

Using the nondeterministic assignments, we can define other, more well-known programming constructs. As examples, we show how the *abort*, *skip* and ordinary (multiple) assignment commands can be defined:

$$\begin{array}{rcl} \mathit{abort} & = & \langle \vee x * y'.\ \mathit{false} \rangle \\ \mathit{skip} & = & \langle \wedge x * x'.\ x' = x \rangle \\ \langle x := e \rangle & = & \langle \wedge x * x'.\ x' = e \rangle \end{array}$$

Arity and free variables. Note that the arity of a command is not the same as the free (program) variables occuring in a syntactical representation. The free variables of the assignment $\langle x := x + y \rangle$ are x and y, but the arity of this command depends on what is the underlying global state space (we can only say that if the arity is $u \leftarrow u$, then x and y are both in u).

Guards and assertions. The *guard* $[P]$ and the *assertion* $\{P\}$ are special cases of the nondeterministic assignments, defined as follows:

$$\begin{aligned} [P](Q) &= P \Rightarrow Q \\ \{P\}(Q) &= P \wedge Q \end{aligned}$$

where P is a predicate. Note that the guard command is not strict. This means that it can be interpreted as *miraculous*, when executed in a state where P does not hold. Dually, the assert command is not total; it may be *nonterminating.*

Conditional composition and iteration. The ordinary if-then-else statement can be written using demonic choice and guards:

$$\textbf{if } g \textbf{ then } S_1 \textbf{ else } S_2 \textbf{ fi} \;=\; ([g]; S_1) \wedge ([\neg g]; S_2)$$

The nondeterministic conditional **if** $g_1 \rightarrow S_1 \,[\!]\, g_2 \rightarrow S_2$ **fi** can be defined in a similar way. The iteration **do** $g \rightarrow S$ **od** is defined using least fixpoints.

Special commands. We shall now define a few commands which will be used in the theories of data refinement and superposition. The angelic and demonic *abstraction commands* $\alpha = \langle \vee + a - c.\ R \rangle$ and $\beta = \langle \wedge + a - c.\ R \rangle$ are special cases of nondeterministic assignments, both with arity $a, u \leftarrow c, u$ (where u stands for the rest of the variables in the space):

$$\begin{aligned} \langle \vee + a - c.\ R \rangle(Q) &= \exists a.\ R \wedge Q \\ \langle \wedge + a - c.\ R \rangle(Q) &= \forall a.\ R \Rightarrow Q \end{aligned}$$

Both abstractions commands change the variables c to the variables a, with R as an invariant. Note that α is total and disjunctive while β is strict and conjunctive.

Special cases of the demonic abstraction command are the (demonic) *initialisation command* $\langle \wedge + x.\ P \rangle$ and the *finalisation command* $\langle -x \rangle$:

$$\begin{aligned} \langle \wedge + x.\ P \rangle(Q) &= \forall x.\ P \Rightarrow Q \\ \langle -x \rangle(Q) &= Q \end{aligned}$$

The initialisation command adds variables x to the state and assigns them values so that P is established (if this is not possible, then it succeeds miraculously). The finalisation command simply removes the variables x from the state without affecting the remaining variables.

Using initialisation and finalisation commands we can define ***initialised blocks***: we consider $|[\ \textbf{var}\ x.\ P; S\ ]|$ to be an abbreviation for the command $\langle\wedge + x.\ P\rangle; S; \langle -x\rangle$. The initialised block will then have the following meaning:

$$|[\ \textbf{var}\ x.\ P; S\ ]|(Q) \quad = \quad (\forall x.\ P \Rightarrow S(Q))$$

2.3 The refinement relation

The partial order on $Mtran_{u \leftarrow v}$ is the *refinement ordering.* Note that if S and S' are predicate transformers, we have

$$S \leq S' \quad \Leftrightarrow \quad (\forall Q.\ S(Q) \leq S'(Q))$$

which corresponds to the usual definition of the refinement relation between programs.

Refinement can also be expressed in terms of total correctness: The relation $S \leq S'$ holds exactly if S' satisfies every total correctness specification that S satisfies (the command S is totally correct with respect to precondition P and postcondition Q if and only if $P \leq S(Q)$ holds). Thus $S \leq S'$ means that S' can be used as a replacement for S in any program. Since we can write very general specifications in our command language, the refinement relation is useful in program development; if we want to show that a specification S is satisfied by a program S', we show that the refinement $S \leq S'$ holds.

Refinement of initialised loops. Our main interest is the refinement of *initialised loops.* These are commands of the form

$$|[\ \textbf{var}\ x.\ P; \textbf{do}\ g_1 \rightarrow S_1\ [\!]\ \ldots\ [\!]\ g_n \rightarrow S_n\ \textbf{od}\ ]|$$

The bodies S_i can be arbitrary commands, but usually we assume that they can be written as demonic assignments. Some of our rules consider initialised loops with only one action (guarded command).

The initialised loop is a basic building block in sequential programs. It can also be given a parallel interpretation (the *action systems* of [4]). Initialised loops can also be interpreted as reactive components [3]. Thus the initialised loop is a very useful program construct.

2.4 Data refinement

For an arbitrary command E we define the indexed refinement relation $\leq_E$ as follows: $S \leq_E S'$ is defined to hold when

$$E; S \quad \leq \quad S'; E$$

This relation is investigated in more detail in [6]. In the special case of the angelic abstraction command $\alpha = \langle\vee + a - c.\ R\rangle$ we call $\leq_\alpha$ the *data refinement relation.*

We shall now give rules for proving data refinements. For proofs and interpretations of these rules we refer to [13]. Let R be an abstraction relation and let α be the corresponding

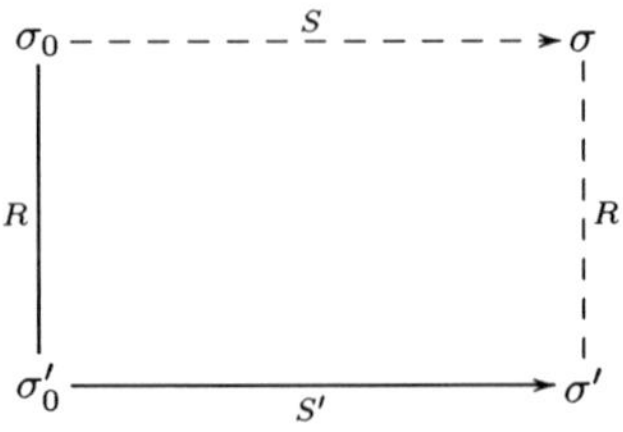

Figure 1: Forwards data refinement

angelic abstraction command, $\alpha = \langle \vee + a - c.\ R\rangle$. The following general rule of data refinement shows how to data refine blocks with local variables:

$$|[\ \textbf{var}\ a.\ P; S\ ]| \ \leq\ |[\ \textbf{var}\ c.\ P'; S'\ ]| \qquad \text{if } P' \leq \alpha(P) \text{ and } S \leq_\alpha S'$$

Furthermore, from the definitions it follows that composed commands can be data refined piecewise:

$$\begin{array}{lcll} S_1; S_2 & \leq_\alpha & S_1'; S_2' & \text{if } S_1 \leq_\alpha S_1' \text{ and } S_2 \leq_\alpha S_2' \\ \bigwedge_i S_i & \leq_\alpha & \bigwedge_i S_i' & \text{if } S_i \leq_\alpha S_i \text{ for all } i \\ \bigvee_i S_i & \leq_\alpha & \bigvee_i S_i' & \text{if } S_i \leq_\alpha S_i \text{ for all } i \end{array}$$

where i ranges over an arbitrary index set.

Finally, we have a rule for data refinement of demonic assignments. This rule has a number of different formulations; we choose one which is easy to use in practice.

$$\begin{array}{l} \langle a, u * a', u'.\ P\rangle \ \leq_\alpha\ \langle c, u * c', u'.\ P'\rangle \\ \quad \text{if } \forall a, c, u, c', u'.\ R(a, c, u) \wedge P'(c, u, c', u') \Rightarrow (\exists a'.\ R(a', c', u') \wedge P(a, u, a', u')) \end{array}$$

This rule can also be used for multiple assignments and guards, since they are special cases of the demonic assignment. Intuitively, the rule says that given initial states related by R and an execution of $S = \langle c, u * c', u'.\ P'\rangle$, there must exist an execution of $S' = \langle a, u * a', u'.\ P\rangle$ such that the final states are related by R (see Fig. 1).

Data refinement of iterations. Data refinement between two iterations can be proved using the following rule:

$$\textbf{do}\ g \rightarrow S\ \textbf{od} \ \leq_\alpha\ \textbf{do}\ g' \rightarrow S'\ \textbf{od} \qquad \text{if } \alpha(g) = g' \text{ and } S \leq_\alpha [g']; S'$$

This rule can be used in situations when the two iterations follow each other step by step, i.e. when the concrete iteration *simulates* the abstract one.

3 The refinement calculus as a theory of HOL

We now show how commands are represented as terms in higher order logic.

3.1 States and predicates

The *state* of a program is a tuple of values, where every component corresponds to a program variable. Thus variables have no global name; they are identified by their position in the state tuple. We will typically use the symbol `s` (possibly primed) for states.

We generally consider states to have the polymorphic type `*s`. In specific cases, this type will be instantiated to a product type with one component for every (global) program variable.

Predicates have type `*s→bool`, which we abbreviate `pred` (actually, the HOL system does not accept abbreviations for polymorphic types; we do it here for notational convenience only). We will typically use `p` and `q` to stand for predicates.

As an example, consider the predicate $x > y$ on the state space

$$(b : bool, x : num, y : num)$$

which is represented as the HOL term
`λ(b,x,y). x>y`
Similarly, any predicate can be written as a term in higher order logic, provided that we know the underlying state space.

Substitution in predicates is represented by a combination of function application and lambda abstraction. As an example, the substitution of 2 for x in the above predicate is represented as

```
λ(b,x,y).  (λ(b,x,y).  x>y)(b,2,y)
```

which by the HOL beta rule is equal to `λ(b,x,y). 2<y`.

Operators on predicates. We define the logical operators on predicates by lifting the corresponding logical operators:

```
⊢def false   = λs. F
⊢def true    = λs. T
⊢def p and q = λs. p s ∧ q s
⊢def p or q  = λs. p s ∨ q s
⊢def p imp q = λs. p s ⇒  q s
```

where `s` has type `*s`.

We also need meets (greatest lower bounds) and joins (least upper bounds) over sets of predicates. Sets are represented by their characteristic functions so we have the following definitions:

```
⊢def glb P = λs. ∀p. P p ⇒  p s
⊢def lub P = λs. ∃p. P p ∧ p s
```

where `P` has type `pred→bool`.

The implication (partial) order on predicates is defined as follows:

```
⊢def p implies q =  ∀s. p s ⇒  q s
```

Note that `and`, `or`, `imp` and `implies` are defined to be infix operators (curried functions).

Chains and limits. A sequence $A = (a_0, a_1, \ldots)$ of elements of some type is formalised as a function `A:num→*`. Thus the element a_i is represented by `A i`.

An ascending sequence of predicates is called a *chain*. The join of such a chain is called its *limit*. Limits of chains will be used in situations where continuity is essential (Section 5). This leads us to make the following definitions:

```
⊢def chain Q = ∀n. (Q n) implies (Q(SUC n))
⊢def limit Q = lub (λq. ∃n. q = Q n)
```

Proving properties of predicates. For proving predicate properties, we have devised a tactic `PRED_TAUT_TAC` that proves tautologous predicate formulas. It is built on the `TAUT_TAC` tactic from the HOL tautology library. Theorems involving quantification usually require rewriting into boolean form and then standard proof techniques.

A least fixpoint operator. For the semantics of iteration, we will need least fixpoints of monotonic functions on predicates. We will not go into details here, we simply assume that `fix f` gives the least fixpoint of an arbitrary monotonic function `f` on predicates. The definition of `fix` and the proofs of fixpoint properties are quite straightforward in HOL.

3.2 Commands in HOL

Commands (predicate transformers) have type `pred→pred'`, abbreviated `ptrans` (note that the two underlying state spaces need not be the same). We will typically use the symbol `c` for commands.

We define constants `skip`, `abort`, `assert` etc, corresponding to the constructors of the command language:

```
⊢def skip q                  = q
⊢def abort q                 = false
⊢def guard p q               = p imp q
⊢def assert p q              = p and q
⊢def demass P q              = λ s. ∀ s'. P s s' ⇒  q s
⊢def angass P q              = λ s. ∃s'. P s s' ∧ q s
⊢def (c seq c') q            = c(c' q)
⊢def Dch C q                 = glb (λq'. ∃c. C c ∧ (q' = c q))
⊢def Ach C q                 = lub (λq'. ∃c. C c ∧ (q' = c q))
⊢def block c q               = λs. ∀x. c(λ(x,s). q s)(x,s)
⊢def if2 (g1,c1) (g2,c2) q = (g1 or g2) and (g1 imp (c1 q)) and (g2 imp (c2 q))
⊢def do1(g,c) q              = fix (λp. ((not g) and q) or (g and (c p)))
```

A few remarks are appropriate here. First, `P` is a state-state relation of type `*s→*s'→bool` and `C` is a set of commands. In the definition of the block, we explicitly say that the local variable(s) must be the first component of the state space (represented by the bound variable `x` in the definition). In the nondeterministic conditional and the iteration, the pair `(g,c)` stands for the action $g \to S$. Another possibility would be to treat such an action as a single command: `(guard g) seq c`. However, it is useful to be able to identify the guard and the body syntactically. This lets us make the additional assumption that c is total, whenever that assumption is needed.

Note that the command notation is open. Commands are abbreviations for predicate transformers and we can define new abbreviations whenever we find that convenient.

The ordinary multiple assignment cannot easily be defined, since we cannot refer to individual variables directly. However, we can define the multiple assignment which assigns to all variables:

$\vdash_{def}$ `assign e q` $=$ `λs. q(e s)`

where `e` is a state-state function of type `*s→*s'`.

Initialised loops. An initialised loop $\|[\ \textbf{var}\ x.\ P; \textbf{do}\ g \rightarrow S\ \textbf{od}\ ]\|$ can be represented as the command

```
block ((guard p) seq (do1(g,c)))
```

Here `p` is the initialisation predicate, `g` is the guard and `c` is the body.

We can also permit loops with an arbitrary number of actions. In order to define this command, we must make a few preliminary definitions which use the HOL facility for making definitions by primitive recursion over lists. We first define the guard of a list of actions:

$\vdash_{def}$ `(lguard NIL = false) ∧ (lguard (CONS (g,c) al) = g or lguard al)`

Here `al` i a list of action, so it has type `(pred#ptrans)list`. Next, we define the list conditional as a recursive binary conditional:

$\vdash_{def}$ `(lif NIL = abort) ∧ (lif (CONS (g,c) al) = if2 (g,c) (lguard al, lif al))`

We can then easily define the general iteration command.

$\vdash_{def}$ `ldo al = do1(lguard al, lif al)`

Reasoning about `ldo` can generally be reduced to reasoning about `do1`.

Junctivity properties of commands. The junctivity properties monotonicity, conjunctivity, disjunctivity and continuity are defined in a straightforward way. As examples, we show the definitions of monotonicity, conjunctivity and continuity:

```
⊢def monotonic c   = ∀p q. p implies q ⇒  (c p) implies (c q)
⊢def conjunctive c = ∀P. c (glb P) = glb (λq. ∃p. P p ∧ (q = c p))
⊢def continuous c  = ∀Q. chain Q ⇒  (c (limit Q) = limit (λn. c (Q n)))
```

where `P` has type `pred→bool` and `Q` has type `num→pred`.

In predicate transformer semantics, it is usually implicitly assumed that all commands are monotonic (or conjunctive). When working with HOL, we must always make such junctivity assumptions explicit. We often put these assumptions in the assumption list, where they indicate what command domain we are working in.

The refinement relation. The refinement relation is defined as an infix constant in the obvious way:

```
⊢def c ref c′ =  ∀q. (c q) implies (c′ q)
```

This relation is then easily proved to be a partial order. A large number of useful refinement rules can be proved directly from the definitions. Examples are:

```
⊢ p implies p′ ⇒  (assert p) ref (assert p′)
⊢ skip seq c = c
conjunctive c ⊢ (c seq (assert p) = (assert(c p)) seq c)
```

The third rule is a rule for *propagating assertions.* Such rules are important in the refinement calculus, since they permit context dependent information to be introduced at different places in a program.

Although the refinement relation is the interesting relation between commands, it is useful to prove equalities whenever they hold, since equalities can be used as rewrite rules directly by the HOL system. In contrast, rewriting using refinements may require extensive manual work.

3.3 Continuity and bounded nondeterminism

In Section 5 we will have a situation where loop bodies are assumed to be continuous. In such cases, we can use another formulation of the semantics of iteration. In our HOL theory, we have proved the theorem

```
continuous c ⊢ do1 (g,c) q = limit (H g c)
```

where the chain of approximations is defined by primitive recursion:

```
⊢def (H g c 0 q = (not g) and q) ∧
     (H g c (SUC n) q = ((not g) and q) or (g and c(H g c n q)))
```

This corresponds to the original definition of the semantics of iteration, given by Dijkstra [9].

Most commands are continuous. However, the demonic assignment can be noncontinuous if it permits unbounded nondeterminism (i.e., if there are an infinite number of possible final state for some initial state). To be able to identify continuous demonic assignments, we have proved the following theorem:

```
⊢ (∀s. finite (λs′. P s s′)) ⇒  continuous (demass P)
```

where finiteness is defined using functions from natural numbers:

```
⊢def finite A =  ∃f n. (∀a. A a ⇒  ∃i. i<n ∧ (f i = a))
```

(i.e., a set A is finite if there there is an onto-function from the set $\{1, 2, \ldots, n\}$ to A, for some n).

3.4 The data refinement relation

The rule of data refinement, given in Section 2.4 can be formulated and proved in HOL. We begin by defining the (angelic) abstraction command as follows:

```
⊢def abst r q = λ(k,u). ∃a. r(a,k,u) ∧ q(a,u)
```

where **r** is the abstraction relation (**a** stands for the abstract variables, **k** for the concrete variables and **u** for the common variables. Then the data refinement relation can be defined:

```
⊢def dataref r c c' = ((abst r) seq c) ref (c' seq (abst r))
```

Rules of data refinement. We will show two rules of data refinement. The first one shows how data refinement is a method for refining initialised loops by replacing the local variables.

```
monotonic c'
 ⊢ (∀k u. p'(k,u) ⇒  ∃a. r(a,k,u) ∧ p(a,u))
    ∧ (∀a k u. r(a,k,u) ⇒  (g(a,u) = g'(k,u)))
    ∧ dataref r c ((guard g') seq c')
    ⇒  (block ((guard p) seq (do1 (g,c))))
        ref (block ((guard p') seq (do1 (g',c'))))
```

The proof of this theorem in HOL is quite extensive. However, it contains no surprises; it is essentially a coding of the ordinary informal proof of the corresponding theorem [13].

Our second rule of data refinement is the rule for data refining demonic assignments:

```
⊢ (∀a c u c' u'. r(a,c,u) ∧ Q(c,u)(c',u') ⇒  ∃a'. r(a',c',u') ∧ P(a,u)(a',u'))
   ⇒  dataref r (demass P) (demass Q)
```

The proof of this rule is straightforward (actually, the implication can be strengthened to equality).

These two rules together give us all we need to verify a data refinement between two initialised blocks, if the bodies are written as demonic assignments. In practice, the bodies may of course contain other constructs, e.g., assignments and conditionals. However, such commands can mechanically be rewritten as demonic assignments. This makes the two rules above quite general.

4 Data refining loops in HOL

We shall now show how data refinements can be verified using HOL. We want to handle the general situation where the iteration in an initialised loop may contain more than one action. Such iterations can be rewritten into iterations with only one action in a standard way:

$$\textbf{do } g_1 \rightarrow S_1 \, [\!] \, \ldots \, [\!] \, g_n \rightarrow S_n \textbf{ od} \;=\; \textbf{do } (\bigvee_i g_i) \rightarrow \textbf{if } g_1 \rightarrow S_1 \ldots g_n \rightarrow S_n \textbf{ fi od}$$

In practice, however, the actions of the two action systems often match each other pairwise, the abstract action being data refined by the concrete one. Thus we want a separate rule for the general case.

4.1 Data refinement with many-branched iterations

In Section 2.4 we gave a rule for data refinement of iterations. This rule can easily be generalised to the case of iterations with more than one action $g_i \rightarrow S_i$ by requiring that the conditions

$$\alpha(g_i) = g_i' \text{ and } S_i \leq_\alpha g_i' \rightarrow S_i'$$

hold for all i. However, this condition is unnecessarily strong. A weaker formulation is the following: the refinement

$$\textbf{do } g_1 \rightarrow S_1 \,[\!]\, \ldots \,[\!]\, g_n \rightarrow S_n \textbf{ od} \quad \leq_\alpha \quad \textbf{do } g_1' \rightarrow S_1' \,[\!]\, \ldots \,[\!]\, g_n' \rightarrow S_n' \textbf{ od}$$

is valid if the following conditions hold:

1. $\alpha(\neg g_i) \leq \neg g_i'$ for all i
2. $S_i \leq_\alpha g_i' \rightarrow S_i'$ for all i
3. $\alpha(\bigvee_i g_i) \leq \bigvee_i g_i'$

4.2 Formalisation of the rule

Using the proof rule for general iterations together with the method for refining initialised blocks we can formulate a proof rule for the data refinement between general initialised loops. This is quite a powerful proof rule that can be used to prove refinements in cases when a concrete program simulates the moves of an abstract one. The proof rule reduces the refinement proof into a proof of data refinement between the constituents of the programs, together with some proofs at the predicate level.

As a HOL theorem, the rule is as follows:

```
⊢  (LENGTH al = LENGTH al')
   ∧ all (λ(g,c). monotonic c) al'
   ∧ (∃r. (∀k u. b'(k,u) ⇒ ∃a. r(a,k,u) ∧ b(a,u))
         ∧ all (λ(g,c)(g',c'). dataref r c ((guard g') seq c')) (lpair al al')
         ∧ all (λ(g,c)(g',c'). (∀a k u. r(a,k,u) ∧ g'(k,u) ⇒ g(a,u))) (lpair al al')
         ∧ (∀a k u. r(a,k,u) ∧ (lguard al)(a,u) ⇒ (lguard al')(k,u)))
   ⇒ (block ((guard b) seq (ldo al))) ref (block ((guard b') seq (ldo al')))
```

where `all P xl` is defined to mean that the property `P` holds for all elements of the list `xl` and `lpair xl yl` is defined as follows:

```
⊢def (lpair xl NIL = NIL)
     ∧ (lpair xl (CONS y yl) = CONS (HD xl,y) (lpair (TL xl) yl)
```

(thus `lpair` pairs up two lists, provided that they are of equal length).

The proof of the data refinement rule is rather tricky. It involves an inductive argument which requires rewriting the general conditional statement as

$$\textbf{if } g_1 \to S_1 \, [\!] \, \ldots \, [\!] \, g_n \to S_n \textbf{ fi} \;=\; \{\bigvee_{i=1}^{n} g_i\}; ([g_1]; S_1 \wedge \ldots \wedge [g_n]; S_n)$$

This means that we have to define `ldch`, a list version of the demonic choice, in order to do the proof in HOL.

4.3 Applying the rule to a case

In order to show how our formalisation can be used in practice, we will show an example refinement proof. Consider two programs searching for the maximal function value within the interval $0..n-1$. The abstract program performs the search by computing the function values step by step, whereas the concrete one computes all the necessary function values in the beginning and then uses the computed values. The concrete program requires more memory but might also be more efficient in some cases.

We let `f` be the function (of type `num→num`). Furthermore, we let `m` be the variable where the maximal value is stored and `i` a counter. Our goal is to prove that the abstract program

$$\begin{array}{l} |[\ \textbf{var}\ a.\ (a = f(0) \wedge m = 0 \wedge i = 0); \\ \quad \textbf{do}\ (i < n \wedge m < a) \to a, i, m := f(i+1), i+1, a) \\ \quad [\!]\ \ (i < n \wedge m \geq a) \to a, i := f(i+1), i+1) \\ \quad \textbf{od} \\ ]| \end{array}$$

is refined by the concrete program

$$\begin{array}{l} |[\ \textbf{var}\ c.\ ((c = \lambda j.\ j \leq n \mapsto f(j)|0)f \wedge m = 0 \wedge i = 0); \\ \quad \textbf{do}\ (i < n \wedge m < c(i)) \to m, i := c(i), i+1) \\ \quad [\!]\ \ (i < n \wedge m \geq c(i)) \to i := i+1) \\ \quad \textbf{od} \\ ]| \end{array}$$

where we use $b \mapsto e|f$ as a notation for a conditional expression. The occurrence of $m = 0 \wedge i = 0$ in the initialisation indicates that both programs assume that both m and i have been initialised to 0, before the blocks are entered.

The abstraction relation used in the proof of data refinement is

$$R: \quad a = c(i) \wedge (\forall j.\ j \leq n \Rightarrow c(j) = f(j))$$

In the HOL formalisation of this problem, the function `f` and the bound `n` are global constants. The abstract program then works with a state tuple `(a:num,i:num,m:num)` and the concrete program has state tuple `(c:num→num,i:num,m:num)`.

The two programs are easily rewritten in the HOL command notation and the goal is set up, saying that the abstract program is refined by the concrete program. We then reduce this goal, matching it against the proof rule and supplying the abstraction relation, which in HOL has the form

```
"(λ (a,c,i,m).(a = c i) ∧ (∀j. (j =< n) ⇒  (c j = f j)))"
```

This yields nine subgoals (verification conditions for the data refinement). Some of them are straightforward while some of them require reasoning over the domain of the program variables (in this case natural numbers). Altogether, the proof is not very long. As an illustration, we show three of the subgoals. The first of these is the initialisation condition, the second one is the guard condition for the first action, while the third one is the body condition for the second action.

```
"(k = (λ j. (j =< n →  f j | 0))) ∧ (i = 0) ∧ (m = 0)
  ⇒  (∃a.(a = k i) ∧ (∀j. j =< n ⇒  (k j = f j))
         ∧ (a = f 0) ∧ (i = 0) ∧ (m = 0))"

"((a = k i) ∧ (∀j. j =< n ⇒  (k j = f j))) ∧ i < n ∧ m < (k i)
  ⇒  i < n ∧ m < a"

"dataref (λ (a,c,i,m). (a = c i) ∧ (∀j. j =< n ⇒  (c j = f j)))
   (assign(λ (a,i,m). (f(i + 1),i + 1,m)))
   ((guard(λ (c,i,m). i < n ∧ (c i) =< m)) seq (assign(λ (c,i,m). (c,i + 1,m))))"
```

The third goal is reduced further using the rule for data refinement of nondeterministic assignments (both commands can easily be rewritten as such assignments). Then, all three goals require some arithmetic reasoning.

The above example could be improved by letting the concrete program put all the function values $f(0), \ldots f(n)$ in a list, rather than in the sequence c. This should make the proof somewhat longer but not essentially more difficult.

5 Backwards data refinement

Generally, a simulation between two initialised loops can be proved using the ordinary rule of data refinement. However, this is not true in situations where a nondeterministic choice is made later in the concrete loop than in the abstract one. In such cases, a dual notion of *backwards data refinement* must be used. This method is investigated in [13], where the results of this section are proved. It is also known under the names of *upward simulation* or *backwards simulation* [10, 12].

5.1 The backwards data refinement relation

The backwards data refinement relation can be defined using the indexed refinement relation: it is the relation $\leq_\beta$ where $\beta = \langle \wedge + a - c.\ R \rangle$ is a demonic abstraction command.

The rules needed for proving refinement of initialised loops by backwards data refinements are similar to the rules for ordinary data refinement. However, we must make the following additional requirements

1. the loop bodies must be continuous (boundedly nondeterministic),
2. R must be total, i.e., every concrete state must correspond to at least one abstract state, and

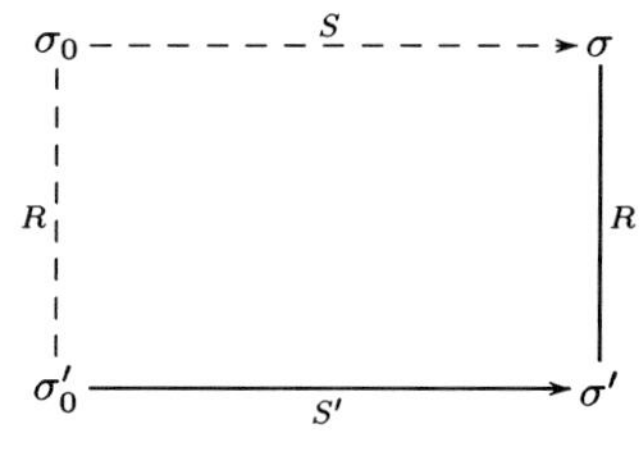

Figure 2: Backwards data refinement

3. R must be image-finite, i.e., every concrete state must correspond to at most a finite number of abstract states.

Backwards data refinement of demonic assignments are proved using the following rule:

$$\langle a, u * a', u'.\ P\rangle \ \leq_\beta \ \langle c, u * c', u'.\ P'\rangle$$
$$\text{if } \forall c, u, a', c', u'.\ R(a', c', u') \wedge P'(c, u, c', u') \Rightarrow (\exists a.\ R(a, c, u) \wedge P(a, u, a', u'))$$

Compare this to the rule for ordinary data refinement (Section 2.4), where we find a final state, given initial states related by R. Here we find an initial state, given final states related by R (see Fig. 2). This justifies calling the method *backwards* data refinement.

5.2 Backwards data refinement in HOL

Since the demonic abstraction command β used in the backwards data refinement is dual to the angelic abstraction command α, we make the following definitions:

```
⊢def ∀c q. dual c q = not(c(not q))
⊢def bwdref r c c′ = ((dual(abst r)) seq c) ref (c′ seq (dual(abst r)))
```

Thus `bwdref` stands for backwards data refinement.

In our HOL theory of backwards data refinement we have proved a number of theorems that show how individual commands can be refined. For instance, the rule for sequential composition is as follows:

```
monotonic c1′ ⊢  bwdref r c1 c1′ ∧ bwdref r c2 c2′
                  ⇒  bwdref r (c1 seq c2) (c1′ seq c2′)
```

The rule for initialised loops. We will consider only the basic case when there is exactly one action in either loop. The rule is then as follows:

```
⊢ continuous c ∧ continuous c′
  ∧ (∃r. (∀k u. ∃a. r(a,k,u))
       ∧ continuous (dual (abst r))
       ∧ (∀a k u. r(a,k,u) ∧ b′(k,u) ⇒  b(a,u))
       ∧ (∀a k u. r(a,k,u) ⇒  (g(a,u) = g′(k,u)))
       ∧ bwdref r c ((guard g′) seq c′))
  ⇒  (block ((guard b) seq (do1(g,c)))) ref (block ((guard b′) seq (do1(g′,c′))))
```

Requiring dual (abst r) to be continuous is the same as requiring the abstraction relation to be image-finite. The proof of the above rule is quite difficult; actually part of our motivation was to check that the pen-and-paper proof reported in [13] was correct.

The proof of the rule for backwards data refinement of a demonic assignment is straightforward:

```
⊢ (∀c u a' c' u'. r(a',c',u') ∧ Q(c,u)(c',u') ⇒  ∃a. r(a,c,u) ∧ P(a,u)(a',u'))
    ⇒  bwdref r (demass P) (demass Q)
```

5.3 Applying the rule to a case

We have applied the rule to a small standard example. Both programs decrement the value of the variable x arbitrarily. In the abstract program, the final value of x is chosen in the initialisation while the decision is postponed in the concrete program.

The refinement we want to prove is as follows:

$$\begin{array}{ll} & |[\ \textbf{var}\ \mathrm{a.\ a} \leq \mathrm{x};\ \ \textbf{do}\ a < x \quad \rightarrow x := x - 1\ \ \textbf{od}\ ]| \\ \leq & \\ & |[\ \textbf{var}\ \mathrm{c.}\ 0 < x;\ \ \textbf{do}\ \mathrm{c} \wedge (0 < x) \ \rightarrow \langle \wedge c, x * c', x'.\ (x' = x - 1)\rangle\ \ \textbf{od}\ ]| \end{array}$$

The abstraction relation is

$$R = (a \leq x) \wedge ((c \wedge (0 < x)) = (a < x))$$

The application of the proof is straightforward: the goal is matched against the proof rule and the abstraction relation is supplied. In this case we get nine subgoals, most of which are straightforward. However, the continuity of dual (abst r) is not trivial. In order to make such proofs simpler, we can use the following result:

```
⊢ (∀k u. finite (λa. r(a,k,u))) ⇒  continuous (dual(abst r))
```

The proof of this theorem uses the fact that dual (abst r) is a special case of the demonic assignment.

6 Superposition refinement

Data refinements changes old local variables for new ones. It also permits new local variables to be added. One important use of this is explored in [5], where action systems are interpreted as parallel programs. The new local variables are used for distributing global information. However, a data refinement cannot add functionality to a program, since it cannot add any global variables. In this section we consider a refinement relation which permits the addition of new global variables.

6.1 A superposition refinement relation

Superposition is a notion which has been used in many different contexts. In this section, we use the word superposition to mean the addition of components to the state space and of computations which affect only the added state components. Given commands $S : u \leftarrow u$ and $S' : x, u \leftarrow x, u$, we say that S' is a *superposition refinement* of S if

$$S \;\leq\; |[\ \mathbf{var}\ x.\ true; S'\ ]|$$

holds. Thus the variables x may give S' some additional functionality, but without affecting (the final values of) the variables u. The superposition refinement relation is easily shown to be reflexive and transitive, if we assume that

$$|[\ \mathbf{var}\ \epsilon.\ true; S\ ]| \;=\; S$$

where ϵ is the empty list of variables.

Formalisation of the relation. We can define the superposition refinement relation directly in HOL:

```
⊢def c sref c′ = c ref (block c′)
```

Here `c'` has one state component (the first) more than `c`. Thus, if the state space of `c` is `*s`, then `c'` operates on the *extended state space* `*#*s`.

It should be noted that our definition does not capture the intended idea completely. For example, neither reflexivity nor transitivity can be proved. This is because HOL does not permit empty tuples and because the two levels of local variables in `block (block c)` cannot be treated as a single component.

6.2 Formalisation of the superposition rule

In proving superposition theorems about our command constructors we need the following definitions:

```
⊢def weaker_cond b b′  =  ∀v. b′ v ⇒  (λ(x, u). b u) v
⊢def weaker_rel P P′   =  ∀u v. P′ u v ⇒  (λ(x, s).λ(y, t). P s t) u v
⊢def same_cond b b′    =  ∀v. (λ(x, u). b u) v = b′ v
```

The first definition states the fact that a condition on an extended state space `*#s` implies a condition on the state space `s`. The second definition contains a similar condition for state relations. The relation `same_cond` says that a condition on a given state space is equivalent to a condition on the extended state space. Dually to `weaker_cond` and `weaker_rel`, we also define `stronger_cond` and `stronger_rel`.

Using the above conditions one can prove some theorems about superposition of different program constructs:

```
⊢ skip sref skip
⊢ weaker_cond b b′ ⇒  (guard b) sref (guard b′)
⊢ stronger_cond b b′ ⇒  (assert b) sref (assert b′)
⊢ weaker_rel P P′ ⇒  (demass P) sref (demass P′)

monotonic c1′ ⊢  c1 sref c1′ ∧ c2 sref c2′
                 ⇒  (c1 seq c2) sref (c1′ seq c2′)
```

The important rule that we have proved is the rule for superposition of initialised loops:

```
⊢ monotonic c ∧ monotonic c′
   ∧ weaker_cond b b′ ∧ same_cond g g′ ∧ c sref c′
   ⇒  (block ((guard b) seq (do1(g,c)))) sref (block ((guard b′) seq (do1(g′,c′))))
```

In practice some additional rules are helpful. In particular, we often use the following "pseudo-transitivity" rules:

```
⊢ c1 sref c2 ∧ c2 ref c3 ⇒  c1 sref c3
⊢ c1 ref c2 ∧ c2 sref c3 ⇒  c1 sref c3
```

6.3 Applying the rule for initialised loops

As a simple example, we consider a superposition which counts the number of times that the loop is executed. The original program is an arbitrary loop:

$$|[\ \mathbf{var}\ x.\ P; \mathbf{do}\ g \rightarrow S\ \mathbf{od}\ ]|$$

and the new program has an added counter t:

$$|[\ \mathbf{var}\ t, x.\ (P \wedge t = 0); \mathbf{do}\ g \rightarrow S; t := t + 1\ \mathbf{od}\ ]|$$

In the HOL formalisation of this example, we assume that the body S is written as a demonic assignment. The old program is simply

```
block ((guard b) seq (do1 (g,demass R)))
```

However, the new program is more complicated, since all predicates must be lifted to the extended state space:

```
block ((guard λ(t,v). b s ∧ (t=0)) seq
       (do1((λ(t,v). g s),(demass λ (t,v)(t′,v′). R v v′) seq (assign λ(t,v).(t+1,v))))
```

Now we can set up the goal. As a first step in solving it, we rewrite the body of the new program as

```
demass λ(t,s)(t′,s′). R s s′ ∧ (t′=t+1)
```

Once this is done, we can match the goal against the proof rule to get the following five subgoals (verification conditions):

```
"weaker_cond b (λ(t,v). b v ∧ (t = 0))"

"monotonic(demass R)"

"monotonic(demass(λ(t,v) (t',v'). R v v' ∧ (t' = t + 1)))"

"same_cond g (λ(t,v). g v)"

"(demass R) sref (demass(λ(t,v) (t',v'). R v v' ∧ (t' = t + 1)))"
```

All these subgoals are straightforward to prove in HOL. In fact, it is not very difficult to compose a strategy, which proves them all (and other similar goals) automatically.

7 Conclusion

We have shown how the refinement calculus can be formalised as a theory in HOL. In particular, we have shown how to formalise three advanced and theoretically interesting refinement methods. For each of these methods there is a proof rule, which has been proved correct within the refinement calculus theory. We have also shown how the proof rules, once they have been proved as HOL theorems, can be used to let the HOL system generate verification conditions.

We intend to go on with this work, proving more rules and applying them to larger case studies of program refinement. It would also be interesting to write semi-automatic strategies which prove the trivial verification conditions automatically (e.g., the monotonicity and continuity conditions) and which simplify the remaining verification conditions as much as possible.

References

[1] R.J.R. Back. On the correctness of refinement in program development. Ph.D. thesis Report A–1978–4, Department of Computer Science, University of Helsinki, 1978.

[2] R.J.R. Back. A calculus of refinements for program derivations. *Acta Informatica*, 25:593–624, 1988.

[3] R.J.R. Back. Refinement calculus, part II: Parallel and reactive programs. In *REX Workshop for Refinement of Distributed Systems,* volume 430 of *Lecture Notes in Computer Science*, Nijmegen, The Netherlands, 1989. Springer–Verlag.

[4] R.J.R. Back and K. Sere. Stepwise refinement of action systems. In *Mathematics of Program Construction,* volume 375 of *Lecture Notes in Computer Science*, Groningen, The Netherlands, June 1989. Springer–Verlag.

[5] R.J.R. Back and K. Sere. Stepwise refinement of action systems. *Structured Programming*, 12:17–30, 1991.

[6] R.J.R. Back and J. von Wright. Refinement calculus, part I: Sequential programs. In *REX Workshop for Refinement of Distributed Systems,* volume 430 of *Lecture Notes in Computer Science*, Nijmegen, The Netherlands, 1989. Springer–Verlag.

[7] R.J.R. Back and J. von Wright. Duality in specification languages: a lattice-theoretical approach. *Acta Informatica*, 27:583–625, 1990.

[8] R.J.R. Back and J. von Wright. Refinement concepts formalised in higher-order logic. *Formal Aspects of Computing*, 2:247–272, 1990.

[9] E.W. Dijkstra. *A Discipline of Programming.* Prentice–Hall International, 1976.

[10] Jifeng He, C.A.R. Hoare and J.W. Sanders. Prespecification in data refinement. *Information Processing Letters*, 25:71–76, 1987.

[11] The HOL System Documentation. Cambridge, 1989.

[12] B. Jonsson. On decomposing and refining specifications of distributed systems. In *REX Workshop for Refinement of Distributed Systems,* volume 430 of *Lecture Notes in Computer Science*, Nijmegen, The Netherlands, 1989. Springer–Verlag.

[13] J. von Wright. The lattice of data refinement. Reports on computer science and mathematics 130, Åbo Akademi, 1992.

[14] J. von Wright and K. Sere. Program transformations and refinements in HOL. In *Proceedings of the 1991 HOL Tutorial and Workshop.* ACM, August 1991.

Higher Order Logic Theorem Proving and its Applications (A-20)
L.J.M. Claesen and M.J.C. Gordon (Editors)
Elsevier Science Publishers B.V. (North-Holland)

Deriving Correctness Properties of Compiled Code

Paul Curzon

University of Cambridge Computer Laboratory, New Museums Site, Pembroke Street, Cambridge, CB2 3QG, England.

Abstract

When developing safety critical software, it is the correctness of the object code that is paramount. However, it is desirable to perform formal verification on the source code. We have combined a derived programming logic with a verified compiler for a generic subset of the Vista structured assembly language. We show how correctness properties of object code can be formally derived from corresponding correctness properties of the source program which have been proved using the programming logic. The work described has been performed using the HOL system and so is machine checked.

Keyword Codes: D.2.4; F.3.1; I.2.3
Keywords: Program Verification; Specifying and Verifying and Reasoning about Programs; Deduction and Theorem Proving.

1. Introduction

Program verification is normally performed on source code. However, it is the object code which ultimately must be correct. As recent standards for safety critical software [22] have recognised, this problem must be addressed when the correctness of code is important. This objective can be achieved in several ways as Figure 1 illustrates.

- The program could be written and validated in the object language. The advantages of high level programs would be lost. Mistakes would be easy to make and hard to find. Formal verification of low level code is much harder than of high level programs.
- The program could be written in a high level language, but validation performed on the compiled code. This is the normal procedure when testing is the validation method used. However, as noted above, it is harder to perform formal verification on object code than on a high level program. The situation is made worse because the object code would be machine generated. Understanding why it was correct, a prerequisite for formal verification, would be much harder especially if an optimising compiler was used.

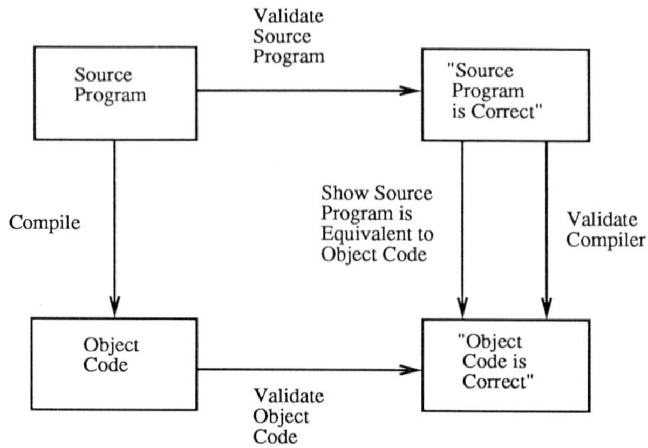

Figure 1. Alternate ways of Obtaining Validated Object Code

- The source program could be formally shown to be equivalent to the compiled code in a one off proof. Validation could then be performed on the source program as the result would be applicable to the object code. This approach has the disadvantage that, in addition to a correctness proof, an equivalence proof must be performed for each program.

- The compiler could be formally verified to be secure. Source programs could be formally verified as the results would apply to the object code. A single proof (that of the compiler) would show that all object programs produced using the compiler correspond to their source programs. However, formally verifying a compiler for a high level language is difficult. Also, care must be taken that the compiler correctness theorem proved is sufficient to ensure that the results really do apply to the low level code.

We adopt the final approach. Our methodology is as follows. We prove a given source program to be totally correct with respect to a specification using a programming logic. We then use an inference rule derived from our compiler correctness theorem to give a theorem stating that the compiled source program is also totally correct with respect to the specification. We have implemented this methodology within the HOL theorem proving system, so all proofs are machine-assisted. Our source language is the structured assembly language, Vista [17]. The instructions correspond directly to machine instructions, but structuring mechanisms such as while loops and procedures are provided for control flow. Verifying an implementation of a structured assembly language is much more tractable than for a high level language, though many of the advantages of using a high level language are retained. This work builds upon previous work to formally verify a compiler for a subset of Vista [9, 7].

A programming logic gives an axiomatic semantics for a language. This is not the most convenient form of semantics for proving a compiler correctness theorem. We have used a relational semantics for that purpose: the possible behaviours of a program are described by a relation between the initial state, final state and input/output behaviour. A disadvantage of using two forms of semantics is that they might not correspond. Thus, the "correct" compiler may only be correct with respect to the relational semantics, not the programming logic. A correctness property of the source program proved using the programming logic may tell us nothing about the object code. We have overcome this by *deriving* the programming logic from the relational semantics. We have formally defined total correctness for the source language in terms of the relational semantics and then derived inference rules about this definition. These rules correspond to the rules and axioms of the programming logic. We have similarly defined total correctness for the object language. We have then derived an inference rule using the compiler correctness theorem which maps theorems about the definition of total correctness for source programs to theorems about the definition of total correctness for object code.

2. Vista

Vista is a structured assembly language designed at the Defence Research Agency (Electronics division), formerly RSRE. The instructions of Vista correspond directly to the instructions available on the underlying machine. For example, the arithmetic instructions are those provided by the ALU. They operate on the registers of the underlying machine, which are visible to the Vista programmer. In this sense it is an assembly language. However, the program counter of the target machine is not directly visible to the Vista programmer. Instead, high level constructs such as while loops and procedures are provided.

Vista was intended for use with the VIPER microprocessor. VIPER is a 32-bit computer, designed for safety critical applications. Some aspects of it have been formally verified [5]. Vista can be re-targeted to other machines by replacing the basic instructions by those available on the new machine. Our work uses a generic version of Vista in which the word sizes, ALU instructions and comparison instructions are left unspecified. The language schema so defined may be instantiated for different machines. It is not a universal schema, however. The target machine must have a similar architecture to that of VIPER. We hope to make future versions of the schema more universal. We only deal with a subset of the language. The syntax of this subset is given in Figure 2. We separate alternatives using the "|" symbol. "$" is used to indicate an empty alternative. The variables `monop`, `aluop` and `compop` are used to represent the monadic ALU operations, binary ALU operations and comparison operations, respectively. They can be instantiated with the operations which correspond specifically to the VIPER microprocessor or alternatively to other machines. The type used to represent a machine word is also left as a variable – `word32`. The name is suggestive of the word size of the VIPER microprocessor, though it could be instantiated to a word type of any length. An example program, written in the subset of Vista, which performs multiplication by repeated addition is given in Figure 3.

```
Variable = string

Register = A | X | Y

Expression = word32 | Variable

Condition = B |                          % Test the B register
            Register compop Expression % comparisons

Command = SKIP |
          STOP |
          Command ; Command |
          Register := Register aluop Expression | % machine functions
          Register := monop Expression |          % monadic functions
          Variable := Register |
          Register := INPUT Variable |
          CALL Variable
          OUTPUT Register Variable |
          WHILE Condition DO Command OD

Vlist = Variable | Variable , Vlist

Declarations = DATA Vlist | $

Proc = PROC Variable BEGIN Command END

Procedures = Proc | Proc ; Procedures

Program = PROGRAM
            Declarations
            Procedures
            Command
          FINISH
```

Figure 2. The syntax of the Vista schema subset

```
PROGRAM
      DATA arg1, arg2, ans, t

      PROC multiply
       BEGIN
       A := 0;
       WHILE X~=0 DO
        A := A + t;
        X := X - 1 OD;
       END

      X := INPUT arg1;
      Y := INPUT arg2;
      t := Y;
      CALL multiply;
      OUTPUT A ans;
      STOP
FINISH
```

Figure 3. A multiplication program in a subset of Vista

3. The Relational Semantics of Vista

To reason formally about the source programs or compiler of a language, we need a formal semantics. We give a relational semantics for the subset of the Vista schema. This is the form of semantics used by Joyce [14] and Gordon [11] extended to include I/O behaviour. The semantics of each construct is defined by a relation between the initial and final states and the I/O behaviour. The I/O behaviour is represented by a single list of events–an *oracle*. All information about the I/O behaviour is held in this data-structure including the relative ordering between different events, whether input or output and on whatever channel [7]. This contrasts with a file based approach, where only the ordering of values input or output to individual files can be specified [13].

For a given command the semantics will relate two states and an oracle if executing the command from the first state could terminate in the second state whilst exhibiting the I/O behaviour of the oracle. For example, for a Skip command the initial and final states must be identical and the oracle must be empty, since Skip does not change the state and performs no I/O. The semantics of the Skip instruction is given by the relation `SemSkip`.

```
SemSkip oracle q1 q2 = (q2 = q1) ∧ (oracle = [])
```

The semantic relations for each command are combined into a single relation, `SemCommand`, defined by structural induction.

```
(SemCommand SKIP rep env penv = SemSkip) ∧
(SemCommand (OUTPUT src chn) rep env penv = ...
```

`SemCommand` takes three arguments in addition to the command, states and oracle. The first, `rep` is a representation tuple [16]. It provides the semantic information of a given instantiation of the schema. For example, the functions corresponding to the operations provided by the ALU in an instantiation are given in this tuple. The second argument to `SemCommand` is the variable environment. It maps variable names to their memory locations. The third argument is a procedure environment. It maps procedure names to the semantic relation corresponding to their body.

A Vista state may be one of three kinds:

```
VistaState = ERROR | HALT Ms Rs | RUN Ms Rs
```

An *Error* state indicates that the program was invalid. A *Halt* state indicates that the program has terminated. A *Run* state indicates that the program is executing normally. The latter two kinds of state contain a memory and a register store, giving the values of memory and registers, respectively.

Three kinds of I/O events are possible. Word values may be input to and from memory mapped channels and a stop light may be set. These are represented by the type `Event`.

```
Event = IN address word32 | OUT address word32 | DONE
```

For example, when executing normally the semantics of the Output command is given by the relation `SemOut`. It outputs a word value to a given address and leaves the state unchanged.

```
SemOut l w oracle q1 q2 = (q2 = q1) ∧ (oracle = [OUT l w])
```

The syntax of the Output command refers to channel names and registers rather than addresses and word values. The full semantics of Output must look up the register's value in the state and the address corresponding to the channel name in the environment. Either of these actions may cause an error; for example, the channel name may not have been declared. If an error occurs the final state must be an error state and the oracle must be empty. If the initial state is not a Run state the state must be unchanged and the oracle empty. The full semantics of Output is given by the relation `SemOutput` defined in terms of `SemOut`. We do not give the details here.

Sequencing is defined inductively in terms of the component commands.

```
SemCommand (c1; c2) rep env penv =
        SemSeq (SemCommand c1 rep env penv) (SemCommand c2 rep env penv)
```

For the relation `SemSeq` to hold, there must be a suitable intermediate state and point at which the oracle can be split. They must be such that the first command results in the intermediate state exhibiting the I/O behaviour of the first part of the oracle. The second command must yield the final state from the intermediate state exhibiting the behaviour of the remainder of the oracle.

```
SemSeq csem1 csem2 oracle q1 q2 =
       ∃q oracle1 oracle2.
        (csem1 oracle1 q1 q) ∧
        (csem2 oracle2 q q2) ∧
        (oracle = APPEND oracle1 oracle2)
```

The semantics of a program is given by **SemProgram**. First the variable environment is determined using SemDec applied to the empty initial environment. The Procedure environment is then determined using SemPDec again with an empty initial environment. The semantics of the command is determined in these environments.

```
SemProgram (PROGRAM d p c FINISH) rep =
       let env = SemDec d EmptyEnv in
       let penv = SemPDec p rep env EmptyPenv
       in
        SemCommand c rep env penv
```

4. A Derived Programming Logic for Vista

We could reason about the correctness of application programs directly using the relational semantics. However, it is more convenient to use a programming logic. To ensure that the programming logic is consistent with the relational semantics, we derive the former from the latter [11]. This involves giving a formal definition of total correctness in terms of the relational semantics, and then proving theorems about the definition. From these theorems we derive rules which correspond to the axioms and rules of the programming logic. The programming logic is thus a toolkit of derived rules for deducing correctness properties about the semantics of programs, which abstract away from the relational definitions.

In the remainder of this section, we first overview the programming logic. It is based on that given previously for a smaller subset of Vista [8]. We then describe its derivation. We also discuss how partial correctness specifications can be deduced from total correctness specifications.

4.1. The Programming Logic

We provide assertions for specifying both the state and the I/O behaviour expected. The assertions {HALT} and {RUN} specify that the state should be a Halt and Run state, respectively. The values in registers may be specified by giving the register name, and similarly for variables. For example, the assertion {A = t} specifies that the value in the A register is the same as the value in the variable t. The I/O behaviour of a program is specified by predicting the I/O events which will occur during the program's execution. Conditions which indicate that a particular sequence of events will take place are defined. The assertion {IN chn w} specifies that a generic word value w is to be input on channel chn. {OUT chn w} specifies that the value w is to be output on channel chn. {DONE} indicates that the stop light will be set (and the program terminate). The assertion {NIL} indicates that no events will take place. Conditions may be sequenced using a temporal conjunction operator (;). Events predicted in the first conjunct must occur prior to those in the second. For example, {IN arg v; DONE} is an assertion which indicates that a single Input event will occur after which the stop light will be set.

We use the notation ⊢ ENV PENV {P} C {Q} to indicate that a command C is totally correct with respect to precondition P and postcondition Q when executed in a variable environment ENV and procedure environment PENV. ENV is a list of the declared variables. PENV is a list which pairs the names of declared procedures with their command bodies. Vista procedures are not recursive so no procedure can appear in its own environment.

The rule for the Skip command is

```
⊢ ENV PENV {P} SKIP {P}
```

We can, for example, use it to deduce that from an initial Run state when the next event predicted is a Done event the final state will be a Run state in which the predicted next event is still a Done event. This holds for all environments.

```
⊢ ENV PENV {DONE ∧ RUN} SKIP {DONE ∧ RUN}
```

The rule for the Output command states that if the channel used has been declared and if an output event is predicted on that channel of the (defined) value in the source register, then the output event will occur leaving the state unchanged.

```
⊢ MEMBER chn ENV
--------------------------------------------------------------------------
⊢ ENV PENV {~(r = UNDEFINED) ∧ ((OUT chn r);P) ∧ RUN} OUTPUT r chn {P ∧ RUN}
```

Thus, we could deduce that from a Run state in which an Output and a Done event remained to occur, after execution only the Done event would remain.

```
⊢ [ans] PENV {~(A = UNDEFINED) ∧ ((OUT ans A); DONE) ∧ RUN}
             OUTPUT A ans
             {DONE ∧ RUN}
```

The sequencing rule is standard. An intermediate assertion must be found.

```
⊢ ENV PENV {P} C1 {R}
⊢ ENV PENV {R} C2 {Q}
---------------------------
⊢ ENV PENV {P} C1; C2 {Q}
```

We use the notation ⊢ {P} p {Q} to indicate that a program p is totally correct. A program is executed by first constructing the environment and procedure environment from empty environments so no environment information is required in the program total correctness definition. We do not give the rules for constructing environments here. The command must have the desired effect in the constructed environments from an initial Run state yielding a Halt state and with no further I/O predicted. This is built into the definition of total correctness for programs. Assertion P is a condition on the oracle only, and assertion Q is a condition on the state only.

```
⊢ {NIL} D {ENV}
⊢ ENV {NIL} PD {PENV}
⊢ ENV PENV {P ∧ RUN} C {Q ∧ NIL ∧ HALT}
----------------------------------------
⊢ {P} PROGRAM D PD C FINISH {Q}
```

An example proof sketch for the multiplication program is given in Figure 4. The specification for the whole program is that its I/O behaviour is to input two values on channels `arg1` and `arg2` and output their product on channel `ans`. The stop light should then be set and no further I/O performed. This is given by the assertion:

```
{IN arg1 v1; IN arg2 v2; OUT ans (v1 × v2); DONE}
```

A state postcondition is given requiring that the A register be left containing the product of the values input.

```
{IN arg1 v1; IN arg2 v2; OUT ans (v1 × v2); DONE}
PROGRAM
 DATA arg1, arg2, ans, t

{~(X = UNDEFINED) ∧ X = w1 ∧ t = w2 ∧ IO ∧ RUN}
 PROC multiply
  BEGIN
  A := 0;
  WHILE X~=0 DO variant{X}
 {~(X = UNDEFINED) ∧ ~(A = UNDEFINED) ∧ t = w2 ∧ A = (w1 - X) × w2 ∧
  IO ∧ (~(X = 0) ⊃ ~(w1 = 0)) ∧ RUN}
   A := A + t;
   X := X - 1 OD);
  END
{~(A = UNDEFINED) ∧ A = w1 × w2 ∧ IO ∧ RUN}

{(IN arg1 v1; IN arg2 v2; OUT ans (v1 × v2); DONE) ∧ RUN}
 X := INPUT arg1;
 Y := INPUT arg2;
 t := Y;
{~(X = UNDEFINED) ∧ X = v1 ∧ t = v2 ∧ (OUT ans (v1 × v2); DONE) ∧ RUN}
 CALL multiply;
{~(A = UNDEFINED) ∧ A = v1 × v2 ∧ (OUT ans A; DONE) ∧ RUN}
 OUTPUT A ans;
 STOP
{A = v1 × v2 ∧ NIL ∧ HALT}
FINISH
{A = v1 × v2}
```

Figure 4. A proof sketch for the multiplication program

4.2. Deriving the Programming Logic

To derive the programming logic, a formal definition of total correctness is needed. Conditions are represented by relations over the oracle, the state and over both, as appropriate. For example, the assertion {`DONE`} is represented by the relation `IsDone` and `RUN` by the relation `IsRun` defined below.

```
IsDone = λoracle.  (oracle = [DONE])

IsRun = λq.  (∃rs ms.  q = RUN rs ms)
```

The assertion {DONE $\wedge$ RUN} is represented by λoracle q. IsDone oracle $\wedge$ IsRun q. Conditions in the second form could be pretty-printed in the first form, though such a pretty-printer has not been implemented.

For programs the definition of total correctness is given by VistaTOTAL.

```
VistaTOTAL P vista Q =
      ∀q1 oracle.
       P oracle ∧ IsRun q1 ⊃
               ∃q2.  vista oracle q1 q2 ∧ Q q2 ∧ IsHalt q2
```

It takes as arguments an oracle assertion P, a state postcondition Q and the semantic relation for the program vista. It states that if the oracle satisfies the precondition and the initial state is a Run state, then there is some Halted final state for which the relational semantics holds and which satisfies the postcondition.

For example, for the multiplication program the desired total correctness theorem would have the following form (where we use the pretty printed version of assertions):

```
⊢ ∀v1 v2.
      VistaTOTAL
       {IN arg1 v1; IN arg2 v2; OUT ans (v1 × v2); DONE}
       <multiplication program semantics>
       {A = v1 × v2}
```

Commands use a different definition of total correctness to programs. The definition must take into account the previously declared variables and procedures. It must also take into account the possibility that some of the I/O behaviour specified in the precondition may be implemented by later commands and so may also appear in the postcondition. Furthermore, the initial state may not always be a Run state nor the final state a Halt state. Apart from these details the form of the definition is similar.

```
VistaCTOTAL rep ENV PENV P vista Q =
      ∀q1 oracle1.
       P oracle1 q1 ⊃
        ∃q2 oracle2.
         oracle2 LLEQ_RIGHT oracle1 ∧
         vista (CONVERT_ENV ENV)
               (CONVERT_PENV rep (EnvOf (CONVERT_ENV ENV)) PENV)
               (UNAPPEND_RIGHT oracle2 oracle1) q1 q2 ∧
         Q oracle2 q2
```

Additional arguments to the definition are the representation tuple, and the environments. The semantic relation now takes an environment and procedure environment argument in addition to the oracle and states. The environment ENV is represented in the programming logic by a list of names. In the relational semantics it is a function. CONVERT_ENV converts between these two representations. Similarly, CONVERT_PENV is used to convert from a list of name–command pairs to a function from names to command semantics. For a command to be totally correct, there must exist both a suitable final state and a suitable point to split the oracle, such that the first part of the oracle satisfies the semantics and the second part satisfies the postcondition. LLEQ_RIGHT tests if its first argument is a final sublist of its second argument. UNAPPEND_RIGHT removes a final sublist from a list.

Deriving the programming logic consists of proving an appropriate theorem about the definition for each programming logic axiom and rule. For example, the derivation of the Skip axiom is based on the theorem:

```
⊢ ∀P rep ENV PENV. VistaCTOTAL rep ENV PENV P SemSkip P
```

HOL inference rules which implement the logic have been derived using these theorems. The rule for Skip is trivially derived as it just involves instantiating the variables. The derivation of the Output rule is based on the theorem:

```
⊢ ∀P rep ENV PENV r chn.
        MEMBER chn ENV ⊃
            VistaCTOTAL rep ENV PENV
                (λoracle q.   ~(MK_REG_ERR r q = UNDEFINED) ∧
                              (IsOseq (IsOut (MK_VAR chn ENV)
                                             (MK_REG r q))
                                      (λoracle.  P oracle q)) oracle ∧
                              IsRun q)
                (SemCommand (OUTPUT r chn) rep)
                (λoracle q.   P oracle q ∧ IsRun q)
```

MK_REG_ERR gives either the value in a register or an indication that the value is undefined. MK_REG gives the value in a register assuming it is defined. MK_VAR gives the address corresponding to a variable. IsOseq is the temporal conjunction operator and IsOut tests if an oracle consists of a single Output event. The inference rule must be given an output command, a membership theorem stating that the channel used is declared in the environment, and an assertion corresponding to P. It performs modus ponens on the above theorem with the membership theorem, specialises the quantified variables and performs beta reduction on the resulting assertions. For example, if the membership theorem was ⊢ MEMBER ans [ans] and the assertion supplied was λoracle q. IsDone oracle we could deduce,

```
⊢ VistaCTOTAL rep [ans] PENV
      (λoracle q.   ~(MK_REG_ERR A q = UNDEFINED) ∧
                    (IsOseq (IsOut (MK_VAR ans ENV)
                                   (MK_REG A q))
                            (λoracle.  IsDone oracle)) oracle ∧
                    IsRun q)
      (SemCommand (OUTPUT A ans) rep)
      (λoracle q.   IsDone oracle ∧ IsRun q)
```

We do not give details of the derivation of the rules for the other commands here. For some the derivation is more complicated than the examples above, though is similar to Gordon's work [11].

4.3. Partial Correctness

As Vista is deterministic up to the values input, we can derive partial correctness statements as defined below from the definition of total correctness for programs.

```
VistaPARTIAL I vista P Q =
      ∀q₁ q₂ oracle.
       I (InValues oracle) ∧ IsRun q₁ ∧ vista oracle q₁ q₂ ⊃
            Q q₂ ∧ IsHalt q₂ ∧ P oracle
```

Partial correctness relates the semantics of the program to an oracle assertion, P, state postcondition, Q, and precondition on the input values, I. The oracle assertion and postcondition are as for total correctness. If the input values of the oracle satisfy the precondition, the initial state is a Run state and the semantics relates the oracle, initial state and final state, then the final state will be a Halt state which satisfies the postcondition and the oracle will satisfy the oracle assertion.

To deduce partial correctness from total correctness, the precondition and oracle assertion must be consistent; that is, there must be at least one oracle matching the oracle assertion for each set of input values matching the precondition.

```
CONSISTENT I P =
      ∀oracle₁.
       I (InValues oracle₁) ⊃
        ∃oracle₂.  P oracle₂ ∧ (InValues oracle₁ = InValues oracle₂)
```

For example, the precondition for the multiplication program might be:

```
I inlist = (inlist = [v1; v2])
```

This is consistent with the oracle assertion:

```
{IN arg1 v1; IN arg2 v2; OUT ans (v1 × v2); DONE}
```

A Partial correctness theorem would be deduced which held for all values of `v1` and `v2`.

```
⊢ ∀v1 v2.
      VistaPARTIAL
       {inlist = [v1; v2]}
       <multiplication program semantics>
       {IN arg1 v1; IN arg2 v2; OUT ans (v1 × v2); DONE}
       {A = v1 × v2}
```

5. A Verified Compiler

The target language of our compiler is Visa – a generic, flat assembly language for VIPER-like microprocessors. Instructions are symbolic and the address space is infinite, consisting of named regions indexed by natural numbers. Further steps in the compilation process will combine the various regions into a single finite store and convert the assembly instructions into machine words. This has not yet been implemented. The instruction set is given in Figure 5. The ALU

```
Operand = LITERAL word20 |                       % Literal word value
          CONTENTS bool address                  % Address holding value

Inst = STP |                                     % Stop
       JMP address |                             % Jump
       JNB address |                             % Conditional Jump
       ALU aluop Register Register Operand |     % Binary ALU operation
       CMP compop Register Operand |             % Comparison
       MON monop Register Operand |              % Monadic ALU operation
       STR Register address |                    % Store register
       INP Register address |                    % Input
       OUT Register address                      % Output
       CLL address                               % Procedure Call
       RTN address                               % Procedure return
       SLK address                               % Store procedure link address
```

Figure 5. The Instruction Set of Visa

and comparison commands act on an operand. This can either be a literal value, or an address where the value can be found. A flag in the operand indicates whether an address refers to the constant store or data store.

A Visa state is represented as a tuple, `(data, p, (a,x,y), b, stop, link)`. It consists of a data store, a program counter, a register store holding the A, X and Y registers, the B condition code, a stop flag and a link memory for storing procedure return addresses. The program and constants are stored in separate read-only stores. The data, link, code, constant and peripheral stores all have distinct address spaces.

The semantics of Visa is given by a relation, `Visa`. It defines a fetch–decode–execute cycle in terms of the semantics for the instructions. The relation will be true if the code can be executed from the initial state using the oracle to reach the final state. In addition, the program should either halt the processor or jump beyond the end address of the program for the first time on the last cycle. The constant store, code store, representation tuple and an end address giving the bound of the fragment of code under consideration must be provided as arguments to the relation. The oracle used at this level differs from Vista in that an extra *no-action* event is provided. Each instruction performs either a normal event or a no-action. This gives a crude form of timing in terms of the number of machine cycles executed.

```
Visa rep cst cs end oracle s1 s2 = ...
```

The compiler is defined in higher-order logic by a function `CompileProgram` which translates Vista programs to Visa assembly code. It returns either compiled code or a compile-time error result. Each basic command is compiled to a single assembler instruction. The structuring constructs compile to jumps of various kinds. Figure 6 shows the compiled version of the multiplication program given in Figure 3. We have omitted the region names from addresses for clarity since they all refer to the default region within the appropriate address space. The start of the program is address 9 in the default region of the code store and the procedure is compiled to address 0. Link address 1 is used to store the return address. The variable `t` is implemented at address 2 in the data store. The channels `arg1`, `arg2` and `ans` are implemented as peripheral addresses 0, 1 and 3, respectively.

```
Address   Object Instruction               Source Command

0:        SLK 1                            PROC multiply BEGIN
1:        MON := A (LITERAL 0)               A := 0;
2:        CMP ~= X (LITERAL 0)               WHILE X~=0
3:        JNB 7                              DO
4:        ALU + A A (CONTENTS F 2)             A := A + t;
5:        ALU - X X (LITERAL 1)                X := X - 1
6:        JMP 2                              OD
7:        RTN 1                            END
8:
9:        INP X 0                          X := INPUT arg1;
10:       INP Y 1                          Y := INPUT arg2;
11:       STR Y 2                          t := Y;
12:       CLL 0                            CALL multiply;
13:       OUT A 3                          OUTPUT A ans;
14:       STP                              STOP
```

Figure 6. The Compiled Multiplication Program

We have proved the following compiler correctness theorem.

```
⊢ ∀p rep.
      (Compiles (CompileProgram p rep)) ∧
      (W32To20ConversionCorrect rep) ⊃
        CompilerCorrect rep (SemProgram p rep) (CompileProgram p rep)
```

This states that under the assumptions that compiling a program does not give a compile error, and the representation tuple used includes valid conversions between 20 and 32 bit words, then all effects exhibited by the semantics of the program are also exhibited by the semantics of the compiled program. The definition of `CompilerCorrect` is given below.

```
CompilerCorrect rep vista code =
      ∀q1 q2 oracle s1.
       Loaded code cs cst ∧
       Ploaded rep s1 (BaseOf code) ∧
       vista oracle q1 q2 ∧
       (CompareStates q1 s1 ∨ (q1 = ERROR)) ⊃
              (q2 = ERROR) ∨
              ∃s2 toracle.
                     Visa rep cst cs (EndOf code) toracle s1 s2 ∧
                     CompareStates s2 q2 ∧
                     (oracle = AbstractOracle toracle)
```

This states that if we assume

- the compiled code is loaded into the code and constant stores,
- the program counter in the initial Visa state s_1 holds the base address of the code,
- the Vista program takes some initial state q_1 to final state q_2 using oracle `oracle`, and
- q_1 corresponds to initial Visa state s_1 as specified by the relation `CompareStates` or is an error state

then we can deduce that either the final state is an error state or there is a Visa state s_2 and Visa oracle `toracle` such that

- the Visa semantics from an initial state s_1 will have I/O behaviour as specified by `toracle` and end in state s_2,
- the final Visa state will correspond to the final Vista state, and
- the Vista and Visa oracles will correspond, as specified by the function `AbstractOracle`. It removes the non-events.

6. Deriving Correctness Theorems about Object Code

Having a proven compiler correctness theorem as described in the previous section increases our confidence that correctness results proved using the programming logic also apply to the compiled code. However, the compiler correctness theorem might not be suitable for this purpose. We show that total correctness theorems about compiled code can be derived from total correctness theorems about the source code using the compiler correctness theorem. In particular we prove, using the compiler correctness theorem, a theorem of the form below, and derive a corresponding inference rule.

```
∀P p Q rep.
      ProgramCompiles p rep ∧
      W32To20ConversionCorrect rep ∧
      VistaTOTAL rep P (SemProgram p rep) Q ⊃
            VisaTOTAL rep P (CompileProgram p rep) Q
```

A different definition of total correctness `VisaTOTAL` is used for the total correctness of Visa programs because it must account for the different kind of states and oracles used at this level. Extra assumptions that the initial Visa state considered corresponds to one which satisfies the precondition, that the program counter is loaded and that the code is loaded into the code store are needed. There must exist a Visa final state which corresponds to some Vista final state which satisfies the postcondition. There must also be a Visa oracle which corresponds to the Vista oracle. The definition of `VisaTOTAL` is given below.

```
VisaTOTAL rep P code Q =
      P oracle q1 ∧
      IsRun q1 ∧
      CompareStates s1 q1 ∧
      Ploaded rep s1 (BaseOf code) ∧
```

```
LOADED code cs cst ⊃
       ∃s₂ q₂ toracle.
          Visa rep cst cs toracle (EndOf code) s₁ s₂ ∧
          CompareStates s₂ q₂ ∧
          (oracle = AbstractOracle toracle) ∧
          Q NIL q₂ ∧
          IsHalt q₂
```

Having proved the correctness of the multiplication program using the derived programming logic, we can deduce a corresponding theorem about the code produced by the compiler, using the derived inference rule.

```
⊢ VisaTOTAL rep
     {IN arg1 v1; IN arg2 v2; OUT ans (v1 × v2); DONE}
     (CompileProgram <multiplication program> rep)
     {A = v1 × v2}
```

We could also obtain a correctness result which includes the text of the compiled code itself. This could be done automatically using constant folding and rewriting techniques on the definitions of the compiler, though this has not been implemented for the subset of Vista considered here.

```
⊢ VisaTOTAL rep
     {IN arg1 v1; IN arg2 v2; OUT ans (v1 × v2); DONE}
     <compiled multiplication program>
     {A = v1 × v2}
```

7. Related Work

Most previous work which has addressed the problem of producing correctness theorems for low level code, has involved systems for reasoning about the low-level code directly. Such systems can be divided into those that perform verification on mnemonic assembly language programs and those which use a bit-representation of the code.

The assembly language systems have typically followed Floyd's approach to program verification [10] modified to deal with low-level code. They provide a verification condition generation program which embodies the semantics of the assembly language. Maurer [20] [21] used this approach to verify IBM 370 code and code for the Litton C4000 airborne computer. Lamb also used this approach in his Intel 8080 Assembly Language Verifier [18]. More recently the approach has been embodied in the SPADE verification environment. It has again been used in the verification of assembly code for the Intel 8080 [4], and also Z8002 code used in the fuel control unit of the RB211-524G jet engine [24].

Verification of bit-level code has typically been based around an operational semantics of the host machine and the use of formal symbolic simulation techniques. MCS was an early system which took this approach. It was used to verify production code for the NASA Standard Spaceborne Computer-2 [3]. A hybrid approach using verification condition generation techniques to verify bit-level microprograms has also been suggested [6]. Recent work of note is that of Boyer and Yu [1] who have verified compiled C and Ada code for the MC68020 microprocessor. Their methodology was to compile the source code using an industrial strength compiler and verify the resulting object code. This was done by first writing a second algorithmic version of the program in the Boyer–Moore logic. This was verified to be equivalent to the object code. The algorithm was then shown to be correct.

There has been much work on the compiler correctness problem from as early as 1967. A good overview of this work is given by Joyce [15]. Notable work, includes that of Polak [25] on a compiler for a Pascal-like language. Generally, compiler correctness proofs have been considered in isolation, with no attempt to make direct use of the resulting theorems. Exceptions to this are Moore's Piton compiler [23] and Young's verified compiler for a subset of Gypsy [26]. They formed part of a stack of verified system components. Thus the compiler correctness proofs were formally linked with correctness proofs for other components of the system. Gypsy programs could be verified with the operational semantics used in the compiler correctness proof. The result then applied to compiled code running on the underlying verified microprocessor. Joyce used the HOL system similarly to link a compiler correctness theorem with the correctness theorem of a simple verified microprocessor [14]. Other compiler correctness work using the HOL system includes that of Martin and Toal [19].

There has also been a vast amount of work on programming logics since they were first conceived by Hoare [12]. For example, an axiomatic semantics was given for the programming language Pascal [13]. Gordon illustrated that a programming logic could be semantically embedded in the HOL system [11].

8. Conclusions

We have formally combined a programming logic with a compiler correctness statement. This allows results about the compiled code to be formally deduced from results proved about source code. All the work has been proven mechanically using the HOL system.

We believe our work is novel in that a formal machine checked link has been made that allows theorems about the total correctness of object code to be deduced from theorems about the total correctness of source code proved in a programming logic. Other work has touched on this issue, but no formal link has previously been made. Gordon's derived programming logic was for a simple imperative language similar but not identical to that used by Joyce in his compiler correctness work. The Gypsy Verification Environment implements a programming logic for Gypsy which complements Young's verified compiler. However, there is no formal link between the semantics used for the compiler correctness proof and that embodied in the Verification Environment, so the two could potentially be inconsistent.

When deriving the programming logic, errors were found in both the original informal version of the programming logic and in the original relational semantics of Vista. These errors were found because the expected programming logic rule could not be derived from the Vista semantics. The formal semantics did not describe the actual semantics of the language. These mistakes were not discovered by the compiler correctness proof because of corresponding errors in the semantics of Visa. This is an additional advantage of using two forms of semantics and deriving one from the other.

We have proved a compiler specification correct rather than an implementation. An interesting case study would be to implement the Vista compiler in Vista itself. It could be verified against the specification using the programming logic. The specification could then be used to compile the implementation using rewriting techniques within the theorem prover. This would give a verified implementation of a Vista compiler in a low level language. A similar idea about bootstrapping a verified compiler implementation, though using an insecure compiler, has independently been suggested by Buth *et al* [2]. They note that a powerful test for correctness of the resulting compiler is to use it to compile its own source program a second time, then compare the two versions for syntactic equivalence.

Visa is still an abstraction of the machine code that is actually run. Work is in progress to verify an assembler from Visa to machine code. This involves loading the disparate address spaces into a single finite memory, and converting the symbolic instructions to machine words. Once this has been done, it could be combined with the Vista compiler described here and theorems about the execution of machine code deduced from Vista total correctness statements. The subset of Vista considered will also be extended. The proofs described are all generic, including that of the multiplication program. No instantiation for particular machines has so far been devised. An instantiation for the VIPER1 processor is planned.

Though intended as a commercial language, Vista is relatively simple and only a subset was considered. Similar results must be obtained for more complex languages.

Acknowledgements

I am grateful to Gavin Bierman, Mike Gordon, John Kershaw, Jeff Joyce, Clive Pygott and the members of the Hardware Verification Group at Cambridge, for their help and advice. This work has been funded by MoD research agreement AT2029/205.

References

[1] Robert S. Boyer and Yuan Yu. Automated proofs of object code for a widely used microprocessor. In *Proceedings of the 11th International Conference on Automated Deduction*, 1992.

[2] Bettina Buth, Karl-Heinz Buth, Martin Fränzle, Burghard v. Karger, Yassine Lakhneche, Hans Langmaack, and Markus Müller-Olm. Provably correct compiler development and implementation. In *Compiler Construction '92*, 1992.

[3] William C. Carter, William H. Joyner, Jr., and Daniel Brand. Microprogram verification considered necessary. In Saki P. Ghosh and Leonard Y. Liu, editors, *AFIPS Conference Proceedings 1978 National Computer Conference*, volume 47, pages 657–664. AFIPS Press, June 1978.

[4] D. L. Clutterbuck and B. A. Carré. The verification of low level code. *Software Engineering Journal*, pages 97–111, 1988.

[5] Avra Cohn. The notion of proof in hardware verification. *Journal of Automated Reasoning 5*, (5):127–138, 1989.

[6] Paul Curzon. A structured approach to the verification of low level microcode. Technical Report 215, University of Cambridge, February 1991.

[7] Paul Curzon. Compiler correctness and input/output. In *Proceedings of the 3rd IFIP Working Conference on Dependable Computing for Critical Applications*, Dependable Computing and Fault-Tolerant Computing. Springer-Verlag, 1992. To be published.

[8] Paul Curzon. A programming logic for a verified structured assembly language. In A. Voronkov, editor, *Logic Programming and Automated Reasoning*, volume 624 of *Lecture Notes in Artificial Intelligence*, pages 403–408. Springer-Verlag, 1992.

[9] Paul Curzon. A verified compiler for a structured assembly language. In *Proceedings of the 1991 International Workshop on the HOL Theorem Proving System and its Applications*. IEEE Computer Society Press, 1992.

[10] Robert W. Floyd. Assigning meanings to programs. In *Proceedings of the Symposium on Applied Mathematics*, volume 19, pages 19–32, 1967.

[11] Michael J. C. Gordon. Mechanizing programming logics in higher order logic. In G. Birtwistle and P. A. Subrahmanyam, editors, *Current Trends in Hardware Verification and Automated Theorem Proving*, pages 387–439. Springer-Verlag, 1989.

[12] C. A. R. Hoare. An axiomatic basis for computer programming. *Communications of the ACM*, 12(10):576–580,583, October 1969.

[13] C. A. R. Hoare and N. Wirth. An axiomatic definition of the programming language PASCAL. *Acta Informatica*, 2:335–355, 1973.

[14] Jeffrey J. Joyce. Totally verified systems: Linking verified software to verified hardware. In M. Leeser and G. Brown, editors, *Specification, Verification and Synthesis: Mathematical Aspects*. Springer-Verlag, 1989.

[15] Jeffrey J. Joyce. A verified compiler for a verified microprocessor. Technical Report 167, University of Cambridge, Computer Laboratory, March 1989.

[16] Jeffrey J. Joyce. Generic specification of digital hardware. Technical Report 90–27, The University of British Columbia, Department of Computer Science, September 1990.

[17] J. Kershaw. Vista user's guide. Technical Report 401–86, The Royal Signals and Radar Establishment, 1986.

[18] Paul Arthur Lamb. *A verification condition generator for Intel 8080 microprocessor assembly language programs*. PhD thesis, George Washington University, 1982.

[19] D. Martin and R. Toal. Case studies in compiler correctness using HOL. In *Proceedings of the 1991 International Workshop on the HOL Theorem Proving System and its Applications*. IEEE Computer Society Press, 1992.

[20] W. D. Maurer. Proving the correctness of a flight-director program for an airborne minicomputer. In *Proceedings of the ACM SIGMINI/SIGPLAN Interface Meeting on Program Systems in the Small Processor Environment*, pages 103–108, 1976. Appeared as a special issue of SIGPLAN Notice V11(4), April 1976.

[21] W. D. Maurer. An IBM 370 assembly language verifier. In P. A. Willis, editor, *Proceedings of the 16th Annual Technical Symposium on Systems and Software: Operational Reliability and Performance Assurance*, pages 139–146. ACM, June 1977.

[22] MoD: Interim defence standard 00-55 on the procurement of safety critical software in defence equipment, 1991.

[23] J. Strother Moore. A mechanically verified language implementation. *Journal of Automated Reasoning*, 5:461–492, 1989.

[24] I. M. O'Neill, D. L. Clutterbuck, P. F. Farrow, P. G. Summers, and W. C. Dolman. The formal verification of safety-critical assembly code. In W. D. Ehrenberger, editor, *Proceedings of the IFAC Symposium on Safety of Computer Control Systems 1988 (Safecomp '88) Safety Related Computers in an Expanding Market*, 1988.

[25] Wolfgang Polak. *Compiler Specification and Verification*, volume 124 of *Lecture Notes in Computer Science*. Springer-Verlag, 1981.

[26] William D. Young. A mechanically verified code generator. *Journal of Automated Reasoning*, 5:493–519, 1989.

Higher Order Logic Theorem Proving and its Applications (A-20)
L.J.M. Claesen and M.J.C. Gordon (Editors)
Elsevier Science Publishers B.V. (North-Holland)

A HOL Mechanization of The Axiomatic Semantics of a Simple Distributed Programming Language[1]

William L. Harrison[a], Myla M. Archer[b], and Karl N. Levitt[b]

[a]Department of Computer Science, University of Illinois at Urbana-Champaign, 1304 West Springfield Ave., Urbana, Illinois, U.S.A.

[b]Department of Computer Science, University of California at Davis, Davis, California, U.S.A.

This paper presents the axiomatic semantics for a simple distributed language, and its mechanization in HOL. The constructs of this language include those basic to a very simple sequential programming language in addition to asynchronous *send* and synchronous *receive* statements. The language has the appearance of a system programming language that supports sequential execution extended with message passing, and would be a suitable basis for coding distributed operating systems and applications. Included in the mechanization are functions which generate the goals associated with the verification of sequences of simple statements in the language. In contrast to Gordon's work on the HOL mechanization of programming logics, the starting point for our work is an axiomatic (rather than denotational definition) of the language, in the style of Schlichting's definition of message passing.

Keyword Codes: D.2.4; F.3.1
Keywords: Program Verification; Specifying, Verifying, and Reasoning about Programs

1 Introduction

In his paper "Mechanizing Programming Logics in Higher Order Logic"[4], Michael Gordon describes his mechanization into HOL of the Hoare semantics for a simple imperative programming language. From the denotational semantics of his language, Gordon's mechanization generates a sentence defining a Hoare triple for each statement; the starting-point interpretation of a programming language statement is its *denotational meaning* — a function on the state of the program which indicates the new state resulting from the execution of the statement. This is in contrast to most approaches to the mechanization

[1]The research reported here was funded in part by contract DOD-MDA 904-91-C-7053 with the National Security Agency's University Program.

of programming logics, where the starting point *is* a Hoare-triple semantics. Gordon's approach is purely *definitional*; that is, it does not introduce axioms (Hoare triples) and, hence, will not introduce inconsistencies into the system.

Because a denotational semantics for our concurrent language would involve the use of powerdomains [12] and hence a complex case structure (at the very least) to obtain a derivation of our inference rules analogous to Gordon's, and since our main purpose was to implement support in HOL for reasoning about programs in our language, we proceeded by defining our language directly by an (extension of Schlichting's) axiomatic semantics. The possibility of defining an appropriate denotational or operational semantics and deriving our axiomatic rules remains an open problem.

Section 2 presents our motivation for this work. Section 3 presents the rules developed by Schlichting for *send* and *receive* without their derivation. In section 4, the distributed target language we are defining is informally described. Section 5 briefly describes the HOL mechanization of the axiomatic semantics of the target language. For more details on the HOL mechanization, please refer to [5]. A discussion of our work and plans for extensions and verified applications of the target language are presented in Section 6.

2 Motivation

The UC Davis system verification group is developing a distributed version of the CLI "stack" [2, 3, 8], which we refer to as the "SILO" project. Among the goals of the SILO project are:

- a formal specification of the R3000 RISC chip as an interpreter,
- extension of the CLI KIT operating system kernel to include distributed services and the correctness verification of the R3000-based object implementation of these services,
- implementation and verification of a generic server,
- implementation and verification of an "enhanced micro-Ada" to include concurrent features taking run-time support from the above operating system.

Verification in HOL related to the second and fourth bullets above requires some means of specifying and reasoning about distributed code in HOL.

The work described in this paper is an attempt to provide the necessary HOL tools to verify the correctness of the extended KIT operating system. Some facility for reasoning about programs in a low-level system language was therefore required. The particular language we chose has the following characteristics:

- it is close to three address code,
- it has asynchronous **send** and **receive** commands (i.e., message passing),
- IF and WHILE control structures instead of jumps.

The choice of language here is incidental to the development of the code reasoning tools—our method is sufficiently flexible to be easily applicable to the mechanization of other programming language semantics.

3 Axiomatic Semantics Of Message Passing

As part of his doctoral dissertation at Cornell University, Rick Schlichting developed Hoare-style proof rules for asynchronous **send** and synchronous **receive** commands [11]. The **send** statement has the following format:

send *expr* **to** *dest*,

and is executed as follows:

1. the value of *expr* is calculated
2. a message with value *expr* is sent to the process named *dest*.

Execution of **send** is *asynchronous*, meaning that the sending process continues execution while the message is transmitted. This is in contrast to the *synchronous* output statement "*dest*!*message*" of CSP [7] which blocks the execution of the sender process until the message is received.

The **receive** statement has the form;

receive m **when** β,

where m is a program variable and β is a (usually first-order) sentence involving m and other program variables.

Execution of a **receive** statement proceeds as follows:

1. the receiving process delays until a message with value MTEXT has been delivered such that $\beta^m_{\mathrm{MTEXT}} = TRUE$ (messages not satisfying β are ignored—but kept around for future **receives**).
2. the program variable m is then assigned the value MTEXT.

This **receive** is *synchronous*; that is, the invoking process is delayed until an appropriate message is delivered. It should be noted that the implementation of such a **receive** is, roughly speaking, as complicated as β is allowed to be. Allowing unrestricted conditions β would be quite difficult, or impossible, to implement. In our case, we assume that $\beta = TRUE$. This implies that no condition can be imposed on the message to be received for our particular **receive** statement.

In the sequel, we summarize the semantics of **send** and **receive** as they appear in [11]. No attempt will be made to derive these rules, but some motivation will be given. The interested reader should refer to [11]. The derivation of the rule for **receive** is particularly complicated.

For any process P, the system state must include information about messages which have been sent to P by other processes, and about those messages which P has received. Associated with each process P are two *bags* or *multisets* which model this aspect of the system state. These bags are referred to as the *send* bag and *receive* bag of P, respectively. *Auxiliary* or *ghost* variables σ_P and ρ_P represent the send and receive bags, respectively, in the semantics. σ_P contains a copy of each message sent to P, while ρ_P contains a copy of each message received by P. Thus, $\rho_P \subseteq \sigma_P$.

When one process sends a message with value *expr* to process P, the result is an assignment to the send bag σ_P. Thus the axiom schema for **send** derives from the standard assignment axiom:

$$\{\, W^{\sigma_P}_{\sigma_P \oplus expr} \} \ \textbf{send} \ \ expr \ \textbf{to} \ P \ \{\, W \,\} \tag{1}$$

where $\oplus$ is the bag insertion operation.

The axiom schema for the **receive** command in [11] is:

$$\{\, R \,\} \ \ r : \textbf{receive} \ \ m \ \textbf{when} \ \beta \ \ \{\, Q \,\} \tag{2}$$

provided that the following sentence is a theorem:

$$(\, R \ \wedge \ \beta^m_{MTEXT} \ \wedge \ MTEXT \in (\sigma_D \ \ominus \ \rho_D)) \implies Q^{m\ ,\ \rho_D}_{MTEXT, \rho_D \oplus MTEXT} \tag{3}$$

where D is the receiving process. We will refer to this last sentence as $Sat(r, R, Q, m)$, or more simply, as $Sat(r)$. A proof of $Sat(r, R, Q, m)$ guarantees the *soundness* of an application of schema 2. Schlichting [11] refers to the above as the axioms and satisfaction condition for "Unreliable Datagrams"; for other models of message passing they are slightly different.

A program proof using this axiomatic semantics requires a "sequential" proof for each process. Proofs of programs consisting of multiple concurrent processes also, in general, require a demonstration of *non-interference* of the processes [11]—that is, some way of showing that assignment actions in one process do not interfere with the *critical assertions*[2] of other concurrently executing processes. Purely as a matter of technique, it is also typically necessary in programs using message passing to include a *program invariant* in the assertions that expresses the possible results of different interleavings of events [11].

4 Our Target Distributed Language

In this section, we present an informal description of the target distributed language. The sequential language constructs are constructed from the instruction set of the AVM-1 microprocessor, designed and verified by Windley [14]. The distributed language constructs are **send** and **receive** as discussed previously, but are specified to take register names as arguments.

As part of his doctoral dissertation [14] at UC Davis, Phillip Windley designed, specified, and verified the correctness of the AVM-1 (*A Verified Microprocessor*). AVM-1 has a load-store architecture and is, in most respects, a RISC machine. The instruction set contains thirty instructions, four of which will be mechanized in our logic.

Our target language contains four AVM-1 instructions: `ADDI` (add immediate), `SUB` (subtract), `JMP_NZ` (jump if not zero), and `JMP_GE` (jump if greater than or equal to). We assume there are two flags – **NZ** and **GE** – which are set if the last arithmetic operation produced a zero result or a result greater than or equal to zero, respectively. A register transfer level (RTL) description of `ADDI` and `SUB` can be found in figure 1 and figure 2.

ADDI

$$R_{dst} \Leftarrow R_{src} + imm$$
$$\mathbf{NZ} \Leftarrow (R_{src} + imm) \neq 0$$
$$\mathbf{GE} \Leftarrow (R_{src} + imm) \geq 0$$
$$PC \Leftarrow PC + 1$$

Figure 1: RTL for ADDI

SUB

$$R_{dst} \Leftarrow R_{src1} - R_{src2}$$
$$\mathbf{NZ} \Leftarrow (R_{src1} - R_{src2}) \neq 0$$
$$\mathbf{GE} \Leftarrow (R_{src1} + R_{src2}) \geq 0$$
$$PC \Leftarrow PC + 1$$

Figure 2: RTL for SUB

Strictly speaking, these instructions are AVM-1-*like*. We ignore *carry* and *overflow* arithmetic conditions for example. Furthermore, we found it convenient to model forward(backward) jumps with IF(WHILE) statements.

Our language differs further from *real* AVM-1 in that instruction operands (i.e., machine registers) are assumed to have type *integer*. Consequently, the distinction between signed and unsigned integers, as well as the finite range of operands, are ignored. These simplifications were made because developing the *means* of reasoning about distributed code in HOL was the primary goal of this project. Using the methods developed here, a language closer to the real AVM-1 instruction set could be mechanized.

The **send** command performs an asynchronous send. The format of the **send** command is:

send R_{msg}, *dst*.

Executing the above statement sends a one word message containing the contents of register R_{msg} to the process named by the constant *dst*. If *dst* refers to an invalid name, **send** R_{msg}, *dst* does nothing. **send** is *asynchronous* — the process executing **send** does not block after executing **send** . (Thus, we are modelling the case where there is either an infinite buffer or some provision for messages being lost.)

Recall that the format of the **receive** command given in [11] is:

receive m **when** β

To simplify our mechanization, we will assume for our **receive** command that β is always TRUE. The format for our **receive** command is then:

receive R_{msg}

[2]For a rigorous definition of *critical assertion* and *process interference*, please refer to [1].

Instruction	Semantics
$\texttt{ADDI}\ \mathtt{R}_{src}, \mathtt{R}_{dst}, imm$	$\{\ \phi^{\mathtt{R}_{dst}}_{\mathtt{R}_{src}+imm}\ \}\ \ \texttt{ADDI}\ \mathtt{R}_{src}, \mathtt{R}_{dst}, imm\ \ \{\ \phi \wedge CF_{\texttt{ADDI}}\ \}$
$\texttt{SUB}\ \ \mathtt{R}_{src1}, \mathtt{R}_{src2}, \mathtt{R}_{dst}$	$\{\ \phi^{\mathtt{R}_{dst}}_{\mathtt{R}_{src1}-\mathtt{R}_{src2}}\ \}\ \ \texttt{SUB}\ \ \mathtt{R}_{src1}, \mathtt{R}_{src2}, \mathtt{R}_{dst}\ \{\ \phi \wedge CF_{\texttt{SUB}}\ \}$
send $\mathtt{R}_{src}, dst$	$\{\ W^{\sigma_{dst}}_{\sigma_{dst}\oplus\mathtt{R}_{msg}}\}\ \ \textbf{send}\ \mathtt{R}_{msg}, dst\ \ \{\ W\ \}$
receive $\mathtt{R}_{msg}$	$\dfrac{\vdash\ Sat(\ r\ ,\ \phi\ ,\ \psi, \mathtt{R}_{msg}, TRUE\ )}{\{\ \phi\ \}\ \ r\!:\textbf{receive}\ \mathtt{R}_{msg}\ \ \{\ \psi\ \}}$

Figure 3: The Instruction Mnemonics and Their Semantics

The process executing this command has the value of a (previously non-**receive**'d) message placed in $\mathtt{R}_{msg}$, if any such message is available. If such a message is not available, the receiving process blocks execution until a message arrives. As indicated previously, **receive** implements *synchronous* receive. Because β is assumed to be the constant TRUE, processes can not put any condition on which message to receive. As a consequence of this, a process cannot "choose" from which process to receive.

The axiomatic semantics for the language can now be given. Axiom schemata for `ADDI` and `SUB` derive from the standard assignment axiom [1, 6, 10] in a straightforward manner. The axiom schemata for **send** and **receive** come directly from the semantics described in [11] which were outlined in the previous section. The semantics for IF and WHILE are simplifications of those found in [1]. The axioms of our programming logic are summarized in figure 3. In figure 3, the symbols $CF_{\texttt{ADDI}}$ and $CF_{\texttt{SUB}}$ state the values of the condition flags after execution of the corresponding instructions. The semantics of IF and WHILE are inference rules, because their consequents depend upon the derivability of other triples in their antecedents, and they are similar to the familiar rules for these constructs. The other inference rules are well-known and can be found in [1, 6, 10], so we do not summarize them here to save space.

5 The HOL Mechanization of the Semantics

This section describes in brief how the axiom schemata and inference rules of the language were integrated into HOL. A predicate `Prov` is defined using the inductive relations definitions package [9] so that

$$\{\phi\}\ \ StatementList\ \ \{\psi\}$$

is a theorem in the logic if and only if

$$\texttt{Prov}\ \ \phi\ \ StatementList\ \ \psi$$

is a theorem of HOL. The pre- and postconditions of a Hoare triple are represented in HOL as terms of the type ":bool", and *StatementList* has HOL type ":(string)list" (i.e., a list of

```
#let th1 = mk_ADDI "['ADDI R1,R2,1']" "R2 = 2" "R2" "R1" "1";;
   th1 = ⊢ Prov(R1 + 1 = 2)['ADDI R1,R2,1'](R2 = 2)
#let th2 = mk_ADDI "['ADDI R1,R1,1']" "R2 = 2" "R1" "R1" "1";;
   th2 = ⊢ Prov(R2 = 2)['ADDI R1,R1,1'](R2 = 2)
#let th3 = DISCH "R2 = 2" (ASSUME "R2 = 2");;
   th3 = ⊢ (R2 = 2) ⟹  (R2 = 2)
#compose th1 th2 th3;;
   ⊢ Prov(R1 + 1 = 2)['ADDI R1,R2,1';'ADDI R1,R1,1'](R2 = 2)
```

Figure 4: `compose` example.

strings). The inductive definition of `Prov` includes a provision for axiom introduction. Our mechanization of these is in terms of axiom schemata: an ML function is provided for each statement type which creates an axiom instance appropriate to the particular components of the statement. For convenience, rather than analyzing the statement itself to discover its components, we simply use the statements as mnemonics (hence the representation of statement lists as strings), and pass the components of a given statement to the ML function for that statement type.

A process in our logic is identified with a uid and a code listing corresponding to a program in execution, and has access to a set of private registers. If desired, a *cobegin* statement may be used to represent the creation of some number of asynchronously executing processes as in [1, 10].

Our mechanization package includes help functions which assist the user in the application of inference rules to complicated Hoare formulas. The method used is sufficiently flexible to be easily applicable to the mechanization of other programming language semantics. Lack of space prevents us from giving a full account of the mechanization (see [5] for further details).

As an example of how inference rules are mechanized in this work, we consider the `compose` function which represents the rule of composition [1, 10]. `compose` takes the three theorems necessary to apply the composition rule; namely, they are $\vdash$ `Prov` $\phi\ sl_1\ \psi$, $\vdash$ `Prov` $\gamma\ sl_2\ \theta$, and $\vdash \psi \Longrightarrow \gamma$. Given these three theorems, `compose` returns:

$$\vdash_{HOL}\ \texttt{Prov}\ \phi\ sl_1\hat{\ }sl_2\ \theta$$

An example application of `compose` can be found in figure 4. In that figure, the HOL and ML commands typed by the user are preceded by a "`#`", and the responses of the HOL interpreter appear directly below each command. In figure 4, the user first creates two axiom instances (i.e., `th1` and `th2`), proves "`R2 = 2` $\Longrightarrow$ `R2 = 2`", and then applies `compose`, from which the desired theorem results. `compose` ensures that `th3` *really* is the necessary theorem by checking that the postcondition of `th1` is the antecedent of `th3`, and that the precondition of `th2` is the consequent of `th3`.

Recall from section 3 that the necessity of demonstrating *Sat(r)* for every `receive` statement *r* complicates the use of the receive axiom, because the user must prove $Sat(r)$ *before* its application. The ML function `mk_receive` takes as arguments a theorem t which

```
#mk_Sat_rec
   "(rhoD = EMPTY_BAG) ∧ (∀M:num. (M IN_B sigD) ⟹  (M = 16))"
   "(R_D=16) ∧ (∀M. M IN_B sigD ⟹  (M = 16))"
   "R_D" "MTEXT" "sigD:(num)bag" "rhoD:(num)bag";;


   "((rhoD = EMPTY_BAG) ∧ (∀M. M IN_B sigD ⟹  (M = 16))) ∧
    MTEXT IN_B (sigD DIFF_B rhoD) ⟹
    (MTEXT = 16) ∧ (∀M. M IN_B sigD ⟹  (M = 16))"
```

Figure 5: mk_Sat_rec example.

proves $Sat(r)$, the destination variable m, and the desired postcondition Q, respectively. In order to demonstrate the soundness of:

$$r : \{ R \} \ \texttt{receive}\ m \ \{ Q \},$$

the following sentence

$$Sat(r) : R \wedge MTEXT \in (\sigma_D \ominus \rho_D) \implies Q^{m\ ,\ \rho_D}_{MTEXT, \rho_D \oplus MTEXT}$$

must be shown to be a theorem.

Constructing the $Sat(r)$ sentence may lead to frustration on the part of the user because one human error can result in considerable wasted time and effort. It is quite easy, for example, for a human to err while performing simultaneous substitution. The first author experienced this problem when first using mk_receive. To avoid similar difficulties in the future, an ML function called mk_Sat_rec was created. mk_Sat_rec takes the precondition R, the postcondition Q, the destination variable m, the message value $MTEXT$, and the send and receive bags σ and ρ, and then constructs the correct satisfaction sentence, and sets the result as a HOL proof *goal*. The user may then attempt to prove the sentence with the HOL tactics package. Similar "help" functions exist for the rules of inference. Figure 5 contains an example application of mk_Sat_rec, and figure 6 contains an example application of mk_receive. In figure 6, sat_r is the theorem resulting from the proof of the goal generated in figure 5.

6 Discussion

This paper has presented a HOL mechanization of the axiomatic semantics of a simple distributed programming language. The language has features in support of sequential program execution in addition to asynchronous *send* and synchronous *receive* through which processes can communicate via message passing. Our semantics for the distributed features of the language derive from Schlichting's axiomatic semantics. Our mechanization package includes help functions which assist the user in applying inference rules to complicated Hoare triples. Among the benefits of our approach are:

```
#let th0_D = mk_receive "[‘ receive R_D ’]" sat_r "R_D"
        "(R_D=16) ∧ (∀M. M IN_B sigD ⟹  (M = 16))";;

   th0_D =
   ⊢ Prov
      ((rhoD = EMPTY_BAG) ∧ (∀M. M IN_B sigD ⟹  (M = 16)))
      [‘ receive R_D ’]
      ((R_D = 16) ∧ (∀M. M IN_B sigD ⟹  (M = 16)))
   Run time: 0.1s
   Intermediate theorems generated: 22
```

Figure 6: `mk_receive` example.

- ease of modification – new rules can be readily added to the system.
- ease of use – help functions relieve the user of the the burden of performing multitudes of repetitive syntactic manipulations, and the HOL tactics package can be applied directly to verification conditions as needed. Also, large libraries (e.g., the bag library in our case) of pre-proven theorems are available to the user in HOL.
- familiarity – someone completely innocent of HOL can look at a HOL session transcript of a derivation and understand what has been demonstrated.

Our semantics for *send* and *receive* include *non-interference* assumptions, stating that the execution of these statements are atomic with respect to the objects they manipulate. The objects in question are machine registers and the *send* and *receive* bags, which record the messages sent to and received by a process, respectively. These assumptions would incur proof obligations on a verified implementation of *send* and *receive*.

The target language will certainly be expanded as necessary to complete the application projects underway, and the tools to do so have been developed and tested in this paper. The help functions (e.g., `mk_Sat_rec`) free the user from performing tedious and complicated syntactic operations like simultaneous substitution, thereby eliminating much human error and effort. Proofs in this system are created and are represented in a manner similar to hand proofs, which renders them more accessible to the HOL novice.

This work described here developed more than an integration into HOL of a particular axiomatic semantics. Rather, we created the necessary HOL machinery for specifying partial correctness conditions as Hoare triples in HOL. That is, axiom schemata such as those of our language may be readily mechanized in HOL with the ML and HOL constructs used here to mechanize our programming logic. State-dependent behavior is frequently specified with Hoare triples, and thus this package could prove quite useful. In particular, we believe the axiomatic semantics for "higher-level" programming language constructs such as the Ada *rendezvous*, the input and output commands of CSP, and the operations of SR may be mechanized readily.

References

[1] G. R. Andrews. *Concurrent Programming : Principles and Practice.* The Benjamin/Cummings Publishing Company, Inc. 1991.

[2] W. R. Bevier. A Verified Operating System Kernel. Technical Report 11, Computational Logic, Inc., October 1987.

[3] W. R. Bevier. KIT: A Study in Operating System Verification. *IEEE Transactions on Software Engineering*, Vol. 15, No. 11, November 1989.

[4] M. Gordon. Mechanizing Programming Logics in Higher Order Logic. *Current trends in hardware verification and automated theorem proving.* G. Birtwistle, P.A. Subrahmanyam, editors. New York : Springer-Verlag, c1989.

[5] W. L. Harrison. Mechanizing the Axiomatic Semantics for a Programming Language with Asynchronous Send and Receive in HOL. Master's Thesis, Dept. of Computer Science, University of California, Davis. September 1992.

[6] C. A. R. Hoare. An Axiomatic Basis for Computer Programming. *Communications of the ACM*, 12, 10(October 1969), pages 576-580, 583.

[7] C. A. R. Hoare. Communicating Sequential Processes. *Communications of the ACM*, 21, 8(August 1978), pages 666-677.

[8] W. A. Hunt. FM8501: A Verified Microprocessor. Inst. Comput. Sci., Univ. Texas at Austin, Tech. Rep. 47, Dec. 1985. 576-580, 583.

[9] T. Melham. A Package for Inductive Relation Definitions in HOL. Technical Report, Cambridge University Computer Laboratory, 1989.

[10] S. Owicki and D. Gries. An Axiomatic Proof Technique for Parallel Programs. *Acta Informatica*, 6, pages 319-340, 1976.

[11] R. D. Schlichting and F. B. Schneider. Using Message Passing for Distributed Programming: Proof Rules and Disciplines. *ACM Transactions on Programming Languages and Systems*, Vol. 6, No. 3, pages 402–431, July 1984.

[12] D. A. Schmidt. *Denotational Semantics: A Methodology for Language Development.* Wm. C. Brown Publishing Co., 1988.

[13] E. T. Schubert. A Methodology for the Verification of Composed Hardware Systems. Ph.D. dissertation, Dept. of Computer Science, University of California, Davis. June 1992.

[14] P. J. Windley. The Formal Verification of Generic Interpreters. Ph.D. dissertation, Dept. of Computer Science, University of California, Davis. June 1990.

Chapter 6

Hardware Description Language Semantics

Higher Order Logic Theorem Proving and its Applications (A-20)
L.J.M. Claesen and M.J.C. Gordon (Editors)
Elsevier Science Publishers B.V. (North-Holland)

A Formalisation of the VHDL Simulation Cycle

John P. Van Tassel†

University of Cambridge Computer Laboratory, New Museums Site, Pembroke Street, Cambridge, CB2 3QG, England.

Abstract
The VHSIC Hardware Description Language (VHDL) has been gaining wide acceptance as a unifying HDL. It is, however, still a language in which the only way of validating a design is by careful simulation. With the aim of better understanding VHDL's particular simulation process and eventually reasoning about it, a formalisation of VHDL's simulation cycle has been developed for a subset of the language. It has also been possible to embed the semantics in the Cambridge Higher-Order Logic (HOL) system and derive interesting properties about specific VHDL programs.

1 Introduction

When the VHSIC Hardware Description Language (VHDL) burst onto the design scene, it was described as the solution to many existing problems. But, when the language was designed very little thought was given to its formal semantics, much less to using such a semantics to reason about individual VHDL programs. The result has been a recent upsurge in research efforts addressing precisely these problems. The following is a discussion of our attempt to formalise the simulation cycle and timing model of VHDL.

The exposition will begin with an overview of VHDL's history and uses and will be followed by a *précis* of other work on the application of formal methods to VHDL. The driving philosophy behind the work will then be explained. The semantics that has been developed is then presented, and the paper will conclude with the description of some experimental results in the application of the semantics to specific VHDL programs.

2 Overview of VHDL

VHDL is a hardware description language (HDL) currently in use by a reasonably large

†Research supported under grant number AFOSR-91-0246 from the United States Air Force Office of Scientific Research

part of the design community. It is an event-driven simulation language whose semantics is defined, at least informally, by the way in which the various language constructs are evaluated by the simulation model [6].

To better understand VHDL and its place in current design practice, we should consider the evolution of the language:

pre-1980's Chaos

This is a time period characterised by the existence of many proprietary hardware description languages and simulators. Furthermore, no one HDL is used throughout the entire design process. In many cases, ad-hoc mechanisms are developed to translate between the tools and languages in use at different stages of the design. This leads to increased confusion and provides a further avenue for the introduction of bugs.

mid-1980's VHSIC programme

The United States Air Force, a large consumer of custom electronics obtained from a variety of vendors, is faced with the need to rationalise its procurement process. The primary concern is that all designs submitted as a part of the bidding process are written in the same non-proprietary HDLand meet certain design and documentation standards. VHDL emerges as the language.

late-1980's IEEE VHDL (STD 1076)

As a part of the VHSIC programme, large vendors such as IBM and Texas Instruments take an interest in VHDL. That interest, coupled with the growing need for an industry-standard HDL, causes VHDL to go through the IEEE standardisation process and emerge as the language in use today.

Since its introduction, VHDL's use has spread to many areas of digital design. While most of these applications are in the areas of design, there has also been recent interest in synthesis directly from VHDL. The language in its current state is not generally amenable to this task, but it is anticipated that modifications to the standard (VHDL'92) will overcome many of the problems.

3 Related Work

Past efforts at applying formal methods to VHDL programming were carried out in the late 1980's. These were intended to give additional simulation-time assurance that particular VHDL programs were performing as anticipated. One of these investigations led to the development of the VHDL Annotation Language (VAL) at Stanford [2], which allowed the user to decorate a program with specifications about its operation. Another of the early projects, conducted by the present author, attempted automatically generate VHDL assertions characterising the behaviour of VHDL programs using syntactic analysis [11, 12].

More recent research has been directed at the verification of VHDL programs. To accomplish their goals, all these projects have necessarily had to build up a notion of VHDL's semantics. Temporal logic is currently in use at the Aerospace Corporation as the framework in which their current research is being conducted [5]. Other useful results are emerging from work being conducted at IMAG in Grenoble on the application of more functional methods to the problem [10].

4 Philosophy

The formalisation of VHDL presented here takes a rather pragmatic view. The semantics of the language, as given informally in [6], emphasizes the way in which the simulation behaves in the presence of events and transactions, and the way in which language constructs are executed by that simulation model. The plan, therefore, is to codify these same principles more formally. The result not only reflects the actual VHDL simulation cycle, but corresponds intuitively to the way in which VHDL users think about the language. For purposes of the current research, it has been necessary to utilise only a subset of full VHDL. This subset is extensible within the context of our semantics by simply formalising the execution of the desired extensions by the simulation model of VHDL. It is, however, sufficient to demonstrate the most salient features of the simulation model.

5 Intuition behind the Semantics

The intention is to formalise as closely as possible the simulation model of [6]. To make this formalisation more understandable, what actually takes place during the course of a VHDL simulation is now explained. A few definitions are first introduced.

Event: A change in the value of a signal.

Transaction: The value that a signal should take on at a particular time in the future. A transaction may (or may not) be converted into an event at that time.

Point of computation: The point at which a particular collection of transactions is processed.

Process: A VHDL concurrent statement equipped with a set of signal names, or sensitivity list, guarding its activation.

Simulation cycle: The evaluation of all the processes in a program at a particular point of computation.

The top-level simulation model of VHDL may be viewed as a four step sequence, and is illustrated in Figure 1. The loop begins by moving forward to the nearest interesting point

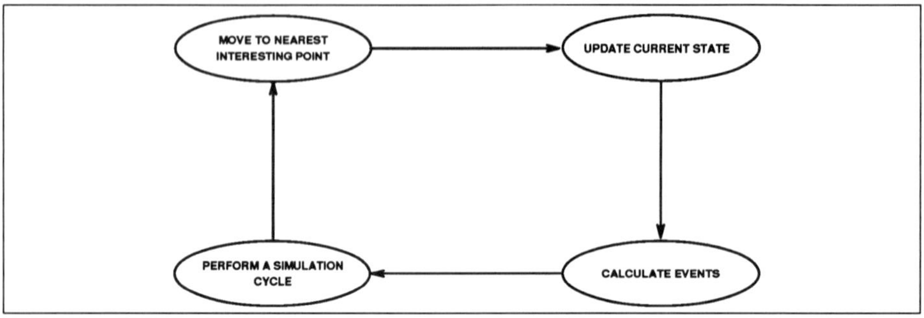

Figure 1: Progression of a simulation

of computation (i.e. one where there are transactions to process), and setting the current time to be the physical time unit associated with that point of computation. The state of the signal values is then updated by those that they are supposed to take on during the present point of computation. We then move on to determine the signals for which this update represents a change in value, or event. A simulation cycle is then performed based on the new state of the world. The cycle is then repeated until there are no more transactions to process. The following pseudo-code is also presented as a more concise description of the simulation loop:

```
while transactions remain to be processed
      go to nearest point of computation with transactions to process
      update the state from the current transactions
      determine which updates represent events
      perform a simulation cycle based on new state and events
end while
```

The state referred to should be understood to contain the values of the individual signals.

Note the use of the phrase "point of computation" above. While one could often equate this concept with that of a time slice, a point of computation is more than that within the context of VHDL's particular simulation model. One could, for instance, say that the points of computation P_1 through P_n represent the time units 1 through n. Alternatively, they could represent 1 through n δ-delays between two major time units.

The concept of δ-delay is the way in which VHDL deals with 0-delay signal assignments. During each iteration through the simulation loop we are using a static state of the world and are scheduling transactions to be performed at some future time point. This scheduling applies to any transaction, whether it is to occur at a delay offset zero units from now, or at some larger one. Each time we go around the simulation loop (Figure 1) in zero time is called a δ-delay. An individual δ does *not*, therefore, represent a quantifiable unit of time. Rather it is simply a simulator artifact.

Figure 2: Progression of a simulation cycle

As mentioned earlier, a component of the simulation loop is the performance of a "simulation cycle". The unfolding of such a cycle as shown in Figure 2 begins with the activation of those processes that are sensitive to any of the current events. We then proceed to run these active processes in parallel. During their execution, each process schedules transactions to be processed in the future. When all the processes have terminated, we gather those separate futures into an amalgamated view of future behaviour. The above steps are again more concisely stated as:

```
determine which processes are active based on current events
run the active processes in parallel
join all activated process' scheduled transactions into a collective whole
```

6 The Femto-VHDL Subset

The VHDL subset that is used in defining the semantics is rather small (hence the name "Femto-VHDL"), but it contains enough constructs of the language to exercise the important features of the simulation model. Only one kind of VHDL concurrent statement is supported, namely `PROCESS` statements with explicit sensitivity lists. The sequential statements addressed are the `IF-THEN-ELSE` conditional and the transport delay signal assignment. It is assumed that:

- Signals may only be Boolean-valued. This is for simplicity, and does not alter the exposition of the simulation model in any way.
- Resolved signals are not allowed. The implication is that constructs such as wired-and's and wired-or's are not permitted.

Given the above listing of supported statements and assumptions, one can ask what actually constitutes a Femto-VHDL program. At the outermost level, such a program is made up of one or more concurrent process statements, each of which comprises some sequential statements. This structuring of constructs is described by to the following abstract syntax for Femto-VHDL:

bexp	::=	*sig* | not *bexp* | *bexp* and *bexp* | *bexp* or *bexp* | *bexp* nand *bexp* | *bexp* nor *bexp* | *bexp* xor *bexp* | *bexp* xnor *bexp* | true | false
ss	::=	*ss* ; *ss* | *bexp* => *ss* I *ss* | *sig* := (*bexp*,*dly*)
cs	::=	*cs* II *cs* | *sl* : *ss*

```
{C} : C := (NOT C,dly)

PROCESS (C)
BEGIN
  C <= TRANSPORT (NOT C) AFTER dly;
END PROCESS;
```

Figure 3: A uniprocess oscillator

```
({H} : H := (NOT H,dly)) || ({H} : C1 := (H,0 NS))

PROCESS (H)
BEGIN
  H <= TRANSPORT (NOT H) AFTER dly;
END PROCESS;

PROCESS (H)
BEGIN
  C1 <= TRANSPORT H AFTER 0 NS;
END PROCESS;
```

Figure 4: A dual process oscillator

bexp ranges over Femto-VHDL Boolean expressions, which are made up of single signal names, or of compound expressions using the various VHDL Boolean operators. Sequential statements (*ss*) may be either two statements in sequence (`;`), a conditional statement guarded by some Boolean expression, or a signal assignment statement that assigns a Boolean value to some signal after some delay expressed in terms of the natural numbers. Finally, *cs* shows the structure of concurrent statements. A concurrent statement may be either two such statements in parallel (`||`), or it may be a single process composed of a sensitivity list (*sl*) and sequential statements (*ss*).

A pair of examples are now presented to demonstrate the abstract syntax of Femto-VHDL and its relationship to real VHDL code. The translation between the abstract syntax and actual VHDL is meant to be obvious and is raised here to avoid confusion when results are presented later. The first example, shown in Figure 3, is a uniprocess program that describes an oscillator with a half-period of `dly`. The second (Figure 4) is a simple extension of the previous device which adds a second process representing a 0-delay buffer. While this second example might seem contrived, it will be used to demonstrate certain properties of the formal semantics later. Both examples are drawn from [1].

7 The Semantic Framework

A brief overview of the formalism to be used in defining the the semantics of Femto-VHDL will now be given. In that semantics use will be made of a collection of labelled transition relations in the style of [9]. Each of these relations is defined inductively as the least relation closed under a set of rules describing the way in which different classes of Femto-VHDL constructs are evaluated by the simulation model.

To better understand the concept of inductively defined relations, let us examine an example taken from [4] using the even natural numbers. One could define the properties necessary to describe even naturals to be:

1. 0 is an even number.

2. if n is even, then $n+2$ is also even.

These statements could be rephrased by the following set of deduction rules holding of a predicate **even**. Such rules are generally made up of *hypotheses* above the line and the *conclusion* below it. As a point of terminology, rule 1 is an *axiom* as it has no hypotheses at all.

$$(1)\quad \frac{}{\mathsf{even}\ 0} \qquad\qquad (2)\quad \frac{\mathsf{even}\ n}{\mathsf{even}\ n+2}$$

Since the rules correspond directly to the minimal properties required of the even numbers, one can define **even** to be the least realation for which the properties hold. It then follows by definition that **even** is satisfied by rules 1 and 2. Furthermore, because **even** is defined to be the least such relation, it is a subset of any other relation satisfying these rules. The net result of all this is that **even** describes only those values that it must describe to characterise the even naturals.

If we wish to express that some relation P is closed under the rules for **even**, the following characteristic statement is made:

$$(P\ 0) \wedge (\forall\, n.\ P\ n \supset P(n+2))$$

Or, in the case of **even** itself the following definition could be made:

$$\forall\, m.\ \mathsf{even}\ m = \forall\, P.\ ((P\ 0) \wedge (\forall\, n.\ P\ n \supset P(n+2))) \supset P\ m$$

Now, if one wished to prove that some property P' holds of **even**, it is required the set $S = \{n \mid P'\ n\}$ be shown to be closed under the rules for **even**. Because **even** is defined to be the least such set, it is known that $\mathsf{even} \subseteq S$. This means that the property in question (P') holds of **even**.

The induction principle just used in reasoning about **even** is known as rule induction, and is necessary for many proofs involving inductive definitions. The way this principle is embedded in the HOL system is discussed in [8].

The rules for Femto-VHDL will be of the same form as those above, but will be made up of more complex components. As mentioned, the formulas will be elements of some transition system. In general, these formulas will look like:

$$env \vdash a \xrightarrow{trans} b$$

where $\xrightarrow{trans}$ is the name of the transition relation, and $\vdash$ separates the environment (env) used by the formula from the items being related (a and b). Such a formula should be read, "in the environment env, a evaluates to b in the relation $trans$".

8 The Semantics of Femto-VHDL

The rules used to describe the semantics of Femto-VHDLare now presented. A good starting point is the rules that specify the evaluation of Boolean expressions. The rules for Femto-VHDL sequential statements will then be given and the exposition will finish with the formalisation of Femto-VHDL concurrent statements. But, before introducing any rules it will be necessary to explain the infrastructure upon which they are built.

8.1 Types and Functions

The types needed to make the semantics work are really very simple. They were developed to correspond with VHDL users' intuition about the structuring of information in VHDL itself. Use will be made of set theory in much of what follows, as it is not only a convenient notation for expressing concepts of interest, but also provides many simple operations that capture the essence of the actions required by the semantics.

Three aliases are used to make what is explained later more readable. The name "value" is used to represent Booleans, "time" for the natural numbers, and "name" for signal names. By employing these aliases, we may define the types used in the semantics as follows:

Type	*Ranges Over*	*Type*
γ	events	(name) set
σ	state	(name $\times$ value) set
τ	transactions	time $\rightarrow$ state
ρ	environment	(time $\times$ state $\times$ events)

Recall that an event is a change in signal value. γ is simply a collection of signal names for which this is the case. State (σ) is used to represent the value of all the signals in a program. It is analogous to the "state" employed in the algorithms of Section 5. Transactions (τ) ensure a way of mapping from future times to state, and will provide the underlying framework for scheduling. Environment (ρ) is used to represent the static view of the world in which a given simulation cycle unfolds, and allows us to keep track of the current time for scheduling purposes, the current state of the signal values, and the signals for which there are currently events.

Three functions are used in defining the semantics. The first is **valcalc**, whose purpose is to calculate the value of a signal sig in a state σ.

$$\mathsf{valcalc}\ sig\ \sigma = \mathsf{choice}\ \{val \mid (sig, val) \in \sigma\}$$

The function choice used in this definition selects an arbitrary element from a non-empty set. The use of choice is justified here because the set from which the choice is made will always be a singleton set (signals have unique values).

Whenever a signal assignment statement is encountered, it becomes necessary to "post", or schedule, a transaction to take place at some future point of computation to effect that assignment. The function post does this, while at the same time implementing the pre-emptive scheduling of transactions demanded by VHDL. post uses the current transactions τ, the signal for which the assignment is to be made *sig*, the value that signal is supposed to take on *val*, and the time at which that assignment is to take place t to create a new version of future transactions by returning a function from time to state. If the time supplied to this function is the one at which the signal assignment is to take place, a set consisting of a "stripped" version of the projected state with the given signal-value pair inserted into it is returned. Otherwise, the function simply returns the stripped version of the projected state. This stripping process is performed by a function local to post called stripped, which in fact does the pre-emption part of scheduling a new transaction. stripped, when given a time t'', will return a version of the starting τ in which all pending transactions on signal *sig* have been removed when t'' is greater than or equal to the time at which the current scheduling is to take place. The function insert below should be understood as simple set insertion.

$$
\begin{array}{l}
\mathsf{post}\ sig\ val\ \tau\ t = \\
\quad let\ \mathsf{stripped}\ t'' = \\
\qquad (\tau\ t'') - \{(x,y) \mid (x,y) \in (\tau\ t'') \wedge (x = sig) \wedge (t'' \geq t)\} \\
\quad in \\
\qquad \lambda t'.\ \mathsf{if}\ (t' = t)\ \mathsf{then}\ ((sig, val)\ \mathsf{insert}\ (\mathsf{stripped}\ t')) \\
\qquad\qquad \mathsf{else}\ (\mathsf{stripped}\ t')
\end{array}
$$

The final function defined is employed during the execution of concurrent statements. As will be seen, each process makes use of the same starting τ when activated and goes on to post new transactions into it. When two processes finish their execution, it is necessary to combine their individual views of future behaviour into an overall one. The function zip does this by returning a function from time to state in which a set union is done at any given future point of computation on the sets of signal-value pairs found there. It is essential to bear in mind that this simple view of transaction resolution falls apart when the restriction that signals not have multiple drivers is removed. In that case the union would have to be replaced by a call to some function which ensures that appropriate user-defined resolution functions are applied to multiply driven signals, while continuing just to add to the global future those which are not [6].

$$\tau'\ \mathsf{zip}\ \tau'' = \lambda t.\ (\tau'\ t) \bigcup (\tau''\ t)$$

8.2 Rules for Boolean Expressions

The rules defining the semantics for the evaluation of VHDL Boolean expressions ($\xrightarrow{Bool}$) are numerous but rather simple. The single unary operator (not) applied to an expression

e is defined to be nothing more than the logical negation of whatever e evaluates to. A similar pattern arises for the binary operators in rules b_{ii} through b_{vii}. It is only when one comes to rule b_{viii} that something interesting happens. Here an axiom is given stating that the value of a signal sig in a particular state σ is obtained by calculating its value using the function valcalc described above.

$$\{\mathsf{b}_i\} \quad \frac{\sigma \vdash e \xrightarrow{Bool} x}{\sigma \vdash (\mathsf{not}\ e) \xrightarrow{Bool} \neg x} \qquad \{\mathsf{b}_{ii}\} \quad \frac{\sigma \vdash e_0 \xrightarrow{Bool} x \quad \sigma \vdash e_1 \xrightarrow{Bool} y}{\sigma \vdash (e_0\ \mathsf{and}\ e_1) \xrightarrow{Bool} (x \wedge y)}$$

$$\{\mathsf{b}_{iii}\} \quad \frac{\sigma \vdash e_0 \xrightarrow{Bool} x \quad \sigma \vdash e_1 \xrightarrow{Bool} y}{\sigma \vdash (e_0\ \mathsf{or}\ e_1) \xrightarrow{Bool} (x \vee y)}$$

$$\{\mathsf{b}_{iv}\} \quad \frac{\sigma \vdash e_0 \xrightarrow{Bool} x \quad \sigma \vdash e_1 \xrightarrow{Bool} y}{\sigma \vdash (e_0\ \mathsf{nand}\ e_1) \xrightarrow{Bool} \neg(x \wedge y)}$$

$$\{\mathsf{b}_{v}\} \quad \frac{\sigma \vdash e_0 \xrightarrow{Bool} x \quad \sigma \vdash e_1 \xrightarrow{Bool} y}{\sigma \vdash (e_0\ \mathsf{nor}\ e_1) \xrightarrow{Bool} \neg(x \vee y)}$$

$$\{\mathsf{b}_{vi}\} \quad \frac{\sigma \vdash e_0 \xrightarrow{Bool} x \quad \sigma \vdash e_1 \xrightarrow{Bool} y}{\sigma \vdash (e_0\ \mathsf{xor}\ e_1) \xrightarrow{Bool} (x \oplus y)}$$

$$\{\mathsf{b}_{vii}\} \quad \frac{\sigma \vdash e_0 \xrightarrow{Bool} x \quad \sigma \vdash e_1 \xrightarrow{Bool} y}{\sigma \vdash (e_0\ \mathsf{xnor}\ e_1) \xrightarrow{Bool} \neg(x \oplus y)}$$

$$\{\mathsf{b}_{viii}\} \quad \frac{}{\sigma \vdash sig \xrightarrow{Bool} \mathsf{valcalc}\ sig\ \sigma}$$

8.3 Rules for Sequential Statements

The relation $\xrightarrow{Seq}$ describes the evaluation of Femto-VHDL sequential statements. The relation makes use of $\xrightarrow{Bool}$ in many places, and the σ that is employed should be understood to be the state component of the environment in rules ss_i through ss_{iv}. Sequencing of Femto-VHDL statements is covered in rule ss_i, and the evaluation of conditionals is defined in rules ss_{ii} and ss_{iii}. The most important statement, from the point of view of describing the simulation model of VHDL, occurs in the postulation of rule ss_{iv}. Here the result of evaluating a transport delay signal assignment is seen. For a signal assignment statement $sig := (e, dly)$ where e evaluates to x in σ, the current time , and the current transactions τ, one says that the result of performing the assignment are new transactions created by using the function post to schedule into τ the value x for signal sig at time $(t + dly)$.

$$\{\mathsf{ss}_i\} \quad \frac{\rho \vdash \langle ss_0, \tau\rangle \stackrel{Seq}{\longrightarrow} \tau' \quad \rho \vdash \langle ss_1, \tau'\rangle \stackrel{Seq}{\longrightarrow} \tau''}{\rho \vdash \langle ss_0 \ \texttt{;}\ ss_1, \tau\rangle \stackrel{Seq}{\longrightarrow} \tau''}$$

$$\{\mathsf{ss}_{ii}\} \quad \frac{\sigma \vdash e \stackrel{Bool}{\longrightarrow} \mathsf{true} \quad (t,\sigma,\gamma) \vdash \langle ss_0, \tau\rangle \stackrel{Seq}{\longrightarrow} \tau'}{(t,\sigma,\gamma) \vdash \langle e \ \texttt{=>}\ ss_0 \ \texttt{|}\ ss_1, \tau\rangle \stackrel{Seq}{\longrightarrow} \tau'}$$

$$\{\mathsf{ss}_{iii}\} \quad \frac{\sigma \vdash e \stackrel{Bool}{\longrightarrow} \mathsf{false} \quad (t,\sigma,\gamma) \vdash \langle ss_1, \tau\rangle \stackrel{Seq}{\longrightarrow} \tau'}{(t,\sigma,\gamma) \vdash \langle e \ \texttt{=>}\ ss_0 \ \texttt{|}\ ss_1, \tau\rangle \stackrel{Seq}{\longrightarrow} \tau'}$$

$$\{\mathsf{ss}_{iv}\} \quad \frac{\sigma \vdash e \stackrel{Bool}{\longrightarrow} x}{(t,\sigma,\gamma) \vdash \langle sig \ \texttt{:=}\ (e, dly), \tau\rangle \stackrel{Seq}{\longrightarrow} \mathsf{post}\ sig\ x\ \tau\ (t + dly)}$$

8.4 Rules for Concurrent Statements

The rules that describe the execution of Femto-VHDL concurrent statements are really those which define how a VHDL simulation cycle works. Rule cs_i shows that if a process' sensitivity list has no signals in common with the current events (γ), then the process does not activate, and the transactions (τ) remain unchanged. If, however, there is an intersection (rule cs_{ii}), then the new transactions (τ') are derived from the execution of the process' sequential statements using $\stackrel{Seq}{\longrightarrow}$. Finally, when two processes are executed concurrently, the resulting transactions are created by an application of **zip** to collect together the transactions derived by running each process separately. Note that the execution is deterministic because signals are note allowed to have multiple drivers (i.e. a signal name for which there is a transaction will appear at any given time in either τ' or τ'', but not both). Syntactically, the requirement is that a signal not appear on the left-hand side of an assignment in more than one process. If the restriction is removed, **zip** will have to be modified as described in Section 8.1.

$$\{\mathsf{cs}_i\} \quad \frac{}{(t,\sigma,\gamma) \vdash \langle sl : ss, \tau\rangle \stackrel{Cyc}{\longrightarrow} \tau} \quad (sl \cap \gamma) = \{\}$$

$$\{\mathsf{cs}_{ii}\} \quad \frac{(t,\sigma,\gamma) \vdash \langle ss, \tau\rangle \stackrel{Seq}{\longrightarrow} \tau'}{(t,\sigma,\gamma) \vdash \langle sl : ss, \tau\rangle \stackrel{Cyc}{\longrightarrow} \tau'} \quad (sl \cap \gamma) \neq \{\}$$

$$\{\mathsf{cs}_{iii}\} \quad \frac{\rho \vdash \langle cs_0, \tau\rangle \stackrel{Cyc}{\longrightarrow} \tau' \quad \rho \vdash \langle cs_1, \tau\rangle \stackrel{Cyc}{\longrightarrow} \tau''}{\rho \vdash \langle cs_0 \ \texttt{||}\ cs_1, \tau\rangle \stackrel{Cyc}{\longrightarrow} \tau' \ \mathsf{zip}\ \tau''}$$

9 Femto-VHDL in HOL

Having given the semantics of Femto-VHDL as a suite of inductively defined relations, that semantics may now be used it to perform some experiments with actual Femto-VHDL programs. Throughout the following examples, use will be made of the rules as they are embedded in the HOL system. The abstract syntax of Femto-VHDL has been coded as a recursive type definition in the logic [7]. The semantics can be viewed as an interpreter which "runs" Femto-VHDL programs by proof. It will then be possible to see what the characteristic future transactions are which express the behaviour of a given program within some environment. Note that the examples are coded in actual VHDL rather than the abstract syntax which we have defined in order to ease readability.

In the first example, we will endeavour to show that a process with an absent (empty) sensitivity list never activates, and therefore does not create any new transactions. Use will be made of the following process (also taken from [1]).

```
PROCESS
BEGIN
  C <= TRANSPORT (NOT C) AFTER dly;
END PROCESS;
```

If the following term is created in HOL, where ρ, τ and τ' are variables:

$"\rho \vdash \langle$
```
PROCESS
BEGIN
  C <= TRANSPORT (NOT C) AFTER dly;
END PROCESS;
```
$, \tau\rangle \xrightarrow{Cyc} \tau'"$

HOL may be asked to "run" the term to generate the following theorem (the symbol |- signifies a theorem in HOL):

|- $\forall \rho\, \tau\, \tau'.$

$\rho \vdash \langle$
```
PROCESS
BEGIN
  C <= TRANSPORT (NOT C) AFTER dly;
END PROCESS;
```
$, \tau\rangle \xrightarrow{Cyc} \tau' =$

$(\tau' = \tau)$

Here one sees that the theorem states that future transactions τ' are the same as the initial transactions τ. The result is what was desired, and demonstrates that for any environment ρ the process never activates.

A sensitivity list is now added that will cause the process to activate whenever there is an event on the signal C. In so doing, one would expect the process to cause a transaction to be posted dly units from now that will give the signal C its inverse value.

```
PROCESS (C)
BEGIN
  C <= TRANSPORT (NOT C) AFTER dly;
END PROCESS;
```

As in the previous example, a term is first created in HOL to set the process up for an execution within the $\xrightarrow{Cyc}$ transition system. A point to note is that we have given a somewhat more concrete environment this time. In particular, it is explicitly stated that there is an event on C.

$$\text{"}(now, \sigma, \{\texttt{C}\}) \vdash \langle \begin{array}{l}\texttt{PROCESS (C)}\\ \texttt{BEGIN}\\ \quad\texttt{C <= TRANSPORT (NOT C) AFTER dly;}\\ \texttt{END PROCESS;}\end{array}, \tau\rangle \xrightarrow{Cyc} \tau' \text{"}$$

Execution of the process leads to the following theorem, and the result of the execution is as expected.

$$\begin{array}{l}\vdash \forall\, now\ \sigma\ \tau\ \tau'.\\ \quad (now, \sigma, \{\texttt{C}\}) \vdash \langle \begin{array}{l}\texttt{PROCESS (C)}\\ \texttt{BEGIN}\\ \quad\texttt{C <= TRANSPORT (NOT C) AFTER dly;}\\ \texttt{END PROCESS;}\end{array}, \tau\rangle \xrightarrow{Cyc} \tau' =\\ \quad (\tau' = (\mathsf{post}\ \texttt{C}\ \neg(\mathsf{valcalc}\ \texttt{C}\ \sigma)\ \tau\ (now + \texttt{dly})))\end{array}$$

It is observed that the future transactions result from posting into the current transactions on signal C its present inverse at `dly` offset from *now*. Special attention should be paid to the fact that the theorem holds for all *now*, σ, τ and τ'. That is, it describes the characteristic behaviour of the process whenever there is an event on the signal C. It should also be noted that `dly` can be just 0 NS. This reinforces the notion that δ-delay is nothing more than a normal occurrence of a standard simulation cycle rather than something special.

As a final example, a second process is added to the Femto-VHDL program and inserts a 0-delay buffer into the system. It would be expected that the behaviour of the new program involved the an application of zip to combine the future transactions arrived at separately by each process. This result should be characterised by the future transactions derived in the previous example being combined with a statement about C1 always getting the current value of C.

```
PROCESS (C)
BEGIN
  C <= TRANSPORT (NOT C) AFTER dly;
END PROCESS;

PROCESS (C)
BEGIN
  C1 <= TRANSPORT C AFTER 0 NS;
END PROCESS;
```

As in the two previous examples, a term is created to express the relationship to be investigated. Here again, an event is currently declared on C.

```
                          PROCESS (C)
                          BEGIN
                            C <= TRANSPORT (NOT C) AFTER dly;
                          END PROCESS;
"(now, σ,{C}) ⊢ ⟨                                                 ,τ⟩ —Cyc→ τ' "
                          PROCESS (C)
                          BEGIN
                            C1 <= TRANSPORT C AFTER 0 NS;
                          END PROCESS;
```

The result after execution is as expected. In particular, it should be noted that the individual future transactions are combined using **zip** to form an overall notion of the future. One also sees that C1 takes on whatever C's value is in the current environment.

```
|- ∀ now σ τ τ'.
                          PROCESS (C)
                          BEGIN
                            C <= TRANSPORT (NOT C) AFTER dly;
                          END PROCESS;
  (now, σ,{C}) ⊢ ⟨                                                ,τ⟩ —Cyc→ τ' =
                          PROCESS (C)
                          BEGIN
                            C1 <= TRANSPORT C AFTER 0 NS;
                          END PROCESS;
  (τ' = (post C ¬(valcalc C σ) τ (now + dly))
                          zip
        (post C1 (valcalc C σ) τ now))
```

10 Conclusions and Future Work

It has been observed that the semantics of the VHDL simulator, in particular a single simulation cycle, is a useful way of expressing the behaviour of individual language constructs. Furthermore, the semantics can easily and naturally be given as a collection of inductively-defined transition relations. In doing so, both unit- and 0- (or δ-) time have each been made to form a part of the same whole, with no special interpretation required in the treatment of one versus the other.

Using the formalisation, it has been possible to animate the semantics as an interpreter within the HOL system. This allows one to examine specific Femto-VHDL programs and automatically prove concise theorems describing their characteristic behaviour. As a result of this embedding in HOL, one may see that the semantics relates closely to the intuition behind the VHDL simulation cycle.

In future, investigations will be made into the ways in which different Femto-VHDL programs purporting to describe the same thing may be proved to be so under some notion of equivalence. A case in point was raised in our discussion of results: how do we know that the uniprocess oscillator and the dual process one represent the same thing? The first task is, of course, to develop the notion of equivalence itself.

11 Acknowledgements

I am grateful to Dr. Juanito Camilleri for his help in coming to grips with this semantics. Further thanks are also due to Mike Gordon, Tom Melham and Monica Nesi for their comments during the development of this report.

References

[1] R. Airiau, J.-M. Bergé, V. Olive, and J. Rouillard, *VHDL: du Langage à la Modélisation*, Presses Polytchniques et Universitaires Romandes (1990).

[2] Larry M. Augustin, Benoit A. Gennart, Youm Huh, David C. Luckham, and Alec C. Stanculescu, 'An Overview of VAL', Technical Report CSL-TR-88-367, Stanford University, Stanford, California (October 1988).

[3] R. Boulton, M. Gordon, J. Herbert, and J. Van Tassel, 'The HOL Verification of ELLA Designs', in: *1991 International Workshop on Formal Methods in VLSI Design*, edited by P.A. Subrahmanyam, (also Univ. of Cambridge Tech Report 199, August 1990) Springer-Verlag (1991).

[4] Juanito Camilleri, 'Symbolic Compilation and Execution of Programs by Proof: A Case Study in HOL', Technical Report 240, University of Cambridge Computer Laboratory, Cambridge, England (December 1991).

[5] Ivan V. Filippenko, 'VHDL Verification in the State Delta Verification System (SDVS)', in: *1991 International Workshop on Formal Methods in VLSI Design*, edited by P.A. Subrahmanyam, Springer-Verlag (1991).

[6] Institute of Electrical and Electronics Engineers, *IEEE Standard VHDL Language Reference Manual*, IEEE Press, New York (1988).

[7] T.F. Melham, 'Automating Recursive Type Definitions in Higher-Order Logic', in: *Current Trends in Hardware Verification and Automated Deduction*, edited by G. Birtwistle and P.A. Subrahmanyam, Springer-Verlag (1988).

[8] T.F. Melham, 'A Package for Inductive Relation Definitions in HOL', in proceedings: *1991 International Workshop on the HOL Theorem Proving System and its Applications*, IEEE Computer Society Press (1991), in publication.

[9] Gordon Plotkin, 'A Structural Approach to Operational Semantics', Technical Report DAIMI FN-19, Computer Science Dept., Aarhus Univ. (September 1981).

[10] A. Salem and D. Borrione, 'Formal Reasoning About Signal Attributes in VHDL', in proceedings: *VHDL Forum for CAD in Europe*, Spring 1991 meeting.

[11] John P. Van Tassel, *The Semantics of VHDL with VAL and HOL: Towards Practical Verification Tools*, M.Sc. Thesis, Dept. of Computer Science and Engineering, Wright State University (1989).

[12] John Van Tassel and David Hemmendinger, 'Toward Formal Verification of VHDL Specifications', in: *Applied Formal Methods For Correct VLSI Design*, edited by Luc Claesen, Elsevier Science Publishers (1990).

Higher Order Logic Theorem Proving and its Applications (A-20)
L.J.M. Claesen and M.J.C. Gordon (Editors)
Elsevier Science Publishers B.V. (North-Holland)

The Formal Semantics Definition of a Multi-Rate DSP Specification Language in HOL [1]

C. Angelo, L. Claesen[2], H. De Man[2]

IMEC vzw, Kapeldreef 75, B3001 Leuven, Belgium
phone: +32/16/281524, fax: +32/16/281515, email: catia@imec.be

Abstract
Silage is an applicative language for describing multi-rate DSP algorithms. It has been used as the high level input language to the CATHEDRAL Silicon Compilers. Hardware implementations of specifications described in Silage are synthesized and inconsistencies in the synthesis trajectory must be avoided. To build the foundations of a verification environment, the formal semantics of Silage must be defined. Our main motivation to formally define the semantics of Silage is to build formal tools, in such a way that the interactivity between the user and the silicon compiler can be increased in a safe way. The designer could then interactively transform the specification, fully exploiting his expertise, but without the risks of introducing inconsistencies. Formally verified transformational design techniques could then be applied at the specification level of the Silicon Compilers. This paper describes the formal semantics definition of a substantial subset of Silage focusing on the multi-rate aspects of the language.

Keyword Codes: F.3.1; F.3.2; J.6
Keywords: Specifying and Verifying and Reasoning about Programs; Semantics of Programming Languages; Computer-Aided Engineering

1 Introduction

The Silage language [1, 2, 3, 4, 5] used as input of the CATHEDRAL [6] system is an applicative language suitable to describe DSP algorithms represented as a data-flow graph. No variables or assignment operators are allowed. A basic concept of the language is the notion of a signal. A signal is an infinite stream of data indexed by discrete and periodic time instants. Signals are defined by function applications. The single assignment principle allows signals to be defined only once. An equation involving signals is indeed a set of equations about the samples of the signals involved. The operations are vectorial and the indexing is implicit. There are operators that can generate signals with rates [3] that are different from the rates of the signals they are derived from. The rates of the signals are implicit and they are derived by the synthesis tool. Previous samples of the signals can be referred to in a signal definition. Arrays of samples can

[1] Work sponsored by CNPq (Brazilian Government) and CHEOPS Project (ESPRIT BRA "3215").

[2] Professors at Katholieke Universiteit Leuven.

[3] Rate is meant to be period, phase or both throughout this text.

be manipulated. No statements are made about the order or concurrency of the operations and the parallelization of computation can be exploited by analyzing the data flow dependencies.

Given different Silage descriptions with the same algorithmical behavior, the ideal silicon compiler should generate the same optimal circuit (according to a certain criterion). However, in practice, the silicon compilers might generate different solutions depending on how the description in Silage is made. Designers often massage Silage descriptions, applying transformations [7, 8, 9] to make them more suitable for a specific synthesis tool. These transformations should *not* change the *input/output behavior* of the DSP algorithm. In [9] we catalog two classes of transformations: those that preserve the *internal behavior* of the algorithm (like commonly used loop transformations) and those that do not preserve it. These transformations have a significant impact on the speed and area of the chip to be synthesized. The organization of loops in a Silage description often influences tasks of the silicon compilation such as memory management [10]. Building the ideal and automatic silicon compiler is difficult. Moreover, the expertise of the designer can be exploited to guide an interactive silicon compiler. The problem with massaging a specification is the possibility of changing its behavior and/or introducing inconsistencies. In order to accurately interpret a Silage description, a definition of the semantics of Silage is necessary [11]. Using the semantics to interpret Silage descriptions, properties of the specification can be proven and the transformations can be formalized and proven correct in a system like HOL [12]. By making available to the user a menu of useful and *correctness preserving* transformations, the designer can take advantage of his creativity and expertise to guide the silicon compiler, choosing appropriate transformations to massage the Silage specification, without changing its behavior. If an automatic tool could be built to generate an optimal Silage description starting from a non-optimal one, the semantics and the formalized transformations could be used to prove that the *behavior* is preserved in the process. However, one should realize that the concept of optimal Silage description is dependent, not only on the criterion to evaluate the optimal circuit to be implemented, but also on the way the description is interpreted by a specific synthesis tool. This paper describes the *multi-rate* semantics of a substantial subset of Silage in HOL, bearing in mind the goal to perform source-to-source transformations. The text is organized as follows: the next section gives an informal description of the Silage language. Since the work described in this paper is built on top of previous works on the semantics of Silage [13, 14, 15], an overview of these contributions is outlined. The HOL definitions and the functions described in ML [16] used in the semantics interpretation are then defined. Next we define the semantics of Silage expressions and definitions and we outline the semantics of Silage functions. Then we discuss the issue of reasoning about Silage programs and we give examples of correctness preserving source-to-source transformations. Finally the status of the work is outlined and some conclusions are drawn.

2 An informal description of Silage

In this section we will informally describe the main Silage constructs and features.

In Silage, signals are periodically sampled and not necessarily defined for all discrete time instants. The time variable and the period of the signals are implicit in Silage programs. For each signal, there is a time instant when samples obeying a certain algorithmic specification start to happen. We will call these samples *algorithmical samples*. These are the interesting samples. Samples happening before the *algorithmical samples* are just initialisation of recursive algorithms. We will call these samples *initialisation samples*. The initialisation samples never depend on input data while the algorithmical samples, except for the case of constant signals, always depend directly or indirectly on input signals. For instance, consider the following piece of Silage code:

```
x@@1 = 0;
x = in + x@1;
```

The first equation initializes the recursive definition in the second equation. The operator @ is used to access previous samples of a signal and the operator @@ is used to define the values of the *initialisation samples.* The first equation sets to the value **0** the first sample of the signal **x** occurring before the beginning of the algorithmical samples. The second equation means that the n-th algorithmical sample of the signal **x** is obtained by adding the n-th algorithmical sample of a signal **in** with the previous sample of the signal **x**.

The time when the first algorithmical sample of a signal happens depends on the phase of the signal. The phase of a signal **sig** is always greater than or equal to the phases of all signals on which **sig** depends on.

The next example of Silage taken from [7] illustrates the use of arrays and iterators.

```
s[0] = 0;
(i: 1 .. 64) ::
        s[i] = s[i - 1] + a[i - 1] * b[i - 1];
return = s[0] + s[32] + s[64];
```

If signals are functions from **time** to **value**, an array is a special kind of signal, whose **value** is a list of data. The definition in the body of the iteration above holds for all values of the iterator **i** in the range **1** to **64**.

The next piece of Silage code taken from [5] illustrates how functions can be defined in Silage:

```
func Adap (a,b : int<24>) c:int<24>; d:int<24,8> =
        z: int<24>;
        begin
                z = a - b;
                c = z@1 + z@2;
                d = c - z@3;
        end;
```

The function **Adap** has as arguments two signals **a** and **b** and returns two signals **c** and **d**. The signal **z** is an internal signal. All signals are integers. The function **Adap** can be called as follows using tuples: `(x, y) = Adap (u, v)`.

Tuples of the form (a,...,z) in Silage aggregates signals that may have different periods and phases. Tuples cannot be nested. On the other hand, the aggregate constructor of the form {a,...,z} aggregates signals with the same period.

Functions returning one signal can be called inside nested expressions. For instance, suppose that **One** is a function returning one signal. Then the next definition is possible: `x = One(u,v) + 2 * One(p,q);`.

There is also in Silage an **if** construct as follows: `if a → x || y;`.

So far, except for the tuples and function definitions, only the Silage operators that handle signals with same periods have been considered. When a signal is generated using these operators, it is implicit that the period of all the signals involved are the same. Next we will consider

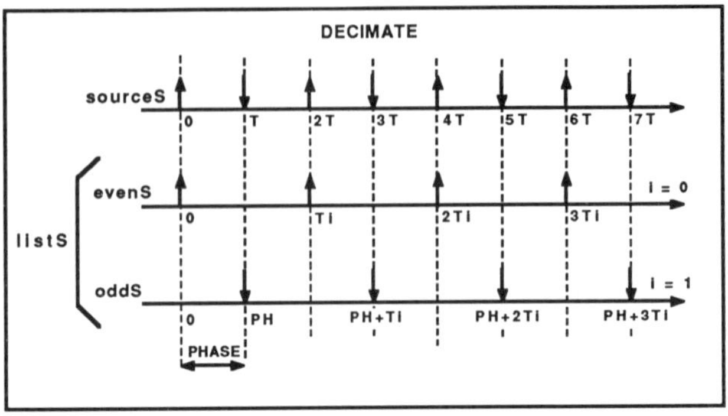

Figure 1: The decimate function

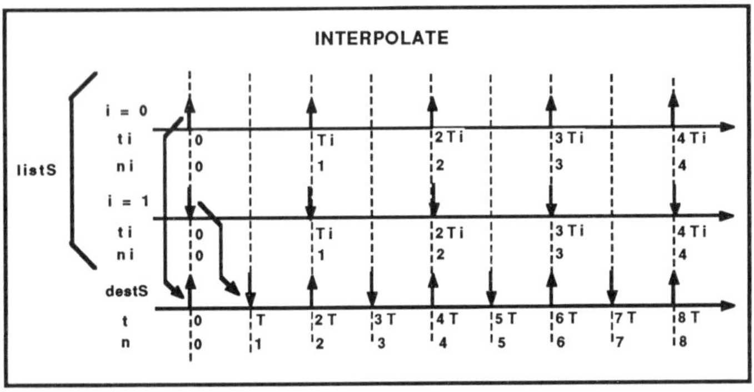

Figure 2: The interpolate function

the **decimate**, the **interpolate** and the **switch** functions, that handle signals with different periods.

Figure 1 illustrates the **decimate** function. The **decimate** function takes as argument a source signal **sourceS** and returns a list of **n** signals **listS**. The **sourceS** is decomposed into **n** signals with a period **n** times greater than the period of the **sourceS**. Each of the first **n** samples of the **sourceS** is the first sample of a signal in **listS**. Similarly, each of the second group of **n** samples of the **sourceS** is the second sample of a signal in **listS**, and so on. The samples of **sourceS** are distributed among the output signals in the order they appear in **listS**. In Figure 1, **listS** has only two signals (n = 2) and the phase of the **sourceS** signal is zero. The horizontal axis represent the absolute time scale. The small vertical arrows represent samples. The signal taking the even samples of **sourceS** is called **evenS** and the signal taking the odd samples of **sourceS** is called **oddS**.

The **interpolate** function takes as argument a list **listS** of **n** signals with equal periods and returns a destination signal **destS**. The samples of the signals in **listS** are interleaved generating the signal **destS** with a period **n** times smaller than the period of the signals in **listS**. Figure 2 illustrates interpolation when **listS** has two signals with same phases.

The **switch** function takes as argument **n** signals and returns **m** signals. The argument signals are interpolated generating a temporary signal and this is decimated into **m** signals.

The time instants in the Figures 1 and 2 are parameterized on the periods **Ti** and **T** of the signals. In a Silage program, the period and the phase of all signals depend on each other. One Silage equation alone does not define the period and the phase of the signals involved, but it defines a relation between them. One has to keep track of the rate information in **every** relation modelling a Silage equation. Even for the operations that handle signals with same periods, this information has to be explicit in the semantic interpretation. Most operations are only meaningful if the periods of the signals involved have some property. For instance, only signals with same periods can be added or interpolated. Each Silage equation relates the periodic samples, the periods and the phases of the involved signals.

In Silage there are two special classes of expressions. *Manifest expressions* are integers, iterator indexes or an arithmetical expression of these. These expressions can be interpreted as scalars. Iterator bounds and array indexes must be manifest. *Selectors* can be a signal variable, an array variable applied to a manifest index or a tuple of selectors. Only selectors can appear on the left hand side of equalities in Silage. Any Silage expression can be interpreted as a signal.

3 Previous work and extensions

The work presented in this paper is built on top of previous works [13, 14, 15]. Next we will summarize their main points.

In [13] two different approaches are discussed to formally define the semantics of a set of Silage definitions:

- *functional model*: a group of equations is interpreted all together to define a signal as a function from time instants to values. For each signal, the equations about its initialisation samples and its algorithmical samples are considered all together to define the signal as a function. Such an approach has to handle functions not defined for all discrete time instants. The definitions take as parameters the input signals and generate the output signals.
- *relational model*: each equation is interpreted separately, stating a relation between the samples, the periods and the phases of the signals. The definitions take signals as parameters (without distinction between input and output signals).

The semantics of Silage is needed to consistently map Silage programs into the HOL logic and then, once in the logic, prove the correctness of source-to-source transformations. The reasoning in the logic is safe but the interpretation of Silage programs into terms of the logic is error-prone. The simpler the semantics interpretation, the safer the whole transformational system. In [13] the relational model is preferred with respect to the functional model because it is simpler to translate Silage programs into the logic. Within that model, issues on how to interpret the multi-rate aspects of Silage into the logic are discussed. In [17] a proposal is made to include in Silage signals with varying period. The idea is that, starting from input signals with fixed periods, an output signal with a varying period can be generated, depending on the values of the input data. In [13] we discuss how the relational model could cope with signals with varying periods in an extended Silage.

In [14, 15] the semantics of a *single-rate* subset of Silage is defined using the relational model. The semantics is divided in three parts:

- the *abstract syntax* defines the production rules of the language without lexical details like precedence of operators.
- the *static semantics* checks if a phrase of abstract syntax is well formed and well typed. Rules are defined to derive the type and attributes of expressions in a given environment. The attributes of an expression can tell in which contexts the expression is allowed. The attributes can indicate if an expression is a selector, a manifest variable or an operator. An environment has information about the types of the variables and it can also tell if a variable is an index, a signal or the name of a previously defined Silage function. The static semantics checks, for instance, if an expression on the left hand side of a Silage equality is a selector and if tuples are nested.
- the *denotational semantics* translates the definitions and expressions accepted by the static semantics into terms of the HOL logic. When expressions with different attributes are acceptable in a given context, the attribute of the expression may decide how it should be interpreted.

The work we are about to describe in the next sections uses the ideas described in [13] and the semantics embedding method described in [14, 15] to define the semantics of a *multi-rate* subset of Silage towards the verification of source-to-source transformations. Our semantics will not be concerned with the executability of Silage programs. We have adapted and extended the abstract syntax and the static semantics of [14, 15] to cope with the multi-rate functions (**interpolate, decimate** and **switch**) and to model Silage functions returning multiple signals. Next follows our version of the abstract syntax of expressions, where **op** is an unary or binary logical or arithmetical Silage operator [5].

1

The abstract syntax of Expressions

$e ::= x$	(variable)
$\mid x(e_1, ..., e_n)$	(function call)
$\mid e_1[e_2]$	(indexing)
$\mid e_1@e_2$	(delay)
$\mid (e_1, ..., e_n)$	(tuple)
$\mid \{e_1, ..., e_n\}$	(vector)
$\mid n$	(numeric literal)
$\mid b$	(boolean literal)
$\mid (op\ e)$	(prefix operator)
$\mid (e_1\ op\ e_2)$	(infix operator)
$\mid (if\ e_1 \rightarrow e_2 \parallel e_3\ fi)$	(conditional)
$\mid (interpolate(e_1, ..., e_n))$	(interpolation)

We use the same manifest interpretation **M** of expressions of [14, 15] as follows:

2

M(x) = (x:(hol_type x))
M(n) = n
M(b) = b
M(op e) = ((hol_op (op e)) M(e))
M(e_1 op e_2) = ((hol_op (e_1 op e_2)) M(e_1) M(e_2))

where:

- **hol_type** maps Silage types into *scalar* HOL types. For instance, if the Silage type of **x** in the environment is **Nat** (naturals) [4] then **(hol_type x)** is the HOL type **num**.
- **hol_op** maps Silage unary and binary logical and arithmetical operators into *scalar* HOL unary and binary logical and arithmetical operators respectively. The mapping depends on the operator and on the types of the operands in the environment. For instance, if e_1 and e_2 have Silage types **Nat** then (hol_op $(e_1 + e_2)$) is the HOL operator $+$.

Similarly to **hol_op, hol_OP** maps Silage unary and binary logical and arithmetical operators into HOL operators on *samples* of signals. The mapping depends on the operator **op** and on the types of the operands in the environment. For instance, if e_1 and e_2 have Silage types **Nat** then (hol_OP $(e_1 + e_2)$) is the operator **SIGPLUS** defined in [14, 15] as:

∀ (A:time→num) (B:time→num). (A **SIGPLUS** B) t = (A t) + (B t)	*3*

Similarly to **SIGPLUS** there are the operators **SIGMINUS** and **SIGMULT** for subtraction and multiplication respectively.

Next follows our version of the abstract syntax of definitions. We choose to handle the interpolation construct at the level of expressions and at the level of definitions.

4

The abstract syntax of Definitions

$d ::= (e_1 @@ e_2 = e_3)$	(initialisation)
$\mid (e_1, ..., e_n) = (e'_1, ..., e'_n)$	(tuple = tuple)
$\mid (e_1, ..., e_n) = x(e'_1, ..., e'_m)$	(tuple = function call)
$\mid (e_1, ..., e_n) = (\mathit{decimate}\ e)$	(decimation)
$\mid x = (\mathit{interpolate}\ (e_1, ..., e_n))$	(interpolation)
$\mid (e_1, ..., e_n) = (\text{switch}\ (e'_1, ..., e'_m))$	(switch)
$\mid (e_1 = e_2)$	(equation)
$\mid (d1; d2)$	(composition)
$\mid (x : e_1..e_2) :: d$	(iteration)
$\mid \mathit{begin}\ d\ \mathit{end}$	(grouping)

In the rest of this text we will assume the results of the static semantics to describe the denotational semantics of expressions and definitions of the *multi-rate* subset of Silage. The multi-rate constructs of Silage changes the signal interpretation of *every* construct of Silage because of the introduction of the concept of rate in the language.

4 HOL definitions and functions described in ML

In this section we will introduce the HOL definitions and the functions described in ML used in the denotational semantics of a *multi-rate* subset of Silage. We use upper case letters for HOL definitions and lower case letters for ML functions.

[4] Actually there is not a Silage type for the natural numbers. Integers instead can be typed in Silage. Currently we are working with the naturals. In the future all Silage types will be accurately modelled by a HOL type.

4.1 HOL definitions

Let **value** be a type variable. The type **sampleTy** defined below is the type of the samples of the signals. [5] Let **rateTy** be the type of naturals **num**. The type of the period and the phase of the signals in the logic is **rateTy**. A signal is a tuple with three components: samples, period and phase. The type **signalTy** is the type of the signals. An array is a special kind of signal, whose sample type is a type instantiation of **sampleTy**, as follows:

```
sampleTy: (time→value)                                   5
signalTy:(sampleTy#rateTy#rateTy)
arraySampleTy: (time→((value)list))
```

The samples, the period and the phase of a signal are obtained as follows:

```
∀ (s:sampleTy) (p:rateTy) (f:rateTy).                    6
  ((SAMP (s, p, f)) = s) ∧
  ((PER (s, p, f)) = p) ∧
  ((PH (s, p, f)) = f)
```

Next follow the definitions of the functions modelling the time scale: **TIME**, **NEGTIME** and **SAMPNUM**. [6] The abstract time of the **n**-th algorithmic sample of a signal is **TIME n**. The abstract time of the **n**-th initialisation sample of a signal is **NEGTIME n**. Given the abstract time of the **n**-th algorithmic sample of a signal **TIME n**, the function **SAMPNUM** returns the sample number **n**.

```
∀ (n:time). (TIME n) = (ORIGIN + n)                      7
∀ (n:time). (NEGTIME n) = (ORIGIN - n)
∀ (time_n:time). (SAMPNUM time_n) = (time_n - ORIGIN)
```

Given a list of natural numbers, the predicate **NOT_ZERO** defined below tells if all elements of the list are different of zero:

```
((NOT_ZERO []) = T) ∧                                    8
(∀ (hd:num) (tl:(num)list). (NOT_ZERO (CONS hd tl)) = ((~(hd = 0)) ∧ (NOT_ZERO tl)))
```

where **T** means true in the logic.

Next follow two definitions about natural numbers: **SYNCLIST** and **MULTIPLE**. Given a list of natural numbers, the predicate **SYNCLIST** tells if all elements of the list are equal. Given two natural numbers **n** and **m**, the predicate **MULTIPLE** tells if **n** is multiple of **m**:

```
∀ (l:(num)list). (SYNCLIST l) = (EVERY (λel.(el = (HD l))) l)      9
∀ (n:num) (m:num). (MULTIPLE n m) = ((n MOD m) = 0)
```

---

[5] We are currently modelling the time with the natural numbers instead of the integer numbers because there is much more support for naturals than for integers in HOL. This temporary hacking forces us to model the origin of the time scale with a natural number **ORIGIN** that must be greater than the greatest delay in a Silage program. In the future **ORIGIN** will be zero and the time will be modelled by the integers.

[6] If **ORIGIN** is zero and the functions **TIME**, **NEGTIME** and **SAMPNUM** are defined on integers, these functions are equal to the identity function and therefore they can be eliminated.

where: **EVERY** is a HOL function that takes as arguments a predicate **P** and a list of elements **l** and returns **true** if **P** holds for every element of **l**; **n MOD m** is the rest of the integer division between **n** and **m**.

Given a list of samples **listSamp**, the samples of a signal **samples**, and the length of the list **leng, (DECSAMP listSamp samples leng 0)** holds if **listSamp** was obtained by decimating **samples** or **samples** is the interpolation of **listSamp**.

```
∀ (hd:sampleTy) (tl:(sampleTy)list) (samples:sampleTy) (leng:num) (shift:num).     10
  ((DECSAMP [] samples leng shift) = T) ∧
  ((DECSAMP (CONS hd tl) samples leng shift) =
        ∀ (ni:time). (((hd (TIME ni)) = (samples (TIME ((leng∗ni) + shift)))) ∧
                    (DECSAMP tl samples leng (SUC shift))))
```

Let a signal **sig** be decimated into a list of signals **listSig**. Given the phases **listPhases** of **listSig** and the period and the phase **period** and **phase** of **sig**, **(DECPH listPhases 0 period phase)** holds where **DECPH** is defined as follows:

```
∀ (hd:rateTy) (tl:((rateTy)list)) (n:num) (period:rateTy) (phase:rateTy).     11
  ((DECPH [] n period phase) = T) ∧
  ((DECPH (CONS hd tl) n period phase) =
          ((hd ≥ (phase + n∗period)) ∧ (DECPH tl (SUC n) period phase)))
```

Silage definitions will be interpreted using the predicates **DEC_SAMP_PH, DECIMATE, INTERPOLATE, SWITCH,** and **EQUAL** defined below. All of them have a boolean parameter **rateInfo** that will be used to relate the periods and the phases of the signals involved. The predicate **SYNC** defined below tells if all elements in a list are equal to a given element.

```
∀ (listSig:(signalTy)list) (sig:(signalTy)) (rateInfo:bool).     12
  (DEC_SAMP_PH listSig sig rateInfo) =
  ((DECSAMP (MAP SAMP listSig) (SAMP sig) (LENGTH listSig) 0) ∧ rateInfo ∧
   (DECPH (MAP PH listSig) 0 (PER sig) (PH sig)))
```

```
∀ (ref:value).     13
  (SYNC [] ref = T) ∧
  (∀ (hd:value) (tl:(value)list). (SYNC (CONS hd tl) ref) = ((hd = ref) ∧ (SYNC tl ref)))
```

```
∀ (listSig:(signalTy)list) (sig:(signalTy)) (rateInfo:bool).     14
  (DECIMATE listSig sig rateInfo) =
              ((DEC_SAMP_PH listSig sig rateInfo) ∧
               (SYNC (MAP PER listSig) ((LENGTH listSig) ∗ (PER sig))))
```

Given a list of numbers **listNum, GREATEST listNum** is the greatest number in the list **listNum**. The function **GREATEST** will be used to model the phase of the signals in the logic. **GREATEST** is defined in terms of the function **GRT** as below:

```
∀ (hd:num) (tl:((num)list)) (ref:num).                                        15
  ((GRT [] ref) = ref) ∧
  ((GRT (CONS hd tl) ref) = (GRT tl ((hd < ref) ⇒ ref | hd)))

∀ (listNum:((num)list)). (GREATEST listNum) = (GRT listNum 0)
```

Consider the Figure 2 illustrating the interpolate function. Given a list of samples of signals **listS**, **(INTERP listS)** models the samples of the signal obtained by the interpolation of the signals whose samples are in **listS**.

```
∀ (listS:((sampleTy)list)).                                                   16
  (INTERP listS) =
  (λ(time_n:time).
              let n = (SAMPNUM time_n) in
              let ni = (n DIV (LENGTH listS)) in
              let i = (n MOD (LENGTH listS)) in
              let theSignal = (EL i listS) in
                    theSignal (TIME ni))
```

where **n DIV k** is the integer division between **n** and **k**, **LENGTH l** is the length of the list **l** and **(EL i l)** is the i-th element of the list **l**.

The predicate **INTERPOLATE** is defined below to interpret the Silage interpolation at the level of Silage definitions. [7] The result of the interpolation of a list of signals **listSig** is the signal **sig**. The argument **listPer** [8] is the list of the non-null periods of the signals in **listSig**.

```
∀ (sig:signalTy) (listSig:(signalTy)list) (listPer:((rateTy)list)) (rateInfo:bool).   17
  (INTERPOLATE sig listSig listPer rateInfo) =
               ((DECSAMP (MAP SAMP listSig) (SAMP sig) (LENGTH listSig) 0) ∧ rateInfo ∧
                (SYNC listPer ((LENGTH listSig) * (PER sig))) ∧
                ((PH sig) ≥ GREATEST (MAP PH listSig)))
```

Next follows the definition of the predicate **SWITCH** used to interpret the Silage function **switch**. The result of the switch of a list of signals **listInSig** is the list of signals **listOutSig**. The argument **listInPer** is the list of the non-null periods of the signals in **listInSig**.

```
∀ (listOutSig:(signalTy)list) (listInSig:(signalTy)list) (listInPer:((rateTy)list)) (rateInfo:bool).   18
  (SWITCH listOutSig listInSig listInPer rateInfo) =
              (∃ (tmp:signalTy).
                 (INTERPOLATE tmp listInSig listInPer rateInfo) ∧
                 (DECIMATE listOutSig tmp T))
```

The Silage equality between two signals **left** and **right** means that the algorithmic samples of the **left** signal are equal to the algorithmic samples of the **right** signal and there is a relation **rateInfo** between the periods of the signals that compose **left** and **right**. This will be defined using the predicate **EQUAL** below:

---

[7] This is not essential since the interpolation is interpreted at the level of expressions.

[8] The argument **listPer** is necessary to be able to interpret the interpolation and switch of constants, without "assigning" zero to the period of **sig**.

∀ (left:sampleTy) (right:sampleTy) (rateInfo:bool). *19*
(**EQUAL** left right rateInfo) =
((∀ (n:time). ((left (TIME n)) = (right (TIME n)))) ∧ rateInfo)

The following definitions of <>, **VAL**, **!!**, **SIGIND**, **VEC**, **SIGCOND** and **TO** were taken from [14, 15]. The <> function delay signals. The **VAL** function define constant signals. The indexing of an array with a manifest index is defined using the **!!** operator. The indexing of an array with a signal is defined using the **SIGIND** operator. The **VEC** definition is used to interpret the aggregate constructor. [9]

∀ (A:sampleTy) (n:time) (t:time). (A <> n) t = A(t - n) *20*
∀ (n:value) (t:time). **VAL** n t = n
∀ (A:arraySampleTy) (n:num) (t:time). ((A **!!** n) t) = (EL n (A t))
∀ (A:arraySampleTy) (B:numSampleTy) (t:time). ((A **SIGIND** B) t) = ((A !! (B t)) t)
∀ (As:(time→value)list) (t:time). (**VEC** As t) = (MAP (λx.x t) As)
∀ (A:boolSampleTy) (B:sampleTy) (C:sampleTy) (t:time).
SIGCOND A B C t = ((A t) ⇒ (B t) | (C t))
∀ (m:num) (n:num). (m **TO** n) = λx. m ≤ x ∧ x ≤ n

4.2 Functions described in ML

In this subsection we will describe the functions defined in ML.

Given a type **ty**, the function **Signal** constructs a signal type as follows:

Signal ty = (time→ty)#rateTy#rateTy *21*

If the semantics of an expression $[\![e]\!]$ is the tuple **(samples, period, phase)** then:

(**samplesOf** $[\![e]\!]$) = samples; *22*
(**periodOf** $[\![e]\!]$) = period;
(**phaseOf** $[\![e]\!]$) = phase.

The **choose** function chooses a period in a list of periods as described below:

choose [periodOf$[\![e_1]\!]$;...;periodOf$[\![e_n]\!]$] = *23*
(if (∀ $period_i$ ∈ [periodOf$[\![e_1]\!]$,...,periodOf$[\![e_n]\!]$]. $period_i = 0$)
then 0
else $period_i$ where:
(($period_i$ ∈ [periodOf$[\![e_1]\!]$,...,periodOf$[\![e_n]\!]$]) ∧ ($period_i \neq 0$) ∧
(∀ j. ($1 \leq j < i$) ∧ $period_j$ ∈ [periodOf$[\![e_1]\!]$,...,periodOf$[\![e_n]\!]$]. $period_j = 0$)))
where ($1 \leq i \leq n$).

The **equalList** function defined below is used to impose that all non-zero elements in a list of periods are equal. **(filter P list)** is the list of the elements of **list** satisfying the predicate **P**.

[9] **numSampleTy** and **boolSampleTy** are the types obtained by instantiating the type variable **value** of **sampleTy** with the types **num** and **bool** respectively as follow: (time→num) and (time→bool).

equalList [periodOf$[\![e_1]\!]$;...;periodOf$[\![e_n]\!]$] = (let filteredList = filter (λ $period_i$. ($period_i \neq 0$)) [periodOf$[\![e_1]\!]$;...;periodOf$[\![e_n]\!]$] in (SYNCLIST filteredList))	*24*

The **makeSigTuple** function defined below takes as argument a list of interpretations of expressions. If the list has only one element, it returns that one. Otherwise it returns a tuple.

makeSigTuple [$[\![e_1]\!]$;...;$[\![e_n]\!]$] = (if (n = 1) then ($[\![e_1]\!]$) else ($[\![e_1]\!]$,...,$[\![e_n]\!]$))	*25*

5 Denotational semantics of expressions

Given an environment and the results of the static semantics, the multi-rate denotational semantics of Silage expressions is defined as in Figure 3. The only Silage expression not interpreted here is the Silage tuple that is interpreted within the denotational semantics of Silage definitions. All the other expressions, except for the Silage function call returning more than one signal, are interpreted in Figure 3 as a signal **(samples, period, phase)**.

If the period of an expression **e** is zero, this just means that this expression does not contain period information and therefore it does not constrain the period of any expression **E** that contains **e** as a subexpression. No signal in the semantics of Silage programs has period equal to zero. When an expression is composed by many subexpressions, its period is chosen from the list of periods of the subexpressions and its phase is the greatest of the phases of the subexpressions. If there is a subexpression whose period is not zero, the period of the expression will not be zero.

6 Denotational Semantics of Definitions

In Silage one cannot define the period and the phase of the signals. However, each Silage definition constrains the possible rates of the signals involved by implicitly imposing a relation between their periods and phases. In Figure 4 the function *check* is defined to model constraints of the periods in the logic. This function is defined for every kind of Silage expression and will be used in the denotational semantics of Silage definitions.

Next follow some comments on the Figure 4. If **(check e)** is true then **e** does not constrain the periods of any signal. If an expression **e** is composed by other subexpressions then the period constrains of the subexpressions are period constrains of the expression **e**. The value of **(check** $e_1[e_2]$**)** does not depend on the value of **(check** e_1**)** because the static semantics guarantees that e_1 is always an array variable and therefore **(check** e_1**)** is true. Similarly, the value of **(check** $e_1@e_2$**)** does not depend on the value of **(check** e_2**)** because the static semantics guarantees that e_2 is manifest and therefore **(check** e_2**)** is true. If an expression **e** is either $e_1[e_2]$, $\{e_1, ..., e_n\}$, (e_1 **op** e_2) or (**if** $e_1 \rightarrow e_2 \parallel e_3$ **fi**) then all the non-zero periods of the subexpressions of **e** must be equal. If the signals e_i of $(e_1, ..., e_n)$ are interpolated then all the non-zero periods of e_i must be equal and multiples of **n**.

Figure 5 shows the denotational semantics of Silage definitions. Next follow some comments about it. The **check** function is used to constrain the periods of the signals on the right hand side of Silage definitions. The period of expressions on the left hand side of Silage definitions do not have to be checked because these expressions are selectors. The interpretations of the Silage constructors **decimate, interpolate, switch** and **equation** relate the periods of the signals on the left hand side of the Silage definitions to the non-null periods of the signals on their right hand side. The semantics of a Silage equation defines that the phase of the signal on the left hand side is greater or equal to the phase of the expression on the right hand side.

26

The Denotational Semantics of Expressions

$[\![x]\!]$ = (if (x is manifest) then ((VAL (x:num)), 0, 0)
else ((SAMP signal), (PER signal), (PH signal)))
where: signal = x:(Signal(Type x))

$[\![x(e_1, ..., e_n)]\!]$ = (if (x returns many signals) then funcCall
else ((SAMP funcCall), (PER funcCall), (PH funcCall)))
where: funcCall = x_FUN (makeSigTuple [$[\![e_1]\!]$;...;$[\![e_n]\!]$]) ;
x_FUN is the HOL function definition of x ;

$[\![e_1[e_2]]\!]$ = (samples, (periodOf $[\![e_1]\!]$), phase)
where: samples = (if (e_2 is manifest) then ((samplesOf $[\![e_1]\!]$) !! M(e_2))
else ((samplesOf $[\![e_1]\!]$) SIGIND (samplesOf $[\![e_2]\!]$))) ;
phase = (if (e_2 is manifest) then (phaseOf $[\![e_1]\!]$)
else (GREATEST [phaseOf $[\![e_1]\!]$; phaseOf $[\![e_2]\!]$])) ;

$[\![e_1@e_2]\!]$ = (((samplesOf $[\![e_1]\!]$) <> M(e_2)), (periodOf $[\![e_1]\!]$), (phaseOf $[\![e_1]\!]$))

$[\![\{e_1, ..., e_n\}]\!]$ = (samples, period, phase)
where: samples = VEC [(samplesOf $[\![e_1]\!]$);...;(samplesOf $[\![e_n]\!]$)] ;
period = choose [(periodOf $[\![e_1]\!]$);...;(periodOf $[\![e_n]\!]$)] ;
phase = GREATEST [(phaseOf $[\![e_1]\!]$);...;(phaseOf $[\![e_n]\!]$)] ;

$[\![n]\!]$ = ((VAL n), 0, 0)

$[\![b]\!]$ = ((VAL b), 0, 0)

$[\![op\ e]\!]$ = (((hol_OP(op e)) (samplesOf $[\![e]\!]$)), (periodOf $[\![e]\!]$), (phaseOf $[\![e]\!]$))

$[\![e_1\ op\ e_2]\!]$ = (samples, period, phase) where:
samples = (hol_OP(e_1 op e_2)) (samplesOf $[\![e_1]\!]$) (samplesOf $[\![e_2]\!]$) ;
period = choose [(periodOf $[\![e_1]\!]$); (periodOf $[\![e_2]\!]$)] ;
phase = GREATEST [(phaseOf $[\![e1]\!]$); (phaseOf $[\![e2]\!]$)] ;

$[\![if\ e_1 \rightarrow e_2 \parallel e_3\ fi]\!]$ = (samples, period, phase) where:
samples = SIGCOND (samplesOf $[\![e_1]\!]$) (samplesOf $[\![e_2]\!]$) (samplesOf $[\![e_3]\!]$) ;
period = choose [(periodOf $[\![e_1]\!]$); (periodOf $[\![e_2]\!]$); (periodOf $[\![e_3]\!]$)] ;
phase = GREATEST [(phaseOf $[\![e1]\!]$); (phaseOf $[\![e2]\!]$); (phaseOf $[\![e3]\!]$)] ;

$[\![interpolate\ (e_1, ..., e_n)]\!]$ = (samples, period, phase) where:
samples = INTERP [(samplesOf $[\![e_1]\!]$);...;(samplesOf $[\![e_n]\!]$)] ;
listPeriod = (choose [(periodOf $[\![e_1]\!]$);...; (periodOf $[\![e_n]\!]$)]) ;
period = (if (listPeriod = 0) then 0 else (listPeriod DIV n)) ;
phase = GREATEST [(phaseOf $[\![e_1]\!]$);...;(phaseOf $[\![e_n]\!]$)] ;

Figure 3: The Denotational Semantics of Expressions

27

The Definition of check

check x = T

check $x(e_1, ..., e_n)$ = (check e_1) $\wedge$ ... $\wedge$ (check e_n)

check $e_1[e_2]$ = (if ((periodOf $[\![e_2]\!]$) = 0) then T
else (((periodOf $[\![e_1]\!]$) = (periodOf $[\![e_2]\!]$)) $\wedge$ (check e_2)))

check $e_1@e_2$ = check e_1

check $(e_1, ..., e_n)$ = (check e_1) $\wedge$ ... $\wedge$ (check e_n)

check $\{e_1, ..., e_n\}$ = ((equalList [(periodOf $[\![e_1]\!]$);...;(periodOf $[\![e_n]\!]$)]) $\wedge$
(check e_1) $\wedge$ ... $\wedge$ (check e_n))

check n = T

check b = T

check (op e) = check e

check (e_1 op e_2) =
(if (((periodOf $[\![e_1]\!]$) $\neq$ 0) $\wedge$ ((periodOf $[\![e_2]\!]$) $\neq$ 0))
then (((periodOf $[\![e_1]\!]$) = (periodOf $[\![e_2]\!]$)) $\wedge$ (check e_1) $\wedge$ (check e_2))
else ((check e_1) $\wedge$ (check e_2)))

check (if $e_1 \rightarrow e_2 \parallel e_3$ fi) =
((equalList [(periodOf $[\![e_1]\!]$); (periodOf $[\![e_2]\!]$); (periodOf $[\![e_3]\!]$)]) $\wedge$
(check e_1) $\wedge$ (check e_2) $\wedge$ (check e_3))

check (interpolate $(e_1, ..., e_n)$) =
(mult $\wedge$ (equalList periods) $\wedge$ (check $(e_1, ..., e_n)$))
where: periods = [(periodOf $[\![e_1]\!]$);...; (periodOf $[\![e_n]\!]$)] ;
mult = (MULTIPLE (choose periods) n) ;

Figure 4: The Definition of check

28

The Denotational Semantics of Definitions

$[\![e_1@@e_2 = e_3]\!] =$
(((samplesOf $[\![e_1]\!]$) (NEGTIME M(e_2))) = M(e_3))

$[\![(e_1, ..., e_n) = (e'_1, ..., e'_n)]\!] =$
($[\![e_1 = e'_1]\!] \wedge ... \wedge [\![e_n = e'_n]\!]$)

$[\![(e_1, ..., e_n) = x(e'_1, ..., e'_m)]\!] =$
x(makeSigTuple [$[\![e'_1]\!]$; $[\![e'_m]\!]$; $[\![e_1]\!]$; $[\![e_n]\!]$]) $\wedge$ (check (e'_1,...,e'_m))

$[\![(e_1, ..., e_n) = \text{decimate } e]\!] =$
(if ((periodOf $[\![e]\!]$) = 0)
then (DEC_SAMP_PH [$[\![e_1]\!]$;...;$[\![e_n]\!]$] $[\![e]\!]$ (check e))
else (DECIMATE [$[\![e_1]\!]$;...;$[\![e_n]\!]$] $[\![e]\!]$ (check e)))

$[\![e = \text{interpolate } (e_1, ..., e_n)]\!] =$
(INTERPOLATE $[\![e]\!]$ [$[\![e_1]\!]$;...;$[\![e_n]\!]$] listPer (check ($e_1, ..., e_n$)))
where: listPer = (filter (λ $period_i$. ($period_i \neq 0$)) [periodOf$[\![e_1]\!]$;...;periodOf$[\![e_n]\!]$])

$[\![(e_1, ..., e_n) = \text{switch } (e'_1, ..., e'_m)]\!] =$
(SWITCH [$[\![e_1]\!]$;...;$[\![e_n]\!]$] [$[\![e'_1]\!]$;...;$[\![e'_m]\!]$] listInPer (check ($e'_1, ..., e'_m$)))
where: listInPer = (filter (λ $period_i$. ($period_i \neq 0$)) [(periodOf $[\![e'_1]\!]$);...;(periodOf $[\![e'_m]\!]$)])

$[\![e_1 = e_2]\!] =$
(EQUAL (samplesOf $[\![e_1]\!]$) (samplesOf $[\![e_2]\!]$) rateInfo)
where: rateInfo = (if ((periodOf $[\![e_2]\!]$) = 0) then setPhase
else (setPeriod $\wedge$ setPhase)) ;
setPeriod = (((periodOf $[\![e_1]\!]$) = (periodOf $[\![e_2]\!]$)) $\wedge$ (check e_2)) ;
setPhase = ((phaseOf $[\![e_1]\!]$) $\geq$ (phaseOf $[\![e_2]\!]$)) ;

$[\![(d1; d2)]\!] = [\![d1]\!] \wedge [\![d2]\!]$

$[\![(x : e_1..e_2) :: d]\!] = \forall$ (x:num) :: M(e_1) TO M(e_2). $[\![d]\!]$
where: $[\![d]\!]$ is evaluated in the current environment enriched by the index x ;
$\forall$ x :: P. B = $\forall$ x. (P x) $\Longrightarrow$ B ;

$[\![\textit{begin} \text{ d } \textit{end}]\!] = [\![d]\!]$

Figure 5: The Denotational Semantics of Definitions

7 The Denotational Semantics of Silage Functions

Our denotational semantics of Silage Functions is similar to the one defined in [14, 15] and therefore is not detailed here. Each Silage function **x** has a relational model **x** and a functional model **x_FUN** in the logic. The relational model can be summarized as follows:

- the arguments of the Silage functions (input and output signals of the DSP algorithm) are universally quantified in the logic;
- internal signals are existentially quantified;
- the body of the Silage functions is interpreted in the current environment enriched by the input, output and internal signals;
- the defined Silage function is included in the environment.

The functional model is defined in terms of the relational model and the Hilbert's operator. For instance, suppose that we have a DSP algorithm that takes a signal **in** and generates a signal **out**. If its relational model is the definition of **x(in,out)** then its functional model is: **x_FUN in = (@x_OUT. ∃ out. (x_OUT = out) ∧ x(in,out))**.

Our definition of the denotational semantics of Silage functions differs from the one in [14, 15] in the following aspects:

- the types of the signals is **signalTy** in order to model the periods and the phases of the signals;
- we impose that every input, output and internal signal has a period different of zero;
- the body of the Silage function is interpreted using the *multi-rate* semantics of definitions;

8 Reasoning about Silage Programs

In this section we are going to discuss and illustrate what kind of reasoning we are able to perform using the models presented in the previous sections. Since we have not modelled the production rules of the Silage language in the HOL logic, we cannot reason about the Silage language in the HOL logic. However, since we have modelled Silage programs in the logic, we can reason about them in the logic. Embedding the production rules of Silage in the logic would not be straightforward.

The main motivation to define the semantics of Silage was to build a formal transformational design system. Whenever possible, we want to model the transformations by generic and pre-proven theorems. However, this will not always be the case. Whenever theorems modelling the transformations cannot be pre-proven, we will develop proof procedures (HOL tactics) that could automatically prove the correctness of the transformations on Silage programs. The granularity of the transformations has to do with the interface of the system with the hardware designer and the complexity of the correctness proof of the transformations. This granularity will be defined bearing in mind the user-friendliness of the system and the construction of fully automatic proof procedures. The library of theorems and proof procedures modelling transformations will always be hidden from the hardware designer. Only a HOL expert could change it.

Next we will illustrate some of the transformations in Silage described in [9]. Examples of transformations that do not change the behavior of the internal signals are loop transformations, commutativity of declarations and properties of arithmetical operators. Transformations that change the behavior of internal signals include the introduction of internal signals, dead code elimination, retiming and the dependency reversal of internal signals. First we exemplify how two Silage programs *P1* and *P2* are interpreted into HOL terms. [10]

P1	⟦*P1*⟧
func f1 (x,y:Nat) (out:Nat) =	∀ x y out. f1(x,y,out) =
begin	(NOT_ZERO[PER x;PER y;PER out]) ∧
y@@1 = 0;	(SAMP y(NEGTIME 1) = 0) ∧
y@@2 = 0;	(SAMP y(NEGTIME 2) = 0) ∧
out = x + (3*y@1 + 7*y@2);	(EQUAL
end;	(SAMP out)
	((SAMP x) SIGPLUS
	(((VAL 3) SIGMULT ((SAMP y) <> 1))
	SIGPLUS
	((VAL 7) SIGMULT ((SAMP y) <> 2))))
	((PER out = PER x) ∧ (PER x = PER y) ∧
	(PH out) ≥ GREATEST[PH x;
	GREATEST[GREATEST[0;PH y];
	GREATEST[0;PH y]]]))

P2	⟦*P2*⟧
func f2 (x,y:Nat) (out:Nat) =	∀ x y out. f2(x,y,out) =
z:Nat;	(NOT_ZERO[PER x;PER y;PER out]) ∧
begin	(∃ z. NOT_ZERO[PER z] ∧
z@@1 = 0;	(SAMP z(NEGTIME 1) = 0) ∧
z = (3*y + 7*y@1);	(EQUAL
	(SAMP z)
	(((VAL 3) SIGMULT (SAMP y)) SIGPLUS
	((VAL 7) SIGMULT ((SAMP y) <> 1)))
	((PER z = PER y) ∧
	(PH z) ≥ GREATEST[GREATEST[0;PH y];
	GREATEST[0;PH y]]])) ∧
y@@1 = 0;	(SAMP y(NEGTIME 1) = 0) ∧
y@@2 = 0;	(SAMP y(NEGTIME 2) = 0) ∧
out = x + z@1;	(EQUAL
end;	(SAMP out)
	((SAMP x) SIGPLUS ((SAMP z) <> 1))
	((PER out = PER x) ∧ (PER x = PER z) ∧
	(PH out) ≥ GREATEST[PH x; PH z])))

Using the *retiming transformation* the program *P1* can be transformed into the program *P2*. The retiming transformation, which uses the distributivity property of the delay operator over other operators, is very important for the optimization of hardware implementations of DSP algorithms [18]. We have proved the following theorem in HOL: ⊢ ⟦*P1*⟧ = ⟦*P2*⟧. This theorem have been proved by a proof procedure developed to prove the correctness of the retiming transformation in general.

[10] Actually we are showing ⟦*P1*⟧ after substituting subterms of the forms "t ∧ T" and "t ∧ (a = a)" for "t". A simple rewriting *in the logic* can prove that ⟦*P1*⟧ is equal to what we call ⟦*P1*⟧ in the table. The same holds for ⟦*P2*⟧.

Consider the Silage programs *P3* and *P4* below. We have proved the following theorem in HOL: $\boxed{\vdash [\![P3]\!] = [\![P4]\!]}$.

Interpolation Transformation	
P3	*P4*

func f1 (a,b,c,d,e,f:Nat)
    out1:Nat; out2:Nat =
begin
  out1 = interpolate (a,b,c,d);
  out2 = interpolate (b,e,d,f);
end;
	func f2 (a,b,c,d,e,f:Nat)
	    out1:Nat; out2:Nat =
	ac:Nat; bd:Nat; ef:Nat;
	begin
	  ac = interpolate (a,c);
	  bd = interpolate (b,d);
	  ef = interpolate (e,f);
	  out1 = interpolate (ac,bd);
	  out2 = interpolate (bd;ef);
	end;

Next we illustrate the *dependency reversal transformation* which was the basic transformation used in the optimization of the Hough algorithm in [8]. We have proved that $\boxed{\vdash [\![P5]\!] = [\![P6]\!]}$ with respect to the single-rate semantics of [14, 15]. Extending the proof procedure to cope with the extensions of the multi-rate semantics should be easy.

Dependency Reversal Transformation	
P5	*P6*

func f1 (in:Nat) (out:Nat) =
    a:Nat[12];
begin
  a[11] = 0;
  (i:1 .. 10) ::
    a[i] = a[i+1] + in[i];
  out = a[1];
end;
	func f2 (in:Nat) (out:Nat) =
	    a:Nat[12];
	begin
	  a[0] = 0;
	  (i:1 .. 10) ::
	    a[i] = a[i−1] + in[i];
	out = a[10];
	end;

The multi-rate semantics of Silage can be used, not only to prove the correctness of transformations, but also it can be used to prove rate properties of Silage specifications. For instance, given the meaning of a Silage program, if the periods of the signals of the program are consistent, one could prove in HOL that there is a period for every signal. If this is the case then, by knowing the periods of some of the signals, one could prove what the periods of all signals are.

9 Status of the work

At the moment we have a ML program that interprets Silage programs described in abstract syntax into HOL terms using the multi-rate semantics of Silage described above. This program is an adaptation and extension of the program developed by Andrew Gordon to interpret Silage programs into HOL terms using the single-rate semantics of Silage described in [14, 15]. We are currently testing the multi-rate semantics and we have proved in HOL properties of the constants used in the semantics. Some theorems have also been inherited from the theories of Andrew Gordon. We are also developing proof procedures to check automatically the correctness of the transformations described in [9]. Hans Samsom is currently developing an interface for the hardware designer to perform source-to-source transformations. The idea is to build a window based interface with a menu of useful transformations for the optimization of DSP algorithms. The correctness of the transformations will be proved using our theorems and proof procedures. We have been working together to build a formal transformational design system with an user friendly interface.

10 Conclusions

We have presented the definition in HOL of the multi-rate semantics of a substantial subset of Silage. The semantics is defined in a relational style and is tuned for source-to-source transformations. A ML program has been developed to translate Silage programs described in abstract syntax into HOL terms. We are currently developing proof procedures to prove automatically the correctness with respect to our semantics of a catalog of commonly used transformations towards the optimization of DSP algorithms. We are also developing with Hans Samsom a window based interface for our formal transformational design system.

11 Acknowledgements

We are grateful to Andrew Gordon for many useful discussions about the semantics of Silage. We also want to thank him for sharing with us all the practical results of his work like ML code and HOL theories.

References

[1] Paul N. Hilfinger, "Silage, a High-level Language and Silicon Compiler for Digital Signal Processing", *Proceedings IEEE CICC-85*, Portland, Oregon, May 1985, pp. 213-216.

[2] Paul N. Hilfinger, *Silage Reference Manual*, December 1987.

[3] D. Genin, P. Hilfinger, J. Rabaey, C. Sheers and H. De Man, "DSP Specification Using the Silage Language", *IEEE International Conference on Acoustics, Speech and Signal Processing*, April 1990, pp. 1057-1060.

[4] Lode Nachtergaele, *A Silage Tutorial*, IMEC, Leuven, Belgium, May, 1990.

[5] Lode Nachtergaele, *User Manual for the S2C Silage to C Compiler*, IMEC, Belgium, May, 1990.

[6] H. De Man, J. Rabaey, P. Six, L. Claesen, "Cathedral-II: A Silicon Compiler for Digital Signal Processing", *IEEE Design & Test of Computers*, December 1986, Vol.3, No.6, pp. 73-85.

[7] Paul Lippens, *Defining control flow from an applicative specification*, Digital VLSI Design, internal report, December, 1988.

[8] J.G. Samsom, L. Claesen, H. De Man, "Correctness Preserving Transformations on the Hough algorithm", *CompEuro 92, IEEE International Conference on Computer Systems and Software Engineering*, The Hague, The Netherlands, May 1992.

[9] C. Angelo, *Transformations in Silage*, IMEC report, November 91.

[10] Ingrid Verbauwhede, *VLSI Design Methodologies for Application-Specific Cryptographic and Algebraic Systems*, Ph.D. Thesis, IMEC, Leuven, Belgium, 1991.

[11] Mark Pauwels, *Requirements for bit-true synthesis on specification languages and simulation, synthesis and verification tools*, IMEC internal report, September, 1990.

[12] M. Gordon, "HOL: A Proof Generating System for Higher-Order Logic", *VLSI Specification, Verification and Synthesis*, Ed. G. Birtwistle and P. A. Subrahmanyam, Academic Press, Boston, 1988, pp.73-127.

[13] C. Angelo, *Issues on the Signal Flow Graph Semantics of Silage in HOL*, IMEC report, October 1991.

[14] A. Gordon, *A Mechanised Definition of SILAGE in HOL*, Cambridge report, February 92.

[15] A. Gordon, "The formal definition of a synchronous Hardware-Description Language in Higher Order Logic", to appear in *ICCD92: International Conference on Computer Design*, Cambridge, Massachusetts, October 1992.

[16] G. Cousineau, M. Gordon, G. Huet, R. Milner, L. Paulson, C. Wadsworth, *The ML Handbook*, INRIA, 1986.

[17] Werner Geurts, Mark Pauwels, Francky Catthoor, Gert Goossens, Lode Nachtergaele, *Proposal for Extensions to the Silage Language*, IMEC internal report, March, 1990.

[18] Miodrag M. Potkonjak, *Algorithms for High Level Synthesis: Resource Utilization Based Approach*, Ph.D. Thesis, University of California, Berkeley, January 1992.

Higher Order Logic Theorem Proving and its Applications (A-20)
L.J.M. Claesen and M.J.C. Gordon (Editors)
Elsevier Science Publishers B.V. (North-Holland)

Design-Flow Graph Partitioning

R.B. Hughes[1] and G. Musgrave[2]

[1] Abstract Hardware Limited, Building 2, The Science Park, Brunel University, Uxbridge, Middx. UB8 3PQ. U.K. Email. roger@ahl.co.uk, Tel. +44 895 258501.

[2] Head of Department, Department of Electronic and Electrical Engineering, Brunel University, Uxbridge, Middlesex. UB8 3PH. U.K.

Abstract

This paper examines the application of rigourous formal techniques to the manipulation of flow graphs used in design partitioning. The concept of a design-flow graph, encompassing the concepts of both data and control-flow is introduced. To aid in partitioning at various levels a set of primitive restructuring operations are defined. The LAMBDA system, based on HOL, is used to provide a strong formal basis to the definition, manipulation and use of such graphs in hardware/software codesign and testing.

Keyword Codes: J.6; G.2.2; F.4.3

Keywords: Computer Aided Engineering; Discrete Mathematics, Graph Theory; Mathematical Logic and Formal Languages, Formal Languages

1 Introduction

Systems level design is becoming of increasing concern since the advent of easy logic synthesis, where silicon structures are generated from a specification of function[1], and the increasing success of behavioural synthesis in automating state assignment. With increasing demands placed on automation of the design *process*, and increasing levels of functional complexity and requirements for high integrity systems, architectural synthesis tools using formal reasoning are required to support system-level design. By *formally* synthesising a system it is not necessary to go to the *simulation* stage which non-formal techniques[2] require.

Formal methods have already been successfully used [3, 4, 5] in a variety of complex case studies. In tackling real-world formal design, the integration of user-interaction and automated synthesis must be considered, as it has been before [6]. In addition, however, we require a system in which high-level specification (including features such as user-defined datatypes and recursively defined functions), and refinement of both data and timing are supported. The LAMBDA/DIALOG design assistant [7] is a system which implements and extends these ideas and by providing VHDL output it can interface with many other, lower level "High-level synthesis" tools[8]. With such useful generic capabilities it can also be used for software engineering design [9, 10, 11].

System-level design, in its more generic sense, requires the design of software within a system to be considered an integral part of the overall system. This is especially important for safety-critical design work, since tradeoffs in one direction imply changes to another area which might affect overall system performance. Thus, the ability to formally design the whole system, bearing in mind various architectural limitations, in a sound logical framework, is considered to be very important. In this paper we present a formal approach to the issue of using *design-flow graphs* (encompassing both data- and control-flow) and their partitioning. Partitioning of these graphs is important if the designer is to be able to implement the design, part in software and part in hardware, in a flexible manner. *Design Partitioning* [12, 13] is also important when implementation in different architectures is considered. We consider that our formal approach to the problem of high-level systems design allows the designer flexibility in exploring the *design space* whilst ensuring that the design always *satisfies the specification.*

It is also important at this stage to observe what is meant by *partitioning.* At the system level, the term 'partitioning' can be used to indicate the partitioning of the design into software and hardware components. It can also be used to describe the architectural partitioning which takes place within the hardware itself. The ability to examine different partitionings, all of which are guaranteed to be behaviourally correct by design, is an area of key interest at present. State-of-the-art systems which can cope with such multiple-level partitioning, to allow the designer to freely explore the design space, are in great need and are becoming reality.

2 Scheduling of Flow-Graphs

A flow graph, in its simplest sense, may be thought of as a connection of nodes which perform some calculation connected to other nodes by means of arcs. There may be delays present on these. If the node has no associated delay then the representation is really that of a signal flow graph (SFG). In an SFG, the computational delay associated with a node can be represented by an initial labelling of all the output arcs of the node. In order to be able to process the data the inputs to the nodes must occur at the same time, i.e. the computations from some of the nodes must be delayed until all sources of input to a node are available. This is called *scheduling* or *balancing* of the flow-graph. In the example that follows all but one of the nodes have been given an initial latency of 1, the exception having a latency of 2. In the scheduling process, delays are added to the arcs so as to correctly balance the graph. Naturally, the minimum delay possible is the best one to use but it is, in general, not unique and several minimal solutions may exist. The scheduling of a such a graph (see Fig. 1) is illustrated in Fig. 2.

It will be noted, on examination of the solution, that the overall latency of the graph is 9 and that this is the same for any path through the graph. The words *any path* in the

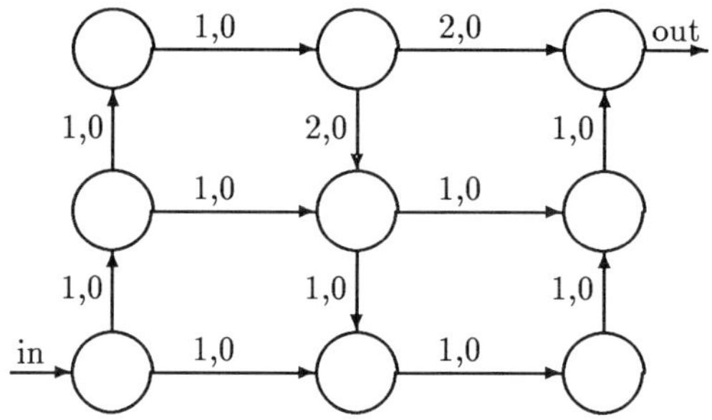

Figure 1: *Simple SFG - before scheduling*

previous statement only has meaning for *acyclic* flow graphs.

These operations have been implemented in the DIALOG interface to the LAMBDA system and the above scheduling was achieved using the scheduling alogorithms available within the existing LAMBDA system. We do not describe the details of the algorithms used here owing to a lack of space. We represent the nodes of the graph as *abbreviations* in the LAMBDA logic, where the specification of an SFG graph is a purely topological description. In this paper, our concern lies with restructuring operations , not on the underlying graph formalism.

3 Graph Restructuring

3.1 Background

Flow graphs, especially data-flow, are currently used mainly by Computer Scientists to describe different systems. Unfortunately, they are not used in a formal manner, but are, in the main, used to pictorially represent different system partitionings, where *system* is usually a set of computer programs. Formalising this work is made more difficult owing to the fact that, unlike Signal Flow Graphs, there is no single universal definition of a DFG. Despite this, such flow-graphs have been used in Kung [14] for real-world designs.Frequently one can think of DFG as a Data Flow Graph (e.g. the DeMarco [15] style), one then needs a separate Control Flow Graph to supply the control to the actual hardware used to implement the system. However, we introduce the more general concept of a *Design Flow Graph*, which encompasses both data and control, so that both aspects can be studied without losing track of the other. By retaining the control-flow information, test pattern generation is made more easily achievable; we will not discuss test-pattern generation here, but it is of interest in various collaborative research projects. Frequently,

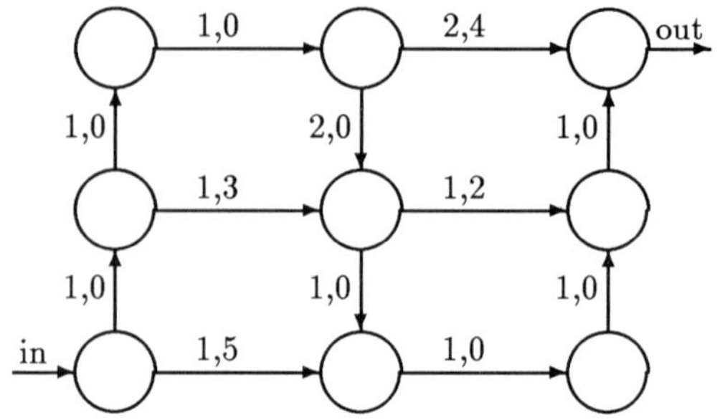

Figure 2: *Simple SFG - after scheduling*

one needs to restructure the data-flow graph to be able to optimise it for a particular implementation or to investigate some particular set of properties. Restructuring is used to :-

- Obtain appropriate partitioning.
- Reduce complexity.
- Present particular system aspects.
- Aid function allocation.

Recently, attempts [16] have been made to try to formalise the operations that take place when a DFG is restructured.

Primitive restructuring operations are considered to be :-

- grouping – folding and merging
- expanding – unfolding and splitting
- translation – raising and lowering

The previously mentioned six operations together with a seventh operation called *Hidden Split* are now provided as a sub-menu in the DIALOG system available with LAMBDA v4.2.

3.2 Formalisation

Description of Operations

It is important to clarify, in natural language terms, what we mean by 'folding' and 'merging'. This is perhaps best achieved by describing the operations in terms of their nearest 'inverse'. We will then describe the HOL formalisation for these operations. In the fold/unfold style of operation a group of nodes are selected to form a new node composed of the sub-nodes previously selected. As no information is lost in this process, an unfold operation will return the user to the same state of DFG as before. Note that the designer will be two steps further on in the design, but at the same state as before when a fold is immediately followed by an unfold. Figure 3 illustrates the effect of the merging and splitting of nodes. It should be noted that the arcs connecting the various nodes are not shown in order to more clearly illustrate the effect on the hierarchy.

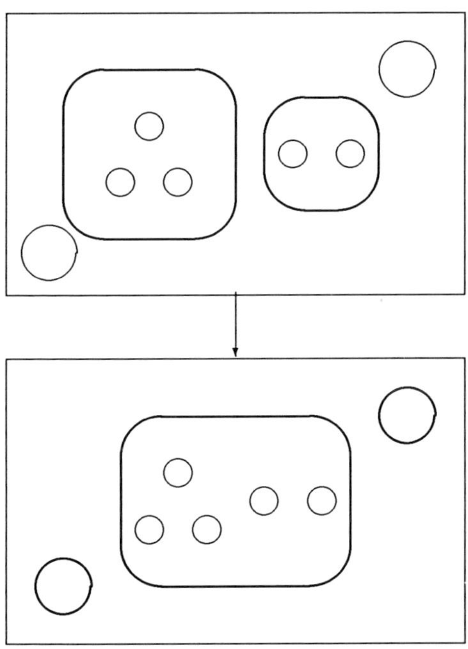

Figure 3: *Merging and Splitting*

It can be seen that the effect of a merge is to combine the effects of two composite nodes, i.e. nodes that have been formed from previous folding operations into a composite node in which the previous nodal definitions are removed, leaving only the sub-nodes in the design. It needs to be emphasised that the resulting node, has *exactly the same functional behaviour* as a node formed by a folding, viz. the nodes formed by the fold and merge operations are *functionally equivalent.* The only difference between the composite nodes formed in each case is in the internal structure of the nodes.

It is also important to realise that the merge operation does not eliminate duplicate nodes. Such elimination is thought possible when no implementation has been carried out since it is then only a matter of comparing the functional specifications of the nodes involved. However, the elimination of an equivalent node after implementation would require that the circuit be split up in such a way as to allow rescheduling to take place. This is a much more difficult operation.

Finally, in the primitive restructuring operations essential for partitioning, the raising and lowering of nodes has been considered. In the DIALOG system, the operations are limited, at present, to movement across one level of hierarchy only.

These operations will only come into their own when they are used in conjunction with equivalence testing operations and rescheduling of the design so as to permit the designer to make implementation decisions for the design. In this, the re-use of existing components, where these may be hierarchical in nature, with associated scheduling and multiplexing with delays is a key requirement of the designers. Despite the NP completeness of many of the problems *in the general case*, it is hoped to provide *specialised* solutions to *specific* problems within this domain.

HOL Description

The operations will be described in some detail. Many of the characteristics of the operations are quite similar to those required for the `foldToComposite` function which provides the *folding* operation. As such, the detail will be in describing this one function with simpler descriptions being given of the other functions.

In the folding operation all the selected components are moved to the conclusion's H list and a conjunction is made out of them and all ocurrences of the selected components are removed from the premises. Thus starting with a rule of the form:-

```
        C3, S4, S5, G // H    |-   R
-------------------------------------
C1, C2, C3, S4, S5, G // H    |-   P
```

where one selects the components S4 and S5, the resulting intermediary sequent is:-

```
        C3,           G // H    |-   R
---------------------------------------
C1, C2, C3,  G // S4 /\ S5, H   |-   P
```

Existential quantification is then carried out on those wires which are connected to outputs that do not occur elsewhere in the conclusion. Thus we are using existential quantification to hide 'hidden' wires in the node to be created. In the following rule form it is assumed that there is just one wire `w` which is common between S4 and S5 that is *not* connected to any other component.

```
        C3,                          G // H   |-   R
--------------------------------------------------
C1, C2, C3,  G // exists w. S4 /\ S5, H   |-   P
```

A name is generated for the new component and a new abbrevation is created for the component.

```
        C3,        G // H   |-   R
-------------------------------------
C1, C2, C3,  G // CN1, H   |-   P
```

The component has been called CN1 in this example. The component data within DIALOG is updated with an entry for the component and a corresponding implementation rule is produced. The implementation rule is craeted by the ML code:-

```
val impRule   = fApplyTr (tacToTr(rawUnify'Tac (gl (gr (gl gI))) foldRule
   thenT unifyTac reflexH)
  thenTr forwardRemoveExists
  thenTr forwardRemoveAnds)
 (mkTrivRuleFromExpn (vbl"G" // newExists $ vbl "H"
     |- newAbbrev))
```

where forwardRemoveExists and forwardRemoveAnds apply to the conclusion of the rule for the composite node abbreviation.

The *merge* operation is similar except that after initial quantification, all the selected components are unfolded. If any of these components are themselves composite nodes then the existential quantifications must be manipulated so as to form a single level conjuction within the one level of quantification. This operation entails recursively removing quantification and conjunction terms from the conclusion followed by forming a conjuction of all the available terms. One then existentially quantifies over all the variables that were previously quantified.

This operation is partly described by the following ML code:-

```
  val innerTr =
       chooseTr(fn rule =>
 let
    val (_,conc) = repRule rule
    val exVars = link (map varsFromExists
    (getHHyps conc))
 in
   forwardRemoveExists thenTr
   forwardRemoveAnds thenTr
   remExistsAndTr thenTr
   addAndsTr thenTr
   foldMapTr makeExistsTr exVars
 end )
```

The *split* operation works within an Edit cycle on an existing composite node. The object must have been formed by using either the merge or the fold facility. A selection of subcomponents is made and a fold operation carried out on these components automatically. An automatic fold is then performed on the remaining components (except IO connector components). The result of these two fold operations is two create two new composite nodes. A implementation rule for the higher-level node is changed in the process of this set of operations and an automatic *unfold* operation then allows the user to see the new nodes connected in place if the old one. In a 'Hidden Split' this unfold is not carried out automatically and thus a subsequent re-edit would reveal the two new nodes instead of the previous component graph.

The *raise* and *lower* operations work in slightly differing ways; 'raise' uses an Edit cycle but 'lower' does not. In the 'lower' operation the aim is to push something down into a node. The resulting node is clearly functionally different to the old one ands, as a result, the name of the node is not preserved to reflect this case. The 'raise' operation lifts a sub-node to the level of hierarchy above, i.e. the level of its parent. This can be represented, similar to 'split' by changing the definition of the node being edited.

All the above Primitive Restructuring Operations at the schematic level are already provided in LAMBDA/DIALOG v4.2, available September 1992. These operations provide the essential basis for further development of higher-order operations.

4 Future Higher-Order Operations

Our work, as part of the PATRICIA[3] project, is also investigating the concepts of neighbourhood and equivalence [17]. In this work we are examining various definitions of node equivalence. It was felt that some form of *finest equivalence*, where 'finest' is taken to mean an equivalence operation which makes the fewest things equivalent would be necessary. That is, the extension of an equivalence over node and edge labels to an equivalence over nodes and edges.

Given, a node equivalence =n=, and a set of successor functions A, a new equivalence == over nodes could be defined as the *finest* equivalence satisfying:-

∀ a in A and nodes x and y : x == y ⇒ x =n= y and a(x) == a(y)

This idea of equivalence states that two nodes are deemed equivalent if they are both *internally* equivalent and connected in the same way to equivalent neighbours. This is stronger than the idea of behavioural equivalence used in our partitioning work, in which the equivalence is restricted to the internal behaviour of the nodes. Equivalence is required for the development of algorithms to compute sub-graph isomorphism.

An alternative, weaker, definition of equivalence was to say that a node $\mathcal{N}$ could be added to an equivalence class $\mathcal{C}$ if it is 'internally' equivalent to the nodes already in the class (i.e. it has an equivalent transfer function AND there is a path, not longer than some number m, from node $\mathcal{N}$ to a node already in $\mathcal{C}$). It was felt that nodes belonging to the

[3]ESPRIT Project No. 5020

same class would probably be closely placed on silicon, favouring a reconfiguration in the case of faults.

The existing design-flow restructuring operations together with an appropriate definition of nodal equivalence will allow for the future development of some higher-order operations to tackle more complex design issues at a very high level of abstraction.

Operations which are likely to be needed, in addition to the essential restructuring operations, are

- the ability to recognise nodes which perform the same task
- manipulation of timing to allow parallel execution of various tasks or to minimise hardware by introducing multiplexing at an appropriate level
- the ability to eliminate redundant nodes
- selection and grouping operations to form redundant nodes
- automation of some recognition operations
- computer assisted 'optimal' node formation, where the definition of what is defined to be 'optimal' may be selected by the user.

The last two operations are likely to be somewhat advanced and will require a good practical knowledge of how to apply the earlier operations. However, they represent achievable long-term goals which will greatly assist in the development of sound systems design approaches based on formal design partitioning. The work is still at an early stage, but the interface between the related work of Politecnico di Milano (I) and Brunel University (UK) and our formal methods [18] will be the key to a sound, yet usable, step in the development of design partitioning tools for total systems design encompassing both hardware and software. Our methods retain the information necessary for *design for testability* which is important, despite behaviourally correct design, to test the physical device for possible manufacturing defects.

5 Acknowledgements

A large part of this work has been supported by the Commission of the European Communities under the PATRICIA project (ESPRIT Project No. 5020). We should also like to thank our project partners for feedback which helped select a common core of operations. Finally, our thanks go to the numerous reviewers whose constructive comments helped refine our abstract into this short paper detailing recent research and development.

References

[1] R.K. Brayton, R. Camposano, G. DeMicheli, R.H.J.M. Otten, and J.T.J. van Eijndhoven. The Yorktown silicon compiler system. In D. Gajski, editor, *Silicon Compilation*. Addison-Wesley, 1988.

[2] R. K. Gupta, C. N. Coello, and G. De Micheli. Synthesis and Simulation of Digital Systems containing interacting Hardware and Software Components. In *Proceedings of 29th ACM/IEEE Design Automation Conference*, pages 225–230. IEEE, 1992.

[3] R. B. Hughes, M. D. Francis, S. P. Finn, and G. Musgrave. Formal tools for tri-state design in busses. In *Proceedings of the 1992 International Workshop on the HOL Theorem Prover and its Applications*, September 1992.

[4] S. Finn, M. Fourman, and G.Musgrave. Interactive synthesis in higher order logic. In *Proceedings of the 1991 International Workshop on the HOL Theorem Prover and its Applications*, Davis, California, August 1991.

[5] E. Mayger and M. Fourman. Integration of formal methods with system design. In A. Halaas and P. B. Denyer, editors, *VLSI 91*, Edinburgh, Scotland, August 1991.

[6] N. D. Dutt, T. Hadley, and D. D. Gajski. An intermediate representation for behavioral synthesis. In *Proceedings of 27th ACM/IEEE Design Automation Conference.* IEEE, 1990.

[7] Abstract Hardware Limited, Brunel University, Uxbridge, Middx. U.K. *The LAMBDA System - Complete Reference Set, v. 4.1*, 1991.

[8] A. Stoll and P.Duzy. High-level synthesis from VHDL with exact timing constraints. In *Proceedings of 29th ACM/IEEE Design Automation Conference*, pages 188–193. IEEE, 1992.

[9] R.B. Hughes. Automatic Software Verification and Synthesis. In *Proc. 2nd. Int. Conf. on Software Engineering for Real-Time Systems*, pages 219–223. Institution of Electrical Engineers, September 1989.

[10] R.B. Hughes and R.M. Zimmer. Automated interactive verification of functional programming languages. In C.M.I. Rattray and R.G. Clark, editors, *The Unified Computation Laboratory*, pages 411–423. Oxford University Press, May 1992.

[11] R. B. Hughes. *Automated Interactive Software Verification and Synthesis.* PhD thesis, Department of Electrical Engineering and Electronics, Brunel University, Uxbridge, Middx, U.K., July 1992.

[12] A. Antola and F. Distante. DFG: a graph based approach for algorithmic flow driven architechture synthesis. In *Proc. of EUROMICRO 91*, Vienna, Austria, September 1991.

[13] P.R. Chang and C.S.G. Lee. A decomposition approach for balancing large scale acyclic data flow graphs. *IEEE TOC*, 39(1), January 1990.

[14] S.Y. Kung. *VLSI Array Processors.* Prentice-Hall, 1988.

[15] T. DeMarco. *Structured Analysis and System Specification.* Prentice Hall, 1979.

[16] M. Chan and C. Chung. Restructuring operations for data-flow diagrams. *Software Engineering Journal*, 6(6):181–195, July 1991.

[17] A.V. Aho, J.E. Hopcroft, and J.D. Ullman. *Data structures and algorithms.* Addison-Wesley, 1983.

[18] R.B. Hughes and G. Musgrave. Strumenti per ragionamenti formali applicati alla progettazione di sistema. *PIXEL*, 13(6):45–47, 1992. ISSN 0392-8217.

Chapter 7

Hardware Verification Methodologies

Higher Order Logic Theorem Proving and its Applications (A-20)
L.J.M. Claesen and M.J.C. Gordon (Editors)
Elsevier Science Publishers B.V. (North-Holland)

Implementation and Use of Annotations in HOL

Saraswati Kalvala Myla Archer Karl Levitt

Department of Computer Science, University of California, Davis
Davis CA 95616, USA

Abstract

The use of formal methods implies a significant leap to be made, both by a human verifier who creates a proof, and by anyone wishing to understand it, when proofs and specifications are expressed in the language of a typical theorem-proving system. This results from the fact that inferences in these systems are made by choosing from a limited set of small logical steps that do not bear an obvious relationship to the way humans derive and understand proofs.

We have developed a system of annotations that can be used to incorporate informal semantic information, concerning the domain being reasoned about, into a formal proof environment, in such a way as to guide proof development and provide proof explanation. These annotations can be used, for example, to transfer a designer's intuitions about why a design is correct to the theorem-proving system that will prove that fact, or to record the steps of the proof in the style of a journal-level proof.

1 Introduction

Mechanized proof systems allow users to discover proofs and test them in a formal context. The user provides the proof steps, and the system contributes in the area of rigor and trustworthiness.

Often, the conjecture to be proved is quite complex. Developing proofs then becomes rather cumbersome, and involves a high degree of familiarity with the formal proof methods available and the logic. Also, the user must access a vast amount of domain knowledge, and relate it to the definitions, lemmas, and proof rules available. That is, to efficiently guide the system in large proofs, users must understand in detail the domain background, as well as have a good understanding of the theorem prover's proof methods. Ideally, all relevant contextual information should be part of the specification of the problem. However, it may not be easy or even possible to express this information in the logic language used for formal description of the problem. We call *extra-formal* information that part of a specification or justification that cannot be expressed formally.

Usually, a formal specification of a design is complemented by a document explaining it in a natural language. This document is used to guide the verifier in developing a proof of correctness. There are many problem with the use of separate documentation written as text:

- More often then not, the process of writing documentation is performed after the design is complete. Such documentation involves a lot of extra effort, and involves re-understanding the design.
- It does not capture the process of development of the design, which involves iterative fine-tuning and re-design.
- Designs may change, and it is difficult to keep documentation up-to-date.
- If the document does not use a schematized notation, it cannot be used for automation.

Proofs are performed to increase assurance of the correctness of some conjecture. Mechanized proof systems may either return yes/no answers, or provide proof scripts. Proofs are evaluated, whether they are successful or not. Failed proofs must provide information on what was wrong with the original statement of the proof goal and/or the domain. In practice, in many reasonably complex instances, the conjecture is rarely correct in its first incarnation.

Attempts at automation of proof are mostly concerned with syntactic scrutiny and manipulation of terms and theorems, and application of proof-theoretical rules and decision procedures. However, this automation does not model important proof decisions particular to the specific problem in hand. The exploitation of this information still depends on human intervention.

Another approach to automation is to make the system use *design knowledge*, instead of proof system heuristics, to suggest proof structuring and application of inferences. This is exactly the type of information that can be present in *annotations*. By an annotation, we mean an informal "piece of information" associated with either a logical expression in a proof or specification or a proof itself. Thus, annotations support clean incorporation of domain heuristics, in the form of design justifications and proof hints.

In sum, the purpose of annotations is to make different types of information accessible in the proof development process, whether to a human user or to the system itself. They can simply guide the user and suggest proof methods, or they can partially automate the

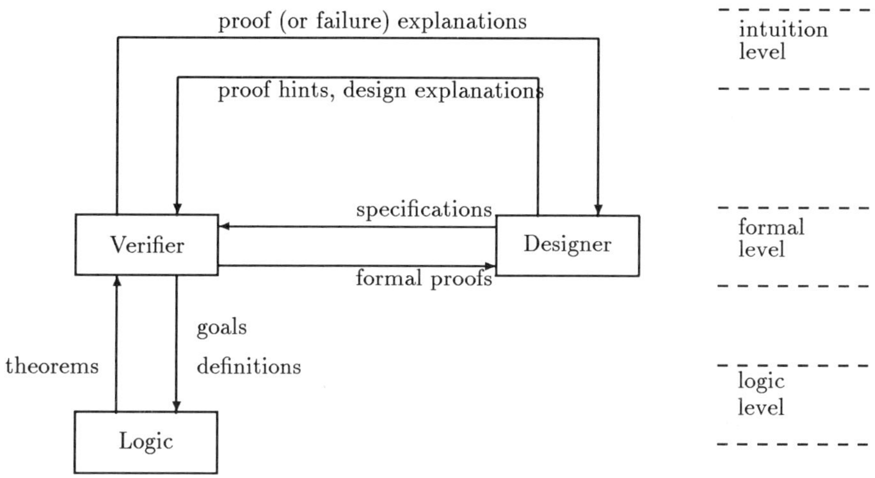

Figure 1: Layered view of proofs

proof process. They ease the process of verification by collecting relevant information and making it accessible when needed. This is especially important when the verification process occurs separately from the design, either after the passage of time or by another group of workers.

Our preliminary results show that, indeed, the proof process requires more information than that contained in formal specifications, and that this information is best captured during the design process, rather than through an *a posteriori* analysis of the design.

We believe that annotations demonstrate the more general concept of stratification of proofs: a more understandable, intuitive, and people-oriented proof that reflects a rigorous, formal proof, as illustrated in Figure 1. While mechanized logics prove a goal by invoking raw inference rules, or by invoking some well-accepted syntactic rules, in practice the target community considers a 'proof' to be the tactic that generates the inference rule or, more broadly, the organization of a *theory*, i.e. the sequence of lemmas and auxiliary definitions that lead to the justification of the primary goal. However, in a more abstract level, a proof is independent of any formal logic, but it is constituted by a more intuitive explanation about the correctness of the original conjecture.

The use of annotations helps bridge the gap between proofs in the formal sense and proofs in the pragmatic sense. Annotations that explain definitions and proofs constitute the informal level, while the manipulation of terms by mechanized inference rules constitutes the formal level.

We have found HOL [Gor88] to be an excellent vehicle for incorporation of such annotations. The annotations we have explored can be used in the HOL environment to guide the path of a proof by being propagated to appropriate subgoals, and in some cases, can even automatically provide the appropriate tactic for proving a subgoal.

In this paper, we first review proof systems currently in widespread use in terms of their facility for control by domain knowledge and explanatory proofs, followed by a description of our implementation of annotations in the HOL system. We then describe examples of how several classes of annotations can be used in the course of a HOL proof, and conclude with comments on work in progress and future directions of the research. The appendix contains the description of a few example verification problems which are targets to the specific taxonomy of annotations we are developing.

2 Related work

Early work in theorem proving was concerned with automatic discovery of proofs through uniform proof methods. A popular application for these tools was program verification. Later, more natural proof systems were developed. Also, the application area for these systems was expanded to include other systems such as mathematics, hardware systems, and process algebras. In this section, we will examine existing proof systems in the context of proof control and understandability.

2.1 Proof control

Uniform proof procedures—such as resolution—provide little direct user control over proof execution, and no readable explanations. They usually proceed by converting lemmas and the conjecture to be proved into terms, in some standard notation, that are then processed algorithmically [Bun83]. User control of proof, when available, is indirect, and consists in changes to parameters that restrict the depth of a uniform search for proofs.

Several researchers have studied uses of annotations to direct the program verification process. Wegbreit developed a methodology for allowing users to supply possible instantiations to quantified variables in assertions [Weg76]. Dershowitz emphasized the use of annotations for development of loop invariants [Der83]. None of these methods, however, is a general enough scheme to cope with all the different types of information that can be brought to bear to the proof process.

The prover developed by Boyer and Moore, known as `NQTHM`, exploits the use of *hints* for guiding the proof process [BM88]. However, `NQTHM` is mostly an automated proof finder, so the purpose of hints is to direct the automatic heuristic search for a proof. The hints are not necessarily a natural aspect of a problem definition, and do not properly annotate a specification or a proof. More often than not, a proof is attempted and then it fails, and only then does a user include hints to then attempt the automatic proof again.

Proof checkers such as HOL allow control of the proof process to be assigned completely to the user, who can then explore different proof strategies. Such systems offer increased user control, while facilitating the execution of higher-level inferences. However, the amplification of the control a user has over the proof makes the proof process more tedious, due to the increase in the details the user must be concerned with.

2.2 Proof explanation

Most explanations of formal proofs relate the conjecture being proved to definitions, lemmas and axioms in the logic. With resolution-based systems, the proof really consists in manipulating objects that have little obvious correspondence to the original problem statement. Proof systems of this type do not really explain reason for correctness, and thus have limited usefulness [Rob91]. When a proof fails, these provers do not help in the subsequent reformulation of the problem specification.

Heuristic-based provers such as **NQTHM** offer better proof explanations than resolution-based systems. Yet, the proofs produced are purely syntactic and, even though they are expressed in natural language, are not real explanations. They provide syntactic sugaring for the machine-driven proof scripts, rather than domain-knowledge based explanations. The way such a proof is executed may differ greatly from a more intuitive proof outline.

HOL users commonly explain a proof with the tactic applied to complete the proof. However, it is usually difficult to isolate the few crucial, insightful steps from the large number of standard, automatic, low-level procedures in which they are interspersed. Also, the conjoined tactic does not illustrate individual intermediate results, which would help in understanding the proof. Often, to understand a proof, one breaks down a tactic and applies it to the goal using the subgoal management system, usually a non-trivial task. Proofs are tied down to the language of the specific prover.

3 Proposed approach

The research described in this paper leads towards a general tool for incorporating methodical annotations into a proof environment. Annotations from problem descriptions are used to help *develop* proofs, and those annotations along with others obtained from proof development are used to *explain* proofs.

Annotations can speed up the search for proofs, by providing support to the creative aspects of proofs. Such information, useful in controlling the development of proofs, is a natural component of problem descriptions, as it represents aspects of the development of the problem. Verification is the flip side of design: the same assumptions that guide design are confirmed when the design is verified. Also, justifications for design can be converted to explanations of the proof.

In verification projects, annotations can be supplied by designers, appended to formal descriptions of the design. For example, a hardware circuit can be conjectured to work correctly because it is assumed that a certain property holds. A program may be thought to be correct because of some relation between values that is expected to remain constant throughout the execution of a loop. Even if the designer does not work out the complete informal proof in detail, the important or complex steps that involve some amount of thought can be captured by annotations.

Proof hints and summaries help one understand a proof and why it is correct. Annotations can be simply propagated from problem specifications to proofs, to make the proofs more understandable by linking them to domain assumptions. If at any point of a proof one finds a contradiction, one can examine the set of annotations that have guided the proof so far, and isolate those that represent erroneous assumptions. A proposed lemma may not apply in the specific case, or all the possible values for a variable may not have been taken in consideration. If on the other hand a proof is correct, the annotations from the specification that gave rise to successful proof steps can be used to increase trust in the design, more than just the (sometimes unreadable) formal proof.

Our proposal is to model extra-formal information in a practical manner, emphasizing the role of proof outlines for formal proof execution. Processing of annotations does not directly create theorems, but generates sequences of tactics that can then be applied to conjectures. However the annotations are processed, they cannot by themselves prove an incorrect conjecture. Wrong hints will lead only to a proof outline that fails to produce a useful proof.

Although we have used HOL as our prototype prover, the use of extra-formal information applies to different proof systems. This is because *structuring* a proof is essential, whatever the proof paradigm. Most natural proofs become more understandable when they are well structured, independent of the method by which proof steps are executed. In fact, annotations present a logic-independent view of proofs, as they capture the intrinsic aspects of a proof that can help a human in understanding and accepting it, and even implementing it in another logic. The same annotations can be used in developing and explaining proofs in different logics—as long as the essential proof outline is the same.

4 Basic structures

Annotations can exist at different places in a HOL session, either in objects of the logic (theorems, constituted by annotated terms) or in the proof management system, in a goal or over a proof. Each of these possibilities will be described separately. Specific examples of annotations will be shown in the next section.

4.1 HOL90

The HOL system developed by Gordon and others [CSD91] is an interactive proof checker that provides a way of manipulating definitions and proofs in higher-order logic. A notation for expressing facts and properties is interfaced to a programming language in which proof procedures and strategies can be encoded. One of the main advantages of using HOL for verification is the safety its approach to strong typing provides: objects of type `theorem` can only be formed by using a very small set of functions—either through proofs that are created from a small set of predefined axioms and inference rules or through conservative extension by new function definitions. Once we trust the set of axioms and rules, we can trust any correctly developed proof in the logic.

The work described here makes use of HOL90 [Sli91], a new implementation of the HOL logic built using Standard ML [Pau91]. The version of the meta-language we use, Standard ML of New Jersey, provides better support for programming the proof system than the version of ML underlying the customary Cambridge implementation (known now as *Classic* HOL). Standard ML allows the use of *structures* for modularization, which we use extensively. It also has a better supported external interface.

Although many parts of the source code for HOL90 have been translated verbatim from Classic HOL, other features specific to HOL90 are very advantageous to our work. It has been very helpful that relevant data types are defined in ML, in contrast to Classic HOL where they are declared in the underlying Lisp. This has resulted in much cleaner coding. There is also a pretty-printing facility—accessible through the meta-language—which makes for good visualization of annotations in terms and proofs.

4.2 Annotated terms

Terms in implementations of HOL are tree-like data structures. Leaves of the tree are variables and constants, while intermediate nodes of the tree are combinations and abstractions, i.e. function applications and variable bindings. In our extension, as illustrated in Figure 2, each branching of this tree structure has a new child node—the `note` field—that stores annotations. The enhancement is also mirrored in the structure of goals and

```
datatype term_plus =    Var_plus of (hol_type * string * note)
     | Const_plus of (hol_type * string * note)
     | Comb_plus of (term_plus * term_plus * note)
     | Abs_plus of (term_plus * term_plus * note);
```

Figure 2: Term data structure with annotations

theorems, resulting in a re-coding of large parts of the implementation of the logic.

In displays of objects of type `term_plus`, annotations will be enclosed by `[*` and `*]`. Note that, whenever this term is displayed, the annotations are displayed too, whether the term appears as part of a theorem, or a proof goal. Standard HOL notation will be used, rather than re-touched logical notation, to demonstrate the actual interaction with the system.

In Figure 3 is an example term, representing the structure of a circuit implementing parity checking. The example has appeared previously in HOL literature: the emphasis of descriptions varies from an explanation of the semantics of the circuit [Gor88] to a description of its proof in a purely syntactic fashion [CSD91]. (The diagram of the circuit is lifted directly from these sources.) By allowing the terms to contain annotations, we have made it possible to always maintain a close link between semantic features and syntactic structures.

4.3 Annotated proofs

While theorems are logically built by applying inference rules to axioms, in practice it is easier to build them in a backward fashion from the goal. Instead of expanding goals with the standard subgoal package, we use an enhanced proof manager that supports an explicit proof structure [Ka[l]91]. Functions exist to apply a tactic to any node in a proof, and to examine the proof developed so far, with intermediate goals and theorems being displayed.

We have added to proof trees an explicit node for annotations, as shown in Figure 4. The proof annotation field can contain annotations created in different forms: some derived from special tactics, others from processing the initial annotations in goals, and others still by the user developing the proof. In Figure 5 we see an example proof that contains annotations. The proof is developed with the use of tactics, as usual. However, the user has also entered comments directly, while performing the proof. When the final proof is displayed, and stored, the comments are part of it.

```
val PARITY_IMP_DEF =
 |-!inp out.
    PARITY_IMP (inp,out)  =
    (?l1 l2 l3 [* out at all but time 0 *]
      l4 l5 [* indicates time 0 *].
     NOT (l2,l1)  /\
     MUX (inp,l1,l2,l3)  /\
     REG (out,l2) [* for next iteration *] /\
     ONE l4  /\
     REG (l4,l5) [* l5(0) = F, else T *] /\
     MUX (l5,l3,l4,out) [* out(0) = T, else l3 *])
     [* definition of parity checker *]
  :thm_plus
```

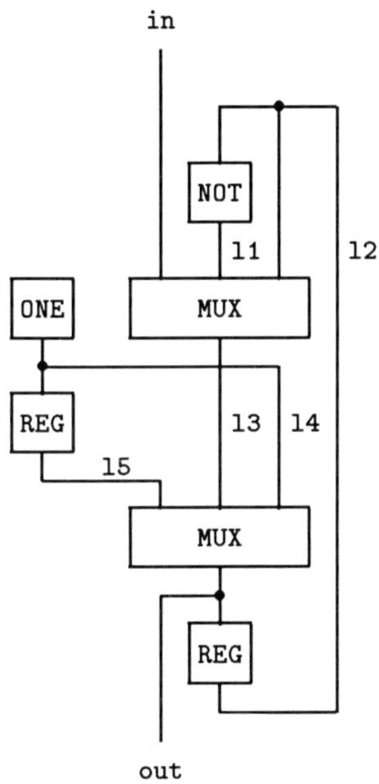

Figure 3: Annotated specification of parity checker

```
datatype prooftree =
  expanded   of ((term_plus list * term_plus)
                 * proof * tactic_plus * note * (prooftree list))
| proved     of (thm_plus *  (term_plus list * term_plus) * tactic_plus
                * proof * tactic_plus * note * (prooftree list))
| unexpanded of (term_plus list * term_plus) ;
```

Figure 4: Proof data structure with annotations

```
- g (--`!a b c. ((c DIVIDES a) /\ (c DIVIDES b)) ==>
        (!m n. c DIVIDES (m*a + n*b)) `--);

...
GOAL PROVED

[]|- !a b c.
       (c DIVIDES a) /\ (c DIVIDES b) ==>
       (!m n. c DIVIDES m * a + n * b)

val it = () :unit


- current_proof();

[]|- !a b c.
       (c DIVIDES a) /\ (c DIVIDES b) ==>
       (!m n. c DIVIDES m * a + n * b)
[* generalizing vars, discharging asms *]

  [c DIVIDES a, c DIVIDES b]|- c DIVIDES m * a + n * b
  [* resolving with: []|- !a b d. b DIVIDES a ==> b DIVIDES d * a *]

    [!d. c DIVIDES d * a,
      !d. c DIVIDES d * b]|- c DIVIDES m * a + n * b
    [* spec'ing asm to: c DIVIDES n * b *]

      [c DIVIDES n * b, !d. c DIVIDES d * a]|- c DIVIDES m * a + n * b
      [* spec'ing asm to: c DIVIDES m * a *]

        [c DIVIDES n * b, c DIVIDES m * a]|- c DIVIDES m * a + n * b
        [* resolving with: []|- (c DIVIDES m * a) /\ (c DIVIDES n * b) ==>
           c DIVIDES m * a + n * b *]

val it = () :unit
-
```

Figure 5: Example proof with annotations

4.4 Semantic hooks

While the HOL system provides some degree of automation by allowing the user to write complex tactics which can examine the current goal, it is still up to the user to build and invoke such tactics when appropriate. The proof manager is inherently passive, it simply presents the current subgoal to the user.

It is our contention that domain-dependent, meta-level hints of a more procedural nature can be effectively incorporated into the proof process by allowing the processing of a goal *before* it is even presented to the user. Thus, we propose a pre-processing phase at which user-defined functions can be automatically integrated into the proof development process.

For this purpose, proof management functions have been re-coded to always invoke a certain function (called a 'hook') whenever they produce a goal. The specific proof functions that have been enhanced in this fashion are the functions for setting goals, expanding goals with tactics, and backing up over previous applications of tactics

While the use of these 'hooks' can help the proof process in the general case, they are even more helpful when associated with the use of annotations. Particular information can be linked to the terms in the goal and the body of the proof being developed, and processed by the hook function at appropriate stages of the proof. The hook function can perform many different tasks: providing hints on how to procede with the proof, printing other relevant information, and actually automating parts of proof. Once the infra-structure to use this hook is added to the system, the particular function can be added without there being a need to reformulate the general proof management functions.

4.5 Instantiation of annotations

The incorporation of the extra information into the term and proof data structures is independent of the particular type of annotations or their use. In fact, the complete proof system can be instantiated differently for distinct applications, without changing the code for the underlying logic basis. This is done by exploiting the existence of *structures* in Standard ML. All relevant modules of the prover are parametrized by annotation structures. A particular instantiation of the proof system for a particular application is generated by building a structure with the *signature* for structures illustrated in Figure 6. Any structure, to match this signature, must provide objects of the types given. The structure must contain:

- a type, with the only restriction being that it must support an equality check (essentially, it cannot have as component a function type);

```
signature justif_sig =
sig
        eqtype note
        val empty_note : note
        val make_note : string list -> note
        val pp_note : note -> unit
        val join_notes : note -> note -> note
        val show_note : note -> unit
end;

signature hook_sig =
sig
        structure trm_inst : t_plus_sig
        val hook_in : (trm_inst.term_plus list * trm_inst.term_plus) ->
                          trm_inst.note -> string -> unit
end;
```

Figure 6: Signature for annotation instantiation

- an element of this type, defined as being an "empty" annotation;
- definitions for functions to create, display, and merge annotation.

An example instantiation is shown in Figure 7. With the given definition of the module `just_of_str`, annotations are simply strings that are included as documentation (for both terms and proofs). The operations to make an annotation out of it simply stores the string verbatim; the only way an annotation is processed is by having it displayed.

Once an instantiation structure is entered, it is simple to build the full proof system around it. Instantiation of modules in Standard ML is simple, and takes the form of a function call, as shown in Figure 8. The `just_of_str` module is used to instantiate the `term_plus` functor, this in turn is used to instantiate other functors, and so on. Finally all these functors are 'opened', so that the (instantiated) functions are made available to the end user.

5 Uses of annotations

In the previous section we have presented the infra-structure for adding extra-formal information to the HOL proof system. However, the form of this information, and the ways it is used, is flexible. In fact, our goal is to allow users to handcraft their own proof support systems, without having to concern themselves with the logic itself. The methodology to use annotations is demonstrated here with applications providing different

```
structure just_of_str =
struct
          datatype note =
                         mk_note of string;
          val empty_note =
                         mk_note "";
          fun make_note x =
                         mk_note (hd x);
          fun join_notes (mk_note c1) (mk_note c2) =
                         mk_note (c1^" "^c2);
          fun pp_note (mk_note note) =
                         if (note = "")
                          then ()
                          else (begin_block consistent 0;
                                add_break (1,0);
                                add_string ("[* "^note^" *]");
                                end_block ());
          fun show_note x =
                         print ((pp pp_note x) ^"\n\n");
end; (*struct *)

structure hook_of_str =
struct
       fun hook_in curr_node glob_note curr_foc =
                          show_note (note_part curr_node);

end; (*struct *)
```

Figure 7: An example instantiation

```
structure term_inst = term_plus (just_of_str);

structure subst_inst = subst_plus (structure TermP = term_inst);

structure thm_inst = thm_p (structure TermP = term_inst
                            structure substP = subst_inst);
...

structure proof_inst = proof_plus (structure tp_inst = term_inst
                                   structure hk_inst = hook_of_str);

open term_inst subst_inst thm_inst proof_inst;
```

Figure 8: Instantiation of proof environment

levels of use.

5.1 Annotations as comments

One simple instantiation is of annotations as supported comments. Generally, whenever a file with comments is loaded into a system for mechanical manipulation, any comments it has are not transferred. One must peruse the original file to examine the commentary. Documentation can be part of a separate medium/object, with many drawbacks. It is very helpful to be able to point directly at parts of the specification. It also makes it easier to have changes in the object being reflected by changes in the documentation.

With our implementation, this use of annotations as text is easy to incorporate into HOL. The annotation structure is illustrated in Figure 7. In this instantiation, annotations are simple strings.

These comments are supported in the sense that they are an integral part of HOL terms; thus whenever a term or a theorem is displayed the annotations are too. They also are carried over with rewriting, so that annotations are inherited by a goal or theorem when it is rewritten with an annotated theorem. During the proof process, annotations from the current goal can be set up to migrate automatically into the node in the proof tree in which it is manipulated, providing automatic explanation of the proof based on design justifications and proof hints from the problem definition. Users can also add comments to the proof structure itself.

5.2 Annotations as labels

One difficulty in running HOL proofs is in accessing subterms and subsets of the assumption lists. This has been considered previously an interface issue, with attempts at using bit-mapped displays for accessing the parts. However, such pointing methods result in structural paths through syntax trees, similar to the ubiquitous sequences of `car`'s and `cdr`'s of Lisp. They do not explain *why* that particular subterm or assumption is relevant.

A more reasonable approach is to attach labels to the subcomponents. These can then be used to document specifications and proofs as well as access subterms and subsets of the assumption list. Specialized functions have been designed, which are variations of existing ones:

`SUBTERM_REWRITE_TAC`: takes as arguments an annotation and a list of theorems, and rewrites all the subterms of the conclusion part of the goal that are annotated by the first argument with the given list of theorems.

SOME_ASM_REWRITE_TAC: takes as arguments an annotation and a list of theorems, and rewrites the conclusion part of the goal with the assumptions that are annotated by the first argument and the given list of theorems.

LABEL_ASM: takes as argument an annotation, and returns the first assumption that is annotated by the argument; this result can be given as an argument to, say, **ASSUME**.

These functions are possible because the type of annotation used must support the equality check. The form of the resulting tactic becomes more explanatory than the structural paths, though they point to the same entities.

5.3 Annotations as proof hints

Some facets of proof can be automated by using proof heuristics such as those found in Boyer and Moore's system and proof functions such as resolution and rewriting. Others, however, have eluded automation. These are the aspects of proofs that we believe can be aided by the use of annotations. This section describes work towards mechanically processing annotations to automatically conduct many parts of proofs.

The following list shows some of the decisions in directing a proof that can be assisted by information gleaned from the design process:

- Cases to split goals: many times a proof needs to test a combination of variables and values.
- Instances: specifications often involve quantified variables; proofs usually involve instantiating these variables.
- Useful properties: some aspects of the domain could provide insight into conjectures to be proved, and could suggest certain proof steps.
- Proof outlines: very often a designer has an intuitive feel for why a design is correct, which can be translated directly into proof steps in a particular logic.
- Analogies: if a design is created in analogy with another one, the proofs may also be similar, at least in a high-level.
- Pruning assumptions: very often it helps to know not only which facts are relevant but also which are not, to reduce complexity of data and avoid unsuccessful paths in the search for proofs.
- Generalizations: general version of conjectures and lemmas that may be easier or more advantageous to prove.
- Partitions: a problem may have several aspects which it may be useful to isolate and prove separately.
- Induction schemes: a problem may be best explained (and proved) inductively, using the indicated inductive expression.

The notion that much of extra-formal information can fit into the above classification suggests that it can be captured with the aid of specification tools, and that their use can be automated. In fact, annotations can actually aid the process of formal specification in certain contexts, by providing templates for specifications that fit the particular type of extra-formal information.

An initial attempt at animating the use of annotations is illustrated in Figure 9. The structuring of annotations into these classes makes it possible to use them in a mechanized way. Another advantage of having a detailed taxonomy of annotations is that having these classes known to specifiers of problems makes it possible for them to clarify their informal justification of the definitions.

Parts of code making up the 'hook' are also shown. One may observe that there are different actions performed by annotations of different types: the `case_analysis` annotation generates a tactic which is executed, the `induct` generates a tactic but does not execute it, only suggests it, and the `analogy` annotation updates the value of a global variable.

Examples of use illustrate the actions. Suppose a specification is reasoned out to be an inductive generalization of a particular variable. When this information is expressed in annotations, it is available to whoever is manipulating it through a prover, as illustrated in Figure 10. Note also that the proof manager was modified to cope with the existence of annotations: in this case, the existence of the `induct` annotation puts a check on the action of other tactics, such as `GEN_TAC`. This is a useful method of dealing with large composed tactics, which may work in general cases but might not be appropriate in each particular case. While stripping off a quantified variable may be the default action, in some cases it is more efficient to make sure the user is aware of the hint.

Similarly, if a design is justified to work when the value of a variable Y is odd and when it is even, then the system performs a case analysis and creates the two corresponding subgoals automatically. Figure 11 shows a proof script where annotations are used to automatically perform this proof step. The goal gives rise to two subgoals based on the annotation that a case analysis on `"b = 0"` is suggested. Observe that the user conducting the proof did not have to supply the tactic for case analysis.

The taxonomy of annotations implemented is not complete, and processing of annotations is not yet fine-tuned. Yet, this instantiation demonstrates the feasibilty of constructing a more complete, rounded off system for automating proof development.

6 Conclusions and further work

Currently, we are applying our annotation-extended version of HOL to numerous examples, to demonstrate the usefulness of annotations for proof development as well as proof

```
     datatype note =  conjecture of string
                    | input_constraint of string
                    | behavioral_factor of string
                    | case_analysis of string
                    | label of string
                    | analogy of string
                    | induct
                    | suggestion of string ;

fun process_each_note (case_analysis x) =
  (print ("case analysis on "^x^"\n";
   eval ("e (ASM_CASES_TAC (--`" ^x^ "`--)THEN REDUCE_NOTE_TAC);"))
   ...
  |process_each_note (analogy x) =
   (global_note := [analogy x] )
  |process_each_note (induct) =
   (print "suggested: INDUCT_TAC\n\n" )
  |process_each_note y = () ;

 ...
```

Figure 9: Animated annotations

```
 new goal:
(--`!p [* induct *] q. (m + n) + p + q = (n + m) + q + p`--)

suggested: INDUCT_TAC

val it = () :unit
- e GEN_TAC;
try induction instead?


 new goal:
(--`!p q. (m + n) + p + q = (n + m) + q + p`--)

val it = () :unit
-
```

Figure 10: Annotations providing suggestions

```
(--'(?c. b = c * a) /\
    (?c. a = c * b) ==>
    (a = b) [* cases: b = 0 *]'--)

current note:
val it = () :unit
- e DISCH_TAC;

 new goal:

(--'a = b [* cases: b = 0 *]'--)

    [(--'?c. b = c * a'--)]
    [(--'?c. a = c * b'--)]

case analysis on b = 0

 new goals:

(--'a = b'--)

    [(--'~ (b = 0) '--)]
    [(--'?c. b = c * a'--)]
    [(--'?c. a = c * b'--)]

(--'a = b'--)

    [(--'b = 0'--)]
    [(--'?c. b = c * a'--)]
    [(--'?c. a = c * b'--)]


val it = () :unit
-
```

Figure 11: Partially automated proof

explanation. Previous verification efforts demonstrate that informal reasoning permeates the whole process from initial design through verification to final publication. Other work, such as the inclusion of hints into Boyer and Moore's system, confirm this conjecture. Thus, the explicit inclusion of annotations into the verification process is clearly justified.

Our work demonstrates that the HOL system provides the right style for the incorporation of annotations. Partial automation of proofs can be well integrated into the tactic-based proof development methodology. Given that the user controls the proof process, it is important to provide easy access to informal knowledge that can provide assistance in proof development.

The use of annotations has been proposed and implemented before. However, these previous efforts do not support the flexibility we see as needed. Our system on the other hand exploits the strong type system of HOL: users can investigate many different classifications of annotations and animate them in different ways, without them ever result in compromising the logic.

An important question is how the design annotations are obtained, and why one should spend time and effort in entering them. In any development project, good explanations of design decisions are expected, and often mandated, in documentation. Furthermore, even when a design is being generated for mechanical processing (verification, compilation, or fabrication, say) it must still be possible for an evaluator to peruse it and understand it. We believe it is much easier to obtain justifications for a design when a design is actually being created. Annotations allow such information to be captured at that time, more efficiently and with less effort. It is no more difficult to enter annotations than to write good comments.

Annotations can also be obtained indirectly. Several studies are being carried out to import proof obligations from other tools, such as CAD and CASE systems, specification languages such as Z, and the parsing of algebras. Translating comments in these inputs into annotations usable in the proof process is a possible outgrowth of our initial effort.

The other thrust of our research is in obtaining meaningful automation of proofs by processing annotations. We have studied initially a fairly restricted domain: hardware circuits that execute a function through iteration, similar to, but more complex than, the canonical parity checker and multiplier examples [Gor88]. We have specified and proved several example circuits, and found that good proofs for many of them share a dependence on certain details of the design: what is the interpretation of a value on a bus (character, integer, etc.), which parts of the circuit are active for the base case and which are active in the inductive case, etc. We are developing an extension to the HOL prover that processes these kinds of annotations.

Proofs in a mechanized system have two desirable properties: rigor and understandability. Formal proofs with a rigorous logic provide the first property; we expect the use of assertions will help support informal proofs geared at understandability.

Acknowledgements

This work was sponsored under Boeing Contract NAS1-18586, Task Assignment No. 3, with NASA-Langley Research Center, and by NSA under agreement DOD–MDA 904–91–C–7053.

References

[BM88] Robert S. Boyer and J. Strother Moore. *A Computational Logic Handbook.* Academic Press, 1988.

[Bun83] Alan Bundy. *The Computer Modelling of Mathematical Reasoning.* Academic Press, 1983.

[CSD91] University of Cambridge, SRI International, and DSTO Australia. *The HOL System*, 1991.

[Der83] Nachum Dershowitz. *The Evolution of Programs.* Birkhauser, 1983.

[Gor88] Michael Gordon. HOL: a proof generating system for higher-order logic. In *VLSI Specification, Verification, and Synthesis.* Kluwer Academic Press, 1988.

[Kal91] Sara Kalvala. Developing an interface for HOL. In *1991 International Workshop on the HOL theorem prover and its applications.* ACM/IEEE, 1991.

[Pau91] Lawrence Paulson. *ML for the Working Programmer.* Cambridge University Press, 1991.

[Rob91] J. A. Robinson. Formal and informal proofs. In Robert S. Boyer, editor, *Automated Reasoning: Essays in Honor of Woody Bledsoe.* Kluwer Academic, 1991.

[Sli91] Konrad Slind. An implementation of higher order logic. Technical Report 91/419/03, Department of Computer Science, The University of Calgary, 1991.

[Weg76] Ben Wegbreit. Constructive methods in program verification. Technical Report CSL-76-2, XEROX Palo Alto Research Center, 1976.

Higher Order Logic Theorem Proving and its Applications (A-20)
L.J.M. Claesen and M.J.C. Gordon (Editors)
Elsevier Science Publishers B.V. (North-Holland)

Towards a Formal Verification of a Floating Point Coprocessor and its Composition with a Central Processing Unit [1]

Jing Pan Karl N. Levitt Myla Archer Sara Kalvala

Department of Computer Science, University of California, Davis
Davis CA 95616, USA

Abstract

In previous work (using HOL and other theorem provers) on the verification of microprocessors, the design is typically represented as a single level (e.g., an electronic block model) or as a linear hierarchy of interpreters (Joyce and Windley). There has been no attempt to verify designs that are in reality a central processor composed with various coprocessors, the typical organization of modern microprocessors. Our work is a step towards the verification (ultimately down to the microcode level) of a microprocessor that consists of a central processing unit that is the master of a floating point coprocessor; the design is drawn from the MC68881 floating point coprocessor slaved to the MC68000, but greatly simplified. The coprocessor in isolation will be verified with respect to a specification that captures the IEEE floating point standard. In our system, CPU and floating point instructions are allowed to execute concurrently, but the appearance to the programmer of the composed system is that of a sequentially executing instruction stream. The CPU and floating point coprocessor communicate through the four-phase handshaking protocol. The verification involves reasoning about a form of behavioral abstraction wherein concurrently executing instruction steams are mapped to a sequential stream.

[1]Work sponsored under Boeing Contract NAS1-18586, Task Assignment No. 3, with NASA-Langley Research Center and NSA Agreement DOD–MDA 904–91–C–7053

1 Introduction

Several microprocessors have been verified using HOL or other theorem provers. What is characteristic of these microprocessors is that they do *not* provide much of the functionality that is common to most microprocessors of commercial interest, e.g., an interrupt controller, a DMA controller, a memory management unit, a floating point coprocessor, or cache memories. As a step towards more interesting systems, several of the microprocessor verification efforts model main memory as a unit separate from the microprocessor, with the four-phase handshaking protocol used to achieve the asynchronous communication [Joy88b]. Through this protocol, systems can be verified where main memory is slaved to a microprocessor and where there is no common clock.

In this paper, we describe work towards the verification of a floating point coprocessor (FPC) and of its interaction with a central processing unit. The designs of the FPC and of the CPU are drawn from, but clearly much simpler than, the MC68881 and the MC68000 coprocessors. The arithmetic functions of our FPC are intended to behave according to the IEEE Floating Point Standard (ANSI-IEEE Std. 754-1085) [IEE85]. Hence, part of our work consisted of writing this standard as a HOL specification.

Previous efforts on the verification of microprocessors used methodologies that model a design as a vertical hierarchy of units. The works of Joyce [Joy88a] and of Windley [Win90] model a microprocessor as a hierarchy of interpreters. The lowest level interpreter is the design to be verified, e.g., the electronic block model. The topmost interpreter in the hierarchy provides the functionality of the microprocessor's instruction sets, and hides all detail not essential to specifying the behavior of instructions. These methodologies provide the basic mechanisms for verifying a "sequence" of interpreters by considering the elements of the sequence in pairs. Generally, an implementing interpreter has a larger state space (reflecting details present in the implementation but hidden in the higher level interpreter), and an instruction execution in the higher-level interpreter is realized as a sequence of instruction executions at the implementation. Thus the pair of interpreters are related by *temporal*, *data*, and *procedural* abstractions [Mel88].

In coping with the interaction between the CPU and FPU, we have developed a start towards a methodology for verifying the horizontal integration of devices—essentially, a technique for composing verified devices that communicate according to asynchronous protocols. In our system, the purpose of the FPC, naturally, is to provide faster arithmetic service. Furthermore, the FPC and CPU can concurrently execute instructions, subject to a requirement that they do not interfere with each other; the CPU assigns floating point instructions to the FPC while assuming responsibility for all other instructions. However, the user's view of the system is that of a single unit that processes instructions sequentially. Hence, as part of our proof, we used a form of behavioral abstraction (by which we model our notion of "implementation") to verify that an interpreter that processes a single stream of instructions is implemented by a pair of interpreters that execute concurrently and communicate according to a particular protocol.

The organization of this paper is as follows. Section 2 summarizes a methodology for the verification of the horizontal composition of verified interpreters that can be viewed as having the behavior of a single interpreter. Section 3 gives an overview of the FPC being verified. The final three sections highlight the proof completed to date, providing a sketch (where appropriate) of the verification of a particular instruction as an example: FLD, which loads a floating point number from main memory into the FPC's accumulator. We do not present the description of our HOL specification of the IEEE FP Standard; see [Pan92] for details.

2 Verifying the composition of verified systems

This section discusses our general approach towards verifying cooperating computer units. It includes an extended interpreter model and abstraction techniques designed to describe and reason about composed systems. In the following sections, the specification and verification of an abstract composed system will be described in further detail.

2.1 The Interpreter model

We use the general interpreter model [Joy88a, Win90] for describing the processes at various abstraction levels. The interpreter model consists of four parts:

- A representation of the state s;
- A representation of the external environment inputs e;
- A set of state transition functions J. Each state transition function $J_k \in J$ relates the state at time $t+1$ to the state and the environment at time t;
- A next state function which select a proper state transition function from J based on the current state and environment.

Thus the interpreter is formalized as the following where $(s\ t)$ and $(e\ t)$ represent the state and the environment at time t respectively, and F is the next state function J_k corresponding to the current stack and environment:

$$I[s,e]: \quad s(t+1) = F(s\ t)(e\ t)$$

In the particular model used here, the environment is used only for input; output to the environment is modeled as part of the state. Each interpreter in the interpreter hierarchy has its own state, environment, and time scale.

Correctness is a relation between a specification and an implementing interpreter where f is the temporal abstraction function that will be discussed later.

$$I_{Imp}[s_{Imp}, e_{Imp}] \Rightarrow I_{Spec}[s_{Spec}\ o\ f, e_{Spec}\ o\ f]$$

2.2 An extended hierarchy of Interpreters

Hierarchical approaches (such as described above) are intended for circumstances where each interpreter in the hierarchy is self-contained. In reality, a computer system often consists of several units which communicate and synchronize with each other. One such example is a system which is composed of a CPU and a FPC—the subject of this work.

Generalizing the single-threaded interpreter hierarchy leads to a tree structure, or even a graph. At the root of the tree we have the top level specification for the whole system. The leaves of the tree correspond to the structural specifications of the physical units.

This model extends the one developed by Windley for AVM-1 [Win90]; our approach allows an interpreter to be implemented by a combination of several lower-level interpreters. Therefore, besides the vertical interaction between the implementing and the implemented interpreter, there is horizontal interaction between the implementing interpreters. The verification that the composition of several interpreters implements a given high-level interpreter is the objective of our work.

In this generalized hierarchy of interpreters, each interpreter can be implemented by an arbitrary but finite number of lower level interpreters. In order to verify the correctness of any interpreter, one needs to verify two properties:

- Each implementing units works properly
- All the implementing units communicate and synchronize among themselves correctly.

At each level, not only are the detailed implementation issues of the next lower level concealed, but so is some or even all of the communication among the lower units. In general, these implementing units may or may not share state. Often there will be a certain degree of concurrency among them.

2.3 Temporal Abstraction for Asynchronous Systems

Formal specifications that describe the behaviors of two interpreters using different notions of discrete time are related by temporal abstraction. This type of abstraction is used when the implementation gives more detail about how a device behaves over time than its abstract specification does. With the mechanism of temporal abstraction, information about the implementation at instances of time that are not of interest can be hidden at the specification (i.e., upper) level.

For synchronous systems, the number of implementation cycles for each instruction on the higher level is a constant for each instruction. However, sometime such deterministic synchrony is not possible. More precisely, in our case of a floating-point coprocessor, the occurrence of an event (such as the completion of an floating-point instruction) not only depends on the type of the instruction, but also on the interaction between the CPU and the FP coprocessor.

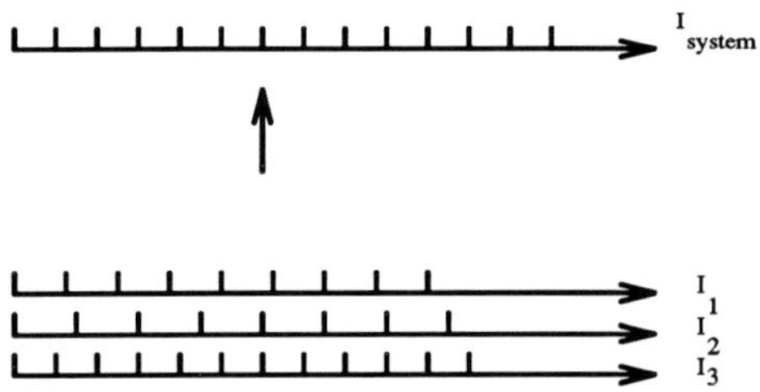

Figure 1: Temporal abstraction for composed systems

For the temporal abstraction model of asynchronous systems, the common approach is to define a predicate g which is true whenever there is a valid abstraction between the lower level and the upper level [Gra92]. Then a generic temporal abstraction function `Temp_Abs` can be defined in terms of g:

```
(Temp_Abs g 0 = @ t:time . First g t) /\
(Temp_Abs g (SUC n) = @t:time . Next g (Temp_Abs (g n) t))
```

where `@` is the Hilbert's choice operator.

The expression `(First g t)` expresses the fact that `t` is the first time that `g` is true, and `Next g (Temp_Abs g n, t)` represents the next time after `(Temp_Abs g n)` that `g` is true.

2.4 Temporal Abstraction for Composed Asynchronous Systems

In a composed system, one often needs to map several independent time scales which belong to the lower (i.e. implementing) level units to the upper level units. Since the lower level units communicate with each other asynchronously, there can be no one common clock shared among them.

Suppose an interpreter I is implemented by a group of interpreters $I_1, I_2, ..., I_n$. Further, assume each of the implementing interpreters I_i has its own time scale. In order to compose all n implementing interpreters together, interpreters $I_1, I_2, ..., I_n$ will all have to be expressed with respect to a *common* time scale. Usually the fastest clock is chosen as the common clock, so the interpreters on other clocks may take more than one common-clock cycle to accomplish what used to take them just one cycle. Assume the clock rate of interpreter I_k is chosen as the common clock rate. Then, the behavior of the remaining interpreters in $I_1, I_2, ..., I_n$ have to be *redescribed* in term of the clock rate of I_k—see Figure 1.

One way to do this redescription is to use *existential quantification.* The specification for interpreter I_i with respect to the common clock will state that there exists some future

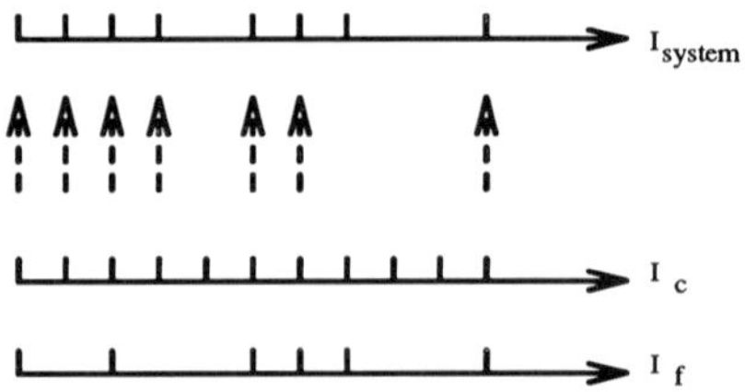

Figure 2: Temporal abstraction for concurrency representation

time t' such that the interpreter finishes executing. If the clock of I_i is of the same speed as the clock of I_k, then t' will be equal to $t+1$. However, if it is slower, t' will exceed $t+1$. We call this type of definition I_i's *behavior from* I_k's *point of view.* Of course, time t' cannot be just any time in the future. Because of the nature of asynchronous communication, t' often is the time at which some signal or signals are raised or lowered.

2.5 Temporal Abstraction to Represent Concurrency

If a system is composed of two concurrently operating processors, its specification has to reflect this property. The so-called concurrent level representation does not have to explicitly describe the existence of two processors. However the concurrency between the two kinds of instruction executed by the two processors has to be specified.

Suppose processor I_c implements instructions of type C while processor I_f implements instructions of type F. As discussed in the previous section, we assume I_f is specified using I_c's clock rate so that the two can be composed together. Concurrency means that C instructions are executed in parallel with F instructions. Naturally, the specification of the system should only be concerned with the start and finish of each instruction, rather than the synchronization detail between the processes. Figure 2 represents one possible mapping between the system and its implementing processors.

Such a representation for the concurrency presents a complication to temporal abstraction. In traditional temporal abstraction schemes, a function that takes several time cycles on the lower level to complete corresponds to the function on the upper level which takes *one* cycle to finish. However, in describing a concurrent system, the corresponding upper level function may also take *several* cycles to complete as well, only the number of cycles must be decreased.

2.6 Proof Structure for our (CPU-FPC) System

The specification and verification of a computer system consisting of a CPU and a FPC will be discussed in detail in the following sections. First, the architecture and design of the coprocessor will be presented.

Our study concentrates on the hiererarchy depicted in Figure 3, which is sufficient

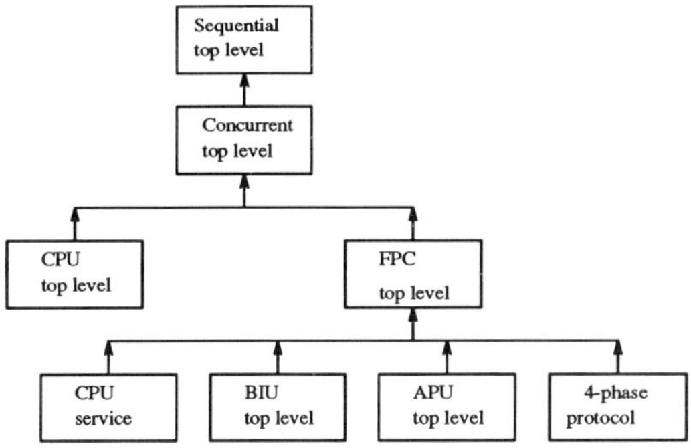

Figure 3: Hierarchy of interpreters for CPU-FPC system

to formalize our technique for modeling communication and concurrency. Other details, such as verification of the CPU (micro-level with respect to the top-level) are omitted. Our verification effort of the system consists of three parts:

- The correctness of the floating-point coprocessor top-level implementation. This level reflects the successful completion of a floating-point instruction. The specification of the coprocessor very much resembles the correctness property of a stand-alone CPU, except it does not increment the program counter, which is part of the CPU. The FPC top level interpreter will be verified to be implemented by three communicating lower-level interpreters, `CPU_Service`, `BIU_Top`, `APU_Top` and the four-phase handshaking protocol. Each interpreter is described in its own time scale, then re-described using the CPU's time scale so that the units may be composed together.

- The correctness of the concurrent top-level implementation. On this level, instructions are categorized into two classes: the CPU instructions and the FPC instructions. Concurrent execution of these two kinds of instructions is specified at this level. The concurrent level is implemented by the CPU and the FPC interpreters jointly. The two processors synchronize with each other by using a busy waiting line.

- The correctness of the sequential top-level implementation. The sequential top level represents the system as seen by an assembly language programmer. All instructions appear to be executed sequentially. Each instruction is completed in one time cycle. There is no knowledge of the existence of two separate processors.

3 The floating-point coprocessor architecture

We now present the design of a floating-point coprocessor (FPC) based on the MC68881 [MC689] to demonstrate our methodology in verifying composed systems.

3.1 The Coprocessor Concept

Our coprocessor is designed to interface with the CPU to provide a logical extension to the CPU's integer data processing capabilities. The coprocessor provides a high-performance floating-point arithmetic unit and a set of floating-point data registers.

The assembly language programmer can view the FPC registers as though the registers were resident in the CPU. Thus, the CPU-and-FPC pair appears from the outside to be one processor that supports both floating-point and integer operations. Moreover, the coprocessor interface allows the FP instructions to execute concurrently with the CPU instructions to improve efficiency. The main features of the coprocessor are:

- The main processor CPU and the coprocessor FPC communicate via standard bus cycles.
- The CPU and the FPC can operate at different clock speeds.
- All effective address calculation and data transfers are performed by the CPU.
- Concurrent (overlapped) instruction execution enhances throughput while maintaining the programmer's model of sequential instruction execution.
- Coprocessor detection of an exception results in a trap that is serviced by the main processor at the request of the FPC.

3.2 The Hardware Overview

The FPC programming model consists of the following:

- Eight general purpose floating-point data registers.
- A control register that contains enable bits for each class of exception trap, and mode bits to set the user-selectable rounding and precision modes.
- A status register that contains floating-point condition codes, quotient bits, and exception status information.
- A instruction address register that contains the main memory address of the floating-point instruction to be executed.

As shown in Figure 4, the coprocessor is internally divided into two processing elements: the bus interface unit (BIU) and the arithmetic processing unit (APU). Though the BIU monitors the state of the APU closely, it operates independently of the APU. The APU operates on the opcode and operands that the BIU passes to it. In return, the APU reports its internal state to the BIU.

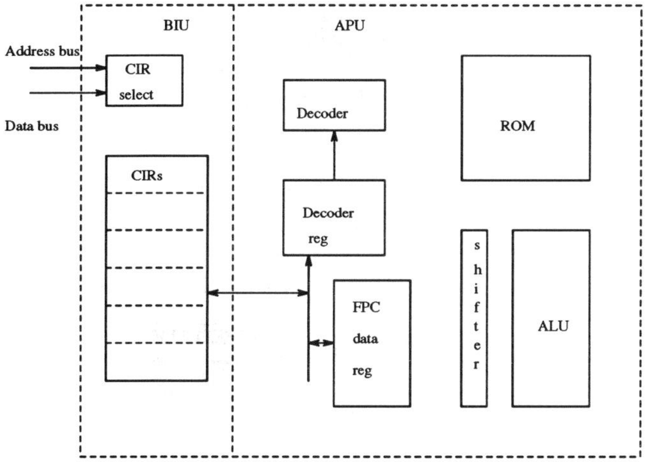

Figure 4: Simplified block design of the FPC

3.2.1 The Bus Interface Unit

The BIU contains the coprocessor interface registers (CIRs) and status flags. These CIRs are addressed by the CPU in the same manner as memory. The coprocessor interface is implemented as a protocol of bus cycles in which the CPU reads and writes to these interface registers. Moreover, the CIR register select and acknowledgment timing control are also contained in the BIU. Since the bus is asynchronous, the FPC need not run at the same clock speed as the CPU.

3.2.2 The Arithmetic Processing Unit

The APU executes all the floating-point instructions. The floating-point data, control, status registers are located inside the APU. In addition to these registers, the APU contains an arithmetic unit used for both mantissa and exponent calculations, and a barrel shifter. The control section of the APU is microcoded.

3.3 The Coprocessor Interface Register (CIR)

There are four CIRs that reside inside the BIU:

- Response CIR. This read-only register is used to communicate service requests from the FPC to the CPU.
- Command CIR. This write-only register is used by the CPU to initiate a dialog for a coprocessor instruction. When the FPC detects a write to this CIR, the data value is latched from the data bus.

- Condition CIR. This write-only register is used by the CPU to initiate the dialog for a conditional type coprocessor instruction.
- Operand CIR. This read/write register is used by the CPU to transfer data to and from the FPC.

Currently we do not have exception handling capabilities built into the coprocessor. We plan to add this feature in the near future.

3.4 The Communication Protocol

As mentioned in the previous section, the FPC contains a number of CIRs which are addressed by the CPU in the same manner as memory. When the CPU fetches a floating-point instruction that is to be executed by the coprocessor, the CPU writes the instruction to the command CIR and reads the response CIR. In this response, the BIU encodes requests for any additional service required by the CPU.

Data read from the FPC-response CIR is referred to as a *primitive*. These primitives and their intended semantics are:

- The FPC is busy. The CPU then checks for interrupts, processes them, and queries the FPC again.
- There is a FPC service request. For example, this request might be to evaluate the effective address and transfer data between the CPU data registers/memory and the FPC.
- The CPU is not needed. Communication is terminated and the CPU is free to execute the next instruction.

3.5 The Instruction Set

The FPC instruction set can be separated into three groups:

1. Data movement instructions. The FLD and FSTR instructions are used to transfer data between the floating-point data registers and the main memory or the CPU data registers. These are the only types of instructions that reference the main memory.
2. Arithmetic operation instructions. One or two operands are specified. The result is always stored into one of the internal FPC data registers. When the FPC is processing one of these instructions, the CPU is free to execute CPU-only instructions.
3. System control instructions.

3.6 Concurrency and Synchronization

Since one FPC instruction usually takes much longer to complete than a CPU instruction, it is desirable for both the CPU and the FPC to be executing instructions simultaneously. When the CPU encounters an FPC instruction (in group 2 above), it (1) waits until the FPC becomes idle, (2) allows the FPC to begin the instruction, and then (3) immediately moves to the next instruction.

The concurrency between the CPU and the FPC must be implemented to maintain a programming model based on sequential instruction execution. There are two possible ways the CPU and the FPC may interfere with one another. They must be disallowed:

- the CPU cannot start another FPC instruction if the FPC is still busy executing the previous instruction.
- the CPU cannot reference a memory location that is being referenced by the previous (but still executing) FPC instruction, e.g. FLD or FSTR instructions.

The first coordination problem is solved by CPU's busy wait feature; while the second one is solved by the assumption underlying the FPC instruction set. There is no arithmetic instruction which can store its result to the memory or the CPU registers. The only way to move data from the FPC to the outside world is through the FSTR instruction, which finishes execution as soon as the communication between the CPU and the FPC is finished.

4 Verifying the FPC top-level from three communicating units

We now discuss the specification and verification of the floating-point coprocessor. For complete details, the reader is referred to [Pan92]. As discussed in the previous section, the coprocessor has two parts: the BIU, which communicates with the CPU, and the APU, which performs the arithmetic duties. Since the coprocessor relies on the CPU to provide instructions and operands, its correct operation *cannot* be proved without the CPU specification. Further, since the communication between the CPU and the coprocessor is carried out asynchronously using the 4-phase handshaking protocol, the correctness of the 4-phase protocol also has to be included when verifying the FPC interpreter. Therefore, in this section we discuss the verification of the FPC interpreter with regard to the three implementing interpreters (the CPU, BIU and APU) and the 4-phase protocol.

4.1 The 4-phase Handshaking Protocol

The FPC executes two different bus cycles, according to the direction of the transfer. They are the asynchronous read and asynchronous write bus cycles. The 4-phase handshaking protocol is used to implement these two types of the bus cycles. For the read cycle, the FPC detects the start of an asynchronous read cycle when the read request line is asserted. The FPC retrieves the data from the register whose address is given, places the data value on the bus and asserts the acknowledgment line (dtack). The acknowledgment remains asserted until the read request line is lowered by the CPU to request the end of read cycle. The BIU consequently lowers the acknowledgment line.

However, we would like to define a more abstract view of message passing between the CPU and the BIU where the handshaking detail is hidden. This more abstract view of communication can be used in the next higher level, that is, in reasoning about the interaction between the CPU and BIU during the execution of an FPC instruction.

```
cir_read_write rep address read write datain dataout
               command response operand_in operand_out condition
               control new_instr operand_ready =

 ! t.(read t) ==>     %read cycle%
    (address t = 1) ==> ((datain(t + 1) = (numtow rep (response t))) /\
                        ~(new_instr t) /\
                        ~(operand_ready t)) |
                        ((datain(t + 1) = (fptow rep (operand_in t)) /\
                            ... )) |
  (write t) ==>      % write cycle
    (address t = 0) ==> (((command(t+1) = wtonum rep (dataout t))) /\
                            ... ) |
            %idle cycle%
                         ((new_instr t = F) /\
                         (operand_ready t = F))
```

The definition `cir_read_write` states that the read or write cycle will be successfully accomplished in one time unit on the more abstract time scale. Also proper signals like `new_instr` and `operand_ready` are raised so that the BIU will know which CIR the data is latched into. Since the data transmitted on the data bus are of type *wordn (abstract type of a word of length n), type conversion is needed in both the reading and writing cycles. When an instruction is written to the command CIR, the signal `new_instr` is raised. Similarly, `operand_ready` will be raised if the data is written into the output operand CIR. To prove that the implementation of the 4-phase handshaking protocol actually implements the more abstract description `cir_read_write`, a temporal abstraction from the lower level time scale to the upper level time scale was used.

4.2 Specification of the Three Implementing Interpreters

This section describes the specification of the three interpreters that implement the FPC top level; they are the interpreters of the CPU service, the BIU and the APU. Each interpreter has a state and an environment. Each element in the state tuple represents

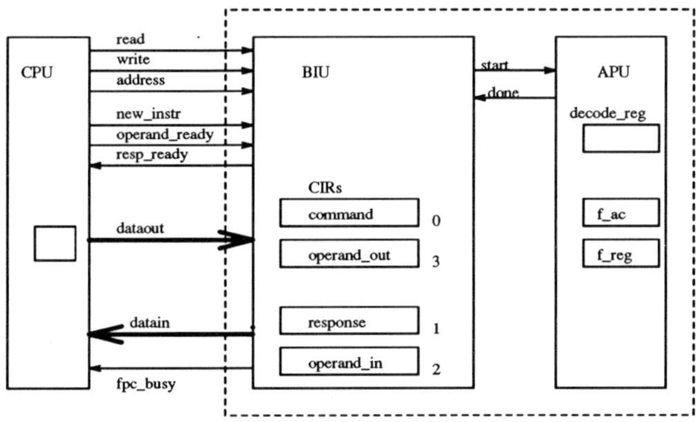

Figure 5: Interaction between ALU, BIU, and APU

objects that can be both read and written by the interpreter, while the elements in the environment tuple can only be read.

The state tuple and the environment tuple for each of the interpreters can be obtained from Figure 5.

The function `cpu_service_NextState` determines the new state for the CPU service interpreter from the current state and the environment according to the current value of the state counter `cpu_state`.

```
cpu_service_NextState n rep cpu_state =
      ((cpu_state = 0) ==> (cpu_begin n rep) |
       (cpu_state = 1) ==> (cpu_wait_for_response n rep) |
       (cpu_state = 2) ==> (cpu_wait_4phase n rep) |
       (cpu_state = 3) ==> (cpu_read_respone n rep) |
       (cpu_state = 4) ==> (cpu_wait_read n rep) |
                    (cpu_put_data n rep))
```

There are four parameters for each state transition function. The second parameter, `rep`, is the parameter for the abstract data types. It represents the "significant" components of a FP machine, including the fields: `fetch`, `store`, `wtonum`, `wtofp`, `numtow`, and `fptow`. The third parameter is the state tuple, and the last one is the environment tuple. The following is the definition of `cpu_begin` where the CPU sends out a new instruction to the BIU. The CPU raises the write request line and puts the instruction on the data bus.

```
cpu_begin n rep
  (c_ac, c_reg, mem, dataout, address, read, write, cpu_state)
  (resp_ready, datain, pc, fpc_busy) =
       ~fpc_busy ==>
           (c_ac, c_reg, mem, (numtow rep ir), 0 F, T, 1) |
           (c_ac, c_reg, mem, dataout, 0, F, F, 0)
```

The top-level specification of CPU service relates the state at time t+1 to the state and environment at t by the `cpu_service_state` function:

```
cpu_service n rep
  (c_ac, c_reg, mem, dataout, address, read, write, cpu_state)
  (resp_ready, datain, pc, fpc_busy) =
     ! t. (c_ac(t+1), c_reg(t+1), mem(t+1), dataout(t+1),
           address(t+1), read(t+1), write(t+1), cpu_state(t+1))
= cpu_service_state n rep (cpu_state t)
       (c_ac t, c_reg t, mem t, dataout t, address t, read t,
        write t, cpu_state t)
       (resp_ready t, datain t, pc t,  fpc_busy t)
```

4.2.1 Specifying the BIU Top Level

In order to allow a FPC instruction to start executing, the BIU communicates with the CPU to obtain the necessary information, and to inform the APU to start processing in the case of an arithmetic instruction. There are four states for the BIU top level state machine. For example, in state 0, the BIU is idle and waits for the next instruction to be sent over by the CPU. The function `biu_top_state` (not shown) describes the state transitions.

As discussed above, the three implementing interpreters (the CPU, the BIU, and the APU) have to be expressed with respect to a common time scale. We chose the `cpu_service` clock rate as the common clock rate. Therefore, the behavior of `biu_top` and `apu_top` have to be redescribed in terms of the clock rate of `cpu_service`.

The following definition `biu_top_cpu` is the APU top-level behavior from the CPU's point of view. The interpreter for the BIU top level relates the state tuple at time `t+1` or `t'` to the state and environment tuple at time `t`, depending on whether the BIU is waiting for the CPU or is doing some work required for the current FPC instruction. For example, in state 1, if the FPC instruction arrives at CPU time `t`, then at CPU time `t'` the instruction will be decoded and the proper response sent. Further, `t'` is the *first* time after time `t` that the response is ready: `(First (t, t') resp_ready)`. The start signal to the APU will remain *stably* low from times `t` to `t'`: `(Stable (start, F, t, t'))`.

```
biu_top_cpu n rep (response, operand_out, decode_reg, resp_ready,
       f_ac, biu_state, start, fpc_busy)
          (command, condition, control, operand_in, done, new_instr,
          operand_ready) =
! (t:num). ? (t':num).((t+1 <= t') /\
    ((biu_state t = 0) \/ (biu_state t = 2)) ==>
      state_tuple_biu (t+1) = biu_top_state n rep (biu_state t)
        (state_tuple_biu t) (env_tuple_biu t) |
    (biu_state t = 1) ==>
       ((First (t, t') (resp_ready)) /\
       (stable (start, F, t, t')) /\  ... )
    (biu_state t = 3) ==>
       ((~operand_ready t)) ==> ...)
```

4.2.2 Specifying the APU Top Level

The definition of the APU top level is straightforward, reflecting updates to the FPU registers as a result of the execution of a floating point instruction, e.g., APU-FADD.

Similar to `biu_top_cpu`, the definition of `apu_top_cpu` specifies the APU top level behavior from the CPU's point of view. It states that if the start signal has not arrived, the APU will remain idle; however if the APU starts to execute at CPU time `t`, then there exists some future CPU time `t'` (where `t'` is the first time since `t` that the APU finishes executing an instruction), at which the APU will finish the execution of the instruction.

```
apu_top_cpu n rep (f_ac, f_reg, cw, sw, done) (start, decode_reg) =
 !t . (~start t) ==>
       ((f_ac(t+1), f_reg(t+1), cw(t+1), sw(t+1), done(t+1)) =
        apu_idle n rep (f_ac t, f_reg t, cw t, sw t, done t)
           (start t, decode_reg t)) |
       (? t' . (t<t') /\
          (First (t, t') (done) /\
         ((f_ac t', f_reg t', cw t', sw t', done t') =
          NextState_apu n rep (Opc n, (decode_reg t))
               (f_ac t, f_reg t, cw t, sw t, done t)
               (start t, decode_reg t))))
```

4.3 Verifying the FPC Top Level

This section focuses on the methodology used to verify that the cooperative effort of the three interpreters implements the FPC top level specification. The key to this proof is the asynchronous communication among the three interpreters. Informally, the proof involves stepping through the states of the three interpreters, starting with the CPU service noting the presence of a FP instruction and terminating with the results being safely stored in the CPU's memory or the APU's accumulator.

4.3.1 Verifying the FLD Instruction

To verify the FPC top-level interpreter, the correctness of *every* FPC instruction has to be proved first, as must the synchronization scheme between the two processors. Theorem `FLD_Correct` states that the FLD instruction and the synchronization schemes are correctly implemented by the three interacting units: the CPU, the BIU and the APU. There are several assumptions that denote the initial states of all the implementing components. More specifically, at time t, the current instruction is FLD (opcode 0), and all the implementing interpreters are in their initial states.

Function FPCstate is used to filter out some elements from the state tuple that might be changed by the CPU after the CPU is freed from its floating-point activity. These elements are the CPU accumulator and registers, the memory, and the instruction register. The same is true for the arithmetic instructions. However, for the FSTR instruction, the entire state will be used since the CPU will not be free until the FSTR is finished execution. Several lemmas, called "waiting lemmas", were needed. These lemmas, internally, indicate that the state tuple remains unchanged while an interpreter awaits a response from another interpreter.

```
FLD_Correct =
 ! rep response, operand_out decode_reg biu_state start command
     condition control operand_in done, new_instr operand_ready
     resp_ready f_ac f_reg cw sw c_ac c_reg mem ir cpu_state
     address read write datain dataout n.
     (biu_top_cpu n rep state_tuple_biu env_tuple_biu ) /\
     (apu_top_cpu n rep state_tuple_apu env_tuple_apu ) /\
     (cpu_service n rep state_tuple_cpu_service env_tuple_cpu_service) /\
     (cir_read_write rep address read write datain dataout command
        response operand_in operand_out condition control
        new_instr operand_ready) ==>
     (! t. (Opc n (ir t) = 0) /\ ~ start t /\
        (biu_state t = 0) /\ (cpu_state t = 0) ==>
         (?t'. (Stable (fpc_busy, T t+1, t')) /\ (fpc_busy t' = F)) /\
               (f_ac t', f_reg t') =
               FPCstate (FLD n rep (f_ac t, f_reg t, c_ac t, c_reg t,
                           mem t, ir t)))
```

5 Verifying the concurrent top level

At the concurrent top level, the instructions are categorized into two classes: the CPU instructions and the FPC instructions. A certain degree of concurrency is allowed between these two classes, provided that there is no interference. The concurrent description of the system defines concurrency possibilities between instruction classes. However, the *detailed* handshaking signals between the CPU and the FPC are hidden at this level.

Instead, the specification of the concurrent interpreter concentrates on the completion of the instructions. There is no explicit statement about the existence of two processors, only the concurrent execution among instructions is described.

The concurrent top-level interpreter is implied by two lower-level interpreters: the CPU and FPC top-level interpreters. The synchronization between the CPU and the FPC is specified by both the CPU and the FPC top-level interpreters through the use of the FPC busy line. This section describes the concurrent-level interpreter and its proof.

5.1 Specifying the Concurrent Top Level Interpreter

The concurrent-level interpreter `Con_Top` is the concurrent specification of the computer system that is composed of the CPU and the coprocessor. The specification indicates that a CPU instruction only needs one clock cycle to complete, while a FPC instruction may take more than one cycles. During the execution of a FPC instruction, several CPU instructions could be executed in parallel, depending on the FPC instruction's type (whether it is a FSTR instruction or one that stores its results into a FP register).

5.1.1 The Specification of the Concurrent Level

The exact time that a FPC instruction completes is unknown. However, it will complete before the next FPC instruction begins. This property is guaranteed by the CPU's busy-wait feature. Specifically, if the FPC is still executing the previous instruction, the CPU will remain idle until the busy line emitted by the FPC is lowered.

There are three cases to the concurrent level definition `Con_Top` based on the current instruction:

- CPU instruction. The CPU state and the joint state are modified in one cycle, while the FPC state remains unchanged.
- FSTR instruction. The CPU state, the FPC state and the joint state are all changed in one cycle.
- non-FSTR FPC instruction. The FPC state is changed once in m+1 cycles, while the CPU state and the joint state may be changed more than once by the CPU instruction executed concurrently with the FPC instruction.

In the third case, the condition $\forall m.((g\ t) \wedge Next1\ g\ (t, t+m+1))$ is used to state that if $t+m+1$ is the next time a FPC instruction is started, then the FPC state will be changed at that time as the result of the current FPC instruction. In reality, the FPC state may be changed before $t+m+1$; that is, the current FPC instruction finishes execution before the next FPC instruction is started. In this case several CPU instructions will be executed following the finishing time of the current FPC instruction. However since CPU instructions never change the FPC state, we can state that at time

$t + m + 1$, the FPC state will only be modified once by the FPC instruction started at time t.

```
Con_Top n rep (f_ac, f_reg, c_ac, c_reg, mem, pc) =
 (!t.
        let instr = wtonum rep ((fetch rep) (mem t) (pc t)) and
        new_cpu = CPU_solo_nextstate n rep (c_ac t, c_reg t,
                                              mem t, pc t) and
        g = (\t : num. ~ (Is_cpu_instr n rep (pc t) (mem t))) in
    (Is_cpu_instr n rep (pc t) (mem t)) ==>
             ((c_ac(t+1), c_reg(t+1), mem(t+1), pc(t+1)) = new_cpu) /\
             (f_ac(t+1), f_reg(t+1)) = (f_ac t, f_reg t) |
    (Is_fstr n rep (pc t) (mem t)) ==>
             (f_ac(t+1), f_reg(t+1), c_ac(t+1), c_reg(t+1), mem(t+1))=
             NextState_fpc n rep instr (f_ac t, f_reg t,
                       c_ac t, c_reg t, mem t) /\
             (pc(t+1) = (pc t) + 1))|
    (! m . ((g t) /\ Next1 g (t, t+m+1)) ==>
             (f_ac(t+m+1), f_reg(t+m+1)) =
             FPCstate (NextState_fpc n rep instr
                       (f_ac t, f_reg t, c_ac t, c_reg t, mem t)) /\
                       (CPU_Parallel n rep (t,t+m+1) (c_ac, c_reg, mem pc))
```

In the above definition, the function `CPU_Parallel` is used to describe the execution of those CPU instructions that are executed in parallel with the FPC instructions between the times t and `t+m+1`.

5.2 The Correctness Statement

The concurrent top level correctness statement, then, is as follows. It is proved by a case analysis based on the type of the current instruction, together with temporal abstraction theorems.

```
 ! (rep : ^rep_ty) c_ac c_reg mem pc f_ac f_reg fpc_busy n .
       let g = \t.(Is_CPU_instr n rep (pc t) (mem t)) \/
                (~fpc_busy t) in
       let abs = (Temp_Abs g) in
       CPU_Top n rep (c_ac, c_reg, mem, pc) (fpc_busy) /\
       FPC_cpu_Top n rep (f_ac,f_reg,c_ac,c_reg,mem,fpc_busy) (pc)
```

6 Verifying the sequential top level

The sequential level description is the highest level in the abstraction hierarchy for our system. This level describes the view of the CPU–FPC system as seen by an assembly language programmer. At this level, the system not only appears as a single unit, but also the instructions appear to be executed in a sequential manner. This sequential execution of instructions is "equivalent" to the underlying concurrent execution in the sense that the final result of any program will be identical in both cases. In this section we discuss the specification of this sequential execution description and the approach to abstracting out the concurrency present in the system.

6.1 Specifying the Sequential Top Level Interpreter

The sequential top level interpreter reflects the assembly language programmer's view of sequential execution. Whatever its type, each instruction takes exactly one cycle to finish. In the following definition of the sequential interpreter, the functions `FPC_Next` and `CPU_Next` define the execution of the instruction pointed to by the program counter, either of type FPC or of type CPU.

```
Seq_Top n rep (f_ac, f_reg, c_ac, c_reg, mem, pc) =
 ! t.  (Is_cpu_instr n rep (pc t) (mem t)) ==>
        CPU_Next n t rep (f_ac, f_reg, c_ac, c_reg, mem, pc) |
        FPC_Next n t rep (f_ac, f_reg, c_ac, c_reg, mem, pc)

CPU_Next n t rep (f_ac, f_reg, c_ac, c_reg, mem, pc) =
        ((c_ac(t+1), c_reg(t+1), mem(t+1), pc(t+1)) =
         CPU_solo_nextstate n rep (c_ac t, c_reg t, mem t, pc t)) /\
        ((f_ac(t+1), f_reg(t+1)) = (f_ac t, f_reg t))

FPC_Next n t rep (f_ac, f_reg, c_ac, c_reg, mem, pc) =
        let instr = wtonum rep ((fetch rep) (mem t) (pc t)) in
        (Is_fstr n rep (pc t) (mem t)) ==>
                (((f_ac(t+1),f_reg(t+1),c_ac(t+1),c_reg(t+1),mem(t+1)) =
                NextState_fpc n rep instr
                        (f_ac t, f_reg t, c_ac t, c_reg t, mem t) /\
                (pc(t+1) = (pc t)+1)) |
        (((f_ac(t+1), f_reg(t+1)) =
                        FPCstate (NextState_fpc n rep instr (f_ac t, f_reg t,
                                c_ac t, c_reg t, mem t)) /\
        ((c_ac(t+1), c_reg(t+1), mem(t+1), pc(t+1)) =
                (c_ac t, c_reg t, mem t, pc t)
```

6.2 The Abstraction

Suppose the sequential execution flow of an instruction stream is represented by the sequence "$F_1C_1C_2C_3F_2F_3C_4F_4C_5C_6$", where F's and C's represent FPC and CPU instructions, respectively. The arrowed line in Figure 6 indicates the mapping points of the abstraction. The instruction labels marked above the line are the starting time of the corresponding instructions, while the labels below the line represent the finishing time. For example, on the concurrent level, instruction F1 starts at time 0 and finishes at time

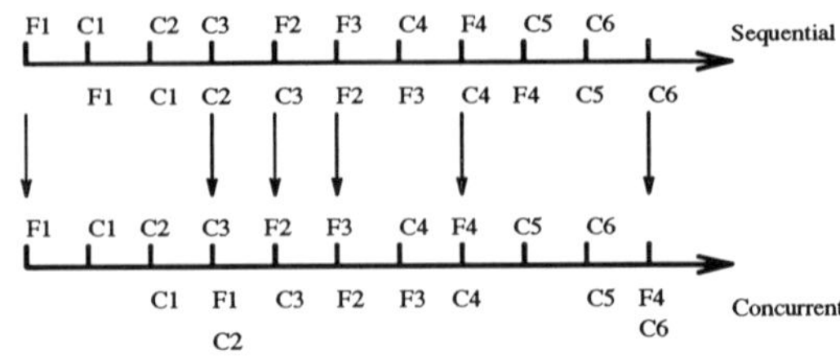

Figure 6: Mapping between sequential top level and concurrent top level

3. A mapping occurs when the states of the concurrent and sequential levels are the same. For example, at time 3 at both levels instructions F_1, C_1, c_2 have completed and C_3 has started. The reader is reffered to [Pan92] for details, where the possible instruction streams are characterized.

7 Discussion

The verification of the coprocessor was not carried out down to the electronic block description. Instead, the coprocessor verified was modeled by three interacting interpreters, representing the behavior of the Bus Interface Unit (BIU), the Arithmetic Processing Unit (APU), and the part of the CPU which communicates with the coprocessor.

The verification of the APU was not carried out to the electronic block diagram mainly because of the difficulties in reasoning about floating-point arithmetic. The current HOL system is not equipped with adequate theories and tactics to deal with bit-vectors, integers, rational numbers, and real numbers. Thus, we decided to concentrate our effort on the modeling of the communication and synchronization between the CPU and the coprocessor.

It should be rather straightforward to carry the proof down to the electronic block level for the BIU. But, since the BIU closely cooperates with the arithmetic unit, it is almost impossible to design the BIU without a electronic block model of the APU. The behavior of both the Bus Interface Unit and the Arithmetic Processing Unit was described as state machines.

In summary, the verification goal for our system is the following:

(APU behavior $\wedge$
BIU behavior $\wedge$
CPU coprocessor service behavior)
$\Rightarrow$ *The sequential description of the system consisting of the CPU and FPC*

Much remains to be done with respect to verification of floating point coprocessors and related areas. The proofs described here could be carried out at the electronic-block-level description of the FPC. The implementation of the APU could be verified with respect to the IEEE standard. Furthermore, the specification of the IEEE standard could be used to prove theorems about floating point arithmetic.

Finally, and most interestingly, a generic theory of communicating interpreters could be developed, extending Windley's theory for stand-alone interpreters. Schubert's thesis [Sch92] describing work on process algebras is a first step towards such a theory.

References

[Gra92] Brian T. Graham. *The SECD Microprocessor, A Verification Case Study.* Kluwer International Series in Engineering and Computer Science. Kluwer Academic Publishers, Boston, 1992.

[IEE85] IEEE Computer Society Press, Washington, DC. *IEEE Standard for Binay*

Higher Order Logic Theorem Proving and its Applications (A-20)
L.J.M. Claesen and M.J.C. Gordon (Editors)
Elsevier Science Publishers B.V. (North-Holland)

Deriving a Correct Computer

Li-Guo Wang[1]

Laboratory for Foundations of Computer Science, Department of Computer Science, University of Edinburgh, JCMB, The King's Buildings, Edinburgh EH9 3JZ, Scotland, U.K.
E-mail Address: lgw@uk.ac.ed.dcs

Abstract

A formal derivation of an implementation at register transfer level of Gordon's computer is presented. The derivation illustrates a proof-driven hierarchical construction process, based on logical inference, which reveals some essential aspects of computer design.

Ketword Codes: F.4.1; I.2.3; J.6
Ketwords: Mathematic Logic; Deduction and Theorem Proving; Computer Aided Engineering.

1 Introduction

Gordon's computer [3] is a simple general-purpose computer invented as a non-trivial example for formal specification and verification. Several papers have reported its verification eg. [3]. Here we report our work about its *formal derivation.*

By formal derivation we mean:

1. Only the formal specification at machine instruction level, and axioms about basic components at register transfer level are given; The implementation is derived.

2. The formal derivation proceeds from the specification, which is manipulated by inference rules. Thus the design process is driven by logical proof and guided by the formal structure of the specification.

   The derivation produces a hierarchical specification series giving a formal development history starting from a behavioral description and ending with a structural one. In the series below microprogram hierarchy and concurrent data path structures are produced:

   - Microprogram hierarchy:

[1]This research is supported by scholarship from Siemens AG Munich and scholarship from SERC GR/F 35890 U.K.

loop microprogram → branch microprogram → linear microprogram → microinstruction → microcode.

- Concurrent data path structure: function units, register, gate and bus.

3. As the result of the derivation we obtain an implementations different from Gordon's, which has 4 buses, 4 registers, shorter microprogram and faster speed.

Derivation differs from verification. For verification both specification and implementation have been given, the task of verification is to prove the their consistency (satisfaction relation). For derivation only the specification is given and the task of derivation is to logically derive implementation. The derivation process is justified by a series of satisfaction relations. Therefore our approach to derivation includes verification as a by-product.

Derivation is a process which is driven by the needs of formal proof on *pure formal structure*. Therefore formal derivation can reveal some essential aspects of the derived goal. The derivation of Gordon's computer shows a quite complicated hierarchical structure. Our approach can lead to different implementations whose variety comes from different selections of proof processes and axioms of basic components. In such a short paper we can only give a rough description.

Our description is based on higher-order logic. For ease of exposition and understanding we mix formal and informal languages. All proofs depend on a theory Γ which is a descendant of the theory HOL [2] and has axioms from [3]. Because Γ is fixed we write $\vdash P$ instaed of $\Gamma \vdash P$. We use italic letters (with subscripts if necessary) to express variables and scheme to range over some sets. Because of space limitations, we omit the proofs of all theorems.

2 Specification

The specification of Gordon's computer [3] defines the state transition semantics of the front panel operation and machine instruction set {halt, jmp, jzr, add, sub, ld, st, skip} of the computer. Here we take its formal description from [3] with a little change and omit its informal explanation (see [3]). Because the specification is at the machine instruction level, a unit of time corresponds to the time interval of the implementation period of a machine instruction.

The following typed variables are used in the specification:
t : num, T, F : bool, knob : num $\rightarrow$ word_2, pc : num $\rightarrow$ word_{13}, VAL_n : word_n $\rightarrow$ num,
idle, button : num $\rightarrow$ bool, memory : num $\rightarrow$ memory_{13_16}, switches, acc : num $\rightarrow$ word_{16},
ADD_{16}, SUB_{16} : $\text{word}_{16} \rightarrow \text{word}_{16} \rightarrow \text{word}_{16}$, CUT_{16_13} : $\text{word}_{16} \rightarrow \text{word}_{13}$,
INC_{13} : $\text{word}_{13} \rightarrow \text{word}_{13}$, OPCODE : $\text{word}_{16} \rightarrow \text{word}_3$,
FETCH_{13} : $\text{memory}_{13\_16} \rightarrow \text{word}_{13} \rightarrow \text{word}_{16}$
STORE_{13} : $\text{word}_{13} \rightarrow \text{word}_{16} \rightarrow \text{memory}_{13\_16} \rightarrow \text{memory}_{13\_16}$

The specification:
$\vdash$ COMPUTER (button, knob, switches, memory, pc, acc, idle) $= \forall u.\ \text{SP}\ (u, u+1) \wedge$
$\text{SP} = \lambda(t, t').\ ((\text{memory}\ t', \text{pc}\ t', \text{acc}\ t', \text{idle}\ t') =$
$(\text{idle}\ t \Rightarrow (\text{button}\ t \Rightarrow$

$(\mathrm{VAL}_2(\mathrm{knob}\ t) = 0 \Rightarrow (\mathrm{memory}\ t,\ \mathrm{CUT}_{16\_13}\ (\mathrm{switches}\ t),\ \mathrm{acc}\ t,\ \mathrm{T})\ |$
$\mathrm{VAL}_2(\mathrm{knob}\ t) = 1 \Rightarrow (\mathrm{memory}\ t,\ \mathrm{pc}\ t,\ \mathrm{switches}\ t,\ \mathrm{T})\ |$
$\mathrm{VAL}_2(\mathrm{knob}\ t) = 2 \Rightarrow (\mathrm{STORY}_{13}\ (\mathrm{pc}\ t)\ (\mathrm{acc}\ t)\ (\mathrm{memory}\ t),\ \mathrm{pc}\ t,\ \mathrm{acc}\ t,\ \mathrm{T})\ |$
$(\mathrm{memory}\ t,\ \mathrm{pc}\ t,\ \mathrm{acc}\ t,\ \mathrm{F}))\ |\ (\mathrm{memory}\ t,\ \mathrm{pc}\ t,\ \mathrm{acc}\ t,\ \mathrm{T}))\ |$
$(\mathrm{button}\ t \Rightarrow (\mathrm{memory}\ t,\ \mathrm{pc}\ t,\ \mathrm{acc}\ t,\ \mathrm{T})\ |$
let $op = \mathrm{VAL}_3\ (\mathrm{OPCODE}\ (\mathrm{FETCH}_{13}\ (\mathrm{memory}\ t)\ (\mathrm{pc}\ t)))$ in
let $addr = \mathrm{CUT}_{16\_13}\ (\mathrm{FETCH}_{13}\ (\mathrm{memory}\ t)\ (\mathrm{pc}\ t))$ in
$(op = 0 \Rightarrow (\mathrm{memory}\ t,\ \mathrm{pc}\ t,\ \mathrm{acc}\ t,\ \mathrm{T})\ |$
$op = 1 \Rightarrow (\mathrm{memory}\ t,\ addr,\ \mathrm{acc}\ t,\ \mathrm{F})\ |$
$op = 2 \Rightarrow (\mathrm{VAL}_{16}\ (\mathrm{acc}\ t) = 0 \Rightarrow (\mathrm{memory}\ t,\ addr,\ \mathrm{acc}\ t,\ \mathrm{F})\ |$
$\qquad (\mathrm{memory}\ t,\ \mathrm{INC}_{13}\ (\mathrm{pc}\ t),\ \mathrm{acc}\ t,\ \mathrm{F}))\ |$
$op = 3 \Rightarrow (\mathrm{memory}\ t,\ \mathrm{INC}_{13}\ (\mathrm{pc}\ t),\ \mathrm{ADD}_{16}\ (\mathrm{acc}\ t)\ \ (\mathrm{FETCH}_{13}\ (\mathrm{memory}\ t)\ addr), \mathrm{F})\ |$
$op = 4 \Rightarrow (\mathrm{memory}\ t,\ \mathrm{INC}_{13}\ (\mathrm{pc}\ t),\ \mathrm{SUB}_{16}\ (\mathrm{acc}\ t)\ \ (\mathrm{FETCH}_{13}\ (\mathrm{memory}\ t)\ addr), \mathrm{F})\ |$
$op = 5 \Rightarrow (\mathrm{memory}\ t,\ \mathrm{INC}_{13}\ (\mathrm{pc}\ t),\ \mathrm{FETCH}_{13}\ (\mathrm{memory}\ t)\ addr, \mathrm{F})\ |$
$op = 6 \Rightarrow (\mathrm{STORE}_{13}\ addr\ (\mathrm{acc}\ t)\ (\mathrm{memory}\ t),\ \mathrm{INC}_{13}\ (\mathrm{pc}\ t), \mathrm{acc}\ t,\ \mathrm{F})\ |$
$(\mathrm{memory}\ t,\ \mathrm{INC}_{13}\ (\mathrm{pc}\ t),\ \mathrm{acc}\ t,\ \mathrm{F})))))$

Let us call memory, pc, acc and idle state signals and call all signals in the specification basic signals.

3 Axioms

The axioms describe basic components at register transfer level: Register, Gate, Bus, Mem, Alu Rom, Mpc and Decode. We take most axioms in [3]. However since our task is derivation some axioms given in [3] will be partly derived in our work, these describe Rom and Decode. We only give the structure of these axioms, and leave their paramenters (which are represented schematically) to be derived. For the sake of saving space we do not repeat these axioms (see [3]) here, but introduce them when they are used.

4 Derivation

In this section we derive, by goal directed proof, a specification series from the most abstract one down to an implementation. The series shows, in rigorous logical form, the development history of our formal design. The key point in the derivation is the hierarchical decomposition which is guided by the formal structure of the specification. The main steps are: 1. time refinement, 2. structural decomposition and 3. term decomposition.

4.1 Time Refinement

In this step we can ignore the internal structure of specification and only abstractly see the specification of Gordon's computer as

Specification 1: $\forall u.\ \mathrm{SP}\ (u, u + 1_u)$

To make "$\forall u$" of the specification 1 become computable we use:

Theorem 1.1 If 1. $\vdash entry\ 0_u$, 2. $\vdash \forall u.\ entry\ u \Rightarrow SP\ (u, u + 1_u) \wedge entry\ (u + 1_u)$

then $\vdash \forall u.\ SP\ (u, u + 1_u)$ $\square$

Here, we use 0_u, 1_u to express the time at machine instruction level and u to range over the time at this level. The *entry* is a new predicate introduced to implement "$\forall u$", which leads to a loop structure. At this stage the *entry* is abstract but later we can see it is about the entry address of microprogram. So we get:

Specification 2: $entry\ 0_u \wedge \forall u.\ entry\ u \Rightarrow \mathrm{SP}\ (u, u + 1_u) \wedge entry\ (u + 1_u)$

To derive an implementation at register transfer level we should refine time to this level. For this we use:

Axiom 1.1 $\vdash \forall s : S_s.\ s\ 0_u = s\ 0$
Axiom 1.2 $\vdash \forall t_1, t_2.\ (ready\ t_1 \Rightarrow ready\ t_2 \wedge \forall t.\ t_1 < t < t_2 \Rightarrow \neg(ready\ t)) \Rightarrow$
$\forall u, s : S_s.s\ u = s\ t_1 \Rightarrow s\ (u + 1_u) = s\ t_2$. ($S_s$ is the set of signals).

Theorem 1.2 If 1. $\vdash ready\ 0$,
2. $\vdash \forall t_1.\ \exists t_2.\ (ready\ t_1 \Rightarrow ready\ t_2 \wedge \forall t.\ t_1 < t < t_2 \Rightarrow \neg(ready\ t)) \wedge F\ (t_1, t_2)$
then $\vdash \forall u.\ F\ (u, u + 1_u)$ $\square$

Corollary 1.1 If 1. $\vdash ready\ 0$,
2. $\vdash \forall t_1.\ \exists t_2.\ (ready\ t_1 \Rightarrow ready\ t_2 \wedge \forall t.\ t_1 < t < t_2 \Rightarrow \neg(ready\ t)) \wedge$
$(entry\ t_1 \Rightarrow SP\ (t_1, t_2) \wedge entry\ t_2)$
then $\vdash \forall u.\ entry\ u \Rightarrow SP\ (u, u + 1_u) \wedge entry\ (u + 1_u)$ $\square$

Here, we use 0, 1 to express the time at register transfer level and t to range over the time at this level. The two axioms give the basic relation between two adjacent time scales. By using a predicate *ready* as mark, any basic signal s can be synchronized between the high and low time scales. Using these theorems and eliminating quantifier we get:

Specification 3:
$\mathrm{ready}\ 0 \wedge entry\ 0 \wedge (\mathrm{ready}\ t_1 \Rightarrow \mathrm{ready}\ t_2 \wedge (\forall t.\ t_1 < t < t_2 \Rightarrow \neg(\mathrm{ready}\ t))) \wedge$
$(entry\ t_1 \Rightarrow \mathrm{SP}\ (t_1, t_2) \wedge entry\ t_2)$
Here, t_1 is arbitrary time constant. t_2 is a time-function of t_1.

From this step we can see that the development of a looping microprogram and the signal "ready" follow logically from the "$\forall u$" component of the specification 1.

4.2 Structural Decomposition

In this step we start to deal with the internal structure of specification. We need only deal with the SP part of the specification 3 and leave others unchanged. We will decompose SP (t_1, t_2) to sub-structures whose controls correspond to linear microprograms. We need a series of transformations but mainly use below two theorems and their reformed or more general forms to decompose specification 3.

Theorem 2.1. $\vdash \forall t_a, t_b, t_c.\ (entry_1\ t_a \Rightarrow r\ t_b = F\ t_a \wedge entry_2\ t_b \wedge$
$K\ (t_a, t_b) \wedge (K\ (t_a, t_b) \Rightarrow (E\ (t_b, t_c)\ (r\ t_b) \Rightarrow E\ (t_a, t_c)\ (r\ t_b))) \wedge t_a \leq t_b) \wedge$

$(entry_2\ t_b \Rightarrow E\ (t_b, t_c)\ (r\ t_b) \wedge t_b \leq t_c)$
$\Rightarrow (entry_1\ t_a \Rightarrow \mathsf{let}\ v = F\ t_a\ \mathsf{in}\ E\ (t_a, t_c)\ v \wedge t_a \leq t_c)$ □

Theorem 2.2. $\vdash \forall t_a, t_b, t_c, t_d.$
$(entry_p\ t_a \Rightarrow P\ t_b = P\ t_a \wedge (P\ t_b \Rightarrow entry_1\ t_c \mid entry_2\ t_c) \wedge t_a \leq t_b \leq t_c \wedge$
$K\ (t_a, t_c) \wedge (K\ (t_a, t_c) \Rightarrow (E\ (t_c, t_d) \Rightarrow E\ (t_a, t_d)) \wedge (F\ (t_c, t_d) \Rightarrow F\ (t_a, t_d)))) \wedge$
$(entry_1\ t_c \Rightarrow E\ (t_c, t_d) \wedge t_c \leq t_d) \wedge (entry_2\ t_c \Rightarrow F\ (t_c, t_d) \wedge t_c \leq t_d)$
$\Rightarrow (entry_p\ t_a \Rightarrow (P\ t_a \Rightarrow E\ (t_a, t_d) \mid F\ (t_a, t_d)) \wedge t_a \leq t_d)$ □

The derived specification 4 is quite complicated, but its two main parts describe the branch structure of the microprogram and each sub-structure respectively.

Specification 4: is same as the specification 3 but, for example,
$entry\ t_1 \Rightarrow \mathrm{SP}\ (t_1, t_2) \wedge entry\ t_2$ is replaced by:
$(entry\ t_1 \Rightarrow$
$(\mathrm{idle}\ t_1 \Rightarrow (entry_{b1}\ t_a \Rightarrow$
$(\mathrm{button}\ t_a \Rightarrow (entry_k\ t_b \Rightarrow$
$(\mathrm{VAL}_2\ (\mathrm{knob}\ t_b = 0) \Rightarrow (entry_{s1}\ t_{c1} \Rightarrow entry\ t_2) \mid$
$\mathrm{VAL}_2\ (\mathrm{knob}\ t_b = 1) \Rightarrow (entry_{s2}\ t_{c1} \Rightarrow entry\ t_2) \mid$
$\mathrm{VAL}_2\ (\mathrm{knob}\ t_b = 2) \Rightarrow (entry_{s3}\ t_{c1} \Rightarrow entry\ t_2) \mid$
$entry_{s4}\ t_{c1} \Rightarrow entry\ t_2)) \mid entry_{s5}\ t_b \Rightarrow entry\ t_2)) \mid$
$entry_{b2}\ t_a \Rightarrow$
$(\mathrm{button}\ t_a \Rightarrow (entry_{s5}\ t_b \Rightarrow entry\ t_2)) \mid$
$entry_{s0}\ t_b \Rightarrow (entry_o\ t_{c2} \Rightarrow$
$(\mathrm{VAL}_3\ (\mathrm{OPCODE}\ (\mathrm{ir}\ t_{c2})) = 0 \Rightarrow (entry_{s5}\ t_d \Rightarrow entry\ t_2) \mid$
$\mathrm{VAL}_3\ (\mathrm{OPCODE}\ (\mathrm{ir}\ t_{c2})) = 1 \Rightarrow (entry_{s6}\ t_d \Rightarrow entry\ t_2) \mid$
$\mathrm{VAL}_3\ (\mathrm{OPCODE}\ (\mathrm{ir}\ t_{c2})) = 2 \Rightarrow (entry_a\ t_d \Rightarrow$
$(\mathrm{VAL}_{16}\ (\mathrm{acc}\ t_d) = 0 \Rightarrow (entry_{s6}\ t_e \Rightarrow entry\ t_2) \mid$
$entry_{s7}\ t_e \Rightarrow (entry_{s8}\ t_{e0} \Rightarrow entry\ t_2))) \mid$
$\mathrm{VAL}_3\ (\mathrm{OPCODE}\ (\mathrm{ir}\ t_{c2})) = 3 \Rightarrow (entry_{s9}\ t_d \Rightarrow (entry_{s8}\ t_{e1} \Rightarrow entry\ t_2)) \mid$
$\mathrm{VAL}_3\ (\mathrm{OPCODE}\ (\mathrm{ir}\ t_{c2})) = 4 \Rightarrow (entry_{s10}\ t_d \Rightarrow (entry_{s8}\ t_{e2} \Rightarrow entry\ t_2)) \mid$
$\mathrm{VAL}_3\ (\mathrm{OPCODE}\ (\mathrm{ir}\ t_{c2})) = 5 \Rightarrow (entry_{s11}\ t_d \Rightarrow (entry_{s8}\ t_{e3} \Rightarrow entry\ t_2)) \mid$
$\mathrm{VAL}_3\ (\mathrm{OPCODE}\ (\mathrm{ir}\ t_{c2})) = 6 \Rightarrow (entry_{s12}\ t_d \Rightarrow (entry_{s8}\ t_{e4} \Rightarrow entry\ t_2)) \mid$
$entry_{s7}\ t_d \Rightarrow (entry_{s8}\ t_{e5} \Rightarrow entry\ t_2))))$
$\wedge \cdots \wedge$
$\mathrm{sub\_structure}_0 = (\mathrm{ir}\ t_{c2} = \mathrm{FETCH}_{13}\ (\mathrm{memory}\ t_b)\ (\mathrm{pc}\ t_b)) \wedge$
$\mathrm{sub\_structure}_1 = (\lambda(t_1, t_2).\ \mathrm{memory}\ t_2 = \mathrm{memory}\ t_1 \wedge$
$\mathrm{pc}\ t_2 = \mathrm{CUT}_{16\_13}\ (\mathrm{switches}\ t_1) \wedge \mathrm{acc}\ t_2 = \mathrm{acc}\ t_1 \wedge \mathrm{idle}\ t_2 = \mathrm{T}) \wedge$
$\mathrm{sub\_structure}_2 = (\lambda(t_1, t_2).$
$\mathrm{memory}\ t_2 = \mathrm{memory}\ t_1 \wedge \mathrm{pc}\ t_2 = \mathrm{pc}\ t_1 \wedge \mathrm{acc}\ t_2 = \mathrm{switches}\ t_1 \wedge \mathrm{idle}\ t_2 = \mathrm{T}) \wedge$
$\mathrm{sub\_structure}_3 = (\lambda(t_1, t_2).\ \mathrm{memory}\ t_2 = \mathrm{STORY}_{13}\ (\mathrm{pc}\ t_1)\ (\mathrm{acc}\ t_1)\ (\mathrm{memory}\ t_1) \wedge$
$\mathrm{pc}\ t_2 = \mathrm{pc}\ t_1 \wedge \mathrm{acc}\ t_2 = \mathrm{acc}\ t_1 \wedge \mathrm{idle}\ t_2 = \mathrm{T}) \wedge$
$\mathrm{sub\_structure}_4 = (\lambda(t_1, t_2).$
$\mathrm{memory}\ t_2 = \mathrm{memory}\ t_1 \wedge \mathrm{pc}\ t_2 = \mathrm{pc}\ t_1 \wedge \mathrm{acc}\ t_2 = \mathrm{acc}\ t_1 \wedge \mathrm{idle}\ t_2 = \mathrm{F}) \wedge$
$\mathrm{sub\_structure}_5 = (\lambda(t_1, t_2).$
$\mathrm{memory}\ t_2 = \mathrm{memory}\ t_1 \wedge \mathrm{pc}\ t_2 = \mathrm{pc}\ t_1 \wedge \mathrm{acc}\ t_2 = \mathrm{acc}\ t_1 \wedge \mathrm{idle}\ t_2 = \mathrm{T}) \wedge$

sub_structure$_6$ = $(\lambda(t_1,t_2)$.
memory t_2 = memory $t_1 \wedge$ pc t_2 = CUT$_{16_13}$ (ir t_1) $\wedge$ acc t_2 = acc t_1) $\wedge$
sub_structure$_7$ = $(\lambda(t_1,t_2)$. memory t_2 = memory $t_1 \wedge$ acc t_2 = acc t_1) $\wedge$
sub_structure$_8$ = $(\lambda(t_1,t_2)$. pc t_2 = INC$_{13}$ (pc t_1) $\wedge$ idle t_2 = F) $\wedge$
sub_structure$_9$ = $(\lambda(t_1,t_2)$. memory t_2 = memory $t_1 \wedge$
acc t_2 = ADD$_{16}$ (acc t_1) (FETCH$_{13}$ (memory t_1) (CUT$_{16_13}$ (ir t_1)))) $\wedge$
sub_structure$_{10}$ = $(\lambda(t_1,t_2)$. memory t_2 = memory $t_1 \wedge$
acc t_2 = SUB$_{16}$ (acc t_1) (FETCH$_{13}$ (memory t_1) (CUT$_{16_13}$ (ir t_1)))) $\wedge$
sub_structure$_{11}$ = $(\lambda(t_1,t_2)$. memory t_2 = memory $t_1 \wedge$
acc t_2 = FETCH$_{13}$ (memory t_1) (CUT$_{16_13}$ (ir t_1))) $\wedge$
sub_structure$_{12}$ = $(\lambda(t_1,t_2)$.
memory t_2 = STORY$_{13}$ (CUT$_{16_13}$ (ir t_1)) (acc t_1) (memory t_1) $\wedge$ acc t_2 = acc t_1)

For every sub_structure$_i$ we have one or more formulae $entry_{si}\ x \Rightarrow$ sub_structure$_i$ (x,y) in the first part of specification 4. Where y occurs again in some $entry\ y$ which is related to $entry_{si}\ x$ by $\Rightarrow$. eg. we have

$$entry_{s9}\ t_d \Rightarrow \text{sub\_structure}_9\ (t_d, t_{e1})$$

and

$$\text{VAL}_3\ (\text{OPCODE}\ (\text{ir}\ t_{c2})) = 3 \Rightarrow (entry_{s9}\ t_d \Rightarrow (entry_{s8}\ t_{e1} \Rightarrow entry\ t_2))\ |\ .$$

In the specification we also have the decomposition of time: the times t_α must be ordered alphabetically (eg. $t_{ci} \leq t_d \leq t_{ej}$); t_1 andt_2 are minimal and maximal, respectively.

Once $entry\ t$ is explained as a microprogram entry address,(ie. refined to address $t = n$), the first part of specification 4 will directly produce the branch structure of the microprogram (connecting linear microprograms by their entry addresses) and the branch part of the device Recode for these branches.

4.3 Term Decomposition

After the specification 4 our main work is to decompose each sub_structure, especially some term equations. The main steps are to unfold each term equation, to find the signals which appear recursively in the term equation and the multiconnection of signals — when several signals are connected to one signal. In addition to introducing function units, Memory and Alu, recursive signals will lead to introducing the Register and multiconnection will lead to introducing a Gate-Bus structure. Meanwhile the microcodes which control these basic components are introduced and concurrently merged along time sequence to form microinstructions and then linear microprograms. Let's take the decomposition of a term equation as example.

term_equation = $\lambda(x,y)$.
acc y = ADD$_{16}$ (acc x) (FEYCH$_{13}$ (memory x) (CUT$_{16_13}$ (ir x)))

The term equation is unfolded to:

$\lambda(x,y).\ \exists\ i_1,i_2,i_3,i_4,i_5,o_1,o_2,o_3.\ \exists\ t_{e1},t_{e2},t_{e3},t_{e4}.$
$\mathrm{acc}\ y = \mathrm{o}_1\ t_{e1} \wedge \mathrm{o}_1\ t_{e1} = \mathrm{ADD}_{16}\ (\mathrm{i}_1\ t_{e1})\ (\mathrm{i}_2\ t_{e1}) \wedge \mathrm{i}_1\ t_{e1} = \mathrm{acc}\ x \wedge \mathrm{i}_2\ t_{e1} = \mathrm{o}_2\ t_{e2} \wedge$
$\mathrm{o}_2\ t_{e2} = \mathrm{FETCH}_{13}\ (\mathrm{i}_3\ t_{e2})\ (\mathrm{i}_4\ t_{e2}) \wedge \mathrm{i}_3\ t_{e2} = \mathrm{memory}\ x \wedge \mathrm{i}_4\ t_{e2} = \mathrm{o}_3\ t_{e3} \wedge$
$\mathrm{o}_3\ t_{e3} = \mathrm{CUT}_{16\_13}\ (\mathrm{i}_5\ t_{e4}) \wedge \mathrm{i}_5\ t_{e4} = \mathrm{ir}\ x \wedge x \leq t_{e4} \leq t_{e3} \leq t_{e2} \leq t_{e1} \leq y$

To introduce registers we note that the signals acc and ir are of appear recursively in this sub_structure e.g. acc t_2 = acc t_1. In addition, for the memory the axiom below for memory deals with its recursive appearances:

$\vdash \mathrm{MEM}\ (memory, addrs, data, wmem, memout) =$
$\forall t.\ (memout\ t = \mathrm{FETCH}_{13}\ (memory\ t)\ (addrs\ t)) \wedge$
$(memory\ (t+1) = (wmem\ t \Rightarrow\ \mathrm{STORE}_{13}\ (addrs\ t)\ (data\ t)\ (memory\ t)\ |\ memory\ t))$
(Note, our MEM is slightly different from that in [3])

Thus we can get:

$\lambda(x,y).\ \exists \mathrm{i}_{acc}, \mathrm{alu}, \mathrm{i}_{alu1}, \mathrm{i}_{alu2}, \mathrm{memout}, \mathrm{i}_{mem1}, \mathrm{i}_{mem2}, \mathrm{mar}, \mathrm{i}_{mar}.\ \exists t_{e1}.\ \ \mathrm{acc}\ y = \mathrm{i}_{acc}\ t_{e1} \wedge$
$\mathrm{i}_{acc}\ t_{e1} = \mathrm{alu}\ t_{e1} \wedge \mathrm{alu}\ t_{e1} = \mathrm{ADD}_{16}\ (\mathrm{i}_{alu1}\ t_{e1})\ (\mathrm{i}_{alu2}\ t_{e1}) \wedge \mathrm{i}_{alu1}\ t_{e1} = \mathrm{acc}\ t_{e1} \wedge$
$\mathrm{acc}\ t_{e1} = \mathrm{acc}\ x \wedge \mathrm{i}_{alu2}\ t_{e1} = \mathrm{memout}\ t_{e1} \wedge \mathrm{memout}\ t_{e1} = \mathrm{FETCH}_{13}\ (\mathrm{i}_{mem1}\ t_{e1})\ (\mathrm{i}_{mem2}\ t_{e1}))$
$\wedge\ \mathrm{i}_{mem1}\ t_{e1} = \mathrm{memory}\ t_{e1} \wedge \mathrm{memory}\ t_{e1} = \mathrm{memory}\ x \wedge \mathrm{i}_{mem2}\ t_{e1} = \mathrm{mar}\ t_{e1} \wedge$
$\mathrm{mar}\ t_{e1} = \mathrm{CUT}_{16\_13}\ (\mathrm{i}_{mar}\ x) \wedge \mathrm{i}_{mar}\ x = \mathrm{ir}\ x \wedge x \leq t_{e1} \leq y$

For each of i_{alu1}, i_{mem1} and i_{mar} there is only one signal connected; thus the connection is direct. But for i_{acc}, i_{alu2} and i_{mem2} there are several signals connected; thus Gate-Bus structures need to be introduced. For this we have:

$\lambda(x,y).\ \exists \mathrm{i}_{acc}, \mathrm{g}_{alu}, \mathrm{alu}, \mathrm{i}_{alu2}, \mathrm{g}_{mem}, \mathrm{memout}, \mathrm{i}_{mem2}, \mathrm{g}_{mar}, \mathrm{mar}.\ \exists t_{e1}.$
$\mathrm{acc}\ y = \mathrm{i}_{acc}\ t_{e1} \wedge$
$\mathrm{i}_{acc}\ t_{e1} = \mathrm{DEST\_TRI}_{16}\ (\mathrm{g}_{alu}\ t_{g1}) \wedge$
$\mathrm{g}_{alu}\ t_{e1} = \mathrm{MK\_TRI}_{16}\ (\mathrm{alu}\ t_{e1}) \wedge$
$\mathrm{alu}\ t_{e1} = \mathrm{ADD}_{16}\ (\mathrm{acc}\ t_{e1})\ (\mathrm{i}_{alu2}\ t_{e1}) \wedge$
$\mathrm{acc}\ t_{e1} = \mathrm{acc}\ x \wedge$
$\mathrm{i}_{alu2}\ t_{e1} = \mathrm{DEST\_TRI}_{16}\ (\mathrm{g}_{mem3}\ t_{e1}) \wedge$
$\mathrm{g}_{mem3}\ t_{e1} = \mathrm{MK\_TRI}_{16}\ (\mathrm{memout}\ t_{e1}) \wedge$
$\mathrm{memout}\ t_{e1} = (\mathrm{FETCH}_{13}\ (\mathrm{memory}\ t_{e1})\ (\mathrm{i}_{mem2}\ t_{e1})) \wedge$
$\mathrm{memory}\ t_{e1} = \mathrm{memory}\ x \wedge$
$\mathrm{i}_{mem2}\ t_{e1} = \mathrm{DEST\_TRI}_{13}\ (\mathrm{g}_{mar}\ t_{e1}) \wedge$
$\mathrm{g}_{mar}\ t_{e1} = \mathrm{MK\_TRI}_{13}\ (\mathrm{mar}\ t_{e1}) \wedge$
$\mathrm{mar}\ t_{e1} = \mathrm{CUT}_{16\_13}\ (\mathrm{ir}\ x) \wedge$
$x \leq t_{e1} \leq y$

To introduce basic components we need the below axioms:

$\vdash \mathrm{REG}_{16}\ (i_{r1}, c_{r1}, o_{r1}) = \forall t.\ o_{r1}\ (t+1) = (c_{r1}\ t \Rightarrow i_{r1}\ t\ |\ o_{r1}\ t)$
$\vdash \mathrm{REG}_{13}\ (i_{r2}, c_{r2}, o_{r2}) = \forall t.\ o_{r2}\ (t+1) = (c_{r2}\ t \Rightarrow \mathrm{CUT}_{16\_13}\ (i_{r2}\ t)\ |\ o_{r2}\ t)$
$\vdash \mathrm{BUS}_n\ (i_{b1}, i_{b2}, \cdots, i_{bn}, o_b) = \forall t.\ o_b\ t = \mathrm{DEST\_TRI}_n\ (i_{b1}\ t\ \cup_n\ i_{b2}\ t \cup_n \cdots \cup_n\ i_{bn}\ t)$
$\vdash \mathrm{GATE}_n\ (i_{g1}, c_{g1}, o_{g1}) = \forall t.\ o_{g1}\ t = (c_{g1}\ t \Rightarrow \mathrm{MK\_TRI}_n\ (i_{g1}\ t)\ |\ \mathrm{FLOAT}_n)$
$\vdash \mathrm{ALU}\ (i_{alu1}, i_{alu2}, inc, add, sub, alu)\ = \forall t.\ alu\ t = (inc\ t \Rightarrow \mathrm{INC}_{16}\ (i_{alu2}\ t)\ |$
$add\ t \Rightarrow \mathrm{ADD}_{16}\ (i_{alu1}\ t)\ (i_{alu2}\ t)\ |\ sub\ t \Rightarrow \mathrm{SUB}_{16}\ (i_{alu1}\ t)\ (i_{alu2}\ t)\ |\ i_{alu2}\ t)$

Then we get:

$\lambda x.\ \mathrm{REG}_{16}\ (\mathrm{i}_{acc}, \mathrm{wacc}, \mathrm{acc}) \wedge \mathrm{wacc}\ (x+1) \wedge$
$\mathrm{BUS}_{16}\ (\cdots, \mathrm{g}_{alu}, \cdots, \mathrm{i}_{acc}) \wedge$
$\mathrm{GATE}_{16}\ (\mathrm{alu}, \mathrm{ralu}, \mathrm{g}_{alu}) \wedge \mathrm{ralu}\ (x+1) \wedge$
$\mathrm{ALU}\ (\mathrm{acc}, \mathrm{i}_{alu2}, \mathrm{inc}, \mathrm{add}, \mathrm{sub}, \mathrm{alu}) \wedge \mathrm{add}\ (x+1) \wedge$
$\neg(\mathrm{wacc}\ x) \wedge$
$\mathrm{BUS}_{16}\ (\cdots, \mathrm{g}_{mem3}, \cdots, i_{alu2}) \wedge$
$\mathrm{GATE}_{16}\ (\mathrm{memout}, \mathrm{rmem}, \mathrm{g}_{mem3}) \wedge \mathrm{rmem}_3\ (x+1) \wedge$
$\mathrm{MEM}\ (\mathrm{memory}, \mathrm{addrs}, \mathrm{data}, \mathrm{wmem}, \mathrm{memout}) \wedge$
$\neg(\mathrm{wmem}\ x) \wedge$
$\mathrm{BUS}_{13}\ (\cdots, \mathrm{g}_{mar}, \cdots, \mathrm{i}_{mem2}) \wedge$
$\mathrm{GATE}_{13}\ (\mathrm{mar}, \mathrm{rmar}, \mathrm{g}_{mar}) \wedge \mathrm{rmar}\ (x+1) \wedge$
$\mathrm{REG}_{13}\ (\mathrm{ir}, \mathrm{wmar}, \mathrm{mar}) \wedge \mathrm{wmar}\ x$

Also using the axioms we get a microcode sequence:

$\mathrm{wacc}\ (x+1) \wedge \mathrm{ralu}\ (x+1) \wedge \mathrm{add}\ (x+1) \wedge \mathrm{rmem}_3\ (x+1) \wedge \mathrm{rmar}\ (x+1) \wedge$
$\neg(\mathrm{wacc}\ x) \wedge \neg(\mathrm{wmem}\ x) \wedge \mathrm{wmar}\ x$

At the time x a microinstruction is constructed to give the above microcodes and followed by another microinstruction which gives these microcodes at $x+1$. Taking the specification as a whole, we have an algorithm to introduce register, gate-bus and time distributions. Using the algorithm for each sub-structure a linear microprogram is introduced.

4.4 Different Implementations

The collection of possible derivations forms a tree: its root is the specification and its leaves are different implementations including Gordon's implementation. Choosing different development paths in this tree leads to different implementations. However in this short paper we only give one example.

5 Implementation

We derived an implementation which has 4 buses, 4 registers and shorter microprogram. The implementation is as Figure 1 and 2.

6 Conclusion and Further Work

In contrast with post design verification, and with *ad hoc* automation of informal design, we formally derive non-trivial hardware (of course still simple), Gordon's computer. Our development is justified by rigorous logical inference and spurred by the inherent logical structure of the hardware. The derivation reveals a general hierarchical structure and various approaches to computer design in a single logical framework. The derivation process naturally shows development histories in rigorous logical form and can produce various implementations. Although derivation (or formal synthesis) is a more difficult task, it accords with a major concern of computer science: to study the active creation of systems. We have tried to give a typical example as evidence to demonstrate this combination of theory and practice. The essential problem of derivation is one of proof methodology.

The work of this paper is one case study in a programme for formal derivation of a class of computers. Further we will work to find a good mathematical structure to describe formal synthesis process and the work will consist of abstract synthesis theory and enginnering practice (down to register transfer level) two respects. We hope the work will help to reveal the inexorable link from specification to implementation, and various possible relations among specification and implementation for hardware in purely logical transparency.

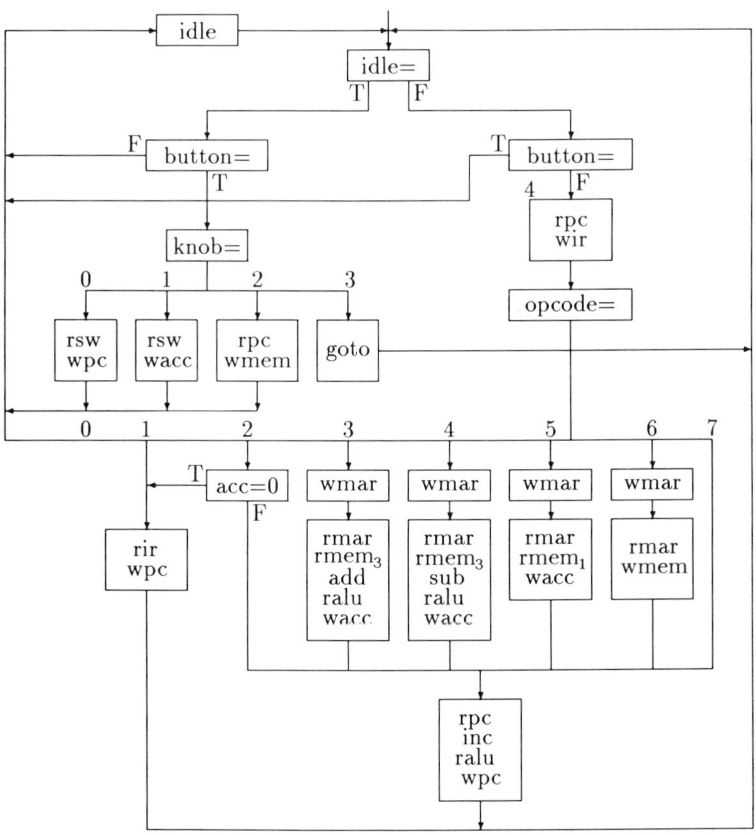

Figure 1: Flowchart of Microprogram

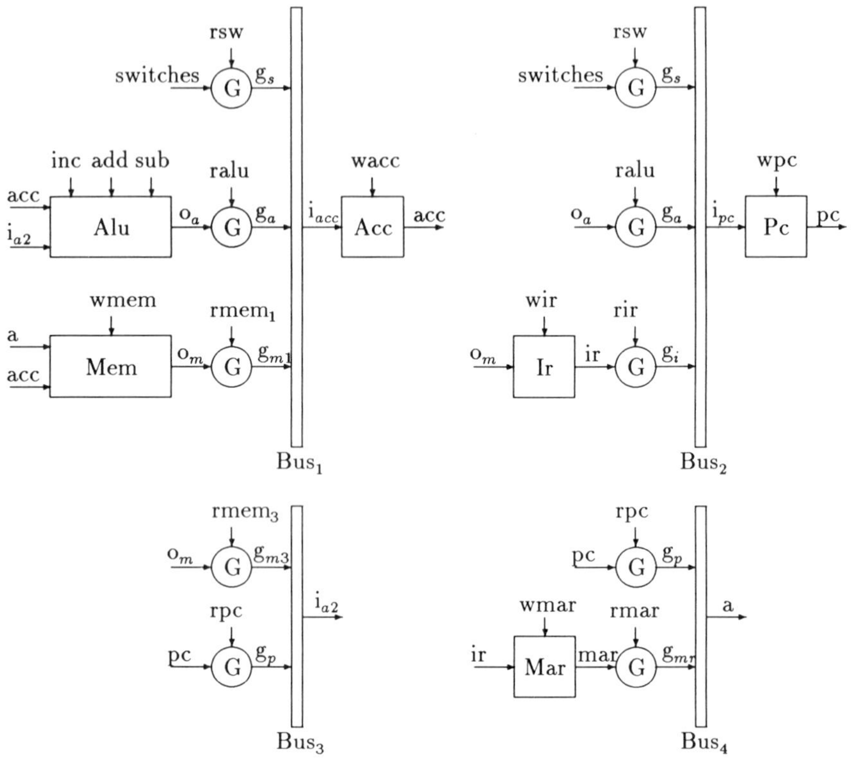

Figure 2: Data Path Implementation.

Note, here $i_{a2} = i_{alu2}$, o_a = alu, a = addrs, o_m = memout.

References

[1] M. P. Fourman, R. L. Harris, Lambda-Logic And Mathematics Behind Design Automation, 26th ACD/IEEE Design Automation Conference, 1988.

[2] M. Gordon, HOL: A Proof Generating System for Higher-Order Logic, University of Cambridge, Computer Laboratory, Tech Report No. 103, 1987.

[3] J. Joyce, G. Birtwistle and M. Gordon, Proving a Computer Correct in Higher Order Logic, Report No. 100, Computer laboratory, Cambridge University, 1986.

[4] Li-Guo Wang, Formal Derivation of A Computer, Correct Hardware Design Methodologies, edited by P. Prinetto and P. Camurati, Elsever science publishers B.V. 1992 ISBN: 0444893679

Higher Order Logic Theorem Proving and its Applications (A-20)
L.J.M. Claesen and M.J.C. Gordon (Editors)
Elsevier Science Publishers B.V. (North-Holland)

Formal Tools for Tri-State Design in Busses

R.B. Hughes[1], M.D. Francis[1], S.P. Finn[1] and G. Musgrave[2]

[1] Abstract Hardware Limited, Building 2, The Science Park, Brunel University, Uxbridge, Middx. UB8 3PQ. U.K. Email. roger@ahl.co.uk, Tel. +44 895 258501.

[2] Head of Department, Department of Electronic and Electrical Engineering, Brunel University, Uxbridge, Middlesex. UB8 3PH. U.K.

Abstract

The LAMBDA[3] system applies formal methods to the system design problem within an engineer-friendly environment. This paper presents our novel approach to the important problem of formal, verified, design using tri-state devices. Applying formal methods to real-world problems requires support for common design paradigms such as the use of tri-state busses. We illustrate our approach by reference to a safety-critical industrial example.

Keyword Codes: J.6; I.1.4; B.6.3

Keywords: Computer Aided Engineering; Algebraic Manipulation, Applications; Logic Design, Design Aids

1 Introduction

Logic Synthesis is now a reality [1] – silicon structures are generated from a specification of function. Behavioural Synthesis is increasingly successful at automating state assignment, scheduling and allocation [2]. With increasing demands placed on automation of the design *process*, and increasing levels of functional complexity and requirements for high integrity systems, architectural synthesis tools using formal reasoning are required to support system-level design.

Formal methods have already been successfully used [3, 4, 5] in a variety of complex case studies. In this paper we present a formal approach to the more complex issue of handling bidirectional busses. This problem has previously been sidesteped or tackled in an *ad-hoc* manner, e.g. VIPER [6] only had unidirectional busses owing to verification

[3] **L**ogic **A**nd **M**athematics **B**ehind **D**esign **A**utomation

difficulties. Bidirectional busses are particularly important since, for the vast majority of applications, they are an essential design component.

Our methodology is illustrated on a large scale, safety-critical, industrial example. Giving the designer freedom to explore the design space, with proof of the implementation being an *integral* part of this process and not a *post-hoc* verification activity, is regarded as essential in the design of the industrial example and is achieved through the use of *abstraction refinement.*

2 The LAMBDA System

In tackling real-world formal design, the integration of user-interaction and automated synthesis must be considered, as it has been before [7]. In addition, however, we require a system in which high-level specification (including features such as user-defined datatypes and recursively defined functions), and refinement of both data and timing are supported. The LAMBDA/DIALOG design assistant [8] is a system which implements and extends [9] these ideas. With such useful generic capabilities it can also be used for software engineering design [10].

Although any step in the synthesis (e.g. scheduling, see [11]) can, in principle, be automated, the toolset also allows interaction. This leaves the designer free to use his own skill and judgement during the course of the design. The system core will record his decisions and ensure that information and constraints, e.g. involving timing, are maintained and propagated correctly throughout the whole design process[12, 13].

Like most Theorem Provers, LAMBDA can make use of equational reasoning to rewrite design requirements into an equivalent, but simpler form. It also supports tactical and inductive proof. In addition, it uses unification, which permits synthesis; the addition of new components in order to achieve a design. The system was designed to be used by engineers rather than logicians, so the user interface is an important element of the toolset. A window interface, called DIALOG, allows the user to interact with the Theorem Prover using a schematic editor.

Our Theorem Prover represents a design state as a logical *rule*. Because the system only allows valid rules to be created only correct designs can be produced. The system creates new rules/designs by transforming[4] old ones. Thus at the start of any design/proof, we need a logically valid rule; this is often a tautology. As the design progresses, this will be transformed to :-

IF	current_design + further_work	ACHIEVES	task_n
AND	...		
AND	current_design + further_work	ACHIEVES	task_1
THEN	current_design + further_work	ACHIEVES	specification

and finally – *If I have this complete implementation and have an external environment then I have implemented this specification.*

[4] see [14] for details of how rules are transformed.

$$\frac{}{\text{completed\_design} + \text{external\_environment} \quad \text{ACHIEVES specification}}$$

Here the external environment will typically be information about how and when inputs and outputs are supplied.

Thus, to start a new design, we need to start with a tautological form of rule solely containing details of the specification – *If I can implement this specification then I can implement this specification.* The proof system ensures that the design, at each and every stage, is correct *with respect to* the given specification.

The formal specification of the tasks to be achieved is thus a very important part of the design process. This usually involves converting an imprecise set of requirements in a natural language into a formal (unambiguous) statement of the goals to be achieved. it should be noted that no one (whether using formal methods or not) can avoid the problems that will occur if a system is incorrectly specified. Higher-order unification, as used in ISABELLE [15], allows our specifications to include parameters (e.g. latency) whose values are determined during the course of the design.

The specification is written in a logical specification language based on the functional programming language ML [16] together with various logical connectives. This has the advantage that functional specifications can be read into an ML compiler to create ML functions, and hence animated.

A *component* (as well as a *specification*) is modelled as a relation between its input and output signals. The relation describes which combinations of inputs and outputs are possible, thus describing the effect of the component on its environment. Note that since the relation is between *signals*, not just the values of those signals at a particular time, we are not restricted to dealing with purely combinational devices – we can relate an output signal to the value of an input signal at any previous time.

INV and DELAY components

For example, INVDELSPEC specifies a circuit whose output is the inverse of its input, delayed by one cycle:

```
val INVDELSPEC#(i : bool signal,          val INVDELIMP#(i : bool signal,
                o : bool signal) =                       o : bool signal) =
  forall t : time.                          forall t : time.
    o (t + 1) == not (i t);                 exists w. INV#(i,w) /\ DELAY#(w,o);
```

Internal wires are 'hidden' using existential quantification, as used in the implementation above. The rule[5] representing the implementation of this specification is:-

```
-------------------------------------------
INV#(i,w), DELAY#(w,o) |- INVDELSPEC#(i,o)
```

[5]In the Logic, ACHIEVES is represented by '|-' ; IF, THEN and AND are assumed

This is a *theorem* of the formal logic, which states that we can use an `INV` and a `DELAY`, appropriately connected, to implement `INVDELSPEC`.

Implementation of INVDELSPEC

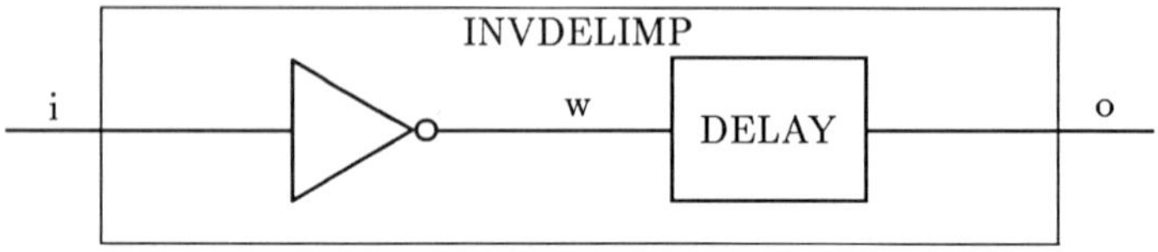

Limitations of the Hardware Model

The DIALOG hardware model is very simple:

- 2-valued (for booleans)
- Synchronous
- One implicit clock (usually)
- Time modelled as natural number

We can use the hardware model to prove, owing to the fact that a *false hypothesis implies anything*, that each of the following circuits also meets the `INVDELSPEC` specification :-

```
val BADCIRC1#(i,o) =
  exists w. PWR#(w) /\ GND#(w)          (* outputs connected together *)
```

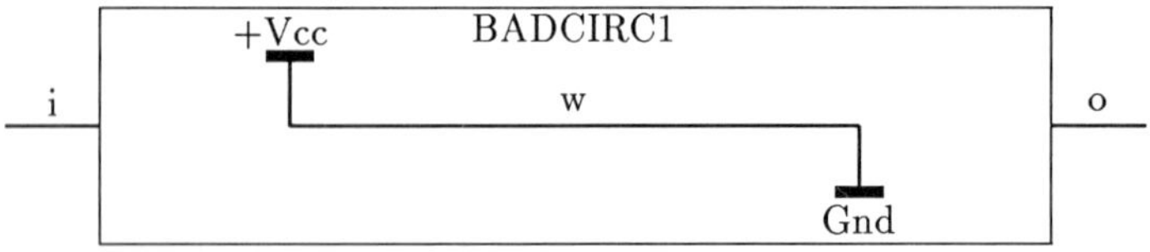


```
val BADCIRC2#(i,o) =
  exists w. INV(w,w)                    (* zero-delay loop *)
```

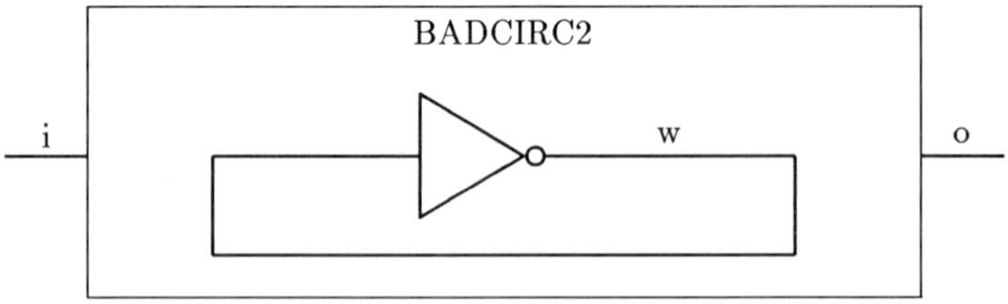

This is clearly ridiculous – real hardware does not behave like this. What has gone wrong?

The problem is that we assume that the relation between the ports of a component, specified by its definition, always holds; in both of the above circuits this is not the case.

Design Rules

How can we avoid this problem? Firstly, the problem only occurs for design errors – well-behaved circuits do not suffer the same problems. Thus, our solution is to keep the model simple, but disallow 'paradoxical' circuits, this also avoids the computational expense of a more sophisticated model.

We keep our existing hardware model, and impose some extra constraints on how we construct circuit descriptions. We achieve this by identifying each port of a component as either an input port or an output port; only an output port is allowed to affect the value on a wire.

The DIALOG description of a component is the LAMBDA description plus:

- Modality (Input or Output) of each port.
- List of zero-delay paths through component.

DIALOG enforces Design Rules to ensure that the naïve, underlying hardware model remains valid:

- Outputs must not be connected together.
- No zero-delay loops.

The first design rule is remarkably effective; although (at the switch level) wires are actually bidirectional devices, the vast majority of them are used to pass information in one direction only. Our design rule checker, implemented as part of DIALOG, will throw out any attempt to construct circuits violating either of the above rules.

3 Industrial Example

As part of the PATRICIA project (ESPRIT Project No. 5020), a formal specification and subsequent synthesis [17] of an industrial-size example was carried out. The example was provided by GEC Avionics and is the PCML (Process Control & Monitor Logic) unit. The unit is intended to provide the interface and timing functions required to support Intel 80386, Motorola 68020 and Inmos T414 microprocessors. The informal specification of this example was given in a GEC Avionics document.

The specification for the PCML is decomposed into parts that can be easily recognised in terms of the informal specification. The sections are: the checkword monitor, the downcounters (for providing the timing for the generation of the sub-frame and frame interrupts), the reset generator, the interrupt generator and the CRC (Cyclic Redundancy Checker). The control and status registers are specified in terms of a special control unit; the address decoder is also handled. We will not go into the details of the whole specification in this paper but we will, instead, concentrate on details pertinent to the use of the formalism for handling the tri-state bus, which is an essential part of the design. The whole of the PCML unit has been synthesised from the specification to a level where the implementations of all the sub-sections of the design have been obtained.

It should be noted that the specification was deliberately abstract, i.e. as uncompromised as possible, in order to give the designer freedom to explore the *design space* by trying out different implementations of the design, all of which formally meet the specification. Flexibility in design and freedom to explore the design space, with proof of the implementation being an *integral* part of the design process and not a *post-hoc* verification activity, are considered essential in this industrial design example. In this *abstraction refinement* process, some parameters, which were left undefined in the specification, were automatically given values during the design process with different implementations resulting in different bindings.

3.1 Overview of Specification

The device is quite complex and has been specified in a hierarchical manner in a traditional "divide and conquer" approach. Each component part of the design has been specified separately to aid both comprehension and verification of these parts. The essential characteristics of the PCML have been captured in the formal specification. The PCML's specification is slightly constrained in order to match the informal requirements as to bus widths, particular bits from registers being used etc. The act of formally specifying the design highlighted some important points in the informal specification that were not apparent.

The basic components of the specification are the Checkword Monitor, the Frame Timers, the Cyclic Redundancy Checker, the Interrupt Generator, Reset Generator and the Address Decoder and control logic. Some of these are further sub-divided into simpler units. Figure 1 shows the top-level LAMBDA specification for this design. In figure 2 the initial steps in achieving the design, by introducing a tri-state component, are shown[6]. The figures show the DIALOG interface to LAMBDA and it will be seen that the figure shows the screen split into three main areas. The larger area to the top left is labelled *Schematic* and is the schematic window which shows the current state of the design in a graphical manner. This also provides a set of menus by which the engineer directs the design process. The window area under this (at the bottom left of the figure) is labelled *Goals* and is the window used to display the current state of the rule being manipulated. Ihe information about the current rule state is used to direct the design. The hypotheses are not shown in this window to simplify the amount of information displayed to the user; hypotheses are usually components, these are shown in the schematic window. Two additional items of information are also displayed in the title bar of this window: *Depth* and *Level*, which indicate the current depth in the hierarchy and design step level respectively. Finally. the window to the bottom right, which is labelled *Netlist*, shows the names of the components at the current depth of the design and the connectivity between these components. The top-level design is concluded in figure 3, which shows the top-level implementation schematic, the bus being shown in bold; the sub-components still require designing.

[6]The figure also illustrates the new Motif style interface available with LAMBDA v4.2, available September 1992 – other figures use v4.1

```
val PCML#(resin_, cs_, we_, oe_, addr, dataIn, databus,
          intack_, intsel, userint, frameint, int1_, int2_, avec_,
          cwmok, reset_) =

  exists subframe, frame, instr, shifting, ftres, ftenable,
         contSel, opSel, preset1Sel, preset2Sel, count1Sel, count2Sel,
         crcDSel, cwdContSel, cwdSel, tcSel, cwrate, crcreset, cwperframeok,
         interlock, resint_, extDriven.
  forall t.

    extDriven t == not (we_ t) && oe_ t

 /\ (extDriven t == true ->> databus t == dataIn t)

 /\ ADDRDEC#(resint_, cs_, we_, oe_, addr, instr,
            contSel, opSel, preset1Sel,
            preset2Sel, count1Sel, count2Sel, crcDSel,
            cwdContSel, cwdSel, tcSel)
 /\ CONTROLS#(resint_, instr, shifting, cwperframeok, contSel, cwdContSel,
             opSel, ftres, ftenable, interlock, cwrate, crcreset, databus)
 /\ CWMONITOR#(resint_, instr, cwdSel, tcSel, databus,
              cwrate, frameint,    cwperframeok, cwmok)
 /\ CRC#(crcreset, databus, instr, crcDSel, shifting)
 /\ INTERRUPT#(intsel, frame, subframe, intack_, userint,
              frameint, int1_, int2_, avec_)
 /\ RESINTERLOCK#(resin_, interlock, cwmok, resint_)
 /\ EXTRESET#(resint_, reset_)
 /\ FRAMETIME1#(ftres, ftenable, databus, instr, count1Sel,
               preset1Sel,   subframe)
 /\ FRAMETIME2#(ftres, ftenable, databus, instr, count2Sel,
               preset2Sel, subframe,   frame);
```

Figure 1: *Top-level PCML Specification*

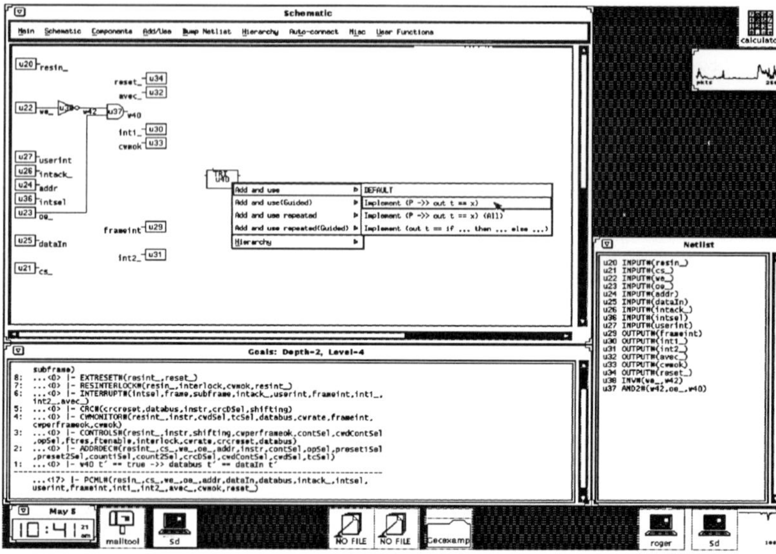

Figure 2: *Introduction of TRI*

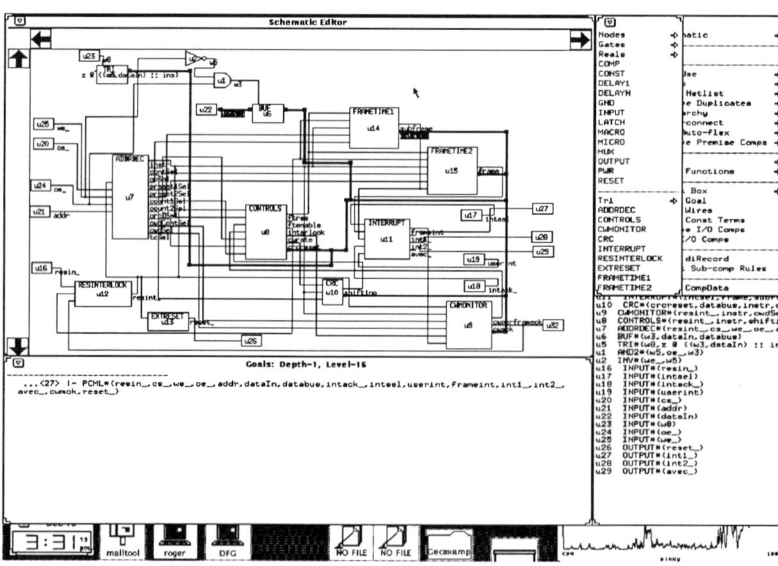

Figure 3: *Top-level Design*

To specify the PCML fully, yet abstractly, the relationship between the control and status registers and the remainder of the PCML must be specified. The control unit provides the missing registers that are not defined in the other units and also supplies various signal lines derived from specific bits of the operations and control registers, viz. *ftres*, *ftenable*, *interlock*, *cwrate* and *crcreset*. It is necessary to drive the bus with the contents of the various data registers; this occurs when the instruction line is a *Read* signal and the appropriate selection signal activated. Thus the bus must be multiply driven, necessitating tri-state logic.

The Need for Tri-state Logic

Tri-state logic is required otherwise busses cannot be multiply driven. Several signals need to be placed onto the bus, but at different points in time in order to avoid the contention problem, viz. a bus clash. Normally, in DIALOG, outputs cannot be simply connected together, this being a Design Rule requirement to allow our formalism to work. Whilst unidirectional busses are not unknown, most designs (including the PCML example given here) incorporate at least one bidirectional bus using tristate drivers. Modelling these busses in logic produces the difficulty explained in section 2 – a design where the bus is simultaneously driven to two different values can be used to implement any specification. Unfortunately, we cannot adopt the same design rule solution – a bidirectional bus is *inherently* bidirectional, so any attempt to impose directionality on it is bound to fail.

Whilst for some safety-critical applications it might be acceptable to forgo the use of bidirectional busses, for the vast majority of applications this would be completely unacceptable. The remainder of this paper explains, and illustrates by means of the industrial example, how we overcome the problem of allowing the use of tristate devices without overstepping the validity of our hardware model.

By way of example, figure 4 shows three remaining sub-goals in the implementation of the CONTROLS section and figure 5 shows the LAMBDA specification for FRAMETIME1 which drives two signals on to the bus. Altogether in the design, 11 different signals need to be driven on to the bus, all but one of these coming from within the PCML. More will be said about this external signal after the model of the ideal bus has been introduced.

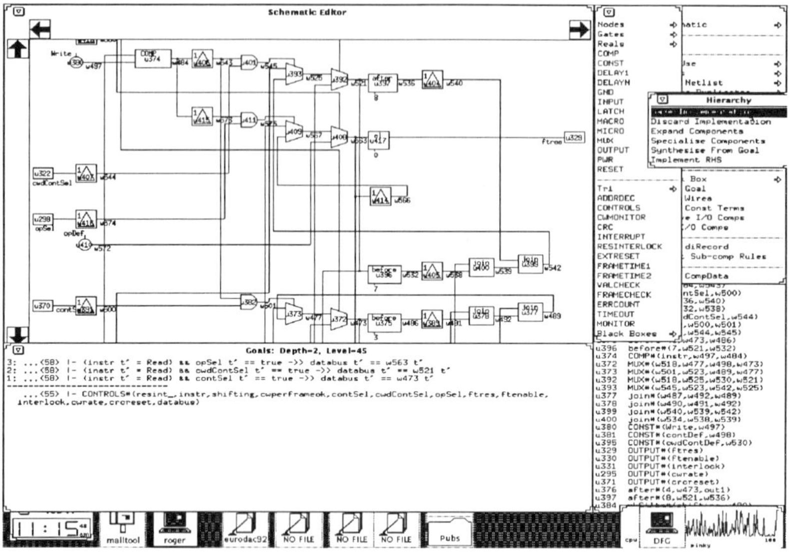

Figure 4: *Control subcircuit with Constraints*

```
val zero16  = zeroN 16;
val FRAMETIME1#(ftres,enable,bus,instr,countSel,presetSel,  interrupt) =
  exists counter, preset.
    forall t.
      (counter (S t) ==
        if (instr t = Write) && countSel t then bus t else
        if ((counter t = zero16) || ftres t) && enable t then preset t else
        decList (counter t))
    /\
      (preset (S t) ==
        if (instr t = Write) && presetSel t then bus t else preset t)
    /\
      (interrupt t == (counter t = zero16))
    /\
      ((instr t = Read) && countSel t == true ->> bus t == counter t)
    /\
      ((instr t = Read) && presetSel t == true ->>  bus t == preset t);
```

Figure 5: *Specification of FRAMETIME1*

Ideal Bus Model

The mechanism required for the handling of lines which are used both for input and output, i.e. in a *bidirectional* manner, is tri-state logic. Since, in order to prove the bus correct, all drive signals must be known, it is essential that the bus be 'internal' to the whole device. Otherwise, a proof of correctness with the bus is only possible if properties are assumed of line drivers external to the circuit being specified.
In our solution, we adopt a two-level approach to bidirectional busses:

1. We implement the circuit using an 'ideal bus'. At this stage we ignore potential bus clashes.
2. We implement the 'ideal bus' using real components. This will generate side conditions. These can be discharged if and only if no bus clashes occur.

To facilitate this approach design rule checking for tristate ports must be relaxed in order to allow several buffer elements to be connected to the same bus. When this is done, any number of such ports may be connected to each other (and to inputs). It should be noted that this would not be necessary, in fact the problem would become moot, if we were to extend our hardware model to a 3 or 4 valued one. We do not do this owing to the need to keep the rules in the rule libraries as simple as possible so as to minimise the time taken in the various rewrite operations. This is especially important when one considers the computational complexity resulting from the numbers of components involved in a 'real-world' industrial design example. As a result, we favour keeping the simple hardware model and using an alternative solution.

Our first approximation uses an unrealisable component – the ideal bus. This hypothetical device transmits data perfectly, even if there is bus contention.

```
type 'a ensig  = bool signal * 'a signal
              (* enable      ,  data    *)

val TRI#(pull : 'a signal,
         ins  : 'a ensig list,
         out  : 'a signal) =              (* Ideal bus          *)
  forall t : time.
    case enabled ins t of
      [w]          => out t == w t        (* Only one enable    *)
    | []           => out t == pull t     (* No enables         *)
    | _ :: _ :: _ => FALSE                (* 2 or more enables *)
```

Each bus takes a list of port pairs, where each pair consists of an enable signal and a data signal. A list is required because it is not known, in advance, how many devices will be connected to the bus. When the bus is first introduced, `ins` will be a 'flexible' variable. As each device is attached to the bus, `ins` becomes further instantiated.

```
G // TRI#(pull,(c,w)::ins,out) $ H |- P#(out t) /\ c t == true
---------------------------------------------------------------
G // TRI#(pull,(c,w)::ins,out) $ H |- P#(w t)
```

The (c,w) port pair in the above represents a connection to the bus. The first element of the pair is an enable signal. When this signal is active the value on the wire w is placed on the bus (out). The bus is used by re-using the TRI component for each extra signal to be connected to the bus, a different (c,w) port pair being added for each additional device. The pull value in the above is the value to which the bus is to be pulled when it is not being driven, i.e. when none of the enable signals in the port pairs list is active.

A critical decision has to be made, at some stage, in the design process. This occurs when it is thought that there are no more components to be connected to the bus, i.e. when the circuit is complete in terms of signals needed to be driven on to the bus. When this stage is reached, the tail of the list is instantiated to nil. Once this decision has been made, it is not possible to add further devices to the bus, which is why we term it a *critical* decision. This 'closed world assumption' is necessary in order to check that no bus clashes occur.

In order to implement the bus, several real components are required. The LAMBDA definitions of these components follow :-

```
val TRIBUF#(cntrl : bool signal,            (* A tristate driver *)
         in1   : 'a signal,
         out   : 'a signal) =
  forall t : time.
    cntrl t == true ->> out t == in t  (* Activated when cntrl active *)

val PULL#(pull : bool signal,               (* To pull to known value *)
          ins  : 'a ensig list,
          out  : 'a signal) =
  forall t : time.
    allDisabled ins t ->> out t == pull t (* Pull bus when not driven *)

val TRIBUFVEC#(pull : bool signal,
            ins  : 'a ensig list,
            out  : 'a signal) =
  PULL#(pull,ins,out) /\
  forall n. n < length ins == true ->>
    TRIBUF#(fst (el n ins), snd (el n ins),out)
```

The key property we need is that a collection of tristate drivers behaves as an ideal bus *provided* that we can show that there are no bus clashes:

```
fun noClashes (ins : 'a ensig) =           (* NoClashes true if  *)
  forall t : time.                         (* just one (or no)   *)
    length (enabled ins t) < 2 == true     (* inputs are enabled *)

G // H |- noClashes ins
----------------------------------------------------
G // TRIBUFVEC#(pull,ins,out) $ H |- TRI#(pull,ins,out)
```

thus allowing the `TRI` component to be eliminated from the design.

It should be noted that `TRIBUFVEC` is never displayed in DIALOG as a component. What happens during implementation of the `TRI` component is that `TRIBUFVEC` is expanded into *n* `TRIBUF` components and an (optional) `PULL` component. In terms of the DIALOG interface, this means that only `TRI` components are ever added manually, with `TRIBUF` components being automatically added each time the `TRI` is re-used. When the `TRI` component is then implemented in DIALOG, it will disappear from the schematic and a `PULL` component (if option necessary) will be added.

The use of these `TRI` components with the associated `TRIBUF` line-driver components is crucial in being able to carry out the synthesis of the design. The reason for this being that many of the registers are designed to be accessible to the external microprocessor device via a single I/O bus and, as such, these values need to be driven on to the same bus, albeit at different times. The use of a `TRI` component for a particular bus allows `TRIBUF` driver units to be introduced, one at a time, for each signal that needs to be output on to the tri-state bus. A new `TRI` component is only required when a different bus needs to be driven.

3.2 The Design

Figure 2 shows the initial step in the design process, where a `TRI` component to represent connection to the non-ideal bus is in position. The same `TRI` component remains in the design until all values that need to be driven on to the bus are included in the implementation. The `TRI` component is then, itself, implemented and this produces a list of *noClashes* which must be met in order to guarantee operation of the circuit. Re-use of the `TRI` component is required to implement a goal of the form :-

```
G //H  |- P ->> databus t == Q
```

which may occur several levels down in the hierarchy, c.f. figure 4 which shows the synthesis of `CONTROLS`, one of the submodules in the `CWMONITOR` module of the PCML (figure 3), in which three goals of the above form are left as a constraints at this level of the hierarchy. When the critical design decision point is reached, the relevant levels of the hierarchy can be expanded and any hidden constraints will appear on the goalstack. As these constraints are discharged, the set of conditions for which no clashes must occur extends on the `TRI` component. The list extends by a length of one for each `TRIBUF` component that is introduced for each signal driven onto the bus. Once all signals that need to be driven on to a particular bus have been accounted for, then, and only then, may the `TRI` component be implemented. Figure 2 also illustrates the use of a `TRIBUF` component for handling an external drive to the databus; this is discussed later.

The similarity of the overall PCML specification (figure refschema) to one presented in the informal specification was refreshingly reassuring.

External Drive – a Practical Approach

In real-world designs, bidirectional busses are not always used in the simple manner described earlier in our 'closed world assumptions'. There is a difficulty when a bidirectional bus is used to communicate with devices external to the device being designed. The problem is currently addressed by conceptually splitting the bus into a 'closed world' *internal* bidirectional bus and an exetrnal bidirectional connection. The bidirectional connection to the external device/s is further split into :-

- Case 1. A unidirectional input.

  The input is connected to the internal bidirectional bus via a tristate driver and the output connection is simply added after the TRI component has been implemented. The main problem here is that no information about the external devices is available, thus there is no timing information to indicate when the line is being driven by an external source.

- Case 2. A unidirectional output.

  There is no problem with tri-state here since it is possible to manually add a single output port after the TRI driver has been implemented.

This solution has been used in the current synthesis approach. Case 1 raised the difficult question of knowing when the external devices are driving the input. However, it is clear that is is not being driven all the time; external devices should only be driving the bus when the internal drivers are inactive. That last comment gives the key to the general solution of this problem, i.e. the inverse of the set of conditions used for driving the internal bus should be taken as the *minimal* set of conditions to be used to drive the external unidirectional input onto the internal bus. This is the solution illustrated in figure 2.

Implementation of TRI Components

Implementation of the TRI component gives rise to a new **noClashes** subgoal. This can either be discharged by proving that no two TRIBUF components are activated at the same time, or provided as a set of conditions which must hold in order for the design to work according to the specification. Naturally, all the various signals which need to be driven onto the bus have already been implemented via buffers, by reselecting the TRI component which was introduced at a very early stage in the design. In this design there is only one TRI component owing to the fact that there is only one bus. The implementation of this component is done by selecting the implementation option on its component menu. In some cases it is necessary to introduce a PULL component to allow for the real-world design case where the bus, if not driven by another signal, is 'pulled' to some value rather than being allowed to float. This is one of the options on the component menu. For this example no PULL was required. These selections become available, as menu options on the TRI component, once the first TRIBUF element has been added. The implementation itself is entirely automatic.

4 Conclusions

A formalism for handling the representation of tri-state components in order to drive bidirectional busses has been given. The practical use of this formalism in various stages during the hierarchical design (synthesis) of a real, industrial-size example, directly from its formal specification, has been clearly illustrated. Once this level of design has been achieved, it is straight-forward to automatically translate LAMBDA's net-list into the format required by the chosen silicon compiler (e.g. VHDL, HILO3, EDIF)

Future work will continue to be example driven. For example, we are improving our representations of drivers to allow bidirectional buffer elements to be represented and to have tristate connector ports as components.

5 Acknowledgements

A large part of this work has been supported by the Commission of the European Communities under the PATRICIA project (ESPRIT Project No. 5020). We should also like to thank GEC Avionics for permission to use their industrial example, the PCML unit.

References

[1] R.K. Brayton, R. Camposano, G. DeMicheli, R.H.J.M. Otten, and J.T.J. van Eijndhoven. The Yorktown silicon compiler system. In D. Gajski, editor, *Silicon Compilation*. Addison-Wesley, 1988.

[2] R. Camposano and R.M. Tabet. Design representation for the synthesis of behavioural VHDL models. In *Proceedings CHDL'89*, Washington DC, June 1989.

[3] S. Finn, M. Fourman, and G.Musgrave. Interactive synthesis in higher order logic. In *Proceedings of the 1991 International Workshop on the HOL Theorem Prover and its Applications*, Davis, California, August 1991.

[4] E. Mayger and M. Fourman. Integration of formal methods with system design. In A. Halaas and P. B. Denyer, editors, *VLSI 91*, Edinburgh, Scotland, August 1991.

[5] E. Mayger, M. Francis, R. Harris, G. Musgrave, and M. Fourman. The need for a core method. In *EUROMICRO 91*, September 1991.

[6] Avra Cohn. Corectness properties of the viper block model: The second level. In *Current Trends in Hardware Verification and Automated Theorem Proving*. Springer Verlag, New York, 1989.

[7] N. D. Dutt, T. Hadley, and D. D. Gajski. An intermediate representation for behavioral synthesis. In *Proceedings of 27th ACM/IEEE Design Automation Conference*. IEEE, 1990.

[8] Abstract Hardware Limited, Brunel University, Uxbridge, Middx. U.K. *The LAMBDA System - Complete Reference Set, v. 4.1*, 1991.

[9] R. B. Hughes and G. Musgrave. Design-flow graph partitioning. In *Proceedings of the 1992 International Workshop on the HOL Theorem Prover and its Applications*, September 1992.

[10] R.B. Hughes and R.M. Zimmer. Automated interactive verification of functional programming languages. In C.M.I. Rattray and R.G. Clark, editors, *The Unified Computation Laboratory*, pages 411–423. Oxford University Press, May 1992.

[11] M.P. Fourman and Eleanor M. Mayger. Formally based systems design—Interactive hardware scheduling. In *VLSI '89*. North-Holland, 1989.

[12] M. Mendler. Constrained proofs: A logic for dealing with behavioural constraints in formal hardware verification. In *Workshop on Designing Correct Circuits, Oxford.* Springer-Verlag, 1990.

[13] H. Busch and G. Venzl. Proof-aided design of verified hardware. In *Proceedings of 28th ACM/IEEE Design Automation Conference*, San Francisco, California, June 1991. IEEE.

[14] S. Finn, M. Fourman, M. Francis, and R. Harris. Formal system design—interactive synthesis based on computer-assisted formal reasoning. In Dr. Luc Claesen, editor, *IMEC-IFIP International Workshop on Applied Formal Methods for Correct VLSI Design, Volume 1*, pages 97–110. Elsevier Science Publishers, B.V. North-Holland, Amsterdam, 1989.

[15] L. Paulson. Natural deduction proof as higher-order resolution. *Logic Programming*, 3:237–258, 1987.

[16] R. Milner, M. Tofte, and R. Harper. *The Definition of ML.* The MIT Press, Cambridge, Massachusetts, 1990.

[17] R.B. Hughes. Final report on GAv Example- the PCML Unit. PATRICIA Report RBH/9112/1, Abstract Hardware Limited, December 1991. Esprit Project No. 5020.

Higher Order Logic Theorem Proving and its Applications (A-20)
L.J.M. Claesen and M.J.C. Gordon (Editors)
Elsevier Science Publishers B.V. (North-Holland)

Specification and formal synthesis of digital circuits

M. Bombana, P. Cavalloro and G. Zaza

ITALTEL SIT, DRSC, 20019 Settimo Milanese, Milan, Italy
email : cavallor@settimo.italtel.it

Abstract

To reduce the 'time to market' and be competitive, new design methodologies must be introduced in order to guarantee rigor of design practice and correctness and reliability of implementations. Formally based synthesis at the most abstract levels of design can be used to reach such goals. In the following we describe the activities carried on in the Design Automation Laboratory of Italtel concerning post design validation using OTTER, and formal synthesis using LAMBDA (*). Particular emphasis is given to the problem of the integration of these design phases into the consolidated Italtel design flow through EDIF and VHDL interfaces. Comments on benefits and drawbacks as perceived by the industrial environment are included.

Keyword Codes: B.7.0; B.7.1; B.7.2
Keywords: Integrated Circuits, General; Types and Design Styles; Design Aids

1. INTRODUCTION

The global quality of the production process and the ability to be on the market before competitors are two factors which allow any industrial company to maintain and increase its consolidated market share. This is more true for the microelectronic market, where the ratio between the design time of a new component and its mean life on the market is so elevated. That is why, in this case more than elsewhere, new design methodologies play such a vital role in reducing effectively the 'time to market' [Go92]. Any innovation and improvement in this area can reduce 'de facto' the design time by applying exact and rigorous design methods, but also help in reducing the implementation time by avoiding the need of redesign steps.

To define a right–first–time design methodology, it is necessary to act in the initial stages of the design flow, where errors and inconsistencies can be introduced into the same specifications because of the complexity of designs. These errors are responsible for a great deal of the further implementation remaking. The earlier the inconsistencies are introduced and the more difficult is the process of their later identification and subsequent correction.

(*) This research is part of the ESPRIT II Project n. 5020 – PATRICIA

The exploitation of formal methods in a design environment, as a way to exhaustively guarantee the correctness of an implementation against its specification, is on progress in Research and Development Laboratories but also in some industrial sites [Ma88] [An91]. A still unsolved problem is the definition of the correct relationship between this phase of the design process and the most traditional ones leading down to layout. Moreover a correct implementation is not enough to satisfy alone the needs of the microelectronic market: a device must not only be correct but also easily testable. This means that the implementer must guarantee the observability and controllability of the I/O ports of the functional blocks which are included in the system. And this must be accomplished at all the hierarchical levels of definition of the design [Bo92].

We intend to cover in the following some of the activities which have been carried on at the Design and Automation Laboratory of Italtel in the last few years on such issues. In more detail we intend to give our contribution to the answers to two of the most crucial questions related to the application of formal methods in a production environment:

1. the identification of the correct phase of the design process where it is more convenient to apply these techniques
2. the implementation of links with the other phases of the consolidated design flow going down to the layout level.

At first we describe the use of a first order logic theorem prover for post–design verification and the approaches used to integrate it into an object oriented proprietary design data–base. Afterwards we explain the reasons why we choose a higher order logic theorem prover to carry on formally based synthesis at the more abstract levels of design. At the end the integration into the Italtel consolidated design flow is considered in detail. Finally comments, results and future activities are discussed.

2. POST DESIGN VALIDATION WITH OTTER

In the past the most widespread use of formal methods has been in the post–design verification of the correctness of the implementation versus the initial specification. The difference between this technique and the traditional simulation phase lies in the fact that in the latter a special set of test signals must be identified and the final result is dependent on this choice (non–exhaustive validation), while in the former all the cases are covered (exhaustive validation). But while formal validation can only guarantee the logical equivalence of the two representations, simulation analyzes details much closer to the physical functionality of the device.

A main role in post–design validation applications is played by the specification language. In fact only logic languages can be formally manipulated by theorem provers. Moreover both behavior and structure must be expressed in the same language to allow the equivalence derivation. Cases have been reported [An89] where the comparison of the two representations highlighted errors which had escaped the simulation phase.

Our choice for implementing post–design validation has fallen on OTTER [Ot89], a first order logic theorem prover developed at the Argonne National Laboratory of Illinois.The entry language is prolog–like, demonstrations are carried on applying rewrite rules based on a set of theorems and lemmas for the structural and functional primitives defined by the user. Different demonstration strategies can be applied [Ca91].

A complete environment has been designed and developed, including:

1. axiomatic definitions of primitive structural cells

2. axiomatic definition of functional primitive operators
3. a set of mixed axioms to express the behavior of the primitive structural cells in terms of the primitive behavioral operators

This environment has been used to validate both combinational and sequential systems, of complexity up to an ALU [Pe90]. Demonstrations are quick and straightforward, once the environment has been correctly defined. The main limitation lies in the fact that being the tool based on first order logic, tactics like induction cannot naturally be defined. They are left to be worked out by the user, increasing the number of working sessions to be carried on to complete one demonstration.

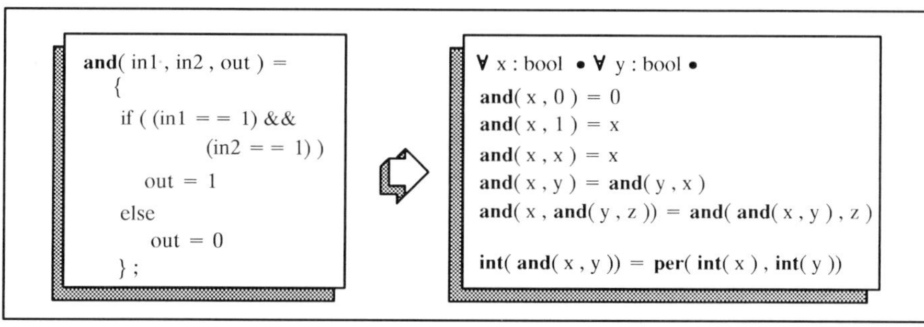

Figure 1. Comparison of the functional and logic models of a simple gate.

The main problem in applying these methodologies in an industrial environment is their complete detachment from the utilized standard methodologies. The problem involves both human and machine aspects:

1. the designer must no longer think in functional, or computational terms (simulation models), as it occurred up to now, but instead in terms of axioms and logic properties (figure 1)
2. different tools must be able to communicate automatically through a common data base or through standard data format.

The first point can be overcome with training or introducing a new professional profile, the mathematician/logic expert in hardware modeling. Analogies can be found in the past, when the expert in simulation was required, or with the knowledge engineer of the area of expert systems.

The second point can be approached in two different ways: the integration approach that is possible in case of proprietary databases, or the encapsulation approach that is feasible whenever a CAD framework allows the use of different vendors tools. We will describe in part 4 our activities in both these lines.

In our opinion post–design validation is a useful technique but doesn't constitute that innovative step in design methodology which is so badly needed to be more competitive on the market. The benefits that it brings are not so overwhelming in comparison to the deal of effort required for its introduction. We believe that these drawbacks make this approach less promising for industrial exploitation than it would appear from a theoretical point of view. Most of the previous issues are instead overcome by the introduction of formally based synthesis tools utilizing higher order logic.

3. FORMALLY BASED SYNTHESIS WITH LAMBDA/DIALOG

At the moment there is no automatic tool/environment able to support completely the first phases of the design process. Still pencil and paper seem to be the most used tools when the designer is trying to derive from an informal specification the best architectural implementation for the device. This phase is really critical, and can heavily influence all the subsequent implementation phases.

The next generation of design environments must fill this gap. Our everyday experience points to some characteristics which will have to be absolutely considered:

1. the designer must be given the possibility to conduct an **exploratory** activity in the first and most abstract design phases
2. it must be guaranteed a quick and automatic generation of correct by construction **qualitative alternatives**
3. **cost/benefit parameters** must be explicitly derived to support the designer in the choice between alternatives

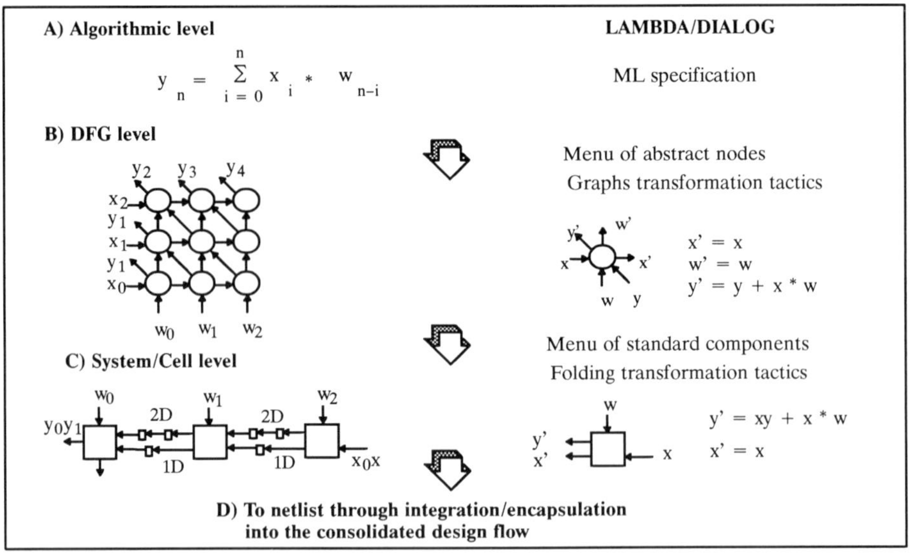

Figure 2. The abstract design levels addressed by formal synthesis.

The LAMDBA/DIALOG [La92] system already includes some of these requirements and looks promising for adding what still is not there [Fi89] [Fo89].

The abstraction levels we have addressed are: the algorithmic level, the Data Flow Graph/Control Flow Graph level, and lastly the System/Cell level where the chosen architecture is implemented (fig. 2) [Pa91a] [Pa91b]. Problems facing the designer at these stages of the design flow are widely dependent on the applications: we have focused on Digital Signal Processors (DSP), and DSP–like devices, for the importance they cover in the communication area. In order to express highly parallel and regular topologies, special tactics have been introduced with the aim to deal with such basic regularity.

On the contrary, when general topologies are concerned, the crucial point lies in the derivation of the overall architecture for the device according to resource optimization, including an efficient use of memories and register files, and to the global scheduling.

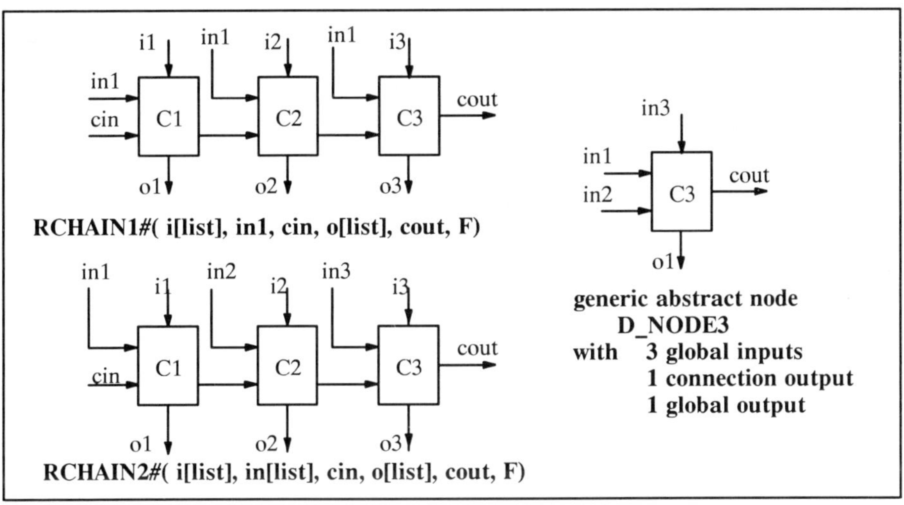

Figure 3. Regular structures and abstract nodes.

The specification language (ML) owns the necessary expressiveness to operate at the required abstraction levels. For example, components are polymorphic, that is they can be applied to signals and busses without modification. No explicit clock is defined. Instead a logic variable, t, synchronizes the whole process. Datatypes include signals, that is functions from time to value, busses and other user defined ones. Recursion is naturally expressed, and so is the partitioning of the specification in different levels made up of groups of related functions.

In order to develop a suitable environment for such applications at this level of abstraction, the following steps have been accomplished:

1 expansion of the standard libraries of components in LAMBDA/DIALOG in order to hook up with the abstract design levels previously referred to, focusing on regular structures
2. definition of abstract computational nodes to implement different cells

3. definition of new chaining tactics which facilitate the task of dealing with recursion and regular topologies

The definition of special tactics to partition and reassemble nodes in a graph are crucial in order to allow graphs manipulation of abstract cells [Hu92].

Nodes describe abstract functional relations with delay 0. They are differentiated on the number of inputs and outputs (S_NODES2 and S_NODES3, for nodes with one output and 2 or 3 inputs respectively, D_NODES2 and D_NODES3 for nodes with two inputs and T_Nodes3 for nodes with three inputs) (fig. 3).
In a similar way recursive structures can be handled by

RCHAIN1#(i[list], in1, cin, o[list], cout, F)

which allows the representation of regular structures with one connection signal between the cells (S_NODES2) (Fig 3). Input signal in1 is common to all the cells. F represents here the abstract function performed by each cell.
Analogously

RCHAIN2#(i[list], in[list], cin, o[list], cout, F)

allows the representation of similar structures with differentiated input lines (the second parameter is a list instead of a single value).

Each cell defined in the structure will be successively synthesized on a different hierarchic level, again as a regular topology or as a general topology using the standard cell menu defined in DIALOG.

The actual limit of the tool is that implementation alternatives are generated by different resource allocation choices. The following step on this line is the broadening of this possibility to include the generation of alternative architectures. Till now the choice of the architecture is implicitly accomplished by the specification style. A recursive function will generally lead to the generation of functionally identical cells (regular topology).

```
fun sumBit x y z = x xor y xor z;
fun carryBit x y z = x && y || y && z || z && x;
fun sumArray [] [] c = c
    | sumArray (h1::t1) (h2::t2) c =
        sumBit h1 h2 c :: sumArray t1 t2 (carryBit h1 h2 c);
```

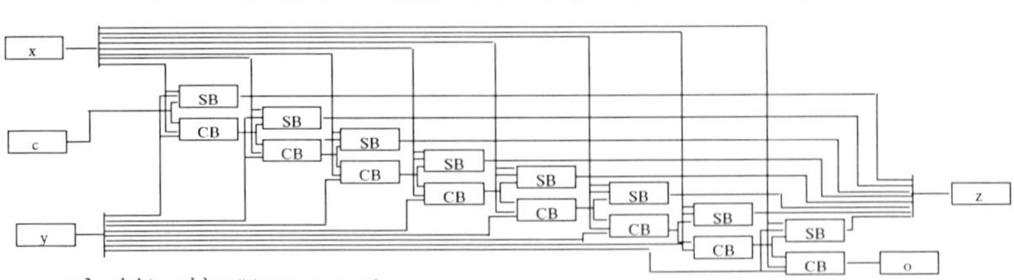


```
val nbit-adder#(x,y,c,z,o)
( forall t:time.exists n.
  z (t + n) = rest (sumArray x t y t c t) /\
  o (t + n) = last (sumArray x t y t c t)).
```

Figure 4. Regular topology generated from a recursive specification.

Let's consider for example the recursive definition of a n–bit adder (fig 4). Here the abstract node role is played by blocks sumBit and carryBit, which will be synthesized using a general topology of AND and NOR gates. The most straightforward way to deal with the recursive function is to unroll it a certain number of times, after having instantiated a bus of the desired length in the input parameters (lists) and then applying rules acting on lists. In this case the obtained regular topology was already implicitly stated in the specification style, and not chosen in a set of provided weighted alternatives. On the other hand, regular structures like systolic arrays need, in order to be generated, a different definition of the global scheduling tactic defined in DIALOG.

The need for formal systems to be more flexible and abstract at the specification level is strongly felt by end–users. This will allow the application of graphs transformations able to produce alternatives also at the architectural level.

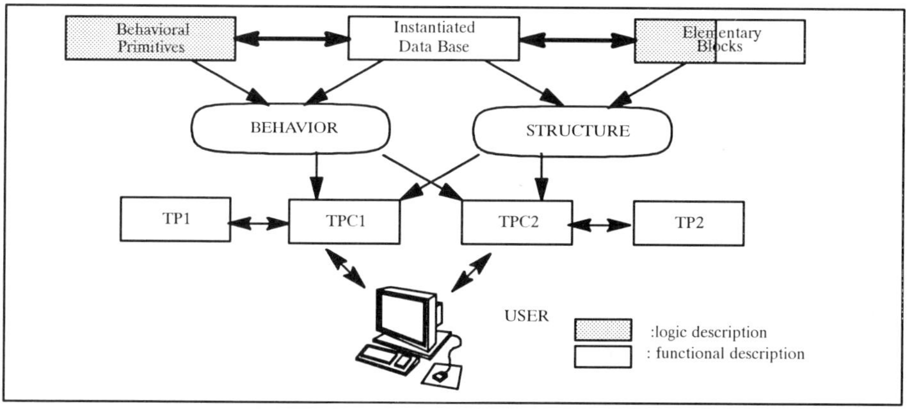

Figure 5. Theorem prover integration into the CAD environment.

4. INTRODUCTION OF FORMAL TOOLS INTO A DESIGN FLOW

In order to fully exploit the benefits deriving from high level formal based synthesis, it is necessary to define and implement integration links between this stage of the design flow and the lower levels, where consolidated, efficient and tested methodologies are used down to the physical level. This aspect is of paramount importance for any application of formal methods to the industrial world, and in a way it is crucial also for the theoretical development in this area.

4.1 Integration

With the term integration we mean a complete share of the database by different tools. This can be accomplished when the database is completely accessible by the user, or when it is proprietary. In our case, we have developed a prototype expansion of an object oriented CAD environment which models structure and behavior [Ap91] adding logical modeling capabilities and the abstract operators needed to the description of logic behavior.

Model–View–Controller paradigm is a good way to handle the complexity of the resulting database. The Model is represented by the complete set of instantiated objects representing the net of the device. The theorem prover is modeled as a specialized View into the database. The translation from semantics into the requested syntax is carried out by special classes, called TPCs (Theorem Prover Controllers). Moreover TPCs handle the interaction of the user with the window running the theorem prover and manage the control flow towards the proof (fig. 5) [Va91].

The OTTER TCP includes several modules devoted to different tasks. Each module is implemented by a set of methods (fig. 6). Automatic steps are involved only when the system can free the user from repetitive operations, but it remains highly interactive when decisions are needed.

4.2. Encapsulation

Encapsulation is a weaker but more realistic way of providing the facility to use tools developed by different vendors and specialized for specific tasks in the same CAD environment. In this way also formal tools can be included in the design flow. No complete share of the database is required in this case, but only a common interface able to call the various tools communicating data in standard description languages and formats (EDIF, VHDL). Commercial interfaces of this kind [Me91] are already on the market and make the integration between different tools easily manageable.

EDIF is a standard description language for schematics (EDIF schematic) and for netlists (EDIF netlist). After completing the analysis at the level of LAMBDA, an EDIF representation of the netlist is generated. As Italtel design flow is based on Mentor environment, we started using this standard description in order to define a connection into AUTOLOGIC environment, where a parametric library (LOGICLIB) of RTL cells can be instantiated and the mapping towards the Italtel library at gate level established (fig. 7).

The encapsulation requires anyway a great deal of customization. The first phase has focused on the definition of a 'ad–hoc' library of components aligned with the standard libraries used by lower level synthesis tools or in the lower levels design phases.

In particular we have established a one to one mapping between the primitive components of the LAMBDA/DIALOG library of cells on which the synthesis process is based, and the components of the LOGICLIB library. The description level of the library (RTL) and its parametrization are widely exploited for this purpose.

The complete design flow starts from the specification in ML, derives the architectural description in LAMBDA/DIALOG and provides a netlist in EDIF format. The modules called 'netlist manipulations', included in the LAMBDA/ITALTEL–FLOW interface, written in C, will operate the necessary manipulation of the EDIF netlist in order to make it readable by the AUTOLOGIC tool which generates the ITALTEL database from which, with standard methods, the layout is generated.

The EDIF approach lacks in generality: such a drawback can be overcome using the VHDL description language (1076–IEEE). A subset of this hardware description language is being used as an entry language also by commercially available synthesis tools. Without entering the debate of the definition of the standard, we think that, as VHDL is going to cover a more and more relevant role in the CAD world, it can represent the most promising link between different levels of the design flow.

In this case the main problem is the alignment of statements and constructs. This means that two requirements must be satisfied:

1. the structural subset of VHDL used by LAMBDA should match the one used by the subsequent commercial synthesis tool
2. the functional description of blocks should also be provided in functional VHDL

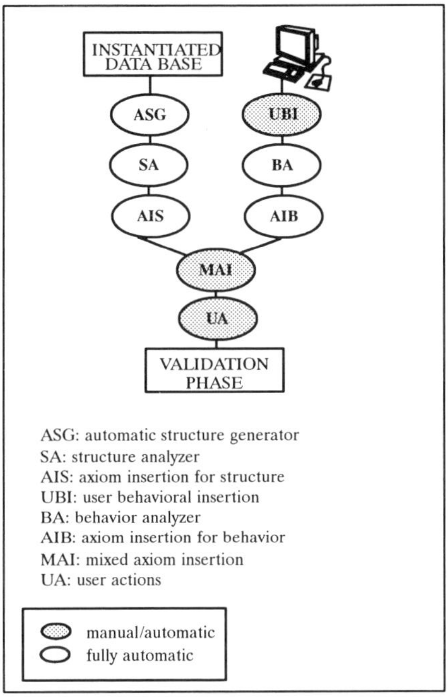

Figure 6. TPC flow diagram.

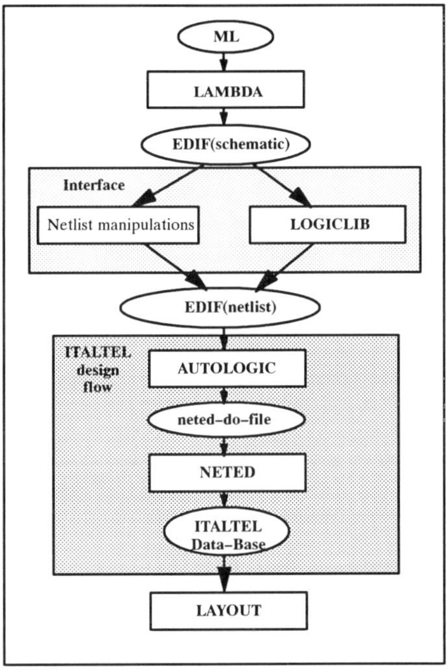

Figure 7. EDIF based design flow.

The first point weakens the approach, as different vendors use different VHDL constructs, and even when they match the syntactic level they differ in terms of semantics. The analysis of verifying the common set used by two commercial vendors, whose synthesis tools will be used in our environment, has led to a guideline in terms of VHDL constructs. No easy solution can be provided at the moment for the description in VHDL of the functionality of blocks not yet structurally synthesized in LAMBDA/DIALOG. An ML/VHDL functional bridge is required for this.

5. CONCLUSIONS

In the last years formal methods have proved more and more reliable and useful as a methodology in order to improve the quality of the design process, by assuring the correctness of the implementation versus a formal specification. In our opinion formally based synthesis tools can fill the current gap existing in automatic support tools at the higher and most abstract design levels. Partitioning and hierarchical structuring of the design can reduce the computational times. Simulation remains a crucial part of the design flow to check the correctness of the design closer to the physical level.

At the moment the use of these techniques remains restricted in most cases to the Research Laboratories but the actual quickly growing developments and the competition from the market are pushing strongly towards their full insertion into the consolidated design flows at pilot industrial sites.

6. REFERENCES

[An89] F. Anceau, Formal verification in industrial environment,Workshop on Formal Methods, L'Aquila, July 1989

[An91] F. Anceau, Panel on formal methods in hardware design, 10th Int. Symposium on Computer Hardware Description Languages, Marseille, April 1991

[Ap91] A. Apostoli, A. Ferrari, S. Loda', L. Plebani, Creazione di un ambiente standard per la progettazione di circuiti integrati, Tesi di laurea, Universita' degli Studi di Milano, 1991

[Bo92] M. Bombana, G. Buonanno, P. Cavalloro, D. Sciuto, G. Zaza, A multi–level testability assistant for VLSI design, EURODAC Conference 1992, Hamburg, September 1992

[Ca91] P. Camurati, T. Margaria, P. Prinetto, Use of the OTTER theorem prover for the formal verification of hardware, G. Jones, M. Sheeran (eds.), Designing Correct Circuits, London Springer 1991

[Fi89] S. Finn, M. Fourman, M. Francis, R. Harris, Formal System design – Interactive Synthesis based on Computer–Assisted Reasoning, in: Proc. IFIP WG 10.2, 10.5 Workshop, North Holland, 1990

[Fo89] M.Fourman, E. Mayger, Formally Based System Design – Interactive Hardware scheduling, G. Musgrave, U.Lauther (eds), VLSI 89, Münich, August 1989, Elsevier 1989

[Go92] G. Gorla, L'automazione del progetto di sistema nell'industria, Workshop on Specification and Synthesis of Digital Systems, CEFRIEL, Milan, May 1992

[Hu92] R.B. Hughes, G. Musgrave, Design Flow Graph Partitioning, HOL '92, IMEC Leuven Belgium, September 1992

[La92] AHL, Lambda Reference Manual - Version 4.1, London 1992

[Ma88] J. C. Madre, J. P. Billon, Proving circuit correctness using formal comparison between expected and extracted behavioral, 25th ACM/IEEE DAC, 1988

[Me91] Getting started with Falcon Framework– Version 8.0–3, Mentor Graphics, 1991

[Ot89] W. W. McCune, Otter 1.0 Users' Guide, Argonne National Laboratory, Illinois, 1989

[Pa91a] ESPRIT PATRICIA Project – Technical Report WP2.2–2 – December 1991

[Pa91b] ESPRIT PATRICIA Project – Technical Report WP2.3–3 – December 1991

[Pe90] U. Pelizzari, Tecniche di theorem proving nella validazione dell'hardware, Tesi di laurea, Universita' degli Studi di Milano, 1990

[Va91] R. Vaghini, Modellazione logica di componenti elettronici in ambiente ad oggetti per applicazioni di verifica formale, Tesi di laurea, Universita' degli Studi di Milano, 1991

Chapter 8

Simulation in Higher Order Logic

Higher Order Logic Theorem Proving and its Applications (A-20)
L.J.M. Claesen and M.J.C. Gordon (Editors)
Elsevier Science Publishers B.V. (North-Holland)

Operational Semantics Based Formal Symbolic Simulation

K. G. W. Goossens

Laboratory for Foundations of Computer Science, Department of Computer Science, University of Edinburgh, The King's Buildings, Edinburgh EH9 3JZ, U.K.

Abstract

This paper describes the development of progressively more powerful and abstract hardware simulators. A small computer hardware design and description language pico-ELLA is then introduced, followed by its formal semantics. Using a number of small examples, we will then show the how this formal semantics may be used within a proof system as a sophisticated simulation tool. Examples include some full adders, a general N bit adder, and two parity checkers.

Keyword Codes: I.2.3; B.7.2; F.3
Keywords: Deduction and Theorem Proving; Integrated Circuits, Design Aids; Logics and Meaning of Programs

1 Introduction

This introduction describes the development of various kinds of hardware simulators. Following this, a small HDL called picoELLA, is introduced in section 2. Its formal semantics, and a brief account of this semantics' embedding in a proof system are described in section 3. Section 4 illustrates the use of the semantics in the capacity of a symbolic simulator, as described in the remainder of this introduction. Finally, integration with other design and verification methodologies will be discussed.

Modern hardware designs are complex. Conventional methods take many iterations to arrive at an acceptable implementation. Though exhaustive testing might allow designers to achieve a high degree of confidence in small circuits, exhaustive testing of modern designs is impossible both on time complexity and cost grounds.

Breadboarding is the process of constructing the design and then using this prototype to perform tests. This is only feasible for small circuits. With greater integration and density of components this method becomes prohibitively expensive. The first move away from using real hardware to test a design is to *simulate* the circuit. Greater complexity encourages the use of structured circuit descriptions, leading to hardware description languages such as ELLA[1] [6] and VHDL [13]. It is possible to design simulators for such languages which model the behaviour of a circuit described in the language.

One of the problems with both breadboarding and value simulation is that for any substantial circuit the number of possible inputs (or *test vectors*) becomes very large. Circuits with internal state are even harder to verify in this manner. By introducing extra values in the value domain, such as don't know and don't care, the number of test vectors may be reduced substantially. If a particular input is irrelevant for a particular test, its value can be set to don't care, instead of having to simulate the test twice, with the value set to true and false respectively.

The MOSSYM simulator [5] is not limited to fixed values as input, but also allows symbolic variables and boolean formulae. That is, we may set an input to the symbolic variable x, say. Wherever x appears we cannot assume anything about its value, so that the result of the operation may be a formula. Note that x is not an extra value in the value domain, but ranges over *all* these values. Of course, this puts an extra burden on the simulator which now needs to be able to handle arbitrary formulae instead of simple values. It may also require algebraic capabilities to simplify intermediate formulae. In theory, we need to do only one simulation; the one with all the input values set to variables. The result would be an expression which would describe the circuit's behaviour. However, this expression may be as complex as the circuit description.

In MOSSYM we have an asymmetry: we are permitted abstraction over data but not over circuits. In other words, we may have symbolic variables ranging over data values, but we are not allowed circuits containing symbolic variables. Symbolic variables are abstract hardware. This idea is not as strange as it may seem; plug-in components are in effect abstract hardware, certainly as long as the circuit is under development. Why would it be useful to have this capability? It would seem that, since we are dealing with the design of a certain circuit, we would only want to simulate that circuit. Consider, however, that large circuits are designed in a modular fashion to allow a number of people to work on separate parts of a circuit at the same time. When a subcomponent is ready, it has to be simulated in a larger context, all of which may not be completed. The availability of an abstract implementation for unfinished parts of the design would enable the component to be simulated in its correct context. A suitable simulator would allow the evaluation of a circuit containing a mixture of concrete and abstract components. Of course, certain properties of the abstract components may be needed to arrive at an output, but these should be available from their specifications. We use the specification of the abstract components to simulate them as long they are not available. Modifying a conventional simulator to deal with these extensions would completely transform it. Mathematical proof systems, in contrast, already have the capability to deal with abstract values of any sort. (This assumes that we work within a suitably powerful logic, such as higher order logic.) If we use a HDL to describe circuits and as input to the simulator, the HDL needs to have a precise mathematical definition to be used in conjunction with a proof system. The approach we advocate here uses a formal definition of the behaviour of a HDL and uses it within a proof system to provide simulation capabilities.

[1]ELLA is a trademark of the Secretary of State for Defence, United Kingdom.

2 picoELLA

picoELLA is derived from ELLA [6]. It contains the 'active ingredients' but lacks its syntactic sugar. picoELLA contains the following constructs:

Type definitions; these are either enumerated types, such as `TYPE Signal = Hi | Lo`, or tuple types such as `TYPE twobool = bool * bool`.

Local declarations. `LET x = e IN e'` defines a local name `x` for a wire (or signal), which may be used in `e'`. Circuit descriptions may be structured using these declarations. Multiple use of a name such as `x` corresponds to a fan-out of the signal. Recursive declarations allow the description of feedback, an example of which is shown at the end of this section.

Constants are built up using constructors such as `Hi`, or `?type` representing the undefined, or don't know value of type `type`, or tuples of constants.

Tuples and indexing are straightforward. $(e_1, e_2)[i]$ behaves the same as e_i, for example. Although, strictly speaking, indexing is not needed it facilitates decomposition of circuit descriptions, as we shall see in the examples.

The `IF` statement, or multiplexor, has the form `IF e MATCHES chooser THEN e1 ELSE e2`. A chooser is a pattern against which the output of circuit `e` is matched. If it matches, the output of the first branch is the result of the `IF`. If the output of `e` and the chooser do not match the `ELSE` part is chosen. A third possibility is that the output of `e` is insufficiently defined to decide between the two branches. In this case, the undefined value `?type` (of the correct type) is the result. For example, consider the NOT gate `IF ?Signal MATCHES Hi THEN Lo ELSE Hi`. If the undefined, or don't know signal `?Signal` turned out to be `Hi` then the output would be `Lo`. On the other hand, if it were `Lo`, the output would be `Hi`. The undefined output `?Signal` therefore reflects our intuition that we don't know what the output should be.

The delay construct introduces a discrete and linear time base into the language, which may be modelled using the natural numbers. The output from circuit `e` at time t will be output by $\mathtt{Delay}(c_t, e)$ at the *next* time step $t+1$. At the *current* time t, the value c_t is the result. This shows that the state of the delay is explicit in its description. This contrasts with other languages, where it resides in a memory or store. picoELLA dispenses with this, and can use a simpler environment instead, as we shall see later. However, as a result of the explicit representation of the state, a new circuit description must be evaluated at each time step. The result of an evaluation consists therefore of a value output together with a description of the circuit at the next time step. The type of the dynamic semantics is therefore *environment* → *expression* → (*value* × *expression*).

As an example, consider the following circuit which implements a parity checker. It returns `Hi` at time $t+1$ if there have been an even number of `Hi`'s on the input signal `input` during the closed time interval $[0, t]$. At time zero it outputs `Hi`.

```
LET INIT ?Signal
    REC xor = DELAY (Hi, IF (input,xor) MATCHES (Hi,Lo)|(Lo,Hi)
                         THEN Hi ELSE Lo)
IN xor
```

We will return to this example in section 4.3.

3 A picoELLA Semantics and Its Embedding in Higher Order Logic

Few HDLs have a precise definition. In practice the simulator serves as this definition, but this leads to problems when different implementations present conflicting outputs. A *formal semantics* may be used to give a mathematical description of the behaviour of a HDL, *i.e.* how a simulator should behave.[2] For example, a structural operational semantics [15] defines the behaviour of a construct in terms of its subexpressions. General properties, such as termination of any simulation within a finite number of steps, may be proved about the semantics (and hence the behaviour of conforming simulators). Various subsets of ELLA [10, 1], Funnel [16] and VHDL subsets [18, 17] are some of the languages that have been given formal definitions.

Although a formal semantics is very useful as a mathematical reference manual, one can also use formal semantics as the basis for design tools. By embedding the semantics in a logical system, supported by a proof assistant such as LAMBDA[3] [8] it is possible to provide support for the formal development of circuit design [11].

The semantics for picoELLA [10] takes the form of a structural operational semantics. It comprises a *static semantics* describing which programs are well-typed, and a *dynamic semantics* defining the run-time behaviour of well-typed programs. We will not discuss the static semantics here; it suffices to say that it is relatively straightforward. picoELLA semantics rules fall into two categories; those dealing with time, and those dealing with the evaluation of expressions within one time step. The ReduceSeqCons rule falls into the first class:

$$\frac{\Gamma' \vdash \ expr_t \Rightarrow \ o_t, \ expr_{t+1} \qquad tl, \ \Gamma \vdash \ expr_{t+1} \Rightarrow \ tl', \ expr_{t+N}}{i_t :: tl, \ \Gamma \vdash \ expr_t \Rightarrow \ o_t :: tl', \ expr_{t+N}}$$

Here $expr_t$ is the program at time t, Γ the environment in which the program runs, and $i_t :: tl$ the input stream. Γ' is Γ with input value i_t adjoined (this will be made more precise later). This rule shows that at time t we run the program $expr_t$ with input value i_t in environment Γ'. The output value o_t is added to the output stream, and the new program $expr_{t+1}$ is evaluated with the remainder of the input stream. As explained previously, since the state of the circuit is explicit in its description we need to evaluate a new circuit at every time step. A typical member of the second category of rules is the ReduceTuple rule:

$$\frac{\Gamma \vdash \ expr_1 \Rightarrow \ v_1, \ expr_1' \qquad \Gamma \vdash \ expr_2 \Rightarrow \ v_2, \ expr_2'}{\Gamma \vdash \ (expr_1, expr_2) \Rightarrow \ (v_1, v_2), \ (expr_1', expr_2')}$$

To evaluate a tuple in environment Γ, both subexpressions must be evaluated in the same environment. The rule for the delay shows the use of the embedded state:

$$\frac{\Gamma \vdash \ expr \Rightarrow \ v, \ expr'}{\Gamma \vdash \ \texttt{DELAY} \ (c, \ expr) \Rightarrow \ c, \ \texttt{DELAY} \ (v, \ expr')}$$

[2]Note that the implementation of the simulator may use any model, as long the input–output relation obeys the definition. The semantic definition should be clear and simple, not necessarily efficient.

[3]LAMBDA and DIALOG are products of Abstract Hardware Limited.

The output from the delay is its latched value *c*. The new description of the delay, to be evaluated at the next time step, contains the output *v* from *expr* at the current time step. In other words, it has latched this clock cycle's output.

The semantics described above, has been embedded in the LAMBDA proof system. The LAMBDA proof system implements a polymorphic constructive higher order logic of partial terms.[4] An existence predicate E is provided to reason about partial terms. Equality == compares two denoting objects, weak equality (or equivalence) === is true if either both objects do not denote, or if both denote and are equal. The functional language ML is used as a command language. A large subset of ML is also used to define new data types and operations on data types *within* the logic. The soundness of the system cannot be compromised through new definitions. LAMBDA returns a number of rules characterising the new ML data type definitions, such as existence of constructors, (in)equality rules and a structural induction principle. In the case of functions rules include a rewrite rule for every function clause. These new rules may be used to define derived rules and tactics. Tacticals can be used to combine tactics into rewrite strategies, or symbolic simulation commands. Examples are `OpSemTac` and `safeOpSemAllTac` in section 4.1.

A type *const* has been encoded using the ML definition system, representing picoELLA constants:

```
datatype const = Cons of natural * natural | CoTuple of const * const;
```

`Cons(i,t)` encodes the $\mathtt{i}^{th}$ constructor of type `t`. `Cons(0,t)` represents `?t` which is the undefined, or don't know, value of type `t`. A constant is therefore a constructor or bottom value, or a tuple containing constants. To illustrate structural induction we consider the `choosers` data type used to encode patterns in the `IF` statement.

```
datatype choosers = C of const
                  | B of choosers * choosers
                  | T of choosers * choosers;
```

`B(ch,ch')` represents the bar, or disjunctive chooser; it matches with a constant if at least one of `ch` and `ch'` does. `T(ch,ch')` is the tuple, or pairing chooser, matching if both subchoosers do. `C c` is the constant chooser. `C (Cons (0,type))` represents the *wild card* chooser `type`; it always matches. It is not allowed to match for the bottom value `?type`, as this would permit non-monotone circuit descriptions. LAMBDA returns the following choosers structural induction rule for this data type.

```
[3] E r1, E r, P#(r1), P#(r) |- P#(T (r1,r))
[2] E r3, E r2, P#(r3), P#(r2) |- P#(B (r3,r2))
[1] E r4 |- P#(C r4)
--------------------
E w |- P#(w)
```

There are three premisses, each containing some hypotheses. `E r4` is an *existence* hypothesis, asserting that `r4` denotes. `P#(r1)` states that the property `P` holds for `r1`. To prove a property `P` of all choosers `w` three subgoals must be proved: in case of the second premise, for example, it must be shown that `P` holds for `B(r3,r2)` provided it holds for `r3`

[4]Note that we use LAMBDA version 3.2. The more recent version 4.0 uses a different logic.

and r2, and r3 and r2 denote. The type representing expressions, or circuits is defined as follows.

```
datatype expr = Const of const
              | Tuple of expr * expr
              | Let of expr * expr
              | Var of natural
              | Delay of const * expr
              | If of expr * expr * expr * choosers
              | Index1 of expr
              | Index2 of expr
              | LetRec of const * expr * expr;
```

Note that no constructor is present for TYPE. Types are dealt with on a meta-level, *i.e.* using LAMBDA's facilities, rather than at the expr object level. To embed the LET operator the de Bruijn encoding of lambda abstractions is used. The bound variables of lambda expressions are encoded as natural numbers indicating the distance (measured in intervening lambdas) away from the defining lambda. Thus $\lambda x.\lambda y.(x,(x,y))\ a\ b$ would be encoded as $\lambda\lambda(1,(1,0))\ a\ b$. In picoELLA this corresponds to encoding `LET x = a IN LET y = b IN (x,(x,y))` by `Let (a, Let (b, Tuple (Var 1, Tuple (Var 1, Var 0))))`. The de Bruijn encoding was sufficient for our purposes because the environment is used only as a stack. Work using the HOL system has usually represented names by strings; the value environment has the type `string -> const` [14, 17]. Finally, the dynamic semantics can be defined as a function Reduce. Its type is `Reduce: const list -> expr -> (const * expr)`, where `const list` represents the value environment.

```
fun Reduce l (Let (e,e')) =
          let val (c, f) = Reduce l e
              val (c', f') = Reduce (c::l) e'
          in  (c', Let (f,f'))
          end |
    Reduce l (Var n) = (elem l n, Var n) |
    Reduce l (If (e,e',e'',ch)) =
          let val (c,d) = Reduce l e
              val (c',d') = Reduce l e'
              val (c'',d'') = Reduce l e''
          in  ( case match ch c of
                tt => c' |
                ff => c'' |
                uu => bottom c', If (d,d',d'',ch) )
          end | ...;
```

The LET statement reduces the defining expression, and pushes the value result on the stack `l` (*i.e.* stores it in the environment). Evaluating a name corresponds to a lookup in the environment Γ in the dynamic semantics, and a lookup in the stack `l` in the embedding.

The IF construct evaluates all of its subexpressions. It then returns (`tt`) the result of the first branch if we have a definite match; or (`ff`) if we have a definite no-match, the result of the second branch; or (uu) a bottom value of the appropriate type if we cannot decide between the two branches.

We have proved a number of properties of the embedded semantics using LAMBDA. For example, the reduction function is monotone, that is, if the input becomes more defined, the output becomes more defined. Also, the reduction function preserves the shape of the program (an adder does not become a multiplier after some time!). In other words, only the contents of delays changes over time. Moreover, we have shown that even in the presence of delayless feedback loops the reduction function terminates in a finite number of steps. In fact, the semantics computes the least fixed point solution of the circuit. It is important to realise that these are results concerning *all* circuits, not particular instances. A more detailed account of the embedding may be found in [11, 9]. For the remainder of this paper, with the exception of subsection 4.3 which deals with feed-back, an embedding without the `LET REC` has been used. The operational semantics rules dealing with `IF` are slightly simpler in the embedding without the `LET REC`. Other rules are identical.

Using these definitions, and derived properties, the operational semantics rules described at the start of this section may be derived within the proof system.[5] These rules encode both the static and dynamic semantics.

The rule `ReduceSeqCons` corresponds to ReduceSeqCons previously. Rules shown in the `typewriter` font denote the embedded rules, those in roman font the 'paper' rules.

```
|- (instream_, env_ |- circ1_ => (outstream_,circ2_))
|- (i1_ :: env_ |- circ_ => (o1_,circ1_) : t_)
--------------------
|- (i1_ :: instream_, env_ |- circ_ => (o1_ :: outstream_,circ2_))
```

To evaluate a program `circ_` with a non-empty input stream `i1_::instream_`, the head of the input stream is pushed onto the environment `env_`. This corresponds Γ' in the paper rule. `circ_` is then evaluated within this time step. The remainder of the input stream is then evaluated using the new circuit `circ1_`. Finally, the output `o1_` is prepended to the resulting output stream. It is a pretty printed version of:

```
E instream_, E env_, E circ1_, E t_
|- ReduceSeq env_ circ1_ instream_ == (outstream_,circ2_)
E (i1_ :: env_), E circ_, E t_
|- typeOfExpr (map typeOfConst (i1_ :: env_)) circ_ == (t_,true)
/\ Reduce (i1_ :: env_) circ_ == (o1_,circ1_)
--------------------
E (i1_ :: instream_), E env_, E circ_, E t_
|- ReduceSeq env_ circ_ (i1_ :: instream_) == (o1_ :: outstream_,circ2_)
```

Henceforth we will only show pretty printed output. Unfortunately, no quotation/anti-quotation system is available, so that any input must still use the raw syntax. The rule for the multiplexor, `ReduceIf'` is similar:

[5] This is in contrast with work by van Tassel [17], which starts directly with the semantic rules. Using the HOL inductive relation package more general relational semantics may be encoded. In LAMBDA version 3.2 we are limited to functional semantics.

```
[6] |- E t_
[5] |- o3_ == (case match chooser_ out_ of
uu => bottom o1_ | tt => o1_ | ff => o2_)
[4] |- chooser_ :  t_
|- (env_ |- branch2_ => (o2_,branch2'_) :  t1_)
|- (env_ |- branch1_ => (o1_,branch1'_) :  t1_)
|- (env_ |- circ_ => (out_,circ'_) :  t_)
--------------------
|- (env_ |- IF circ_ MATCHES chooser_ THEN branch1_ ELSE branch2_ =>
(o3_,IF circ'_ MATCHES chooser_ THEN branch1'_ ELSE branch2'_) :  t1_)
```

There are some extra hypotheses ([4] and [6]) dealing with the static semantics: the choosers must be well-typed and have (denoting) type t_. Note that branch1_ and branch2_ must have the same type t1_, which is also the type of whole IF. The output of the IF is computed in premise five. The three cases (match, no match and don't know) are represented in the case statement by tt, ff and uu respectively. The IF is strict; both branches must always be evaluated. Four other rules dealing with the IF are particular instantiations of this rule, as we shall see later.

It is important to realise that circ_, t_, *etc.* are *meta-variables.* The ReduceIf' rule is really a rule schema, which may be instantiated in an infinite number of different ways. When it is applied to a particular IF statement such as IF Hi MATCHES Hi THEN Lo ELSE Hi, circ_, chooser_, branch1_ and branch2_ will be unified with Hi, Hi, Lo, and Hi respectively. This unification is reflected in every place where these variables occur in the rule. The unification works both ways, meta-variables in a rule are unified to the current goal so that the rule applies (as in the example below). But meta-variables in the goal may also be unified (specialised, made more concrete) for the rule to apply. We will see examples of this later on. In LAMBDA meta-variables may be *flexible* or *rigid.* The former are used to stand for some term to be determined as the proof proceeds, the latter require proofs to be schematic in the variable. Rigid variables ensure that a general result, rather than an instantiation of the result, is proved.

4 Examples

In this section we will illustrate the possible uses of an embedded operational semantics. First we will simulate a simple AND gate in various ways to illustrate the basic principles. Following this, some one bit adders and a general N bit adder, parametrised on the word size and one bit adder subcomponent, will be shown. Finally, two parity checker implementations will be discussed.

4.1 A Simple AND Gate

An AND gate may be described in picoELLA as

```
IF e MATCHES (Hi,Hi) THEN Hi ELSE Lo
```

or, using the syntax of the embedding:

```
If (e, Const Hi, Const Lo, T (C Hi, C Hi));
```

Here e is the input to the circuit. Signal, Hi and Lo have been defined as Cons(0,1), Cons(1,1) and Cons(2,1) respectively. All of Signal, Hi and Lo have type Type 1. We will simulate an AND gate with (Hi,Lo) as input using the ReduceIfFf and ReduceConst rules. The rule ReduceIfFf is comparable to the rule ReduceIf' of the previous page, but always chooses the ELSE branch.

```
> apprl ReduceIfFf;

***** Level 2 *****
[6] |- E t_
[5] |- match (Hi, Hi) out_ == ff
[4] |- (Hi, Hi) :  t_
[3] |- (env_ |- Lo => (Lo, Lo) :  Type 1)
[2] |- (env_ |- Hi => (o1_, Hi) :  Type 1)
[1] |- (env_ |- (Hi, Lo) => (out_, (Hi, Lo)) :  t_)
--------------------
|- (env_ |- IF (Hi, Lo) MATCHES (Hi,Hi) THEN Hi ELSE Lo
=> (Lo,IF (Hi, Lo) MATCHES (Hi,Hi) THEN Hi ELSE Lo) :  Type 1)
```

We now have six subgoals to prove, the first of which deals with the input to the IF. The second and third subgoals compute the THEN and ELSE branches respectively. As stated earlier, both branches must be evaluated, because the result circuit is always used to describe the IF at the next time step. Note, however, that the value output o1_ does not appear in the output. The fourth premise states that the chooser must be well-typed; in this case it has type t_. t_ is an as yet uninstantiated meta-variable. As we shall see below, evaluating premise 1 forces t_ to become a tuple type. We also have to prove that the type denotes in premise 6. Subgoal five expresses the constraint that we choose the ELSE part of the IF; the result out_ of the input circuit must not match with the chooser.

We may now apply ReduceTuple to reduce the tuple in premise 1 to two subgoals. Following this we apply ReduceConst to premises one to four:

```
> applyTacn [1,2,3,4] (doRule ReduceConst);

***** Level 4 *****
|- E (TyTuple (t1_,t2_))
[6] |- match (Hi, Hi) (Hi,Lo) == ff
|- (Hi, Hi) :  TyTuple (t1_,t2_)
[4] |- Lo :  Type 1
[3] |- Hi :  Type 1
[2] |- Lo :  t2_
[1] |- Hi :  t1_
--------------------
|- (env_ |- IF (Hi, Lo) MATCHES (Hi,Hi) THEN Hi ELSE Lo
=> (Lo,IF (Hi, Lo) MATCHES (Hi,Hi) THEN Hi ELSE Lo) :  Type 1)
```

The tactical applyTacn l t applies tactic t to all premises in the list l. doRule converts a rule into the tactic which applies the rule if it is applicable, and fails otherwise. `tryRule`, on the other hand, is the identity tactic if the rule fails to apply. `tryRules` and `doRules` are similar functions operating on lists of rules.

As mentioned earlier, the type of the chooser has been constrained to a less general type `TyTuple (t1_,t2_)`. Evaluating premises one and two will specialise it further to `TyTuple (Type 1,Type 1)`, as the type of `Hi` and `Lo` is `Type 1`. All the subgoals, except `[6]`, are now dealing with the static semantics, or typing of terms. It makes sense to deal with the static and dynamic semantics simultaneously because they are both structural semantics. Moreover, the dynamic semantics only evaluates well-typed expressions.

Using `applyTacn [1,2,3,4] (doRule ReduceHi elseR ReduceLo)` we discharge premises one to four. Using `ReduceMatchTac`, a tactic which rewrites expressions involving `match`, we prove premise six. We also discharge the static typing of the chooser.

```
> applyTac (doRules[ReduceT,ReduceC,ReduceHi]);

***** Level 7 *****
|- E (TyTuple (Type 1,Type 1))
--------------------
|- (env_ |- IF (Hi, Lo) MATCHES (Hi,Hi) THEN Hi ELSE Lo
=> (Lo,IF (Hi, Lo) MATCHES (Hi,Hi) THEN Hi ELSE Lo) :  Type 1)
> applyTac (doRules[ReduceTyTuple,ReduceType,ReduceSn,Reduce0]);

***** Level 8 *****
--------------------
|- (env_ |- IF (Hi, Lo) MATCHES (Hi,Hi) THEN Hi ELSE Lo
=> (Lo,IF (Hi, Lo) MATCHES (Hi,Hi) THEN Hi ELSE Lo) :  Type 1)
> val example1a = popGoal();
val example1a = ?  :  rule
```

Note that `doRules [r1,r2]` is a tactic which applies rule `r1` and then applies `r2` to all resulting subgoals. Thus `ReduceC` and `ReduceHi` are applied to both subgoals resulting from `ReduceT`. The theorem is saved as `example1a` so that we can apply this derivation in one step in the future.

While this is very instructive, it becomes tedious very quickly to this sort of proof by hand. Tactics may be used to great advantage in this sort of regular reasoning. The whole previous example could have been done using one general purpose tactic:

```
val OpSemTac = (repeatT (nonTrivT (tryRules OpSemRules))) thenT
               (tryT (theoremT ReduceMatchTac)) thenT
               (tryT (theoremT ReduceTypeTac));
applyTac OpSemTac;
```

This tactic repeatedly applies one or more of the standard operational semantics rules until none apply. It then applies `ReduceMatchTac` followed by `ReduceTypeTac`, to rewrite any typing subgoals. These last two tactics are applied to a subgoal only if they discharge it.

The circuit as it stands is not very useful, as it deals with only one particular input. Moreover, we had to supply the output from the simulation at the start! We will now quickly redo the example, but using meta-variables as output. These will be flexible, so that they may be instantiated as we compute the output to a fixed answer. We will also use an abbreviation for the AND gate. Unlike the abbreviations for Hi *etc.*, it has an argument. Abbreviations are syntactic functions at the meta-level in the proof system. They are distinct from functions at the object level, such as nadd which we shall see later.

```
val Signal = Cons (0,1);
val Hi = Cons (1,1);
val Lo = Cons (2,1);
val AND#(e) = IF e MATCHES (Hi,Hi) THEN Hi ELSE Lo;
```

When a new goal is to be proven, all meta-variables are rigid; they cannot be (inadvertently) instantiated. In general this is what is required, because the result so proved is then more general. Every operational semantics rule has meta-variables such as env_, circ_ and t_, which are unified with the corresponding expressions in the premise it is applied to. Consider the use of rule ReduceIf below, for example. In this case we want to specialise the meta-variables if required, so we make them flexible using the flex command. A pop-up menu shows the current subgoal, and one selects subterms by clicking on them with a mouse. The flex command is then automatically generated by LAMBDA, so that it may be included in proof scripts for later use. We will now unfold the abbreviation for AND, and apply ReduceIf which is a third rule for the IF statement.

```
> apprl ANDU;

***** Level 3 *****
|- (env_ |- IF circ_ MATCHES (Hi,Hi) THEN Hi ELSE Lo =>
(out_, IF circ_ MATCHES (Hi,Hi) THEN Hi ELSE Lo :  t_)
--------------------
|- (env_ |- AND#(circ_) => (out_,AND#(h)) :  t_)
> apprl ReduceIf;

***** Level 4 *****
|- E t_1
|- T (C Hi,C Hi) :  t_1
|- (env_ |- Lo => (o2_,Lo) :  t_)
|- (env_ |- Hi => (o1_,Hi) :  t_)
|- (env_ |- circ_ => (out_,h) :  t_1)
--------------------
|- (env_ |- AND#(circ_) => (case match (Hi,Hi) out_ of
uu => bottom o1_ | tt => o1_ | ff => o2_,AND#(h)) :  t_)
```

ReduceIf defers the computation of the output of the IF by delivering a *symbolic answer*. In this case, however, we would like to have a concrete value answer rather than an expression describing what happens in the most general case. After undoing everything using undoAll() we prove the result we want by flexing type t_ and circuit newcirc_, expanding all the abbreviations using applyTac (doRules[ANDU,HiU,LoU,SignalU]), finally followed by applyTac OpSemTac. OpSemTac uses the rule ReduceIf' rather than ReduceIf, so that the required answer is obtained.

```
> applyTac OpSemTac;

***** Level 5 *****
|- out_ == (case match (Cons (1,1), Cons (1,1)) out_1 of
uu => bottom (Cons (1,1)) | tt => Cons (1,1) | ff => Cons (2,1))
|- (env_ |- circ_ => (out_1,h) :  TyTuple (Type 1,Type 1))
--------------------
|- (env_ |- AND#(circ_) => (out_,AND#(h)) :  Type 1)
> val ReduceAND = popGoal();
val ReduceAND = ?  :  rule
```

The abbreviations for Hi *etc.* have been expanded so that this derived rule ReduceAND may be used in general contexts without any extra work. This derived rule may be thought of as abbreviating the whole proof tree which was generated to prove this rule. Derived rules may be used very effectively in a hierarchical manner. Simulations may be speeded up by passing rules which reduce subcircuits such as AND gates or adders in one step. An alternative approach, is to write a tactic with the same effect. A tactic would actually *replay* or recreate the proof tree, which would be as slow as rerunning the proof. The application of a derived rule, in contrast, is as fast as a primitive rule.

All of the computations we have shown so far have been within a single clock tick or time step. The following example shows how a delayed AND gate may be simulated during two time steps. At every time step the value at the head of the input stream is put on top of the stack. Var 0 indicates the first value on the stack or environment env_. Delay (c,e) is a unit delay of expression e. Thus the circuit DELAY (Signal,AND#(Var 0)) is an AND gate which takes its input from the input stream, and whose output is delayed by one time step.

```
> apprl ReduceSeqCons;

***** Level 3 *****
|- ([(Signal,Lo)], env_ |- circ1_ => (outstream_,newcirc_))
|- ((Lo,Lo) ::  env_ |- DELAY (Signal,AND#(Var 0)) => (o1_,circ1_) :  t_)
--------------------
|- ([(Lo,Lo), (Signal,Lo)], env_ |-
DELAY (Signal,AND#(Var 0)) => (o1_ ::  outstream_,newcirc_))
> apprl ReduceDelay;

***** Level 4 *****
|- ([(Signal,Lo)], env_ |- DELAY (out_,circ'_) => (outstream_,newcirc_))
[2] |- Signal :  t_
[1] |- ((Lo,Lo) ::  env_ |- AND#(Var 0) => (out_,circ'_) :  t_)
--------------------
|- ([(Lo,Lo), (Signal,Lo)], env_ |-
DELAY (Signal,AND#(Var0)) => (Signal ::  outstream_,newcirc_))
```

Note that the output from circuit is known even though the output from the AND has not been computed yet. We reduce premise 1 using ReduceAND, and the second premise using ReduceSignal.

```
> apprl ReduceSignal;

***** Level 7 *****
|- ([(Signal,Lo)], env_ |-
DELAY (Lo,AND#(Var 0)) => (outstream_,newcirc_))
--------------------
|- ([(Lo,Lo), (Signal,Lo)], env_ |-
DELAY (Signal,AND#(Var 0)) => (Signal ::  outstream_,newcirc_))
```

We have now completed time zero, and can compute the next time step. Note that the description of the delay now has state Lo, which was the output from the AND gate at the previous time step. The second time step may be dealt with in exactly the same manner, resulting in the following:

```
***** Level 10 *****
|- ([], env_ |- DELAY (Lo,AND#(Var 0)) => (outstream_,newcirc_))
--------------------
|- ([(Lo,Lo), (Signal,Lo)], env_ |-
DELAY (Signal,AND#(Var 0)) => (Signal ::  Lo ::  outstream_,newcirc_))
```

The final application of ReduceSeqNil closes the input stream. Note that only at this point do we know what the final circuit looks like, in case we want to continue this simulation.

```
> apprl ReduceSeqNil;

***** Level 11 *****
--------------------
|- ([(Lo,Lo), (Signal,Lo)], env_ |-
DELAY (Signal,AND#(Var 0)) => ([Signal,Lo],DELAY (Lo,AND#(Var 0))))
```

As in the previous example, we could have done all of this with the application of a single tactic safeOpSemAllTac' [ReduceAND]. A list of derived rules may be passed into the tactic. This means that an AND gate, for example, is reduced using one derived rule application, rather a series of primitive rules. This facilitates faster, hierarchical simulation because a circuit does not need to be flattened out into individual gates to be simulated. It is also easier to pinpoint errors in a circuit when it is simulated hierarchically because boundaries of subcircuits are clearer when the subcomponents have not been flattened out. One needs to open up a subcircuit only when it is found to be in error.

4.2 Adder circuits

One of the strengths of the embedding approach used here is that we can manipulate circuit expressions just like any other term in the proof system. This allows us to write functions operating on and delivering circuits. In this subsection we will describe two implementations of a full adder, followed by an N bit adder generator. Formal circuit generators were introduced by Brock *et al.* in [3, 4].

We will first show two implementations of a full adder. ADD1 is composed of two half adders in the following manner:

```
val OR#(e)    = IF e MATCHES (Lo,Lo) THEN Lo ELSE Hi;
val XOR#(e)   = IF e MATCHES (Hi,Lo)|(Lo,Hi) THEN Hi ELSE Lo;
val HA#(e)    = LET e IN (XOR#(Var 0), AND#(Var 0));
val ADD1#(e) = LET e (* ((x,y),c) *) IN
               LET HA#((Var 0)[1]) IN
               LET HA#(((Var 0)[1], (Var 1)[2])) IN
                    ((Var 0)[1],                          (* sum *)
                     OR#(((Var 0)[2], (Var 1)[2])));      (* carry *)
```

The outermost LET is necessary, in case the input expression contains Vars. It also avoids duplication of the input circuit by using a fan-out. For example, without this LET, the second half adder in ADD1#(Var 0) would incorrectly access the first half adder as input. We easily derive ReduceHA and ReduceADD1 using safeOpSemAllTac. In this example we can see quite clearly how we use proof system capabilities to structure our circuits at the object level. AND *etc.* are meta-level syntactic functions.

ADD2 is built directly from three AND gates, two OR gates and two XOR gates.

```
val ADD2#(e) = LET e IN (* ((x,y),c) *)
               LET AND#(((Var 0)[1][2], (Var 0)[2])) IN (* bc *)
               LET AND#(((Var 1)[1][1], (Var 1)[2])) IN (* ac *)
               LET AND#((Var 2)[1]) IN                  (* ab *)
               LET OR#((Var 2, OR#((Var 1,Var 0)))) IN
               LET XOR#((Var 4)[2], XOR#((Var 4)[1])) IN
                   (Var 0,Var 1);
```

Most of the complexity is due to the destruction and construction of tuple wires.

These two adders behave identically on fully defined inputs. However, ADD2 may be more defined than ADD1 on partially defined inputs, such as ((Hi,Signal), Hi). The former outputs (Signal, Hi) while the latter results in (Signal, Signal) for the (sum,carry) pair. For this input we cannot say anything about the sum, but we know that the carry must be Hi. In the case of ADD1 the pessimism is due to non-optimal use of the input; information is consumed piecewise by independent subcomponents. There is no one bit adder implementation whose outputs are more defined than those of ADD2 for partially defined values.

```
> applyTac (safeOpSemAllTac'[ReduceADD2]);

***** Level 4 *****
--------------------
|- ([((Hi,Signal),Hi), ((Signal,Hi),Hi), ((Lo,Signal),Lo),
((Hi,Hi),Signal)], env_ |- ADD2#(Var 0) =>
([(Signal,Hi), (Signal,Hi), (Signal,Lo), (Signal,Hi)],ADD2#(Var 0)))
```

We will now define a N bit adder generating function which is parametrised on the full adder subcomponent.

```
(* onebitadder:  ((x,y),c) -> (s,c) *)
(* nadd:  (((x_{N+1},(..,x_0)),(y_{N+1},(..,y_0))),c_0) -> ((s_{N+1},(..,s_0)),c) *)
fun nadd onebitadder (S 0) x = onebitadder x |
    nadd onebitadder (S (S n)) x =
         LET x IN (* (((x_{N+1},(..,x_0)),(y_{N+1},(..,y_0))),c_0) *)
         LET nadd onebitadder (S n)
             (((Var 0)[1][1][2], (Var 0)[1][2][2]), (Var 0)[2]) IN
         LET onebitadder (((Var 1)[1][1][1], (Var 1)[1][2][1]),
                          (Var 0)[2]) IN
             (((Var 0)[1], (Var 1)[1]),      (* sum *)
              (Var 0)[2])                    (* carry *);
```

nadd is a partial function: there is no such a thing as a zero bit adder. A one bit adder with input $((x_0, y_0), c_0)$ uses the full adder component. A $N+1$ bit adder with input $(((x_N, \overline{x}), (y_N, \overline{y})), c_0)$ uses an N bit adder with input $((\overline{x}, \overline{y}), c_0)$ connected to a full adder with input $((x_N, y_N), c_N)$. As with the ADD1 circuit, virtually all of the complexity is due to the composition of intermediate wires. It is more complicated than in the 'paper version' of picoELLA due to the de Bruijn encoding of variables. The derived rule ReduceNADDSSn, dealing with $N+2$ word size, is quite involved. Premise one evaluates the input circuit; premise two the $N+1$ bit adder, and premise three the full adder. The remaining premises deal with the static semantics.

```
|- E t1_
|- o1_ :  t1_
|- E (Type m2)
|- E t1_3
|- o2_2 :  t1_3
[3] |- (CoTuple (o2_2,Cons (n3,m2)) ::  o1_ ::  env_ |-
add_ (((Var 1)[1][1][1], (Var 1)[1][2][1]), (Var 0)[2])
=> (CoTuple (Cons (n2,m1),Cons (n1,m)),
add_ (((Var 1)[1][1][1], (Var 1)[1][2][1]), (Var 0)[2])) :
TyTuple (Type m1,Type m))
[2] |- (o1_ ::  env_ |-
nadd add_ (S n) ((((Var 0)[1][1][2], (Var 0)[1][2][2]), (Var 0)[2]))
=> (CoTuple (o2_2,Cons (n3,m2)),
nadd add_ (S n) ((((Var 0)[1][1][2], (Var 0)[1][2][2]), (Var 0)[2]))) :
TyTuple (t1_3,Type m2))
[1] |- (env_ |- circ_ => (o1_,circ'_) :  t1_)
--------------------
|- (env_ |- nadd add_ (S (S n)) circ_ =>
(CoTuple (CoTuple (Cons (n2,m1),o2_2),Cons (n1,m)),
nadd add_ (S (S n)) circ'_) :  TyTuple (TyTuple (Type m1,t1_3),Type m))
```

We see that the output of the $N+1$ bit adder is a tuple CoTuple (o2_2, Cons (n3,m2)). Comparing this to the definition of nadd we see that o2_2 represents the partial sum $(s_N, (.., s_0))$, and Cons (n3,m2) the carry c_{N+1}. Decoding the inputs of the final $N+2$nd bit adder add_, we see that its input carry (Var 0)[2] accesses the output carry

Cons(n3,m2) from the $N+1$ bit adder, as expected. The final result of the $N+2$ bit adder consists of (i) the concatenation of the sum bit of add_ (Cons(n2,m1)) concatenated with the partial sum o2_2; and (ii) the carry bit Cons(n1,m) of add_. The derived rule ReduceNADD1 just unfolds the nadd definition to evaluate the full adder. Note that the result circuit must be identical to the circuit we evaluate. This means that the adder is not allowed to have any state.

```
|- (env_ |- add_ circ_ => (out_,add_ circ'_) :  t_)
--------------------
|- (env_ |- nadd add_ 1 circ_ => (out_,nadd add_ 1 circ'_) :  t_)
```

A four bit adder has been simulated, with ADD2 as the subcomponent. For example, binary $1010+1101+1=11000$, that is, a sum of 1000 and a high carry:

```
> applyTac (safeOpSemAllTac'[ReduceNADD4bit']);

***** Level 4 *****
--------------------
|- ([(((Hi,(Lo,(Hi,Lo))), (Hi,(Hi,(Lo,Hi)))), Hi)], env_ |-
nadd (fn e => ADD2#(e)) 4 (Var 0)
=> ([((Hi,(Lo,(Lo,Lo))), Hi)], nadd (fn e => ADD2#(e)) 4 (Var 0)))
```

Note that ADD2 is a meta-level syntactic function, and must therefore be converted into an object level function, using (fn e => ADD2#(e)).

In this section we see most clearly the increased power of our methodology over symbolic simulation as it has been used by Bryant [5] for example. When we remarked that MOSSYM does not allow abstraction over circuits what we intended to convey was that it does not allow the simulation of an N bit adder. Our approach allows more than this; we can even simulate an N bit adder built using any one bit adder onebitadder. As long as we know that the subcircuit onebitadder behaves like a one bit adder, we can simulate any circuit in which it is used. We can simulate a circuit containing abstract hardware, as long as we know what the behaviour of the subcomponent is. Let us consider an ALU, containing an N bit adder. The N bit adder specification will usually be stated at a higher level of abstraction, using natural numbers. The specification for the sum could be $bitsof(natof\ x + natof\ y)\ mod\ 2^N$. *natof* is a data abstraction function, and *bitsof* its inverse. When we simulate the ALU, and arrive at the N bit adder subcomponent, it makes sense to use the specification rather than the implementation. (This assumes we have shown that the implementation specifies the specification.) Rather than simulating the basic gates the adder is composed of, we compute the natural number expressions stating the values *sum* and *carry* have. This is not only faster, also conceptually clearer.

4.3 Two Parity Checkers

Boulton *et al.* illustrate their approach to the verification of ELLA designs with a parity checker [2]. It consists of two multiplexors, two delays and a NOT gate. PCHECK2 below describes the same circuit as PARITY_IMP in the cited paper.

```
val NOT_g#(e) = IF e MATCHES Hi THEN Lo ELSE Hi;
val MUX#(e,b1,b2) = IF e MATCHES Hi THEN b1 ELSE b2;
val REG#(c,e) = DELAY (c,e);
val PCHECK2#(s1,s2,e) = LET e IN
                        LET INIT Signal REC
                        (* Use a LET to avoid duplication of register *)
                        LET REG# (s1,Var 0) IN
                            MUX# (REG# (s2, Hi),
                                  MUX# (Var 2, NOT_g#(Var 0), Var 0),
                                  Hi) IN
                            Var 0;
```

We use NOT_g because NOT is the truth value not operator in LAMBDA. It is worth noting that the state of the parity checker is explicit in the abbreviation. The reason for this is so that the abbreviation may be used in all possible states, and not just the initial state.

```
[5] |- ceq Signal (Cons (n1,1)) == false
|- Cons (n1,1) == (case match Hi (Cons (b,1)) of           (* Outer MUX *)
uu => bottom o1_ | tt => o1_ | ff => Hi)
|- o1_ == (case match Hi (Cons (n,1)) of                   (* Inner MUX *)
uu => bottom o1_1 | tt => o1_1 | ff => Cons (a,1))
|- o1_1 == (case match Hi (Cons (a,1)) of                  (* NOT *)
uu => bottom Lo | tt => Lo | ff => Hi)
|- (env_ |- circ_ => (Cons (n,1),h) :  Type 1)
--------------------
|- (env_ |- PCHECK2#(Cons (a,1),Cons (b,1),circ_) =>
(Cons (n1,1),PCHECK2#(Cons (n1,1),Hi,h)) :  Type 1)
```

The derived rule ReducePCHECK2 contains some points of interest. First note that only the two multiplexors and the NOT gate are present as subgoals; both delays have disappeared. As described in [2], the rôle of the innermost register is to output Lo at time zero, and Hi ever after. This is evident from the conclusion of the rule below, where the state s2 is always Hi after an evaluation. Also note that the output Cons(n1,1) is duplicated in the first register, so that it can be used in the next time step, using the feedback. At time zero, the values in the registers are both Lo. In fact, the value in the first delay at time zero is irrelevant:

```
|- ([Cons (y,1)], env_ |- PCHECK2#(Cons (x,1),Lo,Var 0) =>
([Hi],PCHECK2#(Hi,Hi,Var 0)))
```

This derivation uses an arbitrary input Cons(y,1) and state in the first delay Cons(x,1). The only constraint on these *don't care* values is that they must have the right type. Note that their possible value includes the undefined or *don't know* value. This simulation shows that the state of the new circuit is fully defined no matter what the input at time zero is. In other words, the value of the input at time zero is ignored. This parity checker outputs Hi at time t if there have been an even number of His in the input stream from time *one* to time t inclusive.

An alternative parity checker is listed below.

```
val PCHECK1#(s,e) = LET e IN
                    LET INIT Signal REC
                    REG# (s, XOR# (Var 0, Var 1)) IN
                         Var 0;
```

The initial state must be Hi. PCHECK1 outputs Hi at time $t+1$ if there have been an even number of His in the input stream from time *zero* to time t. The output at time zero is Hi.

```
> applyTac (safeOpSemAllTac'[ReducePCHECK1]);

***** Level 4 *****
--------------------
|- ([Hi,Lo,Hi,Hi,Lo,Lo], env_ |- PCHECK1#(Hi,Var 0) =>
([Hi,Lo,Lo,Hi,Lo,Lo],PCHECK1#(Lo,Var 0)))
```

Using conventional verification techniques we proved that the PCHECK1 circuit does indeed count the number of His in the input stream.

```
(* Number of v's in the input stream from time 0 up to time t.  *)
fun noof v input 0 = 0 |
    noof v input (S t) = if input t = v then (noof v input t) + 1
                                         else (noof v input t);
fun even n = n mod 2 = 0;
fun absinv true = Hi | absinv false = Lo;
fun state x y = absinv (even (noof Hi x y));
```

noof counts the number of vs in the input stream, even returns true if there have been an even number of them, and absinv is the inverse data abstraction function, mapping booleans to constants. state combines these three functions into one, to make the result more readable.

```
|- forall t,l,e,input.  input t == Hi \/ input t == Lo ->>
Reduce l (PCHECK1#(state input t,e t)) ==
(state input t, PCHECK1#(state input (S t),e (S t)))
```

In other words, assuming the input is either Hi or Lo at every time step, the output at time t consists of two parts. The first value is Hi is there have been an even number of His in the input stream. The second part states that the state of the new circuit is given by the state function at time $t+1$. As we discussed at the end of the previous subsection, we can use this specification instead of using the circuit in simulations.

Although it was not shown in the last two examples, the semantics computes the least fixed point of a LET REC. An iterative method is used, and the number of iterations may vary to reach the fixed point. In the case of delayed feedbacks, however, it takes at most one iteration. If the output is not undefined exactly one iteration is needed. In the derived rules for PCHECK1 and PCHECK2 the assumption was made that no undefined values were input to the circuit. (Premise [5] of rule ReducePCHECK2 states this. The assumption is more explicit in the theorem above.) It follows from this assumption that only defined values are output and hence only one iteration is needed. The current tactics do not attempt to deal with recursion.

5 Conclusions

Simulation and verification are usually described as alternative, incompatible approaches. This paper shows that by suitably embedding the operational semantics of a HDL in an appropriate proof tool we are able to integrate simulation and verification within the same framework. We believe the approach taken here is applicable to any HDL. The choice of picoELLA and LAMBDA is not crucial to the discussion.

The strength of our approach is the ability to specify, implement, simulate and reason about a circuit within a single framework. At any stage in this process we may use a conventional HDL notation, the logic supported by the proof system, or a mix of the two. Although the specification will often be expressed using logic, an algorithmic specification, *i.e.* as a high level HDL program, may be useful. An algorithmic specification can also be used to give a more operational intuition by executing it. Logic specifications (and implementations) cannot be animated easily. The common relational hardware description style, which uses existential quantification for hidden wires, is an example. Moreover, structure and behaviour are not properly separated; the form of the behavioural description is used to indicate the intended structure of the circuit. Our approach strictly separates structure and behaviour [9]. Behaviour is given to a purely structural term through a formal embedded semantics, and properties of circuits are derived using this semantics. The ability to reason about and manipulate structural expressions *per se* is very useful. It facilitates interfacing with conventional design tools because they use the same notation. For example, circuits designed using a proof system may be exported directly to layout generators. Alternatively, hardware output by unverified hardware synthesis tools can be validated using the proof system. Hardware may also be synthesised formally using hardware generators such as **nadd** in section 4.2, first introduced by Brock and Hunt in [3]. Formal synthesis [12], and refinement based approaches [7] fit well into our framework. DIALOG is a graphical synthesis package integrated with the LAMBDA proof system. Using DIALOG, it would be possible to synthesise formally verified HDL descriptions without the need to explicitly use the underlying proof system. This could be seen as a HDL interface to the proof system; the user does not need to interact with the underlying proof system. We can also treat circuit optimisations formally. If two structural terms have equivalent behaviours, they may be substituted for one another in any context. A given circuit could be optimised by (possibly context dependent) rewriting, which is certainly possible in LAMBDA. Finally, we can use the embedded semantics to simulate the structural terms. Both data and circuit descriptions may be meta-variables, enabling powerful symbolic simulation. Partial implementations may be simulated by using the specifications of the missing components. In our opinion the main advantage of this approach is the possibility of using a conventional HDL in more formal setting. This bridges the gap between hardware designers and verification engineers.

Future work includes the optimisation of tactics. Tactics must be made to deal with recursion automatically if possible. It will also be helpful to make the use of the system more user-friendly by providing a menu-based X window interface using LAMBDA's built-in browser. Finally, picoELLA was designed to exhibit the ideas outlined here and in [9]. A larger, more readable, subset of ELLA must be used for practical applications.
A longer version of this paper is available as LFCS report number ECS-LFCS-92-231.

References

[1] H Barringer, G Gough, and B Monahan. Operational semantics for hardware design languages. In P Prinetto and P Camurati, editors, *Advanced Research Workshop on Correct Hardware Design Methodologies*, pages 313–334. North Holland, June 1991.

[2] R Boulton, M Gordon, J Herbert, and J van Tassel. The HOL verification of ELLA designs. Technical Report 199, University of Cambridge Computer Laboratory, August 1990.

[3] B Brock and W Hunt, Jr. The formalization of a simple hardware description language. In L Claesen, editor, *Applied Formal Methods For Correct VLSI Design*, pages 778–792, November 1989. Elsevier Science Publishers.

[4] B Brock, W Hunt, Jr, and W Young. Introduction to a formally defined hardware description language. In V Stavridou, T Melham, and R Boute, editors, *Theorem Provers in Circuit Design: Theory, Practice and Experience*, pages 3–35. North Holland, June 1992.

[5] R Bryant and C Seger. Formal verification of digital circuits using symbolic ternary system models. Technical Report CMU-CS-90-131, School of Computer Science, Carnegie Mellon University, Pittsburgh PA 15213, May 1990.

[6] Computer General Electronic Design, The New Church, Henry St, Bath BA1 1JR, England. *The ELLA Language Reference Manual*, issue 4.0, 1990.

[7] M Fourman and E Mayger. Formally based system design – interactive hardware scheduling. In G Musgrave and U Lauther, editors, *Int'l Conf. on VLSI*, 1989.

[8] M Francis, S Finn, and E Mayger. *Reference Manual for the Lambda System.* Abstract Hardware Limited, version 3.2, November 1990.

[9] K G W Goossens. Embedding hardware design and description languages in proof systems. Forthcoming PhD thesis. University of Edinburgh.

[10] K G W Goossens. Semantics for picoELLA. Manuscript, June 1990.

[11] K G W Goossens. Embedding a CHDDL in a proof system. In P Prinetto and P Camurati, editors, *Advanced Research Workshop on Correct Hardware Design Methodologies*, pages 359–374. North Holland, June 1991.

[12] F Hanna, M Longley, and N Daeche. Formal synthesis of digital systems. In L Claesen, editor, *Applied Formal Methods For Correct VLSI Design*, pages 532–548, November 1989. Elsevier Science Publishers.

[13] The Institute of Electrical and Electronics Engineers, Inc. *IEEE Standard VHDL Language Reference Manual*, IEEE std 1076-1987, 1988.

[14] T Melham. Using recursive types to reason about hardware in higher order logic. Technical Report 135, University of Cambridge Computer Laboratory, May 1988.

[15] G Plotkin. A structural approach to operational semantics. Technical Report FN-19, Computer Science Department, Aarhus University (DAIMI), 1981.

[16] V Stavridou, J Goguen, S Elker, and S Aloneftis. FUNNEL: A CHDL with formal semantics. In P Prinetto and P Camurati, editors, *Advanced Research Workshop on Correct Hardware Design Methodologies*, pages 115–137. North Holland, June 1991.

[17] J van Tassel. A formalisation of the VHDL simulation cycle. Technical Report 249, University of Cambridge Computer Laboratory, March 1992.

[18] P Wilsey, T McBrayer, and D Sims. Towards a formal model of VLSI systems compatible with VHDL. In A Halaas and P Denyer, editors, *VLSI '91*, August 1991.

Higher Order Logic Theorem Proving and its Applications (A-20)
L.J.M. Claesen and M.J.C. Gordon (Editors)
Elsevier Science Publishers B.V. (North-Holland)

Simulating Microprocessors from Formal Specifications [1]

Kelly M. Hall and Phillip J. Windley

University of Idaho Laboratory for Applied Logic, Moscow ID 83843

Abstract
Simulation is an important aspect of microprocessor design. In a verified microprocessor design, simulation can be used to ensure that the functionality of instruction set is sufficient and that the microprocessor behaves as intended by the designer. This paper documents and demonstrates a tool written in the HOL theorem proving system for executing microprocessor specifications written in higher-order logic.

Keyword Codes: B.1.4; B.5.3
Keywords: Microprogram Design Aids; Reliability and Testing

1 Introduction

Simulation has long been used by microprocessor designers to ensure functional correctness, check the implementation against the design, ensure that the design behaves as intended, that the instruction set is complete, and as a means of writing and testing code before an implementation actually exists. Recently, there has been much interest in microprocessor specification and verification [Hun87, Coh88, Joy89, Win90]. Microprocessor verification is an analytical method for ensuring functional correctness, but it does not address the other concerns of system designers.

Executable specifications can address many of the simulation needs of computer system designers. An executable behavioral specification can be used to check the specification for completeness and as a means of executing code. An executable structural specification can be used to check an implementation against the design. One approach to integrating simulation is to specify the design in a language common to the verification system and the simulation engine. The work described in [GW92] shows one way this can be done.

Another approach to executable specifications is to actually execute the specification language. The Boyer-Moore logic [BM79] is an example of an executable logic. This approach does not work for higher-order logic, the logic used in this paper, since there

[1]This research was supported in part by NSF under Research Initiation Grant MIP-9109618.

are many constructs in the logic that cannot be executed.[2] There are, however, executable subsets of higher-order logic. Camilleri [Cam88] presents an implementation of an executable subset of higher-order logic in the HOL theorem proving system [Gor88].

The work reported in this paper is an attempt to make Camilleri's work more accessible to microprocessor designers. We have designed a simulator, GSIM, capable of simulating any microprocessor specification meeting some very general structural constraints. For now, the simulator provides a simple user interface similar to that provided by the DOS `DEBUG` command.

Our goal is not to use out simulator for detailed timing analyses. Indeed, at the level of abstraction we are working at (behavioral specifications of microprocessors, time usually ticks once per instruction executed and the details are left to lower levels. The primary goal is the animation of these specification for purposes of inspection and for use as a programming aid.

The remainder of this report describes the simulator. The next section describes the generic interpreter theory since that is the basis for the work. The following section describes the simulator and shows how it operates. Finally, an example using Tamarack-3 [Joy89] is given to demonstrate how the simulator is used. An appendix provides a demonstration of the interactions provided by the simulator.

2 The Generic Interpreter Theory

There have been numerous efforts to formally model microprocessors. The best known of these include Jeff Joyce's Tamarack microprocessor [Joy89], Warren Hunt's FM8501 microprocessor [Hun87], Avra Cohn's VIPER microprocessor [Coh88] and our *AVM-1*. Tamarack is a simple microprocessor with only 8 instructions. FM8501 is larger (roughly the size of a PDP-11), but has not been implemented (a 32-bit version is currently being verified and implemented by Hunt *et al.*). VIPER was designed by Britain's Royal Signals and Radar Establishment (RSRE) at Malvern to provide a formally verified microprocessor for use in safety critical applications and is commercially available. *AVM-1* is a 32-bit microcoded microprocessor with a load-store architecture. *AVM-1* was not implemented, but a pipelined version, designated SAWTOOTH, is currently being specified, designed, and verified for implementation.

2.1 Microprocessor Specification.

The specifications for the microprocessors mentioned above appear very different on the surface; in fact, the specification to FM8501 is even in a different language. On closer inspection, however, we find that each of them (as well as many others) use the same implicit behavioral model. In general, the model uses a state transition system to describe the microprocessor. A microprocessor specification has four important parts:

1. A representation of the state, **S**. This representation varies depending on the verification system being used.

[2] A good example is Hilbert's choice operator, `@`, used in the definition of the logic in HOL.

Table 1: The abstract functions and their types for the generic interpreter model.

Operation	*Type*
`instructions`	`":*key -> (*state->*env->*state)"`
`select`	`":*state->*env->*key"`
`substate`	`":*state'->*state"`
`subenv`	`":*env'->*env"`
`Impl`	`":(time'->*state')->(time'->*env')->bool"`
`count`	`":*state'->*env'->*key'"`
`begin`	`":*key'"`

2. A set of state transition functions, **J**, denoting the behavior of the individual instructions of the microprocessor. Each of these functions takes the state defined in step (1) as an argument and returns the state updated in some meaningful way.

3. A selection function, **N**, that selects a function from the set **J** according to the current state.

4. A predicate, **I**, relating the state at time $t+1$ to the state at time t by means of **J** and **N**.

In some cases, the individual state transition functions, **J**, and the selection function, **N**, are combined to form one large state transition function. Also, a functional specification would use a function for part (4) instead of a predicate. The overall form, however, is the same.

Discovering the similarity in previous efforts provided a common basis for studying microprocessor verification. To make the commonality usable, we have formalized a microprocessor specification model. The formalization provides a step–by–step approach to microprocessor specification by enumerating the important definitions that need to be made for any microprocessor specification. This is the important for the purposes in this paper—the fact that the specification is structured provides the basis for writing a generic simulator.

The formalization models an interpreter. An interpreter is a computing structure with one control point. One of the many available instructions is chosen at this control point based on the current state and inputs. The state is then processed by this instruction and the cycle begins again. Interpreters differ from standard state machine models only in form, not in power; but, their form is particularly amenable to modeling microprocessors and other similar computing devices.

2.2 The Abstract Representation

The formalization is done in the HOL theorem proving system [Gor88] using the abstract theory package [Win92a]. An abstract theory is parameterized so that some of the types and constants in the theory are undefined inside the theory except for their syntax and an algebraic specification of their semantics. The abstract representation is the interface to the generic model and crucial to the work reported here since it the abstract representation that structures the specification. We specify the abstract representation of the generic interpreter by defining a list of abstract objects and operations. Table 1 shows the operations and their types.

The following abstract types are used in the representation.

- `:*state` represents the state.
- `:*env` represents the environment. The environment provides inputs to the interpreter.
- `:*key` is type containing all of the keys. Keys are used to select instructions. For example, the opcodes form the keys at the macro–level.

The operations in the abstract representation are abstract counterparts to those found in specific microprocessor specifications. The two abstract operations that are used in the generic simulator are the instruction set, `instructions` and the selection operator, `select`. `instructions` represents the instruction set as a function from a key (used to identify instructions) to a state transition function. `select` represents the function that picks a key based on the present state and environment.

Using the functions from the abstract representation, we can define an interpreter. The form of this definition is essentially the same as the form of specific microprocessor specifications given in [Joy89, Hun87, Coh88].

```
⊢def INTERP rep s e =
       ∀ t:time.
        let k = (select rep (s t) (e t)) in
        s(t+1) = (instructions rep k) (s t) (e t)
```

The specification of an interpreter is a predicate relating the contents of the state stream at time $t+1$ to the contents of the state stream at time t. The relationship is defined using the functions from the abstract representation.

The generic interpreter theory also contains numerous theorems about the abstract representation. The most important is a correctness proof that shows that `Impl` implements `INTERP` when the theory obligations are satisfied. The details of the generic interpreter theory are beyond the scope of this paper, but interested readers are referred to [Win92b].

3 Generic Interpreter Theory Simulator

The Generic Interpreter Theory Simulator (GSIM) is a program that simulates the execution of microprocessors written using the generic interpreter theory discussed in the last section. GSIM is capable of simulating any specification written as a generic interpreter. This includes not only top–level specifications (programmer's model), but register transfer level simulations as well.

As mentioned in the last section, the generic interpreter theory dictates that the specification have a function that returns an instruction selection key based on the current state and environment and another function from keys to state transition functions. GSIM uses translations of these functions to execute code inside a user-interface patterned after the program DEBUG on CP/M and MS-DOS systems.

Our original goal was to automatically translate HOL-to-ML using Camilleri's executable HOL subset. However, problems with HOL 2.0 inject_input and ML_eval functions prevented the use of Camilleri's code. Currently, an AWK program is used to perform an *ad hoc* translation of the HOL functions to ML for subsequent execution. The translation is not complete and some hand–translation is still required. Work is currently underway to implement a more robust, rule–based translator.

The user interface for GSIM allows the user to modify and view processor state, including memory, registers, flags, and the program counter. GSIM provides a facility for a tiny assembler to make entering source programs tractable. After the program is entered, instructions can be executed one at a time, or the program can be executed without stopping.

The user interface was patterned after DEBUG since it has a small and easily implemented command set. In the future we hope to expand and improve the user interface to include a simulation control language and other features common in dedicated simulation packages.

GSIM achieves its generality by requiring that the user provide instantiations for abstract data types used in the specification. GSIM provides "hooks" for common functions that are used to manipulate the abstract data. For example, most microprocessor specifications use some HOL representation of words in memory. This representation must be associated with a corresponding ML type, and the GSIM hook INT-TO-WORDN-FUNC must be bound to an ML function that coerces an integer into the HOL representation.

Since GSIM is tied closely to the generic interpreter theory, it simulates best those levels of a microprocessor that are most like interpreters: the assembly language level and the RTL level. At these levels of abstraction there is no need to simulate units of time—the machine simply moves from state 1 to state 2. When external events occur at this level of abstraction it is enough to know that the event occurred between state 1 and state 2.

GSIM does make some assumptions about processor architecture over and above the assumptions of the generic interpreter theory; namely, that state is composed solely of four elements: memory, registers, flags, and the program counter. Also, we assume each of these items can be changed solely by the processor (i.e. there is no shared memory or externally controlled processor flags).

3.1 Theory of Operation

There are two main activities that GSIM provides users: altering processor state or simulating processor execution.

Aspects of processor state can either be viewed or modified. GSIM performs these tasks at an abstract level, relying heavily on function hooks to perform the actual work. As stated before, function hooks are used throughout GSIM to make it as general as possible. The hooks are defined by the user to provide a link between the simulator and the specification. The following are examples of the hooks used in the program:

- INT-TO-WORDN-FUNC to coerce an integer into an n–bit word.
- GET-MEM-FUNC to retrieve an n–bit from memory.

- SET-REG-FUNC to update the contents of a register.
- GET-FLAGS-FROM-STATE to extract the processor-flags aspect of processor state.
- MAKE-STATE-FUNC to construct a state-type variable from its memory, registers, and flags aspects.
- SELECT-FUNC to return an opcode given processor state and external environment.

For example, consider modifying the contents of a processor register. Given the current state and a register number to modify, GSIM will return a new state that reflects the user's changes. The algorithm is simple:

- Print the register's current contents using the GET-REG-FROM-STATE, GET-REG-FUNC, and DISPLAY-WORDN hooks.
- Prompt the user for a new value.
- Build the new state using the SET-REG-FUNC and MAKE-STATE-FUNC hooks.

Operations involving memory, flags, and the program counter are very similar.

As a second example, consider the execution of one instruction. Given the current state and environment, GSIM will return a new state that is the result of executing the next instruction. The algorithm is:

- Based on the current state and environment, get the appropriate opcode with the SELECT-FUNC hook.
- Get the appropriate state-transition function from the INSTRUCTIONS-FUNC using the opcode.
- Apply the state transition function to the current state and environment to yield the new state.

The instruction simulation is performed by the trace-execution and begin-execution functions. These apply the various parts of the specification to get the appropriate state transition function according to the current state and environment. Then the state transition function is applied to the current state to yield a new state, and the process repeats.

3.2 Applying GSIM to a Specification

Applying GSIM to an existing specification is a straightforward task.

- If needed, the HOL specification must be made to conform with the GIT. Generally, this means converting processor opcodes into state transition functions, and writing the instructions and select functions.
- The HOL specification is translated into ML. The AWK programs we use handles much of the syntax, but some editing of the ML is still needed to ensure the syntax is correct.

- Data types are defined. These are the types used in the ML code above: `wordn`, `memory`, `registers`, `flags`, etc.
- Subordinate ML functions are written if needed. These are the small functions that are used in the state transition functions: fetching and storing memory locations, adding or subtracting `wordn` values, predicates testing for zero or positive `wordn` values, and so forth. These functions depend upon the representations for data chosen above, and cannot be generated automatically.
- GSIM hooks are bound to appropriate functions. At this point, all of the low-level functions for simulation of the processor exist. All that remains to bind the GSIM hooks to the low-level functions.
- If desired, an assembler can be written and attached to the GSIM tiny-assembler hooks.

These steps are shown in more detail in the Tamarack–3 example.

3.3 Using GSIM

GSIM provides a user-interface patterned after the `CP/M` and `MS-DOS` program `DEBUG`. GSIM command lines are a sequence of words separated by spaces, terminated with a newline. The first word of a command line is a single character which identifies the command, and remaining words are parameters for the desired command. If the first word of a command line is undefined, GSIM prints a short menu of the available commands. GSIM operates in a simple read-eval-print loop until the user exits the program, at which point it returns to the ML read-eval-print loop.

The following commands are currently recognized by GSIM:

- Show or change memory
- Show or change registers
- Show or change the program counter
- Show processor flags
- Set and remove Breakpoints
- Assemble an instruction
- Execute a single instruction
- Begin execution
- Save or Load Memory
- Exit GSIM

The show command for memory takes an optional parameter. If present, that parameter represents the first address to be displayed; otherwise, location zero will be displayed. The show function prints the address and contents of the next twenty memory locations and waits for further input. A "q" will return to the GSIM read-eval-print loop and any other input will cause another twenty locations to be displayed.

The change command for memory is similar to the show command. If given a starting address, the routine uses it; otherwise, the starting address is assumed to be zero. The routine displays the address and contents of a location in memory and prompts for input: an empty newline advances to the next address leaving the current contents unchanged, a "q" returns control to the GSIM read-eval-print loop, and anything else is taken to be the new value for the current location in memory.

The show and change commands for registers, the program counter, and processor flags are similar to the memory versions, with slight differences: registers must be indexed by an integer register number (instead of mnemonic name like ACC or X-reg, the program counter and processor flags commands do not accept a starting address parameter, and there is no change command for the processor flags.

Memory addresses may be specified such that before execution of the instruction at that address is attempted, the simulator halts and a message is printed for the user. These addresses are called breakpoints and are useful when debugging large or complicated programs. Breakpoints are kept in a list of integers. Addresses may be added to the list with the set breakpoints command. Likewise, a breakpoint may be deleted from the list with the remove breakpoint command. After either command is executed, the updated list of current breakpoints is displayed.

The Assemble command has no strict form since the end-user is responsible for defining the calling syntax. GSIM simply passes any parameters to the assemble function as a list of words.

The command to execute a single instruction (called trace) and the command to begin continuous execution (called go) are similar. trace takes an optional starting address (it defaults to address zero) and simulates the execution of a single instruction. go takes an optional starting address and executes instructions until an error occurs or the GSIM hook HALT-GIM-FUNC returns true. In either case, go returns control to the GSIM read-eval-print loop.

Capability to save and load blocks of memory to and from disk files is provided through the write memory and load memory commands. The write command has parameters for the file name to save to, the starting address, and the number of memory locations to write. The load command has parameters for the file name to load from, and the starting address for the loaded data. The write and load commands facilitate the use of external assemblers and compilers with the simulator.

The quit command exits GSIM and returns control to the ML read-eval-print loop.

4 Tamarack–3 Example

This section shows how Joyce's Tamarack–3 can be simulated in GSIM. We have also used GSIM to simulate the operation of *AVM–1*. We anticipate that GSIM will be used extensively in the design and verification of the SAWTOOTH microprocessor.

4.1 Modifying Tamarack–3.

To make Tamarack–3 work with GSIM, it must first be modified to conform to the GIT. To do this, Tamarack's opcode functions must be modified to take the current environment as well as the current state. Also, an opcode selection function and a function to select the appropriate state transition function must be written. For Tamarack–3, the limited environment (a single boolean value) makes the changes to the opcodes very simple. For example, the following shows the original and new specifications for the `NOP` instruction.

The original definition:

```
⊢def NOP (rep:^rep_ty)
        (mem:*memory,pc:*wordn,acc:*wordn,rtn:*wordn,iack:bool) =
        (mem,(inc rep) pc,acc,rtn,iack)
```

The new definition:

```
⊢def NOP (rep:^rep_ty)
        (mem:*memory,(pc:*wordn,acc:*wordn),rtn:*wordn,iack:bool)
        (ireq:bool) =
        (mem, (inc rep) pc, (acc,rtn), iack)
```

There are only two changes, the addition of an environment variable, and the grouping of both processor registers into a single tuple. The single tuple for registers is easier to use with GSIM, but is not required.[3] Other Tamarack–3 instructions are changed in a similar manner and the pseudo-instruction `IRQ` is also made into a state transition function to handle interrupt requests.

In addition to changing the format of the instruction denotations, we must define the two functions from the GIT required for GSIM: the selection function, `select`, and the instruction function, `instructions`.

Joyce does not define an explicit selection function, but this is not difficult to do. The selection function operates on the current state and environment and returns an opcode for the appropriate state transition function. Since `IRQ` is now a state transition function it has an opcode as well. We model this using an injection on boolean 3–tuples (for the actual opcodes) and the unit type `:one` (for the single pseudo–instruction). Functions that correspond to the Tamarack–3 instructions will have as opcodes a left-injection of their Tamarack–3 opcode, and the pseudo-instruction `IRQ` will use a right-injection of the value `one`. The function `GetOpcode` is used as the selection function in the GIT.

[3]The environment variable is not used in the `NOP` instruction, or any other Tamarack–3 instruction, but is necessary for conformance to the GIT.

```
⊢def OpcVal (rep:^rep_ty) (mem,pc) =
        let nextword = (fetch rep) (mem,((address rep)pc)) in
        ((opcode rep) nextword)

⊢def GetOpcode (rep:^rep_ty) (mem,pc,acc,rtn,iack) (irec) =
        ((ireq ∧ ¬iack) →  INR(one) |
         INL(OpcVal rep (mem,pc)))
```

The instruction function simply maps opcodes to state transition functions:

```
⊢def GetFunction (rep:^rep_ty) op =
        ((op = INL(F,F,F)) →  (JZR rep) |
         (op = INL(F,F,T)) →  (JMP rep) |
         (op = INL(F,T,F)) →  (ADD rep) |
         (op = INL(F,T,T)) →  (SUB rep) |
         (op = INL(T,F,F)) →  (LDA rep) |
         (op = INL(T,F,T)) →  (STA rep) |
         (op = INL(T,T,F)) →  (RFI rep) |
         (op = INL(T,T,T)) →  (NOP rep) |
         % (op = INR(one)) %  (IRQ rep))
```

4.2 Data Types

Before translating the HOL definition given above into ML, we must define the types needed in the ML code. The choices are not automatic but depend on the machine being simulated. For Tamarack–3, opcodes have the type `((bool#bool#bool)+int)` where the right injection is arbitrarily chosen to be an integer (we never need the value since simply knowing that its a right injection is enough for our purposes).

Tamarack–3 has 16–bit words. We can think of words as either data words (representing integers) or instruction words (representing an opcode and an address in memory). Data will be stored as a twos-complement 16–bit value, while instructions will be three opcode bits and thirteen bits for the address. Data, then, can take on values from -32768 to 32767 and addresses can range from 0 to 8191. In both cases, we realize the 16–bit and 13–bit words as ML integers:

```
lettype wordn_t = int;;
```

Functions to coerce the opcode, address, and 16–bit values from integers are written in ML.

Memory, `mem_t`, is defined as an association list of integers (the address in memory) and `wordns` (the content). Registers individually have the type `wordn` but we combine all registers `reg_t` into the type `(wordn_t # wordn_t)`. All flags (in both the state and environment) are boolean values. The type `state_t` used to represent state has the type `(mem_t # wordn_t # reg_t # flag_t)` where the `wordn` represents the program counter.

The choice of appropriate data types is very important to the speed of the finished simulator. The difference between implementing the `wordn` data type as an ML integer or a list of boolean values becomes quite apparent in simulator operations. While ML will probably never match the speed of dedicated simulators, with appropriate data types we have had acceptable results using ML.

4.3 Converting HOL to ML

The HOL definitions from the previous section are next translated into ML. The translation is currently done using a simple AWK program.

Joyce makes use of an abstract representation of n–bit words in the Tamarack–3 specification. There are several approaches to dealing with these abstract representations.

- The first, and most straightforward, is to simply define an abstract data type in ML to correspond to them and supply it as an argument to the functions in place of the abstract representation. The primary disadvantage of this approach is that the argument and the selectors will have to be repeatedly executed.

- Another alternative is to treat the abstract representation as a macro and expand the instantiation prior to sending the code to ML for interpretation. This is speedy, but more difficult to implement.

For now, we use the first alternative except when execution speed becomes critical and in those cases, we simulate macro expansion by hand expanding the abstract data types at definition.

The translation process is quite simple: HOL functions are replaced by the corresponding ML functions, if they exist and HOL constants are replaced by similar ML constants. When a corresponding ML function or constant does not exist, the user is responsible for defining them. For example, `NOP` from above becomes:

```
let (NOP':state_t -> env_t -> state_t)
        (mem, pc, (acc, rtn), (iack)) (ireq) =
        (mem, (inc pc), (acc, rtn), (iack));;
```

The ML version of the HOL functions `OpcVal`, `GetOpcode` and `GetFunction` are:

```
let OpcVal (mem, pc) =
  let nwd = fetch (mem, (address pc)) in
  let bit15 = mod (nwd / (exp 2 15)) 2 and
      bit14 = mod (nwd / (exp 2 14)) 2 and
      bit13 = mod (nwd / (exp 2 13)) 2 in
  ( (bit15 = 1), (bit14 = 1), (bit13 = 1) );;

let GetOpcode ((mem, pc, (acc, rtn), (iack)):state_t) (ireq) =
  (ireq & not iack) => inr(1) |
   inl( OpcVal (mem, pc) );;

let GetFunction op =
           (op = inl(false,false,false)) => JZR' |
           (op = inl(false,false,true )) => JMP' |
           (op = inl(false,true ,false)) => ADD' |
           (op = inl(false,true ,true )) => SUB' |
           (op = inl(true ,false,false)) => LDA' |
           (op = inl(true ,false,true )) => STA' |
           (op = inl(true ,true ,false)) => RFI' |
           (op = inl(true ,true ,true )) => NOP' |
           % (op = inr(1)) => %             IRQ' ;;
```

Primes were added to state transition function names to distinguish the HOL theorems from the ML functions.

4.4 Subordinate Functions

Most subordinate functions for Tamarack–3 are simply ML versions of the abstract parts of `rep_ty`, including `iszero`, `inc`, `add`, `sub`, `wordn`, `address`, `val`, `fetch` and `store`. These functions are easy to write once the data types are known. For example, the add and fetch functions are defined as follows:

```
let (add:(wordn_t#wordn_t)->wordn_t) (w1, w2) =
  let op1 = (val w1) and
      op2 = (val w2) in
  wordn (op1 + op2);;

let fetch (mem:mem_t, addr) =
  snd (assoc addr mem) ? failwith `fetch`;;
```

4.5 GSIM Hooks

The next step is to bind the functions hooks from GSIM to the definitions we have made above (some trivial definitions are given directly). For example, the following shows some of the type-hook bindings:

```
lettype wordn_type       = wordn_t;;

lettype memory_type      = mem_t;;

lettype register_type    = reg_t;;

lettype flags_type       = flag_t;;
```

The following show some of the function-hook bindings:

```
let WORDN_TO_INT_FUNC = address;;

let GET_MEM_FUNC mem addr =
  fetch (mem, (address addr));;

let (SET_REG_FUNC:register_type -> int -> wordn_type -> register_type)
    reg number value =
  (number = 1) => ( value, (GET_REG_FUNC reg 2)) |
  (number = 2) => ((GET_REG_FUNC reg 1) , value) |
  failwith `SET_REG_FUNC: bad register number`;;

let SET_NEXT_ADDRESS_FUNC (s:state_type) (addr:wordn_type) =
  ( (fst s) , addr , ((fst o snd o snd) s), ((snd o snd o snd) s) );;

let (MAKE_STATE_FUNC:memory_type   ->
                     register_type ->
                     flags_type    ->
                     wordn_type    -> state_type)
     m r f pc = ( m, pc, r, f );;
```

4.6 A Tiny Tamarack–3 Assembler

GSIM can call an assembler program so that small programs can be conveniently entered. GSIM has no built-in assumptions about the assembler function other than that it returns an updated processor state. Therefore, the assembler can be as full-featured or as sparse as desired.

For testing, our assembler for Tamarack–3 simply took an address, opcode, and data value and stored the new n–bit word into memory. This was sufficient for our needs, but for more advanced processors, more complicated assemblers would be created.

For Tamarack–3, the assembler has the following syntax:

```
A <loc> <opcode> <operand>
```

where `loc`, `opcode`, and `operand` are defined as:

- `loc`—location in memory where the assembled instruction will be stored
- `opcode`—Tamarack–3 opcode
- `operand`—decimal address

The code for the Tamarack–3 assembler is quite simple. The `assm_opcode` function takes a list of parameters, and returns an integer corresponding to the opcode parameter (the second parameter). The opcode is made case-insensitive by mapping an `upper_case` function onto the opcode.

```
let assm_opcode plst =
  snd (assoc
         (implode (map upper_case (explode (el 2 plst))))
         [ ('JZR' , 0);
           ('JMP' , 1);
           ('ADD' , 2);
           ('SUB' , 3);
           ('LDA' , 4);
           ('STA' , 5);
           ('RFI' , 6);
           ('NOP' , 7) ] );;
```

The new_val function takes the integer opcode from assm_opcode and the parameter list and returns the new wornd value to be stored in memory. Since Tamarack–3 has only one addressing mode (direct mode), this function is quite concise.

```
let new_val opcode plst =
  let t = (int_of_string (el 3 plst)) in
  wordn ( (opcode * (exp 2 12)) + t );;
```

The assemble function takes the current state and a list of parameters, and returns a new state with the assembled instruction stored in memory. The function operates on states instead of memory so that it follows the same format as other top-level GSIM functions.

```
let assemble state plst =
  let addr = int_of_string (el 1 plst)
  in
  MAKE_STATE_FUNC
     (SET_MEM_FUNC
        (GET_MEM_FROM_STATE state)
        (wordn addr)
        (new_val (assm_opcode plst) plst))
     (GET_REG_FROM_STATE state)
     (GET_FLAGS_FROM_STATE state)
     (GET_NEXT_ADDRESS_FUNC state);;
```

4.7 Using GSIM with Tamarack–3

Loading GSIM is a two-step process:

1. The first part of GSIM is loaded from the ML top-level via load, loadf, or loadt.

2. The second part of GSIM is loaded via the functions gsim, gsimf, or gsimt which are analogous to the ML functions in step one. Their file name parameter is the name of a data file for the processor-specific ML functions and definitions.

When loaded, control returns to the ML top-level prompt. The command loop for GSIM is the function command-loop, and it is called with the initial state for the processor.

The command-loop prompts the user for input, dispatches the appropriate function to service the user's request, and repeats until the quit option is selected. Unknown user requests cause a short menu of available commands to be displayed.

A detailed script showing how GSIM can be used can be found in Appendix A.

5 Conclusion

Simulation is an important tool in hardware design. We have shown how a simple simulator for microprocessor specifications can be written for the HOL theorem proving system. The simulator is important because it provides a means of checking that the specification is written correctly and is a trustworthy method for executing programs before an implementation is available.

The GSIM program was our first attempt to combine the specification and simulation aspects of microprocessor design. Our goal was to explore the feasibility of combining these tasks within the HOL environment. In the future we hope to build a more complete framework for these tasks including a simulation control language.

The program we have presented is fairly simple, but provides a surprisingly complete tool for microprocessor design. The biggest improvement to the simulator will be a more robust translator for creating ML code from HOL specifications. Since execution speed is important, we are not adamant that the translation be done in HOL. We do feel, however, that a completely mechanical process is necessary—not just for convenience, but as a way to control translation errors and increase trustworthiness.

References

[BM79] R. S. Boyer and J S. Moore. *A Computational Logic.* Academic Press, 1979.

[Cam88] Albert J. Camilleri. *Executing Behavioral Definitions in Higher-Order Logic.* PhD thesis, Cambridge University, February 1988.

[Coh88] Avra Cohn. Correctness properties of the VIPER block model: The second level. Technical Report 134, University of Cambridge Computer Laboratory, May 1988.

[Gor88] Michael J.C. Gordon. HOL: A proof generating system for higher-order logic. In G. Birtwhistle and P.A Subrahmanyam, editors, *VLSI Specification, Verification, and Synthesis.* Kluwer Academic Press, 1988.

[GW92] Jody W. Gambles and Phillip J. Windley. Integrating formal verification with CAD tool environments. Technical Report LAL-92-02, University of Idaho, Laboratory for Applied Logic, April 1992.

[Hun87] Warren A. Hunt. The mechanical verification of a microprocessor design. In D. Borrione, editor, *From HDL Descriptions to Guaranteed Correct Circuit Designs.* Elsevier Scientific Publishers, 1987.

[Joy89] Jeffrey J. Joyce. *Multi-Level Verification of Microprocessor-Based Systems.* PhD thesis, Cambridge University, December 1989.

[Win90] Phillip J. Windley. *The Formal Verification of Generic Interpreters.* PhD thesis, University of California, Davis, Division of Computer Science, June 1990.

[Win92a] Phillip J. Windley. Abstract theories in HOL. Technical Report LAL-92-07, University of Idaho, Laboratory for Applied Logic, June 1992.

[Win92b] Phillip J. Windley. A theory of generic interpreters. Technical Report LAL-92-06, University of Idaho, Laboratory for Applied Logic, April 1992.

A GSIM in Action

This appendix provides detailed scripts showing GSIM in use. The first example will add two numbers and store their sum back to memory. The first step is to load GSIM.

```
#loadf ` /hol/GSIM/main`;;
..
setup = - : (string -> void)

File  /hol/GSIM/setup loaded
...() : void
```

The next step loads the definitions for Tamarack–3. Loading another file at this point would cause GSIM to simulate a different microprocessor:

```
#gsimf ` /hol/GSIM/tam-data`;;

[..lots of definitions removed for brevity..]

type `command_loop INITIAL_STATE' to start gsim() : void
```

Next, we start GSIM and enter a simple program:

```
#command_loop INITIAL_STATE;;

gsim> ?
 (D)ump memory       (E)nter memory
 (R)egisters(show)   (S)et registers
 (B)reakpoint(add)   (C)ancel Breakpoint
 (T)race execution   (G)o (start execution)
 (P)C (view PC)      (O) Set PC
 (L)oad file         (W) Write file
 (Q)uit              (?) this menu
 (A)ssemble to memory
 (F)lags (view flags)

gsim> e 0
0 :  ???  == 0
1 :  ???  == 1
2 :  ???  == q

gsim> a 100 lda 200
gsim> a 101 add 201
gsim> a 102 sta 202
```

The constants 0 and 1 were entered in memory locations 0 and 1 to make later examples more concise, since Tamarack–3 does not allow immediate-mode instructions.

Next, three instructions were assembled to memory, starting at location 100. [4] These instructions cause the value at location 200 to be added to the value at location 201 and the result stored to location 202.

Now we put some values at locations 200 and 201 and see if the program (and the simulator) work as expected.

```
gsim> e 200
200 :   ???   == 17
201 :   ???   == 71
202 :   ???   == q

gsim> g 100
gsim> d 200
200 : 17
201 : 71
202 : 88
203 :   ???
204 :   ???
q

gsim>
```

The simulator added 71 to 17 and stored the result in location 202. The ??? represent memory locations that are not defined, since memory is only partially specified in the associative list. GSIM halts execution when the function-hook `HALT-GIM-FUNC` returns true, or when failure occurs (due to trying to fetch undefined memory). Examining memory starting at location 200 shows that the correct sum was stored in location 202, as expected.

The next program sums the integers 1 to n. Memory location 200 will be set to the desired n, and will be decremented to 0 during program execution. The program starts by storing a 0 in location 202, where the sum will be kept. Next, n is loaded into the accumulator. If it is zero, we branch to exit the loop. Now n is decremented by 1, and stored back into address 200. Since the accumulator now contains $n-1$, we add 1 to get n back, and then add the current sum to the accumulator. We store the updated sum back to address 202 and branch back to address 102 to repeat the loop. Address 110 contains a `NOP`, but it could be left undefined since the error caused fetching from an undefined memory location would halt the simulator anyway. [5]

[4] All numbers used with GSIM are decimal.

[5] This happens anyway trying to fetch from address 111, but the `NOP` makes a convenient target for the branch.

```
gsim> a 100 lda 0
gsim> a 101 sta 202
gsim> a 102 lda 200
gsim> a 103 jzr 110
gsim> a 104 sub 1
gsim> a 105 sta 200
gsim> a 106 add 1
gsim> a 107 add 202
gsim> a 108 sta 202
gsim> a 109 jmp 102
gsim> a 110 nop
```

To demonstrate the program, we load address 200 with the number 10, and start execution at address 100. As expected, location 202 holds the value 55 ($= \frac{1}{2} * 10 * 11$), and location 200 has been decremented to 0.

```
gsim> e 200
200 : 17 == 10
201 : 71 == q

gsim> g 100

gsim> d 200
200 : 0
201 : 71
202 : 55
203 :  ???
q

gsim>
```

Our last program multiplies two integers. It uses the last program as its starting point. The instruction at location 106 (`ADD 1`) performs the task of loading the accumulator with the number we add to the ongoing sum each loop iteration. In the last example, we add the loop variable to the sum; instead, we could load a constant from memory. This would change the program to adding a constant n times to the (initially zero) sum. We keep the constant, called k, in location 201; and we replace `ADD 1` with `LDA 201`. Now when the loop terminates, the sum with be $n * k$.

```
gsim> a 106 lda 201
gsim>
```

To multiply 5 and 25, we store the numbers are locations 200 and 201 respectively and start the program running at location 100.

```
gsim> e 200
200 : 0 == 5
201 : 71 == 25
202 : 55 == q

gsim> g 100
gsim>
```

Examining locations 200, 201, and 202 shows that the program works as expected.

```
gsim> d 200
200 : 0
201 : 25
202 : 125
203 :  ???
q
gsim>
```

As we would expect, location 200 holds 0, 201 is unchanged, and 202 holds 125.

Higher Order Logic Theorem Proving and its Applications (A-20)
L.J.M. Claesen and M.J.C. Gordon (Editors)
Elsevier Science Publishers B.V. (North-Holland)

Executing HOL Specifications: Towards an Evaluation Semantics for Classical Higher Order Logic

P. Sreeranga Rajan

Department of Computer Science, University of British Columbia, Vancouver B.C, Canada, sree@cs.ubc.ca OR sree@cs.stanford.edu

Abstract

Formal methods for hardware design and software engineering have become increasingly essential to avoid costly errors. But, this justification for using formal methods needs to be supported by their ease of use. Classical Higher Order Logic (HOL) has provided a foundation for formal verification. However, specification and verification in HOL is, invariably, fairly involved. Often it is desirable for the specifier to increase one's understanding of the specification by executing the specification. Evaluating specifications prior to embarking on tedious proofs may, at the very outset, lead to uncovering discrepancies, with little effort. This work attempts to provide an environment in HOL in which specifications using a subset of HOL can be executed. We accomplish this by translating HOL terms into ML programs: the translation process being totally automatic. Executing specifications as programs in a formal setting provides a way to guarantee certain properties that need to be satisfied by the specification. Furthermore, we also find that it has a potential to provide a practical insight into the evaluation semantics for classical logic.

Keyword Codes: B.7.2; F.3.2; F.4.1
Keywords: Integrated Circuits, Design Aids; Semantics of Programming Languages; Mathematical Logic.

1 Introduction

In general, theorem proving, as its underlying logic becomes more expressive, becomes less automated. HOL is more expressive among theorem provers, while being less automatic. The specifications, on which tedious proofs are carried out, could turn out to have unintended errors not uncovered by the proof process itself. In this context, the specifications would have to be modified and the proof procedure repeated. Hardware designers and software engineers are used to simulating hardware and execute programs to uncover errors. In this work, we attempt to provide an environment in which HOL terms can be transformed into ML (Meta-Language) programs, which then could be evaluated. This environment, in which HOL is equipped with evaluation, serves to detect specification errors before a tedious verification process is undertaken.

Past research in executing HOL specifications [2] has been very similar to our work. This work was also based on transforming HOL specifications into ML programs. However, transformations such as on types were not totally automated. Moreover, it concentrated on translating relational specification to functional specification in the context of hardware. Our work considers general HOL specifications with very restricted quantification. The aim of our work is also more general, and transformations we carry out are all automatic.

The paper is organized as follows: we discuss the issues in evaluating formal logical specifications in section 2, section 3 gives an overview of the translation process from HOL to ML. In section 4, we walk through the translation process with a simple example. We then give a brief discussion on the evaluation semantics for classical logic in section 5, and finally summarize our contributions in section 6. Sample evaluation sessions are included in the appendix.

2 Issues in Evaluating Specifications

There are several issues that need to be considered in providing an environment for executing formal logical specifications.

- Operational (rewrite rule) semantics for either the logic or the target language.
- The relationship between the metalanguage and the object language of the logic: whether the metalanguage can be used to express the operational semantics of the object language.
- Formalization of the object language interpreter in the object language.

The circularity, apparent in the last two issues above, may suggest a choice of a proper subset of the object language for execution.

Boyer-Moore Logic [1], a type-free quantifier-free first-order logic, resembling pure Lisp has an interpreter in Lisp, that operates on the constants of the logic that represent terms. Tableau based classical deduction has had limited success in program synthesis [6]. Theorem provers based on constructive logics have been highly successful in synthesizing programs from proofs, and thus providing a rich environment for execution of specification and verification [4]. LCF, a precursor of HOL has already been considered as a programming language in the context of relating denotational and operational semantics as part of fully abstract models for functional languages [11]. HOL, based on simple theory of types, getting its expressiveness by quantification over objects of arbitrary order leads to difficulties in designing an interpreter for it.

3 Translation Overview

The HOL subset we have targeted are HOL terms that can be safely skolemized. The execution essentially involves transforming HOL terms into ML programs. The ML programs are then interpreted by the ML interpreter. The transformation process is divided into several steps:

- Transforming HOL types into ML types:

  Translation of HOL type constructors to ML recursive type constructors.

  Definitions of destructors corresponding to each argument of the constructors.

- Skolemization:

  Safe removal of outermost quantifiers.

  Removal of redundant quantifiers.

- Transforming recursive definitions:

  Identifying and splitting the recursion arguments.

  Transforming pattern matching by a tree structure.

- Equational definitions:

  Creation of let, letrec and lambda bindings for the arguments on the left-hand side.

- Variables are translated without change.

- Constants:

  Constants such as numbers are treated as primitive ML constants. Constructors and destructors associated with such constants are defined separately in ML.

  The special constant **ARB** is replaced by failure in ML.

- Combinations:

  Translated as function applications in ML: the evaluation order comes into picture here since HOL-ML is call-by-value, many HOL combinations have to be treated by call-by-need depending on the context.

  Conditionals are explicitly translated as if-then-else statements in ML: this is a key example in considering the evaluation order.

- Abstractions are translated as lambda abstractions in ML.

The translation of partial specification in HOL to escape or failure construct in ML leads to a non-functional program. The partial specification, might be explicitly indicated by introducing the special constant **ARB** in the definitional specifications. Also, partial specifications might come about, in part, due to the axiom of choice and the excluded middle in classical logic. Failure or catch-throw in programming languages is considered[1] a non-functional feature. This hints at classical logic as a foundational basis for the semantics of non-functional programming languages [9].

4 Example Translation: Step by Step

4.1 Translating Type Definitions

Consider the recursive type definition in HOL:

$$
\begin{aligned}
\textit{let}\;\; \textit{stack} \;\; = &\\
& \textit{define\_type}\ \text{`}\textit{stack\_Axiom}\text{`}\\
& \text{`}\textit{stack}\ =\\
& \textit{STACK\_EMPTY}\ |\\
& \textit{STACK\_NODE}\ *\ \textit{stack}\text{`};;
\end{aligned}
$$

It is a polymorphic stack recursive type introducing two constructors: ***STACK_EMPTY*** and ***STACK_NODE***. We first translate it into an ML recursive type definition:

$$
\begin{aligned}
\textit{rectype}\;\; *\,\textit{stack} \;\; = &\\
& \textit{STACK\_EMPTY}\ |\\
& \textit{STACK\_NODE}\ \textit{of}\ \ *\#*\ \ \textit{stack};;
\end{aligned}
$$

We note that the ML type constructors are uncurried unlike their HOL counterparts. Then, we introduce the destructors corresponding to the two arguments to the second (recursive) type constructor STACK_NODE:

$$
\begin{aligned}
&\textit{let}\;\; \textit{dest\_STACK\_NODE}_1\,(\textit{STACK\_NODE}(v,s))\\
&\qquad\qquad\qquad\qquad = v;;\\
&\textit{let}\;\; \textit{dest\_STACK\_NODE}_2\,(\textit{STACK\_NODE}(v,s))\\
&\qquad\qquad\qquad\qquad = s;;
\end{aligned}
$$

[1] by purists

4.2 Translating Equational Definitions

We will consider a recursive definition in HOL using the stack recursive type defined above. It defines the ***POP*** operation on the stack - popping an empty stack is specified by ***ARB***, and popping any other stack gives a stack with the top element removed. The definition in HOL appears as follows:

$$
\begin{aligned}
\textbf{\textit{let POP}} \; = \; & \\
& \textbf{\textit{new\_recursive\_definition}} \\
& \textbf{\textit{false}} \\
& \textbf{\textit{stack}} \\
& \text{`}\textbf{\textit{POP}}\text{`} \\
& \text{``}(\textbf{\textit{POP}}(\textbf{\textit{STACK\_EMPTY}}) = \textbf{\textit{ARB}}) \wedge \\
& \forall \boldsymbol{v}\ \boldsymbol{s}. \\
& (\textbf{\textit{POP}}(\textbf{\textit{STACK\_NODE}}\ \boldsymbol{v}\ (\boldsymbol{s} : (*)\textbf{\textit{stack}})) = \boldsymbol{s})'';;
\end{aligned}
$$

We first skolemize the above term by removing the universal quantifications on the variables $\boldsymbol{v}$ and $\boldsymbol{s}$. We then transform the pattern-matching style definition on the RHS into a tree-style definition in HOL by

- Substituting one constructor variable for all the constructor combinations:

  Cons_Var substituted for ***STACK_EMPTY*** and (***STACK_NODE v s***)

- Substituting destructor combinations for each of the arguments of a constructor - the substitution is actually delayed till execution by creating ***let*** declarations for the arguments:

  let v = dest_STACK_NODE1 (***Cons_Var***)

  let s = dest_STACK_NODE2 (***Cons_Var***)

We arrive at this HOL term by using the above transformations:

$$
\begin{aligned}
\boldsymbol{POP}\ \ \boldsymbol{Cons\_Var}\ \ =\ \ & \\
& ((\boldsymbol{Cons\_Var} = \boldsymbol{STACK\_EMPTY}) \Rightarrow \\
& \boldsymbol{ARB} \mid \\
& \boldsymbol{let\ s}\ =\ \boldsymbol{dest\_STACK\_NODE2}\ (\boldsymbol{Cons\_Var})\ \boldsymbol{in} \\
& \boldsymbol{let\ v}\ =\ \boldsymbol{dest\_STACK\_NODE1}\ (\boldsymbol{Cons\_Var})\ \boldsymbol{in} \\
& \boldsymbol{s};;
\end{aligned}
$$

We then carry out the final transformations into an ML program:

- Transform the conditional HOL combination into an *if-then-else* statement and replace the constant ***ARB*** by failure:

$$
\begin{array}{c}
\boldsymbol{if}\ (\boldsymbol{Cons\_Var}\ =\ \boldsymbol{STACK\_EMPTY})\ \boldsymbol{then} \\
\boldsymbol{failwith}\ \text{`}\boldsymbol{partial\ specification}\text{`}\ \boldsymbol{else} \\
\boldsymbol{let\ s}\ =\ \boldsymbol{dest\_STACK\_NODE2}\ (\boldsymbol{Cons\_Var})\ \ \boldsymbol{in} \\
\boldsymbol{let\ v}\ =\ \boldsymbol{dest\_STACK\_NODE1}\ (\boldsymbol{Cons\_Var})\ \ \boldsymbol{in} \\
\boldsymbol{s}
\end{array}
$$

- Introduce the lambda binding for the variable on the RHS:

$$
\begin{aligned}
& \boldsymbol{\lambda Cons\_Var.} \\
& \qquad \begin{array}{c}
\boldsymbol{if}\ (\boldsymbol{Cons\_Var} = \boldsymbol{STACK\_EMPTY})\ \boldsymbol{then} \\
\boldsymbol{failwith}\ \text{`}\boldsymbol{partial\ specification}\text{`}\ \boldsymbol{else} \\
\boldsymbol{let\ s} =\ \boldsymbol{dest\_STACK\_NODE2}\ (\boldsymbol{Cons\_Var})\ \ \boldsymbol{in} \\
\boldsymbol{let\ v} =\ \boldsymbol{dest\_STACK\_NODE1}\ (\boldsymbol{Cons\_Var})\ \ \boldsymbol{in} \\
\boldsymbol{s}
\end{array}
\end{aligned}
$$

- Introduce ***letrec*** binding for the function name on the LHS:

$$
\begin{aligned}
\boldsymbol{letrec}\ \ \boldsymbol{POP}\ \ =\ \ & \\
& \boldsymbol{\lambda Cons\_Var.} \\
& \boldsymbol{if}\ (\boldsymbol{Cons\_Var} = \boldsymbol{STACK\_EMPTY})\ \boldsymbol{then} \\
& \boldsymbol{failwith}\ \text{`}\boldsymbol{partial\ specification}\text{`}\ \boldsymbol{else} \\
& \boldsymbol{let\ s} =\ \boldsymbol{dest\_STACK\_NODE2}\ (\boldsymbol{Cons\_Var})\ \ \boldsymbol{in} \\
& \boldsymbol{let\ v} =\ \boldsymbol{dest\_STACK\_NODE1}\ (\boldsymbol{Cons\_Var})\ \ \boldsymbol{in} \\
& \boldsymbol{s}
\end{aligned}
$$

Several examples involving compiler specifications, microprocessor specifications, boolean arithmetic functions have been executed successfully.

5 Limitations

The limitations of the translation come about due to the expressiveness of the HOL logic itself and the evaluation order of the ML language associated with the HOL environment:

- Type definitions: currently there is no logical way to identify the type constructors apart from other constants in a theory:
  We adopt a convention that the ML key for a type definition to be the <Type_name>_Axiom.
- Quantification:

  Hilbert choice operator ϵ cannot be translated.

  User introduced quantifiers are not handled.

  Arbitrary quantification using $\forall$ and $\exists$, may not be safely skolemized.
- "Lazy" combinations: several combinations in HOL whose rewrite rules correspond a call-by-need evaluation order, though are translated as applications, do not evaluate in ML to the expected result. This is due to HOL ML being call-by-value. One such special case is the translation of the special constant **ARB** by failure.

6 Towards Evaluation Semantics

The target language of our translation being ML and that the interpreter being written in ML itself has several advantages:

- ML being the metalanguage for HOL, the specifications can be executed in the HOL environment itself.
- ML has a well-defined formal dynamic and static semantics: the interpreter can be formalized in HOL.

A set-theoretic denotational semantics has been defined for HOL [5]. However, its operational semantics has not yet been fully investigated.

It is well known that functional programs can be extracted from proofs in constructive logics. One way to understand this is that the Curry-Howard isomorphism underlying constructive type theories identifies propositions as types with an associated semantics of evidence [3]. Thus, informally, proofs can be viewed as programs and evidence viewed as values. Furthermore, constructive logics provide foundations for functional programming languages.

On the other hand, classical logic cannot be directly interpreted to have existence properties. However recent work [7, 8, 10, 12] has shown that classical proofs by appropriate translations to constructive logic can, indeed, be associated with computations. This work was done in the context of semantics for imperative programming languages, using appropriate translations to functional languages and constructive logics.

7 Summary and Future Work

We believe that our environment would be useful to HOL researchers and practitioners as an aid to quicker development of defect-free specifications, leading to a better understanding of HOL specifications. Furthermore, evaluating specifications as programs in a formal setting serves to guarantee certain properties intended to be satisfied by the specification. Besides providing a useful tool for practitioners, this work could provide further insight into formulating an evaluation semantics for HOL, leading to HOL proofs as programs. On the other hand, it also provides an insight into classical logic forming the foundations of non-functional programming languages.

Acknowledgements

The author thanks Jeff Joyce and Carl Seger for their constant encouragement and reviews. Many of the ideas in this paper could not have come about without a long association with, and constant mentoring by Carolyn Talcott. The author also thanks Chet Murthy for providing a better understanding of his work on classical proofs as programs.

References

[1] Boyer, Robert S. and Moore J Strother , 'A Computational Logic Handbook', Academic Press, 1988.

[2] Camilleri, Albert J., 'Executing Behavioural Definitions in Higher Order Logic', PhD Thesis, University of Cambridge 1988.

[3] Constable, Robert, 'Semantics of Evidence', *Department of Computer Science, Cornell University Technical Report, 1986*

[4] Constable, Robert, et. al. , 'Implementing Mathematics with the Nuprl Proof Development System' Prentice Hall, 1986.

[5] Gordon, M. J. C. , 'The HOL Description', Dec. 1990.

[6] Manna, Zohar and Waldinger, Richard, 'A Deductive Approach to Program Synthesis', *Manuscript: Stanford University CS257 Course*

[7] Murthy, Chetan R., 'An Evaluation Semantics for Classical Proofs', in: *Proceedings of the Fifth Annual Symposium on Logic in Computer Science*, IEEE, 1991.

[8] Murthy, Chetan R., 'Classical Proofs as Programs: How, What and Why' in: *Technical Report TR 91-1215*, Department of Computer Science, Cornell University, July 1991.

[9] Murthy, Chetan R., 'Extracting Constructive Content from Classical Proofs', PhD Thesis, Department of Computer Science, Cornell University, 1990

[10] Murthy, Chetan R., *Personal Communication*

[11] Plotkin, G. D. , 'LCF Considered as a Programming Language', in: *Theoretical Computer Science, 5:223-255, 1977* .

[12] Talcott, Carolyn L., *Personal Communication*

Appendix: Example Evaluation Sessions

```
#eval "3+2";;
` FIX_PLUS  ( 3 ) ( 2 );;
` : string

#new_theory `foo`;;
() : void

#let fact = new_prim_rec_definition (
           `fact`,
           "(fact 0 = 1) /\
            (fact (SUC n) = (SUC n) * (fact n))");;

###fact = |- (fact 0 = 1) /\ (!n. fact(SUC n) = (SUC n) * (fact n))

##exec `foo` [] [];;
Translating HOL definitions in foo and its parents ...

Loading derived ML definitions in foo_hol.ml ...
.................................
Type a function name to evaluate or
Type "eval <hol_term>;;" to evaluate a new <hol_term> or
Type "exit();;" to exit
`#` : string

hol-exec> fact 10;;
3628800 : int

hol-exec> exit();;
`hol-exec> ` : string

#quit();;
Bye.
```

Chapter 9

Extended uses of Higher Order Logic

Higher Order Logic Theorem Proving and its Applications (A-20)
L.J.M. Claesen and M.J.C. Gordon (Editors)
Elsevier Science Publishers B.V. (North-Holland)

Linking Other Theorem Provers to HOL Using PM: Proof Manager[1]

Myla Archer, George Fink, and Lie Yang

(archer@cs.ucdavis.edu, gfink@cs.ucdavis.edu, yangl@cs.ucdavis.edu)

Department of Computer Science, University of California, Davis, California, USA

Abstract.

The tool PM (for Proof Manager), initially developed as an improved interface for HOL, has been previously described in [FAY91]. At that time, there were plans to improve PM as a HOL interface, and to extend the PM interface to permit additional theorem provers to be used to supplement HOL. A preliminary design for a translation algorithm was described. In this paper, we describe how these plans have been realized, and our continuing work to improve PM as a multiple-prover interface. We illustrate the use of PM with multiple provers with an example.

Keyword Codes: I.2.3, F.4.1, D.2.2
Keywords: Deduction and Theorem Proving; Mathematical Logic; Tools and Techniques

1. Introduction.

PM (Proof Manager) is a theorem-prover interface that was originally designed to be an improved interface to the HOL system [Gordon87], providing a facility for recording and manipulating the structure of a proof, easy travelling to arbitrary unproved subgoals in an ongoing proof, and convenient collecting of the parts of a completed proof into a single proof.

Many of the proof-management features of PM are useful for nearly any theorem prover: the maintenance of a proof tree, the marking of nodes as validated or not, the invariant that the assertion at a validated node follows from the assertions at its children, the effect of tree-editing operations on node validation information, and so on. The ideas for these features owe much to an earlier proof manager called TED, developed at the University of Illinois [Ham85]. Other features of PM, such as the facility for ''squashing'' contiguous segments of a proof tree into a single node with a single tactic as validation, are aimed at HOL. However, the general design of PM gives it the potential to provide an interface to many different theorem provers.

We reported in [FAY91] on our intention to develop PM into a multiple-prover interface, and described the methods we expected to use in doing so. We also presented there our rationale for doing so, which we briefly recapitulate here. The strength of a higher-order-logic theorem prover such as HOL is its the ability to express high-level results in its logic and to make high-level inferences. However, low level proofs that could be done automatically by many first-order provers typically are very tedious to do in a higher-order system such as HOL. Rather than creating new supplementary decision procedures and so on for, say, HOL, we believe advantage should be taken of the years of polishing that have gone into other systems such as the Boyer-Moore prover [BM88], various resolution provers, and other systems. Beyond supplementing a higher-order system such as HOL, it

[1]This research is supported in part by the Department of Defense under contract grant DOD-MDA 904-91-C-7052.

is also worth noting that different theorem provers have often been polished to work especially well on certain problem formulations: hence, a multiple-prover system could be useful for combining the strengths of different provers.

This paper reports on our progress towards turning PM into a multiple-prover interface, and other improvements to PM.

We have continued to develop the general interface features of PM; some of our new features, such as support for forward proofs, are more aimed at provers other than HOL, although they can be used with HOL as well. We give more details in section 2 below.

In addition, the translation algorithm outlined in [FAY91] has been implemented for the Boyer-Moore prover and two resolution provers. We are currently working on extensions to the translation capability and a better user interface to the translator that could, among other things, allow the user to choose among possible translations. The current and future translation capabilities of PM are discussed in section 3.

Permitting PM to be used with multiple provers within a single proof requires implementing not only translation algorithms, but a design for maintaining the proof tree structure and translation information during a proof. Our initial implementation of such a design and our intended improvements to it are described in section 4. A small example of multiple-prover use of PM is given in section 5. We expect to present a more extended example in the full paper.

Although informal justifications can be given, we do not yet have formal proofs of soundness for PM tree and translation operations. This is an important direction for future work.

2. Improvements to PM as a proof manager.

In this section, we briefly review the nature and origins of PM, and then describe the new features that we either have added or are in the process of adding to PM.

We began our development of PM by concentrating on creating a HOL interface and those features of proof tree maintenance that are most useful for users of HOL. In transforming it to a multiple-prover system, we have continued using HOL's as the principal interface language, because of its expressive power.

PM is written in E-lisp (Emacs lisp) for GNU emacs [Sta87], and is built on top of a tree editor Tree-mode [Ham88] [Kamin90], also designed for GNU emacs. Tree-mode was developed at the University of Illinois as a means of interfacing theorem provers—most notably NUPRL [Con86] [Swarup88]—and other software systems.

PM maps proof trees into a Tree-mode tree. The tree-mode tree keeps track of the state of a theory at the root of the tree in a special ''theory'' node. Other nodes in the tree are called ''proof'' nodes, and store the assertion, tactic, and type information for their associated goal. The mapping between HOL proof and PM tree is maintained in an automatic fashion by PM. As the HOL proof grows, so does the PM tree. Given a valid HOL tactic at a childless node, PM will validate the node, attaching the appropriate subgoals as children, and record the tactic as the ''method'' of proof that the node assertion follows from its child assertions.

Various tree operations are provided that are intended to permit convenient manipulation of the proof structure without causing the validation status of proof nodes to become incorrect. One simple such operation is the copying of a subtree to a point elsewhere in the proof where the assertion at its root node might be useful as a lemma. Probably the most useful tree operations for users of HOL are the various ''squashing'' operations that combine contiguous parts of the proof tree into a single node and simultaneously combine the tactics that validate the squashed nodes into an appropriate tactic to validate the single node in terms of its children.

As we indicated in [FAY91], we intended to add support for forward proof to PM. A simple version of such support using SUBGOAL_THEN, ASSUME_TAC, and so on, has been implemented that validates a node with children by adding a new child obtained from the parent node by including the children appropriately on the assumption list. This is accomplished with a simple one-keystroke ''lemma'' command.

For a prover that is not tactic-and-subgoal-driven, but is, say, automatic, forward proof has a different style: the assertion at a node is packaged up with the assertions at its children as assumptions, and given as a complete problem to the prover; if the prover succeeds, the node is validated. We are in the process of implementing such forward proof support for the Boyer-Moore prover and the resolution provers. We also intend to implement an alternative, analogous forward proof mechanism to be used with HOL that would take an unvalidated node with children, accept a HOL tactic as a proposed validation, and, if the node can be proved using the tactic with the children added to the assumptions, adds the new subgoals as additional children, while modifying the proposed tactic to reflect the use of additional assumptions. We expect that all of these forward proof steps will be invoked by simply including arguments with the lemma command.

A method of using a proof tree as a proof script that can be rerun has been implemented for the case that HOL is the sole prover. Prior to the formal proof of soundness of tactic squashing operations, this provides a method for checking their validity. It also provides a method for re-checking a stored proof. We intend to extend this rerun feature to include multiple-prover trees. Until formal proof of soundness of our translation schemes exists, rerunning multiple-prover proofs could be used, say, to raise one's confidence level in a stored proof, but cannot formally guarantee proof validity.

A few new user conveniences have been added to PM, such as a command to list the remaining unvalidated nodes, and a limited ''undo'' feature for operations such as squashing.

We have a number of new features planned for the future, one of which is the ability to make different prover calls simultaneously on a node, providing one possible way to speed proofs when multiple processors are available.

3. Translation from HOL to other theorem provers.

Our initial goal has been to supplement HOL with existing provers that can automatically handle simple first-order formulae, assertions from linear arithmetic and other theories for which there exist decision procedures, and many cases of induction. With this in mind, we have not attempted to translate all HOL assertions into first order form, but only those of a certain simplicity (although—provided a weakened notion of validity is sufficient for the intended application—it is, in principle, possible to formulate any HOL assertion as a first order formula [Monk76]).

It should be noted that even for first-order HOL formulae, translation for other provers is not always straightforward. One clear reason for this is the matter of types, which are largely implicit in HOL, and are handled in various ways in other systems. However, as indicated below, there are additional complications with respect to choosing best translations.

In [FAY91], we described a translation algorithm by which we proposed to translate a subset of HOL formulae to first-order form. This algorithm has now been implemented, and extended to sequents by the expedient of, in effect, discharging the assumptions. The algorithm decides whether it can translate a formula, and if it can, produces an intermediate-form translation in the form of an expression tree in which type information is preserved, and in which a decision for all boolean nodes has been made about whether to consider them to be of meta-boolean or low-boolean type. (For example, a boolean-valued function in a HOL formula will either remain such—the ''low-boolean'' case—or be translated as a predicate—the "meta-boolean" case; correspondingly, an "=" node with boolean children will either remain such, equating two low-booleans, or become an "iff" node, connecting two meta-booleans.) The low-boolean meta-boolean distinction in the expression tree is used to decide whether a simple translation is feasible: for example, an application of a low-boolean operator to a term of the form $\forall x. P(x)$ would not be allowed.

For particular first-order provers, the expression tree can then be further processed into the appropriate form. The low-boolean meta-boolean distinction turns out to be a useful intermediate step in producing a "most efficient" form of input for certain provers; in particular, it is very useful with resolution provers, since the meta-booleans affect the breakdown into clauses.

Currently, we are focusing our attention on supplementing HOL with interfaces to the Greenbaum resolution prover [Gre86] and the Boyer-Moore prover with Kaufmann's skolemizer [BM88, Kauf88] as our target first-order provers, but are considering including an interface to FAUST [SKK91] that would use the HOL-FAUST interface described in [SKK91], and take advantage of the PM proof tree structure when appropriate. The resolution interface is now implemented for two versions of the Greenbaum prover: the latest version currently under development [Gre89], which we will continue to refer to as the Greenbaum prover, and an earlier version, which we will refer to as the TED prover [Ham85] [HKC85]. We are in the process of implementing the Boyer-Moore interface. The translation aspect of the latter is essentially done; we are attempting to develop an appropriate format for calls to the prover that could make it as automatic as possible—for example, by encoding hints in the call. Such a hint would be needed for a one-step Boyer-Moore proof in the example shown below in section 5.

There are many possible directions for improving the translation facility of PM. For example, it should be noted that there are often several candidate translations of a HOL formula to first-order form. For example, the formula:

$$(\exists\, p\, q.\ (p = a \wedge b) \wedge (q = c \wedge d) \wedge (out = p \vee q)) = (out = a \wedge b \vee c \wedge d)$$

if translated according to the principle that as many nodes as possible should be meta-boolean, becomes:

```
(IFF (E (p q) (AND (= p (and a b)) (= q (and c d)) (IFF (out) (Or p q))))
     (IFF (out) (OR (And a b) (And c d))))
```

Here, "IFF", "AND", and "OR" are the boolean connectives, "out" is treated as a 0-ary predicate, and "Or" and "And" are treated as binary predicates; "and" is simply a function. However, this formulation requires including axioms for Or and And, at least, to obtain a proof. A more natural translation would minimize the number of distinct boolean operators and attempt to maximize common sub-expressions:

```
(IFF (E (p q) (AND (= p (and a b)) (= q (and c d)) (IFF (out) (Or p q))))
     (IFF (out) (Or (and a b) (and c d))))
```

which proves in a fraction of a second with the TED resolution prover, and requires no additional axioms. Our translation algorithm uses the simplicity-plus-consistency criterion that we have indicated to produce the latter translation. However, what will be the best translation in the general case is not yet clear. We plan eventually to provide the user the opportunity to examine the possibilities and make a choice.

In the example given above, we were able to ignore the problem of types, since the only type present is boolean. In more complex examples, the type information in the intermediate-form tree must be used. Translating such type information using predicates is a relatively simple task. With resolution proving, it is possible to use a type-tag scheme (in which each expression is imbedded in a declaration of its type) instead of predicates; such a scheme prevents terms of different types from unifying. Experience has shown that this increases proving speed. We are implementing such a scheme with our resolution prover interfaces. The next example translation below illustrates how type tags can be used in conjunction with type variables and polymorphic types.

We have done some experimentation that has shown that it may well be worthwhile extending the subset of HOL formulae that we translate. The following HOL assertion arose in connection with the proof of a delay-line:

```
∀ (i: num -> bool) (out: num -> bool).
    ∃ p. (∀ time: num. p (time + 1) = i time) ∧ (∀ time: num. out (time + 1) = p time)
         =
    ∀ time: num. out (time + 2) = i time
```

Proving this in HOL required the time needed to develop the appropriate tactic, plus time to execute the tactic—on the order of forty seconds.

We have performed a hand translation of this assertion aimed at the TED resolution prover, in which the type information is encoded using type tags, and an APPLY function is used to express function application:

```
(A (i out)
   (IFF (E (p)
            (AND (A (time)
                      (= (TYPE bool (APPLY p (TYPE num (SUC (TYPE num time)))))
                         (TYPE bool (APPLY i (TYPE num time)))))
                  (A (time)
                      (= (TYPE bool (APPLY out (TYPE num (SUC (TYPE num time)))))
                         (TYPE bool (APPLY p (TYPE num time)))))))
        (A (time)
           (= (TYPE bool (APPLY out (TYPE num (SUC (TYPE num
                                                    (SUC (TYPE num time)))))))
              (TYPE bool (APPLY i (TYPE num time)))))))
```

The forward implication proves with TED automatically in two or three seconds; the reverse implication proves in roughly thirty seconds when supplemented by the following axiom about APPLY:

```
(A (f g type1 type2 type3)
   (E (h) (A (x) (= (TYPE type3 (APPLY h (TYPE type1 x)))
                    (TYPE type3 (APPLY f (TYPE type2 (APPLY g (TYPE type1 x)))))))))
```

which essentially expresses the fact that the composition of two functions is a function, provided that domain and range types match correctly. (This reverse implication proves in two seconds or less with the new Greenbaum prover, but requires some interactive choices concerning rewrites by the user.)

In order to more fully take advantage of the induction capabilities Boyer-Moore prover, we wish to handle the problem of translating data type and recursive function definitions from HOL form to Boyer-Moore form. Doing so in an automatic fashion is not an easy problem for recursive function definitions, since HOL-form definitions tend to be given in ''constructor form'', while Boyer-Moore definitions are in ''destructor form'': for example, a HOL definition of a function on lists would be given in terms of nil and append, while a Boyer-Moore definition would use head and tail. One solution is to require the user to prove an appropriate theorem about the HOL definition. Automatic construction of such theorems in some cases—perhaps using additional information from a theory—is not inconceivable. The problem of translating data type definitions is also of interest to us, but is work for the future.

As indicated in the introduction, formal proofs of soundness of our translations have yet to be done. Prior to that, however, we believe our multiple-prover tool still has utility in raising confidence in the validity (or lack of it) of proposed assertions—in particular, it could speed the discovery of errors.

4. Using PM with multiple provers.

Our current implementation of PM uses HOL as the assertion language. Calls to other provers are done by first calling the appropriate version of the translation algorithm for that prover, which validates the translated node, giving it the translation as its single child. Currently, the assertion at the child node is in the language of the target prover; once this child node is created, the PM user can then attempt to validate its assertion using the target prover. Developing a single common interface language is a problem we may address in the future.

Our current goal has been to supplement HOL with (essentially) automatic proof support from the additional provers. Developing an appropriate suite of prover calls and interpreting the calls for individual provers has turned out to be important. For example, if a call to the Boyer-Moore prover were to incorporate hints, it would be desirable for a PM user to not be asked to provide the names of events that would be required by the standard Boyer-Moore interface. For simple hints at least, event names can be generated invisibly to the PM user.

We do not yet have support for changing into the environment of a target prover, except as that change is accomplished during the invocation of that prover on a particular problem. Thus, PM does not yet support switching control to the Boyer-Moore system, although it does allow the interactive queries of the Greenbaum prover to be answered by the user. If sufficient reasons appear for true prover-context-switching, we may eventually provide the support for doing so. Short of doing this, we hope to develop convenient methods for communicating supplementary information from the HOL theory to a different target prover.

5. An example of using multiple provers with PM.

The following simple example, shown in Figures 1-3, is an illustration of how our implementation of the design for multiple-prover use of PM described in section 4 actually looks in practice.

It should be noted that the ''Prover :'' fields in the nodes refer to the language in which the assertion is expressed. The ''Method'' at each node gives the justification for the assertion at a node following from those at its children; the Method may be a HOL tactic, a call to an automatic prover, or a call to a translation scheme.

```
emacs@ent
/*  Theory trial
  Definitions : inverter,even,ninv,invspec
  /1*  Proved Wed Sep  9 16:33:27 1992
    Method : INDUCT_TAC THEN REPEAT STRIP_TAC
    Assertion : ! n in out. ninv n in out = invspec n in out
    /1,1  unproven
      Assertion : ninv 0 in out = invspec 0 in out
    /1,2  unproven
      Assertion : ninv(SUC n)in out = invspec(SUC n)in out
      Assumptions : "!in out. ninv n in out = invspec n in out"

-----tree-mode: fig1.pm        (at root)           (Tree Basic)-----
let inverter = new_definition(`inverter`,
        "! in out. inverter in out = (out = (~in))");;

let even = new_definition(`even`,
        "! n. even n = (? m. n = 2 * m)");;

let ninv = new_prim_rec_definition(`ninv`,
        "(ninv 0 in out = (out = in)) /\
         (ninv (SUC n) in out = (? a. (inverter in a) /\
                                      (ninv n a out)))");;

let invspec = new_definition(`invspec`,
        "! n in out. invspec n in out =
                     (even n) => (out = in) | (out = ~in)");;

--%%-Emacs: Recedit: definitions    12:48am     (Hol)----All-------
Type C-c C-c to enter this text into the tree
```

Figure 1: Definitions and first proof step in the proof of correctness of an implementation of an inverter.

```
emacs@ent
/* Theory trial
 Definitions : inverter,even,ninv,invspec
 /1* Proved Wed Sep  9 16:33:27 1992
   Method : INDUCT_TAC THEN REPEAT STRIP_TAC
   Assertion : ! n in out. ninv n in out = invspec n in out
   /1,1* Proved Wed Sep  9 23:03:21 1992
     Method : REWRITE_TAC[ninv;invspec;even;inverter]
     Assertion : ninv 0 in out = invspec 0 in out
     /1,1,1* Proved Wed Sep  9 23:04:16 1992
       Method : SUBGOAL_THEN "(? m. 0 = 2 * m):bool" ASSUME_TAC
       Assertion : (out = in) = ((?m. 0 = 2 * m) =>
                                  (out = in) | (out = ~in))
       /1,1,1,1 unproven
         Assertion : ?m. 0 = 2 * m
       /1,1,1,2 unproven
         Assertion : (out = in) = ((?m. 0 = 2 * m) =>
                                    (out = in) | (out = ~in))
         Assumptions : "?m. 0 = 2 * m"
   /1,2 unproven
     Assertion : ninv(SUC n)in out = invspec(SUC n)in out
     Assumptions : "!in out. ninv n in out = invspec n in out"

-----tree-mode: fig2.pm       (at root)          (Tree Basic)-----
```

Figure 2: The proof tree after the definitions have been expanded and a lemma suitable for the Boyer-Moore prover has been isolated. The sibling of the lemma is suitable for translation for a resolution prover.

```
emacs @ toadflax
/1*  Proved Wed Sep  9 16:33:27 1992
  Method : INDUCT_TAC THEN REPEAT STRIP_TAC
  Assertion : ! n in out. ninv n in out = invspec n in out
  /1,1*  Proved Wed Sep  9 23:37:38 1992
    Method : REWRITE_TAC[ninv;invspec;even;inverter]
    Assertion : ninv 0 in out = invspec 0 in out
    /1,1,1*  Proved Wed Sep  9 23:38:17 1992
      Method : SUBGOAL_THEN "(? m. 0 = 2 * m):bool" ASSUME_TAC
      Assertion : (out = in) = ((?m. 0 = 2 * m) =>
                         (out = in) | (out = ~in))
      /1,1,1,1*  Proved Wed Sep  9 23:38:22 1992
        Prover : HOL
        Method : Translation to bm
        Assertion : ?m. 0 = 2 * m
        /1,1,1,1,1  unproven
          Prover : bm
          Assertion : (exists m (and (numberp m)
                                (equal 0 (times 2 m))))
      /1,1,1,2*  Proved Wed Sep  9 23:38:28 1992
        Prover : HOL
        Method : Translation to gpr
        Assertion : (out = in) = ((?m. 0 = 2 * m) =>
                                  (out = in) | (out = ~in))
        Assumptions : "?m. 0 = 2 * m"
        /1,1,1,2,1  Proved Thu Sep 10 00:20:19 1992
          Prover : gpr
          Assertion :
          (IMPLY [E (M) (AND (NUMBERP M)
                 (= ZERO (TIMES (SUC(SUC ZERO)) M]
                 (IFF (IFF (OUT) (IN))
                    (OR (AND [E (M) (AND (NUMBERP M)
                             (= ZERO (TIMES (SUC(SUC ZERO)) M]
                             (IFF (OUT) (IN)))
                        (AND (NOT [E (M) (AND (NUMBERP M)
                                  (= ZERO
                                    (TIMES (SUC(SUC ZERO)) M])
                             (IFF (OUT) (NOT (IN))))))))
  /1,2 unproven
      Assertion: ninv(SUC n)in out = invspec(SUC n)in out

-**--tree-mode: fig3.pm        (child 1 of 1)       (Tree Basic)-----
```

Figure 3: The proof tree after the translations have been performed and the resolution proof completed.

6. References.

[BM88] Robert S. Boyer and J Strother Moore, A Computational Logic Handbook, Academic Press (1988).

[Con86] R. L. Constable, et al., Implementing Mathematics with the Nuprl Proof Development System, Prentice-Hall (1986).

[FAY91] G. Fink, M. Archer, and L. Yang, *PM: A proof manager for HOL and other provers*, Proceedings of the 1991 International Tutorial and Workshop on the HOL Theorem Proving System and Its Applications, IEEE Computer Society Press, 1992, to appear.

[Gordon87] Mike Gordon, *HOL: A proof generating system for higher-order logic*, in: VLSI Specification, Verification and Synthesis (G. Birtwistle and P. A. Subrahmanyam, eds.), Kluwer (1987).

[Gre86] Steven Greenbaum, Input Transformations and Resolution Implementation Techniques for Theorem Proving in First-Order Logic, Doctoral Dissertation (Technical Report No. UIUCDCS-R-86-1298, Department of Computer Science), University of Illinois at Urbana (1986).

[Gre89] Steven Greenbaum, Private communication.

[HKC85] David Hammerslag, Samuel N. Kamin, and Roy H. Campbell, *Tree-oriented interactive processing with an application to theorem-proving*, Proc. of the ACM/IEEE Conf. on Software Development, Tools, Techniques, and Alternatives (SoftFair II) (December, 1985).

[Ham85] David Hammerslag, TED: A Tree Editor with Applications for Theorem Proving, Master's Thesis, University of Illinois at Urbana (1985).

[Ham88] David Hammerslag, Treemacs Manual, Technical Report No. UIUCDCS-R-88-1427, Department of Computer Science, University of Illinois at Urbana (1988).

[Kamin90] S. Kamin, *Differences between Treemacs and Tree-mode*, University of Illinois at Urbana-Champaign, 1990.

[Kau90] Matt Kaufmann, *An extension of the Boyer-Moore theorem prover to support first-order quantification*, Computational Logic, Inc., 1990.

[Monk76] J. D. Monk, Mathematical Logic, Springer-Verlag (1976).

[SKK91] Klaus Schneider, Ramayya Kumar, and Thomas Kropf, *Integrating a first-order automatic prover in the HOL environment*, Proceedings of the 1991 International Tutorial and Workshop on the HOL Theorem Proving System and Its Applications, IEEE Computer Society Press, 1992, to appear.

[Stall87] Richard Stallman, GNU Emacs Manual, Free Software Foundation (1987).

[Swarup88] Vipin Swarup, The UIPRL Proof Development System Master's Thesis, University of Illinois at Urbana (1988).

Higher Order Logic Theorem Proving and its Applications (A-20)
L.J.M. Claesen and M.J.C. Gordon (Editors)
Elsevier Science Publishers B.V. (North-Holland)

Adding New Rules to an LCF-style Logic Implementation

Konrad Slind

slind@cpsc.ucalgary.ca

Abstract

By encapsulating the primitive rules of inference as the constructors of the abstract type of theorems, LCF-style systems provide a successful answer to the problem of how to soundly add derived inference rules to a theorem prover. We suggest a supplementary method which uses formal proof of program equivalence to justify the dynamic addition of new primitive rules. The method does not increase the strength of the logic; rather, it is a means of increasing the computational strength of the implementation. The method can be used to enforce rigour in the process of iteratively improving the performance of a system. This work uses first class environments, a recent addition to the SML/NJ compiler.

Keyword Codes: F.4.1;I.2.3

Keywords Mathematical Logic; Deduction and Theorem Proving.

1 Introduction

The problem of how to add new inference rules to a theorem prover is important, for if one can not add new rules, all proofs must be done by using the fixed set of primitive rules of the logic. The ability to define derived rules provides an abstraction facility, allowing proofs comprising many small steps to be regarded as single leaps. This gives the user a means of raising the level of discourse with the theorem prover. Of course, there is a problem: how to ensure that a derived rule, as implemented in the theorem prover, is sound. The LCF[GMW79] family of theorem provers has been particularly successful in this regard, allowing derived rules of inference to be defined by the user in the programming metalanguage of the logic, and guaranteeing their soundness (relative to the fixed base of primitive rules) by means of strong typing.

What is an LCF-style logic? There are many aspects of LCF that were adopted by other systems, but the one we focus on is the encapsulation of the rules of inference as the only constructors of the abstract datatype `thm`.[1] To be more precise, let the set of primitive inference rules be $\mathcal{R}$. Then the typical LCF-style `thm` abstract type is defined by the implementations of the rules in $\mathcal{R}$ plus a destructor function `dest_thm`, which breaks a theorem down into its user-accessible components. For example, in hol90, `dest_thm` has type `thm -> term list * term`.

The set of "LCF-definable" rules of inference is the closure of the primitive rules under SML function construction, *i.e.*, all ML functions with range type `thm`. Call this closure $\overline{\mathcal{R}}$. The most important property of the abstract type of theorems is that the *only* way to get an SML value of type `thm` is by fully evaluating a function in $\overline{\mathcal{R}}$. Since $\mathcal{R}$ is *exactly* the programs implementing the primitive rules of inference in the logic, there is a close correspondence between the execution of SML programs in $\overline{\mathcal{R}}$ and proofs in the logic.

The purpose of this paper is to point out that more can be said: there are SML definable functions that "effectively do inference" but are not in $\overline{\mathcal{R}}$. Consider a simple example from hol90: the addition of a term of type[2] `` `:bool` `` to the assumptions of a theorem. The following two programs[3]

```
fun ADD_ASSUM1 (t,th) = MP (DISCH t th) (ASSUME t);

fun ADD_ASSUM2 (t,th) = if (has_boolean_type t)
                        then (union [t] (hyp th), concl th)
                        else raise NOT_A_PROPOSITION
```

have the following types

```
val ADD_ASSUM1 : (term * thm) -> thm
val ADD_ASSUM2 : (term * thm) -> (term list * term).
```

`ADD_ASSUM1` performs three primitive inferences (`MP`, `DISCH`, and `ASSUME` are members of $\mathcal{R}$):

1. t is discharged from the assumptions of th. This requires checking t for α-convertibility with each assumption of th.

2. t is assumed.

3. *modus ponens* gives a theorem whose conclusion is that of th, and whose assumptions are the assumptions of th unioned with the singleton set t. The union operation requires a traversal of the assumptions. (Depending on the formulation of the logic, union or may not need to be done modulo α-convertibility.)

---

[1]The implementation of LCF did not itself define the type of theorems in ML; the type `thm` was built in Lisp, the meta-meta-language. However, Gordon[Gor82] gives a careful exposition of how to develop what we call an LCF-style logic in ML.

[2]Notational clarification—HOL types are quoted, SML types are not.

[3]The functions giving the assumptions and conclusion of a theorem are `val hyp = #1 o dest_thm` and concl = #2 o dest_thm

The execution of ADD_ASSUM1 requires that the assumptions of th be traversed twice. We turn now to `ADD_ASSUM2`. It returns a pair (A, c) where c is the conclusion of the original theorem, and A is the union of the assumptions of the original theorem with the list consisting solely of t. The execution of `ADD_ASSUM2` requires that the assumptions of th be traversed once. Now consider the function

```
dest_thm o ADD_ASSUM1
```

which decomposes the result of `ADD_ASSUM1` into a pair that *in every situation* will be identical to (A, c). If we could somehow "coerce" `ADD_ASSUM2` to return a value of type `thm`, we are assured that it will do the same job as `ADD_ASSUM1`, except that it will be twice as fast at computing the new assumptions. Therefore, we should be able to use this new rule of inference to speed up future proofs. Suppose that the coercion is achieved by `mk_thm : term list * term -> thm`. Then `mk_thm o ADD_ASSUM2` is equivalent to `ADD_ASSUM1`, but there is a big problem with `mk_thm`. It must be a constructor for the abstract type `thm`, but it doesn't correspond to a rule of inference in the logic. The correspondence with formal proof has been lost; so much so that in both hol88 and hol90 any term can be coerced to be a theorem since `mk_thm` is user-accessible and does no argument checking at all! Another problem with this not-so hypothetical scenario is that non-formal reasoning about the equivalence of programs is apt to be wrong.

This paper proposes a method that allows the use of `mk_thm` for efficiency in reasoning while at the same time preserving the correspondence between execution of a member of $\overline{\mathcal{R}}$ and formal proof. The essence of the method is to require the user to supply

1. Two formalized programs $\ulcorner p_1 \urcorner$ and $\ulcorner p_2 \urcorner$, where p_1 is a member of $\overline{\mathcal{R}}$.
2. A theorem in the HOL system that says $\ulcorner \mathit{dest\_thm} \circ p_1 \urcorner$ is behaviourally equivalent to $\ulcorner p_2 \urcorner$.

Then, provided various checks are passed, we accept $\mathit{mk\_thm} \circ p_2$ as a new member of $\overline{\mathcal{R}}$, *i.e.*, we dynamically add it into the system and bind it to a name. This is a meta-logical principle for an HOL implementation; we are going to call it $\mathcal{MM}$. The essential property of $\mathcal{MM}$ is that the use of it does not change the set of provable formulas. For any i, where i is the number of times $\mathcal{MM}$ has been invoked, $\overline{\mathcal{R}_i} \vdash th$ if and only if $\overline{\mathcal{R}_{i+1}} \vdash th$.

2 The formalization of programs

$\mathcal{MM}$ depends on a formal, *i.e.*, HOL, theory of the syntax and semantics of SML programs. On top of that, the HOL implementation is itself formalized. This gives the following picture of a *reflective* HOL implementation. An item delimited by the quotation marks ($\ulcorner$_$\urcorner$) means that it has been formalized in the HOL logic:

⌜hol90⌝
⌜SML⌝
hol90
SML

In this picture, the top two layers (the formalization[4]) purport to "talk about" the bottom two layers (the implementation); if we believe that they do so accurately, we can use proof of program equivalence as justification to soundly and dynamically change the implementation of the logic.

The formalization of SML programs in HOL is a form of Gödelization, the coding up of a formal system inside itself. The interest of Gödelization in logic is often to see how weak a system it can be done in (so as to get correspondingly strong incompleteness results, for instance); our interest is in making it as transparent as possible, so we use datatypes in HOL to represent programs. The syntax of programs is given by terms of type `':ast'`, which is an exact copy of `ast`, a type of untyped SML abstract syntax trees provided by the SML/NJ compiler. Environments (HOL type `':env'`) may be represented in the logic by the type `':ast list'`, with the proviso that every member of the representing list be a declaration. State, formalized as a finite mapping between addresses and values, is modelled by terms of type `':state'`.

The following diagram illustrates how we move programs between the formal and the real. Types and functions provided by the compiler are italicized.

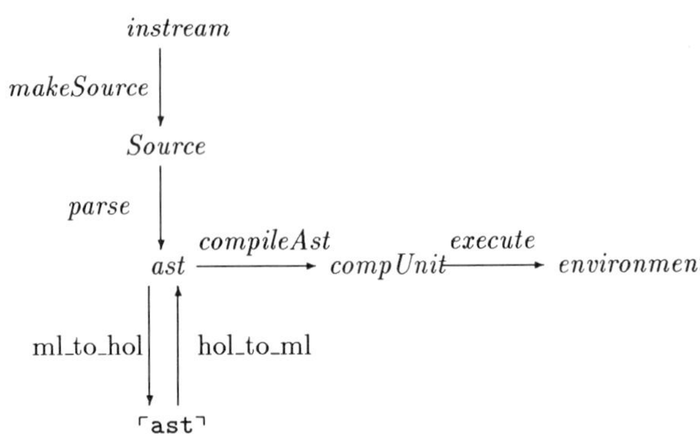

The SML/NJ implementation provides a function `parse` to map from SML concrete syntax to SML abstract syntax. We need merely to write trivial translation routines (`ml_to_hol` and `hol_to_ml`) between `ast` and ⌜`ast`⌝. The mapping from formalized objects to real objects is accomplished by using `hol_to_ml` to map from ⌜`ast`⌝ to `ast`, followed by compilation from `ast`. The HOL parser and prettyprinter need to be

[4]Neither $\mathcal{MM}$ nor **mk_thm** is included in the formalization

upgraded to recognize formalized SML programs, since dealing with abstract syntax trees is not satisfactory for the user.

Question: Does the embedding language have to be the same as the embedded language? Yes, if one is to be truly self-referential. I suppose that the embedded language could be a "subset" of the embedding language, but then to be self-referential, the HOL implementation (or at least the parts that get reasoned about) would need to be coded solely in the subset.

3 First class environments

The SML/NJ compiler has recently been augmented with a user-accessible type of environments [Mac92]. Both static and dynamic kinds of environments are accessible. There are some pre-defined environments available, which are used as backgrounds against which to build up new environments. Static environments are built up by elaboration and dynamic environments by evaluation of declarations. Environments are incremental: the evaluation of a declaration against a background environment will result in an environment consisting only of the declaration. Other operations allow environments to be concatenated together (•) and also thinned (−).

To model the SML environment in HOL, we use a function from theory identifiers to environments:

```
new_constant{Name = "HOL_ENV", Ty = `:thid -> env`}
```

Theory identifiers (*thid*s) need to be unique: no two different theories can have the same thid. $\ulcorner HOL\_ENV_{thid} \urcorner$ represents the state of the SML environment after an application of $\mathcal{MM}$ in the theory designated by *thid*. $\ulcorner HOL\_ENV_{HOL} \urcorner$ is a formalization of the hol90 system when it starts up, *i.e.*, the SML pervasive environment on top of which is the standard HOL environment.

$\mathcal{MM}$ requires the compilation of the prospective rule of inference in an environment that has `mk_thm`; however, `mk_thm` is not allowed to occur in any environment available to the user. We meet this requirement by having two environments, E_{sys} and E_{user}. In E_{sys} compilation using `mk_thm` is allowed; use of E_{sys} (by invoking $\mathcal{MM}$) is rare and guarded. E_{user} is the ordinary top level environment available to the user. They are defined by the following equations

$$E_{sys_{th}} = HOL\_ENV_{th} \bullet PERV\_ENV$$

$$E_{user_{th}} = (HOL\_ENV_{th} - mk\_thm) \bullet PERV\_ENV$$

We reiterate that $\mathcal{MM}$ and `mk_thm` are in HOL_ENV_{HOL}, but not in $\ulcorner HOL\_ENV_{HOL} \urcorner$; we are allowed to use them in the system, but not to reason about them.

4 Program equivalence

There are a bewildering number of formalisms for proving programs, including operational, axiomatic, and denotational methods. In the world of ML-like languages, there has been some work on proving equivalence in functional languages with reference variables[MT92, FF89, Pit91], but these attempts all make simplifying assumptions (a subset of ML, no local state, no user defined datatypes, *etc.*) and are all still very much research-in-progress. It is clearly much too early to commit ourselves to a particular semantics for ML. Therefore, we treat the notion of program equivalence *abstractly*, only requiring that it explicitly use the supplied formalizations of abstract syntax, environment, and state[5]. Behavioural equivalence is a predicate named `Equiv` over pairs of programs ((environment,syntax) pairs) in a common state.

```
val program = `:(env * ast)`
new_constant{Name = "Equiv",
             Ty = `:state -> ^program -> ^program -> bool`}
```

5 Adding a new rule

The implementation of $\mathcal{MM}$ takes a theorem stating the equivalence of two programs and a name to bind the new rule with.

```
val new_rule:{Name:string, Witness:thm} -> unit
```

The argument theorem is of the form

$$\vdash \; Equiv \ulcorner s \urcorner\ (\ulcorner E_1 \bullet HOL\_ENV_{th} \urcorner, \ulcorner dest\_thm \circ p_1 \urcorner)$$
$$(\ulcorner E_2 \bullet HOL\_ENV_{th} \urcorner, \ulcorner p_2 \urcorner)$$

The incremental environment E_1 is not used, except that it figures in the checks; otherwise, it could be an existentially quantified variable.

What needs to be checked?

- *th* equals the *thid* of the last ancestor theory $\mathcal{MM}$ was invoked in.
- $E_j \bullet HOL\_ENV_i$ closes p_j, *i.e.*, the values of all free variables in a program can be found in its environment.
- `dest_thm` must not have been redefined in $\ulcorner E_1 \urcorner$ or $\ulcorner E_2 \urcorner$. `mk_thm` must not have been redefined in $\ulcorner E_2 \urcorner$. $\circ$ (composition) must not have been redefined in either $\ulcorner E_1 \urcorner$ or $\ulcorner E_2 \urcorner$.

[5] Reasoning about state is essential in order to reason about `hol90`, since the implementation of theories depends on a reference variable.

- the program $dest\_thm \circ p_1$ must be well-typed in the environment $E_1 \bullet HOL\_ENV_{th}$.
- the program $mk\_thm \circ p_2$ must be well-typed in the environment $E_2 \bullet HOL\_ENV_{th}$.
- p_1 and p_2 must have the same domain type. We assume that the two programs are not curried.
- The current theory is in draft mode.

Once these constraints have been checked, the formalized environment $\ulcorner E_2 \urcorner$ is compiled against the environment HOL_ENV_{th} into a real environment E_2, and `val Name = mk_thm o` p_2 is compiled with respect to $E_2 \bullet HOL\_ENV_{th}$. The following equation summarizes what happens in the (real) SML environment in an invocation of `new_rule`, taking the system from $\overline{\mathcal{R}_{th}}$ to $\overline{\mathcal{R}_{current\_theory}}$.

$$HOL\_ENV_{current\_theory} = (val\ Name = mk\_thm \circ p_2) \bullet E_2 \bullet HOL\_ENV_{th}$$

The above compilation makes E_2 and the new rule known to the compiler; the formalized SML environment is moved to the new level by asserting the following axiom

$$`HOL\_ENV(current\_theory) = \ulcorner val\ Name = mk\_thm \circ p_2 \urcorner \bullet \ulcorner E_2 \urcorner \bullet HOL\_ENV\ th`$$

5.1 Effect on the implementation of theories

Now we address the question of how HOL_ENV_{th} gets used in theories. The theory graph gives a form of persistence to the system. How does this view of the history of the user's (or community's) proof activity interact with the history of the implementation's development (as given by the growth of HOL_ENV)? In other words, do theories get stored in environments, or *vice versa*?

We decided to make environments subsidiary to theories. This is effected by the addition of an **env** field to the theory representation. As noted above, we store the new axiom for HOL_ENV in the current theory. The addition of environments to theories has the following consequences for the theory operations:

`new_theory`: No change.

`new_parent`: Each new ancestor of the theory is traversed recursively, and the incremental env stored there is concatenated with the current environment as the recursion unwinds. For each such env, the associated axiom is added to $HOL\_ENV_{current\_theory}$ (we note that this requires another index for HOL_ENV, but that would render the description even more opaque).

`load/extend_theory`: Essentially the same as `new_parent`.

Thus we can improve the efficiency of the implementation merely by loading in a theory where $\mathcal{MM}$ had been invoked.

6 Methodology, or Modelling the Growth of an Implementation

The style of reasoning used on `ADD_ASSUM1` and `ADD_ASSUM2` is often used when developing an implementation of a logic; the informal process goes as follows:

1. Implement logic;
2. Profile the implementation on demanding examples; find bottleneck derived rule r;
3. Replace r by composing `mk_thm` with a computationally cheaper "alogical" equivalent and binding it to name of r; this amounts to calling `new_rule` with a name identical to the name currently binding the supplied derived rule of inference (p_1).
4. Now we recognize that there may have been a lot of development between the definition of the pure rule and the invocation of $\mathcal{MM}$. We want to take advantage of the efficiency of the new rule in the building of those various tools; hence, we *bootstrap* by re-doing the system development. Tools defined in this intermediate area that use r will retain their original semantics, but will now run faster because they incorporate the new rule.
5. go to 2

How to properly implement bootstrapping is not yet clear. So far, it is very much an *ad hoc* process: various bits of code are included in one pass of compilation, and excluded (by being commented out, usually) in the next pass of compilation. This is not satisfying, because it works by obliterating the past: the history of the development of the system should be extractable from the current source text of the system (comments don't count).

An alternative way to look at the problem and temporarily sidestep the proof obligation of $\mathcal{MM}$ is to place the statement of the witness theorem on the assumption list of every theorem proved using the new rule. This effect might cleanly be accomplished by using abstract theories.

By either method, $\mathcal{MM}$ provides a novel and rigourous way to optimize an LCF-style system.

7 Related Work

To pun, one could call this an "Iterative Progression of Implementations"[Fef62] approach because $\mathcal{MM}$ has many parallels with the idea of a "reflection principle". Reflection refers to the introduction of an unprovable, but true (in some external sense), formula as a theorem to a logic. The canonical example is the addition of a statement of consistency to the theory of first order arithmetic. Philosophically, a reflection principle is an expression of trust in a logic system. In contrast, $\mathcal{MM}$ is an expression of trust in the computer implementation of the logic and the formalization of the SML semantics, as embodied in the definition of program equivalence. The unrestricted availability of `mk_thm` as in the current system can be seen as an expression of confidence in every possible user.

A reflection principle increases the set of formulas that are provable in a system; we are much less ambitious logically since $\mathcal{MM}$ merely increases the number of definable inference rules in the system. $\mathcal{MM}$ also increases the number of proofs that will run to completion, in that more proofs can go through, *i.e.*, terminate successfully, simply because machine resoures will not get exhausted, since proofs can be more efficient.

Related work on the Computer Science side has been done by Weyhrauch[Wey80] and Boyer and Moore[BM81]. Weyhrauch uses a technique called *semantic attachment*, which associates programs to theories: the program is intended to be a partial model of the theory. As such, the program really lives in the metatheory of the theory. When an object (term, formula, or theorem) of a theory needs to be manipulated in the course of doing proof, it can either be manipulated according to the usual logical rules in the theory, or lifted to the meta-theory, where the associated program may be called directly on that object. Thus the FOL slogan is that proof in the theory is evaluation in the metatheory. The weakness of this methodology is that it leaves open the question of correctness of the program that implements the model.

Boyer and Moore require the user to prove a form of validity maintenance. For them, a *metafunction* is a mapping between terms in their logic. If, under all variable assignments, the input and output terms are identical, then the metafunction can be shown not to affect the soundness of the implementation, and hence can be used for inference.

The NUPRL group are investigating reflection in a constructive logic[ACHA90]. They investigate "logical" reflection; proofs about a logical type of proofs are used in conjunction with the "proofs-as-programs" interpretation to (among other things) soundly build up the power of their proof system.

$\mathcal{MM}$ is closest to the method of Boyer and Moore. Both approaches require proofs that the new rule preserves validity; the difference is that we focus on program equivalence, which, since one of the programs is already a rule of inference, implies (relative) validity. Whether this is a fruitful trade-off must be borne out by examples.

Partial evaluation [JSS89] is another means of transparently gaining efficiency, one that doesn't require as much formalization. It should not be viewed as a competitor to the logic-based approaches, so much as a supplement. It operates mostly syntactically, while the approach here depends essentially on formal derivations of program equivalence.

8 Conclusions and Future Work

$\mathcal{MM}$ gives us a way to rigourously control a facet of the life-cycle of an LCF-style logic implementation. We can soundly combine (initial) simplicity with efficiency. This has application in alleviating the tension between the LCF school and the so-called "decision procedure" implementations of theorem provers. For example, using $\mathcal{MM}$ (and I'm not saying it will be immediately feasible!) it should be possible to derive an efficient (modulo the inherent complexity of the problem) rule of inference that provides a limited Presburger arithmetic decision procedure. The same alleviation of tension should be possible between simulation and proof-based verifications of hardware (BDDs come to mind). The intent is not to compete with these proof methodologies in speed, but to soundly incorporate them in HOL in a form reasonably close to their descriptions in the literature.

None of the requirements for $\mathcal{MM}$ are particularly HOL- or even logic-dependent, other than requiring an implementation of a sufficiently expressive logic to perform proofs of equivalence in. I believe that the use of formal proof to guard a "loophole" function (`mk_thm` in our case) in order to safely optimize an implementation has general application to abstract datatypes, but it does seem to depend on a particular relationship between the representation and abstraction functions of the type.

In view of this logic-independence, it might be interesting to consider this work in the context of a logical framework such as Isabelle[Pau90]. It is not clear that the motivation for $\mathcal{MM}$ (efficiency) would still be compelling in that context, since there is basically a "unit" cost per rule application; both primitive and derived rules are just lambda terms that need to be (higher-order) unified with the goal in order to be applied.

The point of this paper can be expressed as a sort of implication: *if* one has a notion of equivalence between programs, and it can be expressed in terms of abstract syntax, environment, and state, *then* $\mathcal{MM}$ will give a way to soundly build up the computational power of the theorem-prover.

9 Acknowledgements

For advice and encouragement thanks are due Elsa Gunter, Tim Griffin, Amy Felty, Pierre Cregut, Bill Aitken, and Graham Birtwistle.

References

[ACHA90] S.F. Allen, R. Constable, D. Howe, and W. Aitken. The semantics of reflected proof. In *Fifth annual IEEE symposium on Logic in Computer Science*, pages 95–107, Philadelphia, USA, June 1990.

[BM81] R.S Boyer and J.S. Moore. Meta-functions: proving them correct and using them as new proof procedures. In R.S. Boyer and J.S. Moore, editors, *The Correctness Problem in Computer Science*, pages 103–184, Liege, Belgium, 1981.

[Fef62] Solomon Feferman. Transfinite recursive progressions of axiomatic theories. *Journal of Symbolic Logic*, 27(3):259–316, 1962.

[FF89] Matthias Felleisen and Daniel Friedman. A syntactic theory of sequential state. *Theoretical Computer Science*, 69:243–287, 1989.

[GMW79] Michael Gordon, Robin Milner, and Christopher Wadsworth. *Edinburgh LCF: A Mechanised Logic of Computation*, volume 78 of *Lecture Notes in Computer Science*. Springer-Verlag, 1979.

[Gor82] Michael Gordon. Representing a logic in the LCF metalanguage. In D. N'eel, editor, *Tools and Notions for Program Construction*, pages 163–185. Cambridge University Press, 1982.

[JSS89] N.D. Jones, P. Sestoft, and H. Sondergaard. Mix: a self-applicable partical evaluator for experiments in compiler generation. *LISP and Symbolic Computation*, 2(1):9–50, 1989.

[Mac92] David MacQueen. First class environments for sml/nj. Preliminary system documentation, 1992.

[MT92] Ian Mason and Carolyn Talcott. References, local variables, and operational reasoning. In *Proceeding of the Seventh Annual Symposium on Logic in Computer Science*, San Jose, California, USA, June 1992. Computer Society Press.

[Pau90] Lawrence Paulson. Isabelle: The next 700 theorem provers. In P.G. Oddifreddi, editor, *Logic and Computer Science*, pages 361–386. Academic Press, 1990.

[Pit91] Andrew Pitts. Evaluation logic. In Graham Birtwistle, editor, *IV Higher Order Workshop, Banff 1990*. Springer-Verlag, 1991.

[Wey80] Richard Weyhrauch. Prolegomena to a theory of mechanized formal reasoning. *Artificial Intelligence*, 13(1-2):133–170, 1980.

Higher Order Logic Theorem Proving and its Applications (A-20)
L.J.M. Claesen and M.J.C. Gordon (Editors)
Elsevier Science Publishers B.V. (North-Holland)

Why We Can't have SML Style datatype Declarations in HOL

Elsa L. Gunter

AT&T Bell Laboratories, Rm. #2A-432, Murray Hill, NJ, 07974-0636, USA

Abstract

The type descriptions that `define_type` is capable of handling are noticeably more limited than those allowed by SML. In particular, `define_type` requires of a type description it is given that the type being defined should not occur within any compound type. While this restriction is more severe than is necessary for there to be a solution in HOL to the description, we show that some restriction on the nature of the compound types within which the type being defined may occur is necessary. Not all descriptions allowable in SML will have a solution in HOL. Moreover, owing to the nature of the basic principle of type definition in HOL, no purely syntactic non-ad hoc test of a recursive type description will be sufficient to allow us to extend `define_type` to compound types involving "safe" type constructors such as `list` while at the same time barring all descriptions for which no solution is possible. Any general extension to the `define_type` package that allows the types being defined to occur within compound types in the type description will need to take as additional arguments theorems about the type constructors used in the compound type that justify their being so used. Finally, we show that an extension to the case where all the type constructors used in compound types involving the types being defined are essentially recursive type constructors themselves; the type constructors must satisfy an "initiality" theorem of the form returned by `define_type`, which must be supplied as an argument to this extension of `define_type`.

Keyword Codes: F.4.1; I.2.3
Keywords: Mathematical Logic; Deduction and Theorem Proving.

1 Introduction

In the early days of HOL, people using the system rarely made new types. This was considered to be in the domain of only the most sophisticated of user, and among them it was left to those who were working on extending the basic system (*e.g.* adding the type of lists or integers). With the advent of Tom Melham's `define_type` package, this situation has changed dramatically. Now that it is roughly as easy to define new types, acquire properties of elements of those types and define new functions over those types as to create a `datatype` declaration in SML, it has become a commonplace aspect of setting up a new problem in HOL to create several types via `define_type` to ease the process. But the type descriptions that `define_type` is capable of handling are noticeably more limited than those allowed by SML. In particular, `define_type` requires of a type description it is given that the type being defined should not occur within any compound type. As people have become more accustomed to using `define_type` to create new types to solve their problems, they have come to want the greater flexibility that would come from allowing the type being defined to occur within a compound type in its description. The question has been asked "Why can't I do the following:

```
define_type "x1" "x1 = A1 | B1 ((x1)list)";
```

Is there some intrinsic reason for the limitation which forbids this?" In this paper we intend to show that in fact there is an intrinsic reason why we must have some such limit. The particular description given above, using the type constructor `list` in fact causes no problem. However, we intend to show that there is no syntactic test, shy of an ad hoc enumeration of some favorite "safe" type constructors, that will allow us to be able to determine if it is possible to define in HOL a type allowing the appropriate principle of recursive function definition. That is, while we could explicitly extend `define_type` to handle `list`, `pair` and a few others, in general it will not be possible to generically construct a solution for a description such as

```
define_type "x1" "x1 = A1 | B1 ((x1)foo)";
```

or even to determine if one exists.

2 Differences Between SML Types and HOL Types

Although the languages of SML and HOL look very similar, there are several important differences that HOL users should be aware of. One difference of which most users are aware is that every HOL type must have an element of that type, whereas it is possible to define a type in SML which has no elements. The SML declaration

```
datatype Empty = Nothing of Empty
```

causes no problems, whereas the HOL command

```
define_type "Empty = Nothing Empty"
```

is rejected because of the lack of a non-recursive operator.

There is another significant way in which the types of SML and HOL differ, and that is with regard to the meanings of function spaces. In SML the function space between two types is taken to mean the "computable functions" between two domains which are the meanings of the two types. There are various ways of defining "domain" and "computable", but it is important to note that we cannot use sets for domains and the full set of functions between them for the meaning. On the other hand, we can with HOL; HOL admits a set-theoretic model. Since we can interpret types as sets and functions as all functions between the sets, we cannot have a solution, for example, to the description

```
"lambda = Var num | App lambda lambda | Abs (lambda -> lambda)"
```

For a type to be a solution to this description would require that there be a one-to-one mapping `Abs` of the function space over the type into itself. Moreover, there would need to be a one-to-one mapping `Var` of the naturals into it as well. However, given a set A, the set $A \to A$ can be injected into it if and only if A is a singleton set. For any set A of cardinality greater than 1, the cardinality of $A \to A$ is always strictly greater than that of A. This is usually shown using the Cantor Diagonal argument, but can also be shown by a variation of Russell's Paradox. The reason this problem does not arise in SML is because the meaning of the function space is necessarily not the full function space between sets, but rather something like monotone continuous functions between domains of algebraic directed-complete partial orders.

The fact that we cannot have a set-theoretic model for HOL if it has a solution to

```
"lambda  = Var num | App lambda lambda | App (lambda -> lambda)"
```

tells us that the current principles of type and term definition will not allow us to create a type and terms that satisfy this description. It does not tell us that there are no principles of definition that could be added to HOL that would be consistent and would admit a solution to this description, at least not directly. However, no such such principles can exist because any solution to the above description is inconsistent with HOL. The next two sections are devoted to proving a generalization of this fact and discussing some of the consequences for attempts to extend the `define_type` package.

3 The HOL Proof that the Function Space Can't Be Embedded

The precise result we show is that the following is a theorem of HOL:

$$\vdash (\exists \mathsf{x}{:}\beta\ \mathsf{y}{:}\beta.\ \neg(\mathsf{x} = \mathsf{y})) \Longrightarrow \forall \mathsf{f}{:}(\alpha \to \beta) \to \alpha.\ \neg(\mathsf{ONE\_ONE\ f})$$

The reason this shows the non-existence of a solution of the description of the type `(==':lambda'==)` is because `(==':lambda'==)` must have at least two distinct elements since it has two distinct operators, and hence the function `(--'App:(lambda -> lambda) -> lambda'--)` cannot be one-to-one.

In order to prove

$$\vdash (\exists \mathsf{x}\ \mathsf{y}{:}\beta.\ \neg(\mathsf{x} = \mathsf{y})) \Longrightarrow \forall \mathsf{f}{:}(\alpha \to \beta) \to \alpha.\ \neg(\mathsf{ONE\_ONE\ f})$$

we factor it through a specific instance, namely:

$$\vdash \forall \mathsf{f}{:}(\alpha \to \mathsf{bool}) \to \alpha.\ \neg(\mathsf{ONE\_ONE\ f})$$

To see that we can acquire the general result, suppose that β is a type with at least two elements $\mathsf{x}{:}\beta$ and $\mathsf{y}{:}\beta$, and fix $\mathsf{f}{:}(\alpha \to \beta) \to \alpha$. Then we know by the special case

$$\vdash \neg(\mathsf{ONE\_ONE}\ \lambda \mathsf{g}.\ \mathsf{f}(\lambda \mathsf{z}.\ \mathsf{gz} \Rightarrow (\mathsf{x}{:}\beta) \mid \mathsf{y})).$$

Therefore, there exist functions $\mathsf{g}{:}\alpha \to \mathsf{bool}$ and $\mathsf{h}{:}\alpha \to \mathsf{bool}$ such that

$$\vdash (\neg(\mathsf{g} = \mathsf{h})) \wedge (\mathsf{f}(\lambda \mathsf{z}.\ \mathsf{g}\ \mathsf{z} \Rightarrow \mathsf{x} \mid \mathsf{y}) = \mathsf{f}(\lambda \mathsf{z}.\ \mathsf{h}\ \mathsf{z} \Rightarrow \mathsf{x} \mid \mathsf{y})).$$

To show that $\neg(\mathsf{ONE\_ONE}\ \mathsf{f})$, we therefore need to show

$$\begin{bmatrix} \neg(\mathsf{x}=\mathsf{y}) \\ \mathsf{f}(\lambda \mathsf{z}.\ \mathsf{g}\ \mathsf{z} \Rightarrow \mathsf{x} \mid \mathsf{y}) = \mathsf{f}(\lambda \mathsf{z}.\ \mathsf{h}\ \mathsf{z} \Rightarrow \mathsf{x} \mid \mathsf{y}) \\ \neg(\mathsf{g}=\mathsf{h}) \end{bmatrix} \vdash \exists \mathsf{x}_1\ \mathsf{x}_2.\neg((\mathsf{f}\ \mathsf{x}_1 = \mathsf{f}\ \mathsf{x}_2) \Longrightarrow (\mathsf{x}_1 = \mathsf{x}_2))$$

Let x_1 be $\lambda \mathsf{z}.\ \mathsf{g}\ \mathsf{z} \Rightarrow \mathsf{x} \mid \mathsf{y}$ and let x_2 be $\lambda \mathsf{z}.\ \mathsf{h}\ \mathsf{z} \Rightarrow \mathsf{x} \mid \mathsf{y}$. Then we must show that

$$[\neg(\mathsf{x}=\mathsf{y}),\ \neg(\mathsf{g}=\mathsf{h})] \vdash \neg((\lambda \mathsf{z}.\ \mathsf{g}\ \mathsf{z} \Rightarrow \mathsf{x} \mid \mathsf{y}) = (\lambda \mathsf{z}.\ \mathsf{h}\ \mathsf{z} \Rightarrow \mathsf{x} \mid \mathsf{y}))$$

To show that these functions are not equal, it suffices to show there is some term $\mathsf{w}:\alpha$ on which they differ. Since g and h differ, there exists some term on which they differ. This is the term we need. Thus we need to show

$$[\neg(\mathsf{x}=\mathsf{y}),\ \neg(\mathsf{gw}=\mathsf{hw})] \vdash \neg((\mathsf{g}\ \mathsf{w} \Rightarrow \mathsf{x} \mid \mathsf{y}) = (\mathsf{h}\ \mathsf{w} \Rightarrow \mathsf{x} \mid \mathsf{y}))$$

This follows by a simple case analysis. Therefore,

$$\vdash \neg(\mathsf{ONE\_ONE}\ (\mathsf{f}:(\alpha \rightarrow \beta) \rightarrow \alpha))$$

Hence, we are left showing the specific case

$$\vdash \forall \mathsf{f}:(\alpha \rightarrow \mathsf{bool}) \rightarrow \alpha.\neg(\mathsf{ONE\_ONE}\ \mathsf{f})$$

That is, we must show that the powerset of a type cannot be injected into that type.

Fix a function $\mathsf{f}:(\alpha \rightarrow \mathsf{bool}) \rightarrow \alpha$. To show

$$\vdash \neg(\mathsf{ONE\_ONE}\ \mathsf{f})$$

is equivalent to showing

$$\vdash \exists \mathsf{s}_1\ \mathsf{s}_2:\alpha \rightarrow \mathsf{bool}.\ (\mathsf{f}\ \mathsf{s}_1 = \mathsf{f}\ \mathsf{s}_2) \wedge \neg(\mathsf{s}_1 = \mathsf{s}_2))$$

Let s_1 be $\lambda \mathsf{z}:\alpha.\ \exists \mathsf{s}.\neg((\mathsf{z} = \mathsf{f}\ \mathsf{s}) \Longrightarrow \mathsf{s}\ \mathsf{z})$. Before determining what s_2 should be, let us show that $\mathsf{s}_1(\mathsf{f}\ \mathsf{s}_1)$, *i.e.*

$$\vdash \exists \mathsf{s}:\alpha \rightarrow \mathsf{bool}.\ \neg((\mathsf{f}\ \mathsf{s}_1 = \mathsf{f}\ \mathsf{s}) \Longrightarrow \mathsf{s}(\mathsf{f}\ \mathsf{s}_1))$$

which is the same as showing

$$\vdash \neg\forall \mathsf{s}:\alpha \rightarrow \mathsf{bool}.\ (\mathsf{f}\ \mathsf{s}_1 = \mathsf{f}\ \mathsf{s}) \Longrightarrow \mathsf{s}(\mathsf{f}\ \mathsf{s}_1)$$

To see this, let us suppose to the contrary and derive a contradiction. Then we are assuming $\forall \mathsf{s}:\alpha \rightarrow \mathsf{bool}.\ (\mathsf{f}\ \mathsf{s}_1 = \mathsf{f}\ \mathsf{s}) \Longrightarrow \mathsf{s}(\mathsf{f}\ \mathsf{s}_1)$ (which is the same thing as $\neg(\mathsf{s}_1(\mathsf{f}\ \mathsf{s}_1))$). Therefore, we have it for the particular case of $\mathsf{s}_1 = \lambda \mathsf{z}:\alpha.\exists \mathsf{s}.\ \neg((\mathsf{z} = \mathsf{fs}) \Longrightarrow \mathsf{sz})$. This specific case, after beta-reduction, simplifies to $\neg\forall \mathsf{s}:\alpha \rightarrow \mathsf{bool}.\ (\mathsf{f}\ \mathsf{s}_1 = \mathsf{f}\ \mathsf{s}) \Longrightarrow \mathsf{s}(\mathsf{f}\ \mathsf{s}_1)$ (that is, $\mathsf{s}_1(\mathsf{f}\ \mathsf{s}_1)$), which contradicts our original assumption. (This argument is the part that I am referring to as a modification of Russell's Paradox.) With this contradiction we have established that

$$\vdash \exists \mathsf{s}:\alpha \rightarrow \mathsf{bool}.\ \neg((\mathsf{f}\ \mathsf{s}_1 = \mathsf{f}\ \mathsf{s}) \Longrightarrow \mathsf{s}(\mathsf{f}\ \mathsf{s}_1))$$

Let s_2 be such an s. Then we need to show

$$[\mathsf{s}_1(\mathsf{f}\ \mathsf{s}_1),\ \mathsf{f}\ \mathsf{s}_1 = \mathsf{f}\ \mathsf{s}_2,\ \neg(\mathsf{s}_2(\mathsf{f}\ \mathsf{s}_1))] \vdash (\mathsf{f}\ \mathsf{s}_1 = \mathsf{f}\ \mathsf{s}_2) \wedge \neg(\mathsf{s}_1 = \mathsf{s}_2)$$

Simplifying, we need to show

$$[\mathsf{s}_1(\mathsf{f}\ \mathsf{s}_1),\ \mathsf{f}\ \mathsf{s}_1 = \mathsf{f}\ \mathsf{s}_2,\ \neg(\mathsf{s}_2(\mathsf{f}\ \mathsf{s}_1))] \vdash \neg(\mathsf{s}_1 = \mathsf{s}_2)$$

However, this follows immediately since we have $\mathsf{s}_1(\mathsf{f}\ \mathsf{s}_1)$ and $\neg(\mathsf{s}_2(\mathsf{f}\ \mathsf{s}_1))$. Therefore $\mathsf{f}:\alpha \rightarrow \mathsf{bool}$ is not one-to-one.

4 Constraints on type constructors that don't work

In the previous section, we showed that a certain class of functions in HOL cannot be one-to-one. This puts limitations on possible extensions to the class of specifications to which `define_type` might be extended. Specifications of the form

$$\mathsf{ty} = \ldots \mid \mathsf{C_i} \ldots (\mathsf{ty} \rightarrow \tau) \ldots \mid \ldots$$

are obviously ruled out by the result in the previous section, since $\mathsf{C_i}$ could not be a one-to-one function (assuming τ has at least two elements). On the other hand, there is nothing to rule out specifications of the form

$$\mathsf{ty} = \ldots \mid \mathsf{C_i} \ \ldots (\tau \rightarrow \mathsf{ty}) \ldots \mid \ldots$$

provided that ty does not occur in τ, and indeed it is possible to extend `define_type` to allow such cases. It is tempting to generalize these results to the conclusion that positive occurrences of the type being defined in the argument types of the constructors for the type being defined are allowable. However, the result from the previous section can be used to see that this is not the case. Consider a specification of the form

$$\mathsf{ty} = \ldots \mid \mathsf{C_i} \ ((\mathsf{ty} \rightarrow \tau_1) \rightarrow \tau_2) \mid \ldots$$

where both τ_1 and τ_2 are types not involving ty and each having at least two elements. Then ty has only a positive occurrence. However, if there were a solution to this specification, the constructor $\mathsf{C_i}$ would have to be a one-to-one function. Since τ_2 has at least two elements, there exists a one-to-one function, $\mathsf{f}:(\mathsf{ty} \rightarrow \tau_1) \rightarrow ((\mathsf{ty} \rightarrow \tau_1) \rightarrow \tau_2)$. Thus, if $\mathsf{C_i}$ were a one-to-one function, the composition $(\mathsf{C_i} \circ \mathsf{f}) : (\mathsf{ty} \rightarrow \tau_1) \rightarrow \tau_1$ would be a one-to-one function, contradicting the result of the previous section, since τ_1 is assumed to have at least two elements.

There exists another problem with trying to use a condition such as saying that a specification is acceptable provided that there are no negative occurrences of the type being defined among the argument types of the constructors. A "negative occurrence" is only referring to occurrences within the function space type constructor. What conditions are to be placed on occurrences of the type being defined within other type constructors? One possible attempt is to say that the recursive occurrences of the type being defined cannot be within a type constructor if the polymorphic variable for which the type being defined is being substituted occurred under a function space arrow in the definition of the type constructor. This is either way too much or not sufficient, depending upon how you interpret the statement. If only explicit occurrences of type variables under function space arrows are counted when examining the definitions of type constructors, this is not sufficient. Using the principle of type definition in HOL, we can create new type constructors that are as problematic as the function space arrow. (For example, we can create a type isomorphic to, but distinct from, the function space.) When this troublesome type constructor is used in defining another type constructor, we cannot allow recursive occurrences of the type being defined within it, even though it is not the function space type constructor itself. Thus, the restriction is not sufficient to guarantee that we will be able to solve specifications satisfying it.

Alternately, we could recursively calculate whether a type variable occurs "implicitly" within the function space arrow in the definition of a type constructor. However, if pursued uniformly, this approach is destined to failure. All non-vacuous polymorphism for type constructors currently in the core HOL system is derived from the polymorphism of the function space arrow. If we attempt to rule out all type constructors whose type variables have implicit occurrences within function space arrows, we will end up ruling out all type constructors currently in main use, leading us to no real extension whatsoever. For example, $(\alpha, \beta)\mathsf{prod}$ is defined in terms of a predicate on $\alpha \rightarrow \beta \rightarrow \mathsf{bool}$ and $(\alpha, \beta)\mathsf{sum}$ is defined in terms a predicate on the type $\mathsf{bool} \rightarrow \alpha \rightarrow \beta \rightarrow \mathsf{bool}$, so even the product and sum constructors would be ruled out. To a certain extent, the function space constructor is a red herring. It is the source of all difficulties, but then again it is the principle source of polymorphism. Clearly, some other solution needs to be sought for the problem of within which type constructors to allow recursive occurrences of a type being define.

5 What we can build

To get a handle on how to extend `define_type` to allow recursive occurrences of the type being define within some type constructors, let us recall another possible extension, namely to mutually recursive types. With mutually recursive type definitions, the result we need to use the mutually recursive types is the initiality theorem, which allows us to define functions over the mutually recursive types by mutual recursion and by cases. Tom Melham gives the following example in an email message sent to `info-hol`, 26 April 1992. Consider the following mutually recursive type specification:

$$\mathsf{a} = \mathsf{A}_1 \mid \mathsf{A}_2\ \mathsf{b} \quad \text{and} \quad \mathsf{b} = \mathsf{B}_1 \mid \mathsf{B}_2\ \mathsf{a}$$

For this type specification, the initiality theorem we need is

$$\begin{array}{l} \vdash \forall \mathsf{x}_1{:}\alpha\ \ \mathsf{x}_2{:}\beta\ \ \mathsf{f}_1{:}\beta \rightarrow \mathsf{b} \rightarrow \alpha\ \ \mathsf{f}_2{:}\alpha \rightarrow \mathsf{a} \rightarrow \beta. \\ \quad \exists!(\mathsf{fn}_1, \mathsf{fn}_2).(\mathsf{fn}_1\ \mathsf{A}_1 = \mathsf{x}_1) \wedge \\ \qquad (\forall \mathsf{b}.\ \mathsf{fn}_1\ (\mathsf{A}_2\ \mathsf{b}) = \mathsf{f}_1\ (\mathsf{fn}_2\ \mathsf{b})\ \mathsf{b}) \wedge \\ \qquad (\mathsf{fn}_2\ \mathsf{B}_1 = \mathsf{x}_2) \wedge \\ \qquad (\forall \mathsf{a}.\ \mathsf{fn}_2\ (\mathsf{B}_2\ \mathsf{a}) = \mathsf{f}_2\ (\mathsf{fn}_1\ \mathsf{a})\ \mathsf{a}) \end{array}$$

It would not be sufficient to give back only the two theorems:

$$\vdash \forall \mathsf{x}_1{:}\alpha\ \ \mathsf{f}_1{:}\mathsf{b} \rightarrow \alpha.\ \ \exists!\mathsf{fn}_1.(\mathsf{fn}_1\ \mathsf{A}_1 = \mathsf{x}_1) \wedge (\forall \mathsf{b}.\ \mathsf{fn}_1\ (\mathsf{A}_2\ \mathsf{b}) = \mathsf{f}_1\ \mathsf{b})$$

and

$$\vdash \forall \mathsf{x}_2{:}\beta\ \ \mathsf{f}_2{:}\mathsf{a} \rightarrow \beta.\ \ \exists!\mathsf{fn}_2.(\mathsf{fn}_2\ \mathsf{B}_1 = \mathsf{x}_2) \wedge (\forall \mathsf{a}.\ \mathsf{fn}_2\ (\mathsf{B}_2\ \mathsf{a}) = \mathsf{f}_2\ \mathsf{a})$$

We need to be able to make mutually recursive function definitions over our mutually recursive types.

The same situation holds for the cases of types defined with recursive occurrences within type constructors. To see this, let us consider another example:

$$\mathsf{Tree} = \mathsf{LEAF}\ \mathsf{bool} \mid \mathsf{NODE}\ (\mathsf{Tree}\ \mathsf{list})$$

and suppose we want to define the function that takes the conjunction of all the leaves of a tree. If the only theorem allowing us to define functions over trees is

$$\vdash \forall \mathsf{f}:\mathsf{bool} \to \alpha\ \ \mathsf{g}:\mathsf{Tree\ list} \to \alpha.\\ \quad \exists!\mathsf{fn}.(\forall \mathsf{b}.\ \mathsf{fn}\ (\mathsf{LEAF\ b}) = \mathsf{f\ b})\wedge\\ \quad\quad (\forall \mathsf{tl}.\ \mathsf{fn}\ (\mathsf{NODE\ tl}) = \mathsf{g\ tl}),$$

then it will not be possible for us to make our definition. We need a stronger principle, namely:

$$\vdash \forall \mathsf{x}:\beta\ \ \mathsf{f_1}:\mathsf{bool} \to \alpha\ \ \mathsf{f_2}:\beta \to \mathsf{Tree\ list} \to \alpha\ \ \mathsf{f_3}:\alpha \to \beta \to \mathsf{Tree} \to \mathsf{Tree\ list} \to \beta.\\ \quad \exists!(\mathsf{fn_1},\mathsf{fn_2}).(\forall \mathsf{b}.\ (\mathsf{fn_1}\ (\mathsf{LEAF\ b}) = \mathsf{f_1\ b})\wedge\\ \quad (\forall \mathsf{tl}.\ \mathsf{fn_1}\ (\mathsf{NODE\ tl}) = \mathsf{f_2}\ (\mathsf{fn_2\ tl})\ \mathsf{tl})\wedge\\ \quad (\mathsf{fn_2\ NIL} = \mathsf{x})\wedge\\ \quad (\forall \mathsf{t\ tl}.\ \mathsf{fn_2}\ (\mathsf{CONS\ t\ tl}) = \mathsf{f_3}\ (\mathsf{fn_1\ t})\ (\mathsf{fn_2\ tl})\ \mathsf{t\ tl})$$

With this principle, we can make the desired function definition by

$$(\forall \mathsf{b}.\ \mathsf{tree\_conj}(\mathsf{LEAF\ b}) = \mathsf{b})\wedge\\ (\forall \mathsf{tl}.\ \mathsf{tree\_conj}(\mathsf{NODE\ tl}) = \mathsf{map\_tree\_conj\ tl})\wedge\\ (\mathsf{map\_tree\_conj\ NIL} = \mathsf{T})\wedge\\ (\forall \mathsf{t\ tl}.\ \mathsf{map\_tree\_conj}(\mathsf{CONS\ t\ tl}) = ((\mathsf{tree\_conj\ t}) \wedge (\mathsf{map\_tree\_conjtl})))$$

In order to be able to derive such an initiality theorem, precisely what we need is the initiality theorem for lists. More generally, if we are given an initiality theorem for a recursively defined polymorphic type constructor (possibly a type constructor defined by mutual recursion with other types) then we will be able to solve further mutually recursive type specifications which make free use of this constructor. The construction in general makes use of the ability to make mutually recursive type definitions. To see how, we will sketch the construction for the specific case of the tree example.

To solve the tree specification, using the initiality theorem for lists, we transform it into the mutually recursive specification:

$$\mathsf{Tree} = \mathsf{LEAF'\ bool} \mid \mathsf{NODE'}\ (\mathsf{Tree\_list})$$

and

$$\mathsf{Tree\_list} = \mathsf{Tree\_NIL} \mid \mathsf{Tree\_CONS\ Tree\ Tree\_list}.$$

From here, we show that Tree_list is isomorphic to Tree list, and use this isomorphism to define LEAF, NODE, and to translate the initiality theorem for this mutual recursion into the initiality theorem we desire for Tree.

This last summer Healfdene Goguen and myself implemented in `hol90` an extension to `define_type` that would solve recursive specifications that involve recursive type constructors by using this general principle to translate them into mutually recursive specifications not involving the recursive type constructors. The mutually recursive specifications are then solved using a package written by Myra VanInwegen and myself following along the lines laid out by Tom Melham in his email message of 26 April 1992. This work currently only allows single recursive type specifications involving recursive type constructors. Also it takes as input an SML datatype description rather than a string to be parsed into such a description. In future work it will be extended to allow for mutually recursive type specifications involving recursive type constructors, and to include a parser to allow concrete syntax for specifications. Because the extensions described above is fundamentally built out of the existing `define_type` package, type specifications of the form

$$\mathsf{ty} = \ldots \mid \mathsf{C_i}\ \ldots(\tau \rightarrow \mathsf{ty})\ldots \mid \ldots$$

still cannot be handled. To allow these kinds of specifications would require changing `define_type` to be based on arbitrarily branching trees, instead of only finitely branching trees. Once such an extension is incorporated into the basic `define_type` package, it should be routine to change the mutually recursive definitions involving recursive type constructors to also include such specifications.

References

[1] M. J. C. Gordon. *The HOL System.* Cambridge Research Centre, SRI International, and DSTO Australia, 1989.

[2] T.F. Melham. 'Automating Recursive Type Definitions in Higher Order Logic', in: *Current Trends in Hardware Verification and Automated Theorem Proving*, edited by G. Birtwistle and P.A. Subrahmanyam (Springer-Verlag, 1989), pp. 341–386.

[3] T.F. Melham. Email correspondence. `info-hol` email, 26 April 1992.

IFIP

The INTERNATIONAL FEDERATION FOR INFORMATION PROCESSING is a multinational federation of professional and technical organisations (or national groupings of such organisations) concerned with information processing. From any one country, only one such organisation – which must be representative of the national activities in the field of information processing – can be admitted as a Full Member. In addition a regional group of developing countries can be admitted as a Full Member. On 1 October 1992, 45 organisations were Full Members of the Federation, representing 62 countries.

The aims of IFIP are to promote information science and technology by:

- fostering international co-operation in the field of information processing;
- stimulating research, development and the application of information processing in science and human activity;
- furthering the dissemination and exchange of information about the subject;
- encouraging education in information processing.

IFIP is dedicated to improving worldwide communication and increased understanding among practitioners of all nations about the role information processing can play in all walks of life.

Information technology is a potent instrument in today's world, affecting people in everything from their education and work to their leisure and in their homes. It is a powerful tool in science and engineering, in commerce and industry, in education and adminstration. It is truly international in its scope and offers a significant opportunity for developing countries. IFIP helps to bring together workers at the leading edge of the technology to share their knowledge and experience, and acts as a catalyst to advance the state of the art.

IFIP came into official existence in January, 1960. It was established to meet a need identified at the first International Conference on Information Processing which was held in Paris in June, 1959, under the sponsorship of UNESCO.

Organisational Structure
The Federation is governed by a GENERAL ASSEMBLY, which meets once every year and consists of one representative from each Member organisation. The General Assembly decides on all important matters, such as general policy, the programme of activities, admissions, elections and budget.

IFIP ORGANISATION CHART

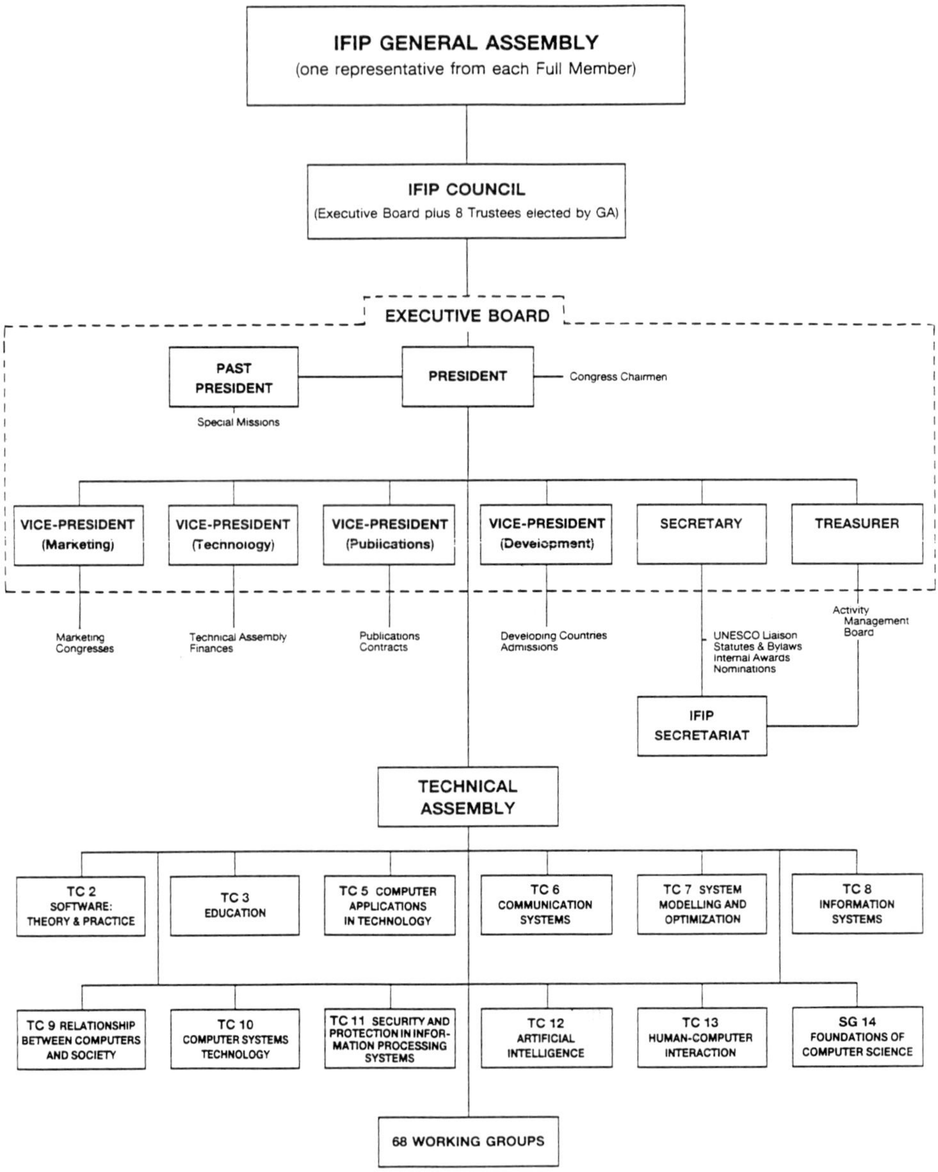

September 1992

The day-to-day work of IFIP is directed by its Officers: the President, Vice-Presidents, Secretary and Treasurer, who are elected by the General Assembly and together constitute the EXECUTIVE BOARD.

The COUNCIL, consisting of the Officers and up to eight Trustees elected from the General Assembly, meets twice a year and takes decisions which become necessary between General Assembly meetings.

The headquarters of the Federation are in Geneva, Switzerland where the IFIP Secretariat administers its affairs.

For further information please contact:

IFIP Secretariat
attn. Mme. GWYNETH ROBERTS
16 Place Longemalle
CH-1204 Geneva, Switzerland
telephone: 41 (22) 28 26 49
facsimile: 41 (22) 781 23 22
Bitnet: ifip@cgeuge51

IFIP's MISSION STATEMENT

IFIP's mission is to be the leading, truly international, apolitical organisation which encourages and assists in the development, exploitation and application of Information Technology for the benefit of all people.

Principal Elements

1. To stimulate, encourage and participate in research, development and application of Information Technology (IT) and to foster international co-operation in these activities.
2. To provide a meeting place where national IT Societies can discuss and plan courses of action on issues in our field which are of international significance and thereby to forge increasingly strong links between them and with IFIP.
3. To promote international co-operation directly and through national IT Societies in a free environment between individuals, national and international governmental bodies and kindred scientific and professional organisations.
4. To pay special attention to the needs of developing countries and to assist them in appropriate ways to secure the optimum benefit from the application of IT.
5. To promote professionalism, incorporating high standards of ethics and conduct, among all IT practitioners.
6. To provide a forum for assessing the social consequences of IT applications; to campaign for the safe and beneficial development and use of IT and the protection of people from abuse through its improper application.

7. To foster and facilitate co-operation between academics, the IT industry and governmental bodies and to seek to represent the interest of users.
8. To provide a vehicle for work on the international aspects of IT development and application including the necessary preparatory work for the generation of international standards.
9. To contribute to the formulation of the education and training needed by IT practitioners, users and the public at large.

Note to Conference Organizers

Organizers of upcoming IFIP Working Conferences are urged to contact the Publisher. Please send full details of the Conference to:

Mrs. STEPHANIE SMIT
Administrative Editor – IFIP Publications
ELSEVIER SCIENCE PUBLISHERS
P.O. Box 103, 1000 AC Amsterdam
The Netherlands
telephone: 31 (20) 5862481
facsimile: 31 (20) 5862616
email: s.smit@elsevier.nl

IFIP TRANSACTIONS

IFIP TRANSACTIONS is a serial consisting of 15,000 pages of valuable scientific information from leading researchers, published in 35 volumes per year. The serial includes contributed volumes, proceedings of the IFIP World Conferences, and conferences at Technical Committee and Working Group level. Mainstream areas in the IFIP TRANSACTIONS can be found in Computer Science and Technology, Computer Applications in Technology, and Communication Systems.

From 1993 onwards the IFIP TRANSACTIONS are only available as a full set.

IFIP TRANSACTIONS A:
Computer Science and Technology
1992: Volumes A1–A19
1993: Volumes A20–A40
ISSN 0926-5473

IFIP Technical Committees that are involved in IFIP TRANSACTIONS A
Software: Theory and Practice (TC2)
Education (TC3)
System Modelling and Optimization (TC7)
Information Systems (TC8)
Relationship Between Computers and Society (TC9)

Computer Systems Technology (TC10)
Security and Protection in Information Processing Systems (TC11)
Artificial Intelligence (TC12)
Human-Computer Interaction (TC13)
Foundations of Computer Science (SG14)

IFIP TRANSACTIONS B:
Applications in Technology
1992: Volumes B1–B8
1993: Volumes B9–B14
ISSN 0926-5481

IFIP Technical Committee that is involved in IFIP TRANSACTIONS B
Computer Applications in Technology (TC5)

IFIP TRANSACTIONS C:
Communication Systems
1992: Volumes C1–C8
1993: Volumes C9–C16
ISSN 0926-549X

IFIP Technical Committee that is involved in IFIP TRANSACTIONS C
Communication Systems (TC6)

IFIP TRANSACTIONS FULL SET: A, B & C
1992: 35 Volumes. US $ 1892.00/Dfl. 3500.00
1993: 35 Volumes. US $ 2340.00/Dfl. 3885.00
The Dutch Guilder prices (Dfl.) are definitive. The US $ prices mentioned above are for your guidance only and are subject to exchange rate fluctuations. Prices include postage and handling charges.

The volumes are also available separately in book form.

Please address all orders and correspondence to:

ELSEVIER SCIENCE PUBLISHERS
attn. PETRA VAN DER MEER
P.O. Box 103, 1000 AC Amsterdam
The Netherlands
telephone: 31 (20) 5862602
facsimile: 31 (20) 5862616
email: m.haccou@elsevier.nl